■ 材料科学经典著作选译

非线性光学晶体
——一份完整的总结
Nonlinear Optical Crystals: A Complete Survey

[俄罗斯] David N. Nikogosyan

王继扬 译
吴以成 校

高等教育出版社

内 容 简 介

本书是根据施普林格出版社(Springer)出版的 D. N. Nikogosyan 著《非线性光学晶体——一份完整的总结》(Nonlinear Optical Crystals: A Complete Survey) 2005 年版译出。原著作者是施普林格出版社 1999 年出版的 V. G. Dmitriev 等著《非线性光学晶体手册》(Handbook of Nonlinear Optical Crystals,其中译本与本书中译本由我社同时出版)一书的作者之一。

本书继承了《非线性光学晶体手册》(以下简称《手册》)的特色,同时根据近年来非线性光学晶体研究和应用的进展,只收入了《手册》所选的 77 种晶体中的 34 种,其余 43 种未收入而代之以近年发展的 30 种新的非线性光学晶体。并对七个非线性光学晶体相关领域的最新进展,如深紫外激光的产生、通过差频产生太赫兹辐射、超短脉冲压缩、自倍频晶体、周期极化晶体、光子晶体和三阶非线性产生三次谐波等作了介绍。所以本书称得上是当前主要的非线性光学晶体数据大全的最新版本。但本书不包括非线性光学晶体的基本理论和概念的介绍,因为《手册》一书已总结和反映了自 20 世纪 60 年代以来非线性光学晶体的概貌,并且提供了与非线性光学晶体及其应用相关的最基础和较为完整的理论基础。因此,这两本书的结合,可使读者全面、系统地了解非线性光学晶体研究和发展的全貌。

本书实际应用性强、数据翔实准确,实为从事与非线性光学晶体及激光技术相关的科研人员、技术人员和教学人员必备的工具书和重要的参考书。

  David N. Nikogosyan 教授、博士，是爱尔兰科学基金会(SFI)的研究员，在爱尔兰 Cork，Cork 学院大学物理系工作。他在非线性光学、激光物理和量子电子学领域从事了 35 年的科学研究工作，发表了 133 篇文章、综述和专著，其中包括 11 篇综述和 8 本专著。

中译本序

自 20 世纪 60 年代激光问世以来，非线性光学得到了快速发展。它是研究激光与物质相互作用所产生的各种非线性光学现象的一门学科。非线性光学有许多应用：开辟新的激光波段，压缩激光脉宽，提高光谱分辨率，消除光在传播中的畸变等。非线性光学晶体是非线性光学的重要基础。利用非线性光学晶体可以制作改变激光波长频率的变化装置：光开关等。由于它们的特殊重要性，非线性光学晶体已经成为晶体材料基础上发展起来和激光技术相关的新型交叉学科。

由 V. G. Dmitriev，G. G. Gurzadyan 和 D. N. Nikogosyan 三人编著的《非线性光学晶体手册》历经三版，收集了自 20 世纪 60 年代开始到 1999 年为止被广泛研究和应用的 77 种非线性光学晶体的基本数据资料，并完整而简明地介绍了非线性晶体光学和非线性光学晶体应用相关的基础理论，有很重要的学习和参考价值。其后，《非线性光学晶体手册》作者之一的 D. N. Nikogosyan 又在该手册的基础上，重新编著了《非线性光学晶体——一份完整的总结》一书。这本书从体例上基本上继承了《非线性光学晶体手册》的特色和优点，翔实地、有分析和选择地给出了述及晶体的重要数据，并对晶体的特点及应用作出了自己的评价。这本书根据研究和应用工作的进展，保留了《非线性光学晶体手册》中仍然有重要应用和参考价值的 34 种晶体（剔除了 43 种），又代之以新的 30 种晶体，并对所有的数据进行了更新和校对。这两本书的出版，从一个侧面反映了非线性光学晶体研究领域的快速发展。

从上述两本书所收集的晶体，我们可以欣慰地看到中国晶体生长工作者在非线性光学晶体领域的成就和贡献。自 20 世纪 80 年代以来，我国非线性光学晶体的研究从跟踪仿制到走上独立研究和发展新材料的历史，是我国科学发展的一个缩影。其中值得指出的是，我国科学工作者运用分子设计学方法相继发现了 β－BBO、LBO 和 KBBF 等重要的非线性光学晶体。2009 年 2 月《自然》杂志专门撰文介绍了中国深紫外非线性光学晶体研究在国际上的领先状况，认为"其他国家在晶体生长方面的研究目前来看是无法赶上中国的"。从第一本书修订出版的 1999 年，到后一本书出版的 2005 年，新发展的 30 种晶体中，至少有一半以上的晶体是由中国科学家首先发现或已经做了大量工作的。这两本书从一个角度反映了中国在非线性光学晶体研究领域的成就。

这两本书既有相通之处又各有特点。《非线性光学晶体手册》一书总结和反映了自 20 世纪 60 年代以来非线性光学晶体的概貌，并且为读者提供了与非线性光学晶体及其应用相关的最基础和较为完整的理论基础；《非线性光学晶体——一份完整的总结》注重自 21 世纪以来在非线性光学晶体研究和应用方面的成就，并为读者提供了最新的发展领域和发展趋势。两本书都提供了在出版时能够找到的全面、可靠和有出处的资料。因此两本书的结合，为我们基本了解非线性光学晶体研究和发展的全貌提供了重要的基础。

正如 D. N. Nikogosyan 教授所指出的那样，尽管今天我们可以在互联网上查到大量的信息，但真正要获得有用和可靠的数据，却要消耗大量的时间和精力。这两本手册类的专著为我们提供了可以信赖和实用的数据。

中国有一支高水平的晶体研究队伍，同时中国也有一支高水平的激光及非线性光学的研究和应用队伍，因此非常需要一套可供参考的非线性光学晶体手册。这两本书中译本的出版将满足读者的这一需要。

非线性光学晶体的研究是我一生的追求和钟爱，能为非线性光学晶体这一领域作点贡献，是我人生中的最大快乐。值此《非线性光学晶体手册》和《非线性光学晶体——一份完整的总结》（中译本）出版之际，我也愿意和每一位读者分享这一快乐。

陈创天

作者自序

许多许多年以前，当我还是一个 12 岁的孩子时，我的父亲、著名的雕塑家 Nikalai Nikogosyan，带了我和我的母亲一起去拜访著名的苏联外交家、苏联驻英国大使 Ivan Maisky 教授。在那时候，我父亲正在创作教授的雕像，正如通常那样，他开始了和他的艺术创作对象的友谊。我们应邀到大使的夏季住所（俄语称为"dacha"）去共进晚餐，他的住所离莫斯科大约有 25 英里。我今天再也很难详细地描述那个六月的夜晚，但是我仍然真切地记得：那是一个分外明亮的夜晚，在其住所前面圆弧形的路边上，深红色的玫瑰盛开，散发着芬芳的香气。到今天，我仍然能够穿过 45 年时间的距离，清楚地看到大使的书房。在这间大屋子里，在许许多多书籍的旁边，放置着刚刚出版的豪华黑色皮面精装第二版《苏联大百科全书》整套书卷。我打开了其中一卷，马上被其各类深邃的知识所倾倒：彩色的地图、图表、照片、说明和参考书目，等等。"这是多么宝贵的财富！"我当时想。在我们回家的途中，我问父亲，我们能不能买一套这样神奇的书籍，哪怕是没有豪华装饰的普通版。但是他并没有理解我的激情，我母亲则很能体谅我，她告诉我：对我们来说这套书籍太昂贵了，最好还是当我自己有办法和有需要时，亲自去买一套这样的书籍。

后来，到了 20 世纪 60 年代中期，我在莫斯科大学学习物理学，我订购了新一版（第三版，也是最后一版）的《苏联大百科全书》，在其后 7 到 10 年期间，当它们一卷一卷出版时，我都把它们买了回来。我记得它们每一卷的价格是 5.5 卢布（在那个年代，官方汇率是 1 卢布换 8 美元），这对我每月 35 卢布的工资来说是一笔不小的开支。

当今，按照人们流行的想法来看，任何大百科全书都是毫无意义的了。我常常从我的学生那里听到，任何信息都可以在因特网上得到。然而，用学术语言来说，这只是一个粗浅的近似。首先，在因特网上任何一点小小有用信息的种子都会溶化在无用信息浩瀚的大海中，置于无回应或无控制的状态中。在因特网获得的参考文献资料往往是不完整的、过时的，并且常常是与其他来源获得的材料相互矛盾的。任何一位不同意我观点的人可以尝试一下，将任何一种常用非线性光学晶体的名称（例如 BBO、KTP、铌酸锂等）输入 www.google.com，并且将出现于屏幕上的资料予以比较。结果表明，因特网的用户必须是博学的，才可能区分不同数据值之间的差异。现代计算机的电脑，尽管思维敏捷、博学多闻，但仍是愚钝异常，并不具有对不同系列数据进行逻辑比较、选择最可靠系列的能力；换言之，在我们的因特网世界里，对于科学书籍仍然有显著的需求。

从孩提时代到现在，贯穿我整个人生（我现在57岁了），我都是一个热切的收藏者。首先是集邮，再是钱币，还有书籍，然后是密纹唱片，还收集古玩，再有杜鹃花（在我们的爱尔兰花园里就收集了优良的50个品种），等等。这本晶体的总结可看成是我的数据收藏品，这是我在过去25年中予以策划并完成的。我关于非线性光学晶体的第一次综述[1]发表于1977年，在20年后被SPIE选为在光学参量振荡领域中出版物的一个里程碑[2]。这样的个人经历也许是我为什么决定在非线性光学晶体方面再编撰一本书的原因，这件事花费了我近来的每一天（实际上是每一个晚上）时间。在这一年半时间内，每晚我都面对着我的计算机屏幕。换言之，我钟情于这一过程（再也不可能作其他解释）。

这样一本手册的明显特征是，它同时属于许多人，我愿意与每一位读者同享。我希望通过使用（阅读）我这本小小的百科全书，至少将会使读者分享我的巨大的感受，它就是编撰这本书所带给我的快乐和享受。

David N. Nikogosyan
Tower, Blarney
Co. Cork, Ireland
2003 年 12 月 20 日

■ 参考文献

[1] D. N. Nikogosyan: Nonlinear optical crystals (review and summary of data). Kvant. Elektron. 4(1), 5–26 (1977) [In Russian, English trans.: Sov. J. Quantum Electron. 7 (1), 1–13(1977)].

[2] D. N. Nikogosyan: Nonlinear optical crystals (review and summary of data). In: *Selected Papers on Optical Parametric Oscillations and Amplifiers and Their Applications*, *SPIE Milestone Series*, *Vol. MS140*, ed. by J. H. Hunt (SPIE Optical Engineering Press, Bellingham, Washington, 1997) pp. 191–203.

目 录

缩略语

第 1 章 引言 1

第 2 章 基本的非线性光学晶体 5
2.1 β-BaB$_2$O$_4$，偏硼酸钡（BBO） 5
2.2 LiB$_3$O$_5$，三硼酸锂（LBO） 22
2.3 LiNbO$_3$，铌酸锂（LN） 42
2.4 KTiOPO$_4$，磷酸钛氧钾（KTP） 64

第 3 章 主要的红外材料 91
3.1 AgGaS$_2$，硫镓银（AGS） 91
3.2 AgGaSe$_2$，硒镓银（AGSe） 104
3.3 ZnGeP$_2$，磷锗锌（ZGP） 115
3.4 GaSe，硒化镓 130

第 4 章 常用晶体 139
4.1 KH$_2$PO$_4$，磷酸二氢钾（KDP） 139
4.2 NH$_4$H$_2$PO$_4$，磷酸二氢铵（ADP） 161
4.3 KD$_2$PO$_4$，氘化磷酸二氢钾（DKDP） 175
4.4 CsLiB$_6$O$_{10}$，硼酸锂铯（CLBO） 186
4.5 MgO：LiNbO$_3$，氧化镁掺杂铌酸锂（MgLN） 193
4.6 KTiOAsO$_4$，砷酸钛氧钾（KTA） 201
4.7 KNbO$_3$，铌酸钾（KN） 208

第 5 章 周期性极化晶体及"衬底"材料 221
5.1 LiTaO$_3$，钽酸锂（LT） 221
5.2 RbTiOAsO$_4$，砷酸钛氧铷（RTA） 228
5.3 BaTiO$_3$，钛酸钡 234

5.4　$MgBaF_4$，氟化钡镁 …… 240

5.5　$GaAs$，砷化镓 …… 243

第6章　新发展及有前景的晶体 …… 257

6.1　BiB_3O_6，三硼酸铋（BIBO） …… 257

6.2　$K_2Al_2B_2O_7$，硼酸铝钾（KABO） …… 261

6.3　$KBe_2BO_3F_2$，氟硼铍酸钾（KBBF） …… 265

6.4　$BaAlBO_3F_2$，氟硼酸铝钡（BABF） …… 267

6.5　$La_2CaB_{10}O_{19}$，硼酸钙镧（LCB） …… 269

6.6　$GdCa_4O(BO_3)_3$，硼酸氧钙钆（GdCOB） …… 271

6.7　$YCa_4O(BO_3)_3$，硼酸氧钙钇（YCOB） …… 278

6.8　$Gd_xY_{1-x}Ca_4O(BO_3)_3$，硼酸氧钙钇钆（GdYCOB） …… 287

6.9　$Li_2B_4O_7$，四硼酸锂（LB4） …… 292

6.10　$LiRbB_4O_7$，四硼酸铷锂（LRB4） …… 295

6.11　$CdHg(SCN)_4$，硫氰酸汞镉（CMTC） …… 297

6.12　$Nb:KTiOPO_4$，铌掺杂 KTP（$Nb_xK_{1-x}Ti_{1-x}OPO_4$ 或 NbKTP） …… 300

6.13　$RbTiOPO_4$，磷酸钛氧铷（RTP） …… 305

6.14　$LiInS_2$，硫铟锂（LIS） …… 309

6.15　$LiInSe_2$，硒铟锂（LISe） …… 315

6.16　$LiGaS_2$，硫镓锂（LGS） …… 317

6.17　$LiGaSe_2$，硒镓锂（LGSe） …… 319

6.18　$AgGa_xIn_{1-x}Se_2$，硒铟镓银（AGISe） …… 320

6.19　Tl_4HgI_6，碘汞铊（THI） …… 323

第7章　自倍频晶体 …… 325

7.1　$Nd:MgO:LiNbO_3$，掺钕掺氧化镁铌酸锂（NdMgLN） …… 325

7.2　$Nd:YAl_3(BO_3)_4$，掺钕四硼酸铝钇（$Nd_xY_{1-x}Al_3(BO_3)_4$ 或 NYAB） …… 329

7.3　$Nd:GdAl_3(BO_3)_4$，掺钕四硼酸铝钆（$Nd_xGd_{1-x}Al_3(BO_3)_4$ 或 NGAB） …… 337

7.4　$Nd:GdCa_4O(BO_3)_3$，掺钕硼酸氧钙钆（$Nd_xGd_{1-x}COB$ 或 Nd:GdCOB） …… 341

7.5　$Nd:YCa_4O(BO_3)_3$，掺钕硼酸氧钙钇（$Nd_xY_{1-x}COB$ 或 Nd:YCOB） …… 347

7.6　$Nd:LaBGeO_5$，掺钕锗酸硼镧（$Nd_xLa_{1-x}BGeO_5$ 或 NdLBGO） …… 351

7.7　$Nd:Gd_2(MoO_4)_3$，掺钕钼酸钆（$Nd_{2x}Gd_{2-2x}(MoO_4)_3$ 或 NdGMO） …… 355

7.8　$Yb:YAl_3(BO_3)_4$，掺镱四硼酸铝钇（$Yb_xY_{1-x}Al_3(BO_3)_4$ 或 Yb:YAB） …… 359

7.9　$Yb:GdCa_4O(BO_3)_3$，掺镱硼酸氧钙钆（$Yb_xGd_{1-x}COB$ 或 Yb:GdCOB） …… 364

7.10　$Yb:YCa_4O(BO_3)_3$，掺镱硼酸氧钙钇（$Yb_xY_{1-x}COB$ 或 Yb:YCOB） …… 367

第8章　很少用的和传统的晶体 …… 373

8.1　$KB_5O_8 \cdot 4H_2O$，五硼酸钾四水合物（KB5） …… 373

- 8.2 CsB_3O_5，三硼酸铯（CBO） ... 380
- 8.3 $C_4H_7D_{12}N_4PO_7$，氘化左旋磷酸精氨酸一水合物（DLAP） ... 383
- 8.4 α-碘酸（α-HIO_3） ... 387
- 8.5 $LiCOOH \cdot H_2O$，甲酸锂一水合物（LFM） ... 392
- 8.6 CsH_2AsO_4，砷酸二氢铯（CDA） ... 395
- 8.7 CsD_2AsO_4，氘化砷酸二氢铯（DCDA） ... 400
- 8.8 RbH_2PO_4，磷酸二氢铷（RDP） ... 404
- 8.9 $CsTiOAsO_4$，砷酸钛氧铯（CTA） ... 410
- 8.10 $Ba_2NaNb_5O_{15}$，铌酸钡钠（BNN） ... 414
- 8.11 $K_3Li_2Nb_5O_{15}$，铌酸钾锂（KLN） ... 419
- 8.12 $CO(NH_2)_2$，尿素 ... 422
- 8.13 $LiIO_3$，碘酸锂 ... 426
- 8.14 Ag_3AsS_3，硫砷银（淡红银矿） ... 437
- 8.15 $HgGa_2S_4$，硫镓汞 ... 444
- 8.16 $CdGeAs_2$，砷锗镉（CGA） ... 447
- 8.17 Tl_3AsSe_3，硒砷铊（TAS） ... 453
- 8.18 $CdSe$，硒化镉 ... 456

第9章 一些最新的应用 ... 465
- 9.1 深紫外光的产生 ... 465
- 9.2 通过 DFG 产生太赫兹波 ... 467
- 9.3 通过 SHG 的超短激光脉冲压缩 ... 468
- 9.4 自倍频晶体 ... 470
- 9.5 周期性极化晶体 ... 473
- 9.6 光子带隙晶体 ... 478
- 9.7 通过 $\chi^{(3)}$ 非线性过程产生的 THG ... 479

第10章 闭卷的话 ... 481

附录 A 所列举杂志的全称 ... 482

附录 B 在最后审稿时加入的最新参考文献 ... 488

名词索引 ... 491

译者后记 ... 499

缩略语

BPM，双折射相位匹配（birefringent phase matching）
CW，连续波（continuous wave）
FiHG，五次谐波发生（fifth-harmonic generation）
FoHG，四次谐波发生（fourth-harmonic generation）
HeXLN，六方极化铌酸锂（hexagonally poled lithium niobate）
MOPA，主振荡功率放大器（master oscillator power amplifier）
NCPM，非临界相位匹配（non-critical phase matching）
OPA，光学参量放大器（optical parametric amplifier）
OPG，光学参量发生（optical parametric generator）
OPO，光学参量振荡器（optical parametric oscillator）
PPKTP，周期极化磷酸钛氧钾（periodically poled potassium titanyl phosphate）
PPLN，周期极化铌酸锂（periodically poled lithium niobate）
PPLT，周期极化钽酸锂（periodically poled lithium tantalate）
PPRTA，周期极化砷酸钛氧铷（periodically poled rubidium titanyl arsenate）
QPM，准相位匹配（quasi phase matching）
SFD，自倍频（self-frequency doubling）
SFG，和频发生（sum-frequency generation）
SH，二次谐波（second harmonic）
SHG，二次谐波发生（second-harmonic generation）
THG，三次谐波发生（third-harmonic generation）
YAG，钇铝石榴石（yttrium aluminum garnet）
YAP，正铝酸钇（yttrium orthoaluminate）
YLF，氟化钇锂（yttrium lithium fluoride）
YSGG，钇钪镓石榴石（yttrium scandium gallium garnet）

在国家点火装置(NIF)由 Natalia Zaitseva(照片中人)生长的 54 cm×54 cm×55 cm 的 KDP 晶体。这是为在美国加利福尼亚州 Lawrence Livermore 国家实验室建设的世界最大的激光器频率转换应用所制备的。承蒙 Lawrence Livermore 国家实验室提供照片。

第 1 章
引 言

在过去 25 年中，我出版了 9 本关于非线性光学晶体的综述和数据库一类的著作[1-9]。因此，在介绍这本书时，我很乐意回答将来读者最普遍的疑问和要问的问题："我们为什么需要这样一本新书？如我们将它和 1995 年编撰，于 1997 年和 1999 年由 Springer 出版的手册[8,9]来比较的话，这本晶体的总结与前面那些手册有什么重要的变化？"

写这本新书的原因，首先是在过去十年中，激光技术已经获得了巨大的发展。有三个明显的发展趋势：

（1）向更短的脉冲方向发展，脉冲宽度压缩到了几百（或几十）飞秒。更短的脉冲可以增加辐射强度，这就使得所发生的非线性光学过程有更高的效率。同时，这种发展趋势大大提高了对非线性光学晶体激光脉冲引起的损伤阈值的要求。短脉冲宽度使人们必须考虑群速色散的影响。这种效应甚至可能是有利的，它可以在二次谐波产生过程中形成对激光脉冲的压缩。

（2）向能发射可见、紫外和近红外波段激光的小型二极管泵浦连续波（CW）激光光源的方向发展。发明了一种称作准相位匹配的新相位匹配方法，使我们在任何非线性光学晶体中都可采用其二阶非线性系数的最高值，而且可以在任何需要的方向上（例如非临界相位匹配）获得相位匹配。这就极大地提高了二阶三波相互作用（即 SHG）的效率，使激光二极管的频率变换输出功率相当高。另一个途径是在非线性晶体中掺入稀土离子（通常是 Nd 或 Yb），在激光二极管泵浦下，这样的掺杂材料能产生基频，同时将其转变为二次谐波。所以，它们被称为自倍频晶体。

（3）积极地探索新的非线性光学晶体，特别是在低对称晶体中探索新晶体。在过去 10 年中，引入和发展了许多新晶体，诸如 GdCOB、YCOB、YAB、BIBO、CLBO、KBBF、LB4、$MgBaF_4$、GaAs 和许多其他晶体。而且它们已经在准位相匹配、自倍频和深紫外发生等方面成功得到了应用。

这本书与以前的手册[8,9]不同。首先从结构上来说，省略了理论部分，因为现在关于非线性晶体中二阶三波相互作用的理论已经很好地确立了，关于这一题目已有其他书籍出版[10]。我还决定在本书中删去所有关于非线性光学材料的传统应用（SHG、SFG、DFD 和 OPO 等）方面的内容，否则会很容易地增加本书的厚度，使其超出人们可以接受的水平。

第二个不同在于这本新手册的内容。在书中剔除了 43 种旧的非线性光学材料，现在代之以 30 种新的晶体。而且首次特别考虑了周期性极化及自倍频材料。对每种晶体的资料的编排结构也作了根本改变，加上了大量的晶体学、热物理、光谱学、电光学和磁光学的资料。

这一新编的对 63 种非线性光学晶体的总结包括了 1 500 份具有完整标题的参考文献。所有引用资料中有 15% 是引自 2000—2003 年的文献；其中有 41% 是过去 9 年中的（在所收集上一本书资料后到现在的时间）。最频繁的引用源是《应用物理快报》（占所有参考文献的 11.3%）、《光学快报》（占所有参考文献的 10.0%）和《光学通信》（占所有参考文献的 9.8%）。

在晶体性质总结（2—8 章）之后的第 9 章，通过 7 篇小型综述讨论了一些普通的和新型的非线性材料（包括自倍频和准相位匹配）的最新应用，完成了整书。

最后，我乐意在此提到我的朋友和同事，我将他们的名字按字母表的顺序排列如下，衷心感谢他们中肯的批评、有益的讨论，以及送给我相关文章抽印本和电子文件等帮助。他们是：Gerard Aka 教授（法国），Vladimir Alshits 教授（俄罗斯），Ladislav Bohaty 教授（德国），Patrick Mc Carthy 博士（爱尔兰），Subhasis Das 博士（印度），Katia Gallo 博士（英国），Helmut Görner 博士（德国），Sergey Grechin 博士（俄罗斯），Alexander Gribenyukov 博士（俄罗斯），Stas Ionov 博士（美国），Ludmila Isaenko 博士（俄罗斯），Mitsuru Ishii 博士（日本），Kiyoshi Kato 教授（日本），Hideo Kimura 博士（日本），Takayoshi Kobayashi 教授(日本)、Lev Kulevskii 教授(俄罗斯)、Nikolay Leonyuk 教授(俄罗斯)、Wenju Liu 教授(中国)、Alla Makarova 女士(俄罗斯)、Nikolai Merzliakov 博士(俄罗斯)、Kiminori Mizuuchi 博士(日本)、Yusuke Mori 教授(日本)、Eugene Moskovets 博士(美国)、Tatiana Perova 博士(爱尔兰)、Katalin Polgár 博士(匈牙利)、Mariola Ramirez 博士(西班牙)、Martin Richardson 教授(美国)、Eugenii Ryabov 教授(俄罗斯)、Mark Saffman 教授(美国)、Solomon Saltiel 教授(保加利亚)、Ichiro Shoji 博士(日本)、Yuji Suzuki 博士(日本)、Eiko Takaoka 博士(日本)、Daniel Vivien 教授(法国)、Richard White 博士(澳大利亚)、Alexander Yelisseyev 博士(俄罗斯)、Masashi Yoshimura 博士(日本)、Natalia Zaitseva 博士(美国)和 Anatoly Zayats(英国)。

我还特别愿意再单独感谢 Cork 学院大学网上图书馆 Loans 服务中心的 Eileen Heathy 女士、Phil O'Sullivan 女士和 Garret Cahill 先生。

我特别要感谢 Springer-NY 书店的编辑，Hans Koelsch 博士，感谢他建设性的建议和卓有成效的合作。同时，我总是无限感激我亲爱的夫人 Danielle，感谢她对我的鼓励和宽容。

■ 参考文献

[1] D. N. Nikogosyan: Nonlinear optical crystals (review and summary of data). Kvant. Elektron. 4(1), 5 – 26 (1977) [In Russian, English trans.: Sov. J. Quantum Electron. 7(1), 1 – 13 (1977)].

[2] D. N. Nikogosyan, G. G. Gurzadyan: Crystals for nonlinear optics. Biaxial crystals. Kvant. Elektron. 14(8), 1529 – 1541 (1987) [In Russian, English trans.: Sov. J. Quantum Electron. 17(8), 970 – 977 (1987)].

[3] G. G. Gurzadyan, V. G. Dmitriev, D. N. Nikogosyan: *Nonlinear Optical Crystals. Properties and Applications in Quantum Electronics. Handbook* (Radio i Sviyaz, Moscow, 1991), pp. 1 – 160 [In Russian].

[4] V. G. Dmitriev, G. G. Gurzadyan, D. N. Nikogosyan: *Handbook of Nonlinear Optical Crystals*. Springer Series in Optical Sciences, Vol. 64, ed. by A. E. Siegman (Springer, Berlin, 1991), pp. 1 – 221.

[5] D. N. Nikogosyan: Beta barium borate (BBO). A review of its properties and applications. Appl. Phys. A 52(6), 359 – 368 (1991).

[6] D. N. Nikogosyan: Lithium triborate (LBO). A review of its properties and applications. Appl. Phys. A 58(3), 181 – 190 (1994).

[7] D. N. Nikogosyan: *Properties of Optical and Laser-Related Materials. A Handbook* (John Wiley & Sons Ltd., Chichester, 1997), pp. 1 – 594.

[8] D. N. Nikogosyan: Properties of Nonlinear Optical Crystals. In: V. G. Dmitriev, G. G. Gurzadyan, D. N. Nikogosyan: *Handbook of Nonlinear Optical Crystals, Second, Revised and Updated Edition.* Springer Series in Optical Sciences, Vol. 64, ed. by A. E. Siegman (Springer, Berlin, 1997), pp. 67 – 288.

[9] D. N. Nikogosyan: Properties of Nonlinear Optical Crystals. In: V. G. Dmitriev, G. G. Gurzadyan, D. N. Nikogosyan: *Handbook of Nonlinear Optical Crystals, Third Revised Edition.* Springer Series in Optical Sciences, Vol. 64, ed. by A. E. Siegman (Springer, Berlin, 1999), pp. 67 – 288.

[10] R. L. Sutherland: *Handbook of Nonlinear Optics* (Marcel Dekker, New York, 1996), pp. 1 – 685.

第2章
基本的非线性光学晶体

本章包括了四种最为广泛应用的非线性光学晶体的信息：偏硼酸钡(BBO)、三硼酸锂(LBO)、铌酸锂(LN)和磷酸钛氧钾(KTP)。和它们的周期极化同系物一起，包括周期极化铌酸锂(PPLN)和周期极化磷酸钛氧钾(PPKTP)，这些晶体的应用至少占目前实际应用的75%。

所有在这一章以及以下各章给出的关于角度、温度和光谱接受度(带宽)的值适用于1 cm长的相应非线性光学晶体。

2.1 $\beta-BaB_2O_4$，偏硼酸钡(BBO)

负单轴晶：$n_o > n_e$
分子量：222.950
密度：3.84 g/cm^3[1]；3.849 g/cm^3[2]；在T = 293 K时，3.85 g/cm^3[3]
点群：$3m$

晶格常数：$a = 12.532$ Å[4]；(12.532 ± 0.001) Å[2]；12.547 Å[5]
$c = 12.717$ Å[4]；(12.726 ± 0.001) Å[2]；12.736 Å[5]

莫氏硬度：4[6,7]；4.5[2]

熔点：1 368 K[2,8]

线性热膨胀系数 α_t[3]

T/K	$\alpha_t \times 10^6 / K^{-1}$, $\parallel c$	$\alpha_t \times 10^6 / K^{-1}$, $\perp c$
293	0.36	-2.54

线性热膨胀系数平均值[5]

T/K	$\alpha_t \times 10^6 / K^{-1}$, $\parallel c$	$\alpha_t \times 10^6 / K^{-1}$, $\perp c$
298 ~ 1 173	36	4.0

比热容 c_p，$p = 0.101\ 325$ MPa

T/K	$c_p / (J \cdot kg^{-1} \cdot K^{-1})$	参考文献
298	490	[2]
	496	[9]

热导率

$\kappa / (W \cdot m^{-1} \cdot K^{-1})$, $\parallel c$	$\kappa / (W \cdot m^{-1} \cdot K^{-1})$, $\perp c$	参考文献
0.8	0.08	[5]
1.6	1.2	[10]

室温下的直接带隙能量：$E_g = 6.2$ eV[11]，6.43 eV[12]

透明范围：

以 0.5 透过计：对 0.8 cm 长晶体，0.198 ~ 2.6 μm[13]；对 0.3 cm 长晶体，0.196 ~ 2.2 μm[2]

以 "0" 透过计：0.189 ~ 3.5 μm[8,14]

以 0.5 透过计：0.198 ~ 2.6 μm[1]

线性吸收系数 α

$\lambda / \mu m$	α / cm^{-1}	参考文献	备 注
0.193 4	1.39	[15]	$T = 295$ K
	0.29	[15]	$T = 91$ K

续表

$\lambda/\mu m$	α/cm^{-1}	参 考 文 献	备 注
0.213	<0.21	[1]	最佳晶体
0.264	0.04±0.01	[16]	∥c
	0.06±0.003	[16]	⊥c, o 光
	0.10±0.003	[16]	⊥c, e 光
0.266 1	<0.17	[1]	最佳晶体
	0.04~0.15	[2]	
0.532 1	0.01	[17]	
	<0.01	[9]	
1.0	0.001~0.002	[2]	
1.064 2	<0.001	[9]	
2.09	0.008 5	[2]	e 光
	0.07	[2]	o 光
2.55	0.5	[18]	

双光子吸收系数 β

$\lambda/\mu m$	τ_p/ns	$\beta \times 10^{11}/(cm \cdot W^{-1})$	参 考 文 献	备 注
0.211	0.000 9	243±85	[19]	$\theta=30°, \phi=0°$
0.264	0.000 8	93±33	[19]	$\theta=30°, \phi=0°$
	0.000 22	68±6	[20]	∥c
		66±7	[20]	⊥c, o 光
		47±5	[20]	⊥c, e 光
	0.000 2	61	[21]	$\theta=48°$
0.266 1	0.015	90±10	[11]	∥c
0.354 7	0.017	1.0±0.2	[11]	∥c

折射率的实验值[5]

$\lambda/\mu m$	n_o	n_e	$\lambda/\mu m$	n_o	n_e
0.404 66	1.692 67	1.567 96	0.589 30	1.670 49	1.552 47
0.435 83	1.686 79	1.563 76	0.643 85	1.667 36	1.550 12
0.467 82	1.681 98	1.560 24	0.818 90	1.660 66	1.545 89
0.479 99	1.680 44	1.559 14	0.852 12	1.659 69	1.545 42
0.508 58	1.677 22	1.556 91	0.894 35	1.658 62	1.544 69
0.546 07	1.673 76	1.554 65	1.014 00	1.656 08	1.543 33
0.579 07	1.671 31	1.552 98			

293～353 K 温度范围内折射率的温度微商

$\lambda/\mu m$	$dn_o/dT \times 10^6/K^{-1}$	$dn_e/dT \times 10^6/K^{-1}$
0.4～1.0	-16.6	-9.3

最佳色散关系方程（λ 以 μm 为单位，$T = 293$ K）[13]：

$$n_o^2 = 2.7359 + \frac{0.01878}{\lambda^2 - 0.01822} - 0.01354\lambda^2$$

$$n_e^2 = 2.3753 + \frac{0.01224}{\lambda^2 - 0.01667} - 0.01516\lambda^2$$

近红外吸收边处精确度更高的 Sellmeier 方程（λ 以 μm 为单位，$T = 293$ K）[22]：

$$n_o^2 = 2.7359 + \frac{0.01878}{\lambda^2 - 0.01822} - 0.01471\lambda^2 + 0.0006081\lambda^4 - 0.00006740\lambda^6$$

$$n_e^2 = 2.3753 + \frac{0.01224}{\lambda^2 - 0.01667} - 0.01627\lambda^2 + 0.0005716\lambda^4 - 0.00006305\lambda^6$$

在[1]、[5]、[8]、[23]、[24]、[25]、[26]中给出了其他的 Sellmeier 方程组。

非线性折射率 γ

$\lambda/\mu m$	$\gamma \times 10^{15}/(cm^2 \cdot W^{-1})$	参考文献	备 注
0.2661	0.025 ± 0.008	[11]	$\parallel c$
0.3547	0.36 ± 0.08	[11]	$\parallel c$
0.5321	0.55 ± 0.10	[11]	$\parallel c$
0.780	0.40 ± 0.05	[27]	[100]方向
	0.32 ± 0.05	[27]	[010]方向
0.850	0.37 ± 0.06	[28]	$\theta = 29.2°, \phi = 0°$
1.0642	0.29 ± 0.05	[11]	$\parallel c$

室温下在低频区（远低于BBO晶体声共振频率，即"自由"晶体）测量的线性电光系数

$\lambda/\mu m$	$r_{22}^T/(pm \cdot V^{-1})$	$r_{51}^T/(pm \cdot V^{-1})$	参考文献	备 注
0.5145	2.5 ± 0.1		[29]	$T = 296$ K
0.6328	2.5	<0.04	[30]	
	2.2 ± 0.1		[31]	

在高频区（远高于BBO晶体的声学共振频率，即对于"夹持"晶体）测量的线性电光系数

$\lambda/\mu m$	$r_{22}^S/(pm \cdot V^{-1})$	参 考 文 献	备 注
0.514 5	2.1 ± 0.3	[29]	$T = 296$ K
0.632 8	2.1 ± 0.1	[31]	

在通常情况下,二阶有效非线性系数的表达式(Kleinman 对称条件成立, $d_{31} = d_{15}$)[32]:

$$d_{ooe} = d_{31}\sin(\theta+\rho) - d_{22}\cos(\theta+\rho)\sin 3\phi$$
$$d_{eoe} = d_{oee} = d_{22}\cos^2(\theta+\rho)\cos 3\phi$$

二阶有效非线性系数的简化表达式(小双折射角近似,Kleinman 对称条件成立, $d_{31} = d_{15}$)[33]:

$$d_{ooe} = d_{31}\sin\theta - d_{22}\cos\theta\sin 3\phi$$
$$d_{eoe} = d_{oee} = d_{22}\cos^2\theta\cos 3\phi$$

二阶非线性系数的绝对值[32]:

$$|d_{22}(0.532\ \mu m)| = 2.6\ pm/V$$
$$|d_{22}(0.852\ \mu m)| = 2.3\ pm/V$$
$$|d_{22}(1.064\ \mu m)| = 2.2\ pm/V$$
$$|d_{22}(1.313\ \mu m)| = 1.9\ pm/V$$
$$|d_{15}(1.064\ \mu m)| = 0.03\ pm/V$$
$$|d_{31}(1.064\ \mu m)| = 0.04\ pm/V$$
$$|d_{33}(1.064\ \mu m)| = 0.04\ pm/V$$

其他二阶非线性系数的值 d_{22}:

$$|d_{22}(1.064\ \mu m)| = (2.1 \pm 0.1)\ pm/V^{[34]};\ (2.2 \pm 0.2)\ pm/V^{[35]};$$
$$(2.23 \pm 0.16)\ pm/V^{[36]};\ (2.23 \pm 0.18)\ pm/V^{[37]}$$
$$|d_{22}(1.319\ \mu m)| = (1.89 \pm 0.15)\ pm/V^{[37]}$$

d_{22} 和 d_{31} 的符号相反[8,25,34]。

相位匹配角的实验值($T = 293$ K)

相互作用的波长/μm	$\theta_{exp}/(°)$	参 考 文 献
SHG, o + o ⇒ e		
0.409 6 ⇒ 0.204 8	90	[13]
0.41 ⇒ 0.205	90	[38]
0.411 52 ⇒ 0.205 76	82.8	[13]
0.415 46 ⇒ 0.207 73	79.2	[13]
0.418 ⇒ 0.209	77.3	[39]
0.429 ⇒ 0.214 5	71	[40]

续表

相互作用的波长/μm	$\theta_{\exp}$/(°)	参 考 文 献
0.476 5⇒0.238 25	57	[41]
0.488⇒0.244	54.5	[41]
0.496 5⇒0.248 25	52.5	[41]
0.510 6⇒0.255 3	50	[42]
	50.6	[43]
0.514 5⇒0.257 25	49.5	[41]
0.532 1⇒0.266 05	47.3	[5]
	47.5	[13],[44],[45]
	47.6	[46],[47]
	48	[23],[48]
0.589⇒0.294 5	41.5	[49]
0.604⇒0.302	40	[50]
0.615 6⇒0.307 8	39	[51]
0.616⇒0.308	38	[52]
0.709 46⇒0.354 73	32.9	[53],[54]
	33	[44],[55],[56]
	33.1	[47]
	33.3	[18]
	33.7	[57]
0.78⇒0.39	31	[58]
	30	[59]
0.8⇒0.4	26.5	[60]
0.946⇒0.473	24.9	[61]
1.064 2⇒0.532 1	22.7	[5]
	22.8	[4],[13],[44],[47],[62],[63]
SFG,o+o⇒e		
0.738 65+0.257 25⇒0.190 8	81.7	[64]
0.727 47+0.263 25⇒0.193 3	76	[65]
0.592 2+0.296 1⇒0.197 4	88	[66]
0.596 4+0.298 2⇒0.198 8	82.5	[67]
0.599 1+0.299 55⇒0.199 7	80	[66]
0.604 65+0.302 33⇒0.201 55	76.2	[67]
0.532 1+0.325 61⇒0.202	83.9	[13]

续表

相互作用的波长/μm	$\theta_{exp}/(°)$	参 考 文 献
0.609 9 + 0.304 95 ⇒ 0.203 3	73.5	[66]
0.532 1 + 0.346 91 ⇒ 0.21	71.9	[13]
0.773 6 + 0.257 87 ⇒ 0.193 4	70.7	[15]
0.532 1 + 0.354 73 ⇒ 0.212 84	70	[48]
0.515 67 + 0.386 75 ⇒ 0.221	64.7	[68]
0.804 + 0.268 ⇒ 0.201	64	[69]
0.75 + 0.375 ⇒ 0.25	61.7	[70]
1.064 2 + 0.266 05 ⇒ 0.212 84	51.1	[13]
0.78 + 0.373 ⇒ 0.252 3	47.4	[71],[72]
1.064 2 + 0.298 ⇒ 0.232 81	46.1	[73]
0.578 2 + 0.510 6 ⇒ 0.271 15	46	[74]
0.590 99 + 0.532 1 ⇒ 0.28	44.7	[75]
0.78 + 0.43 ⇒ 0.277 2	43.4	[76]
1.064 2 + 0.354 73 ⇒ 0.266 05	40.2	[13]
1.064 1 + 0.532 05 ⇒ 0.354 7	31.3	[77]
1.064 2 + 0.532 1 ⇒ 0.354 73	31.1	[5]
	31.3	[13]
	31.4	[57]
2.688 23 + 0.571 2 ⇒ 0.471 1	21.8	[25]
1.418 31 + 1.064 2 ⇒ 0.608	21	[78]
SHG, e + o ⇒ e		
0.532 1 ⇒ 0.266 05	81	[13]
0.709 46 ⇒ 0.354 73	48	[55]
	48.1	[44]
1.064 2 ⇒ 0.532 1	31.6	[79]
	32.4	[5]
	32.7	[4],[44]
	32.9	[13]
SFG, e + o ⇒ e		
1.064 2 + 0.354 73 ⇒ 0.266 05	46.6	[13]
1.064 2 + 0.532 1 ⇒ 0.354 73	38.4	[5]
	38.5	[13]
SFG, o + e ⇒ e		
1.064 2 + 0.532 1 ⇒ 0.354 73	59.8	[13]

在 $T = 293$ K 时内角带宽、温度带宽和光谱带宽的实验值

相互作用的波长/μm	θ_{pm}/(°)	$\Delta\theta^{int}$/(°)	ΔT/℃	$\Delta\nu$/cm^{-1}	参考文献
SHG, o+o⇒e					
1.064 2⇒0.532 1	22.8	0.021	37	9.7	[4]
	21.9	0.028			[23]
	22.7	0.030	51		[5]
0.532 1⇒0.266 05	47.3	0.010	4		[5]
0.53⇒0.265	47.6(298 K)	0.006			[80]
SFG, o+o⇒e					
1.064 1+0.532 05⇒0.354 7	31.3	0.011			[77]
1.064 2+0.532 1⇒0.354 73	31.1	0.015	16		[5]
2.447 02+0.571 2⇒0.463 1	22.1	0.026			[25]
2.688 23+0.571 2⇒0.471 1	21.8	0.028			[25]
SHG, e+o⇒e					
1.064 2⇒0.532 1	32.7	0.034		8.8	[4]
	32.4	0.046	37		[5]
SFG, e+o⇒e					
1.064 2+0.532 1⇒0.354 73	38.4	0.020	13		[5]
SFG, o+e⇒e					
1.064 2+0.532 1⇒0.354 73	58.4	0.050	12		[5]

在 $T=293$ K 时相位匹配角的温度变化[5]

相互作用的波长/μm	θ_{pm}/(°)	$d\theta_{pm}/dT$/[(°)·K^{-1}]
SHG, o+o⇒e		
0.532 1⇒0.266 05	47.3	0.002 50
1.064 2⇒0.532 1	22.7	0.000 57
SFG, o+o⇒e		
1.064 2+0.532 1⇒0.354 73	31.1	0.000 99
SHG, e+o⇒e		
1.064 2⇒0.532 1	32.4	0.001 20
SFG, e+o⇒e		
1.064 2+0.532 1⇒0.354 73	38.4	0.001 50
SFG, o+e⇒e		
1.064 2+0.532 1⇒0.354 73	58.4	0.004 21

BBO 中 SHG 过程逆群速失配的计算值

相互作用的波长/μm	$\theta_{pm}/(°)$	$\beta/(fs\cdot mm^{-1})$
SHG, o+o⇒e		
1.2⇒0.6	21.18	54
1.1⇒0.55	22.28	76
1.0⇒0.5	23.85	104
0.9⇒0.45	26.07	141
0.8⇒0.4	29.18	194
0.7⇒0.35	33.65	275
0.6⇒0.3	40.47	415
0.5⇒0.25	52.34	695
SHG, e+o⇒e		
1.2⇒0.6	29.91	103
1.1⇒0.55	31.46	130
1.0⇒0.5	33.73	164
0.9⇒0.45	36.98	210
0.8⇒0.4	41.67	276
0.7⇒0.35	48.74	373
0.6⇒0.3	60.91	531

激光诱导的体损伤阈值

$\lambda/\mu m$	τ_p/ns	$I_{thr}/(GW\cdot cm^{-2})$	参考文献	备 注
0.266 1	10	0.3	[81]	10 Hz
	8	>0.12	[46]	
		2.0	[82]	由提拉法生长(CZ-BBO)
		3.0	[82]	由熔盐法生长(flux-BBO)
		3.4	[82]	CZ-BBO，在1 193 K退火(50小时)
0.308	12	>0.2	[83]	
0.354 7	10	0.9	[81]	10 Hz
		5	[18]	
	8	25	[84]	1个脉冲
		19	[84]	1 800个脉冲
	0.03	>0.4	[85]	10 Hz
	0.015	>3	[53]	
0.400	0.000 2	>150	[60]	10 Hz
0.510 6	20	>0.25	[86]	4 kHz
0.51~0.58	20	10	[87]	
0.514 5	CW	>0.000 4	[88]	
0.532 1	10	2.3	[81]	10 Hz

续表

$\lambda/\mu m$	τ_p/ns	$I_{thr}/(GW \cdot cm^{-2})$	参考文献	备注
	8	48	[84]	1个脉冲
		32	[84]	1 800个脉冲
	1	7	[23]	
	0.25	10	[6]	
	0.075	>7	[14]	
	0.025	>4.2	[54]	10 Hz
		>4	[63]	
0.62	0.000 2	>50	[89]	
	0.000 1	1 000(?)	[90]	
0.694 3	0.02	10	[8]	
0.8	0.000 025	>3 400	[91]	1~5 kHz
0.85	0.000 25	>93	[92]	1 kHz
1.054	0.005	50	[93]	
1.064 2	14	50	[84]	1个脉冲
		23	[84]	1 800个脉冲
	10	4.5	[81]	10 Hz
		5	[6]	
	1.3	10	[4]	
	1.1	14	[94]	
	1.0	13.5	[6]	
	0.1	10	[6]	
	0.035	>5	[54]	

激光诱导的表面损伤阈值

$\lambda/\mu m$	τ_p/ns	$I_{thr}/(GW \cdot cm^{-2})$	参考文献	备注
0.266	10	0.15	[81]	10 Hz
0.355	10	0.50	[81]	10 Hz
0.51~0.58	20	1	[95]	4~14 kHz
0.532	10	1.3	[81]	10 Hz
0.539 8	0.015	120~150(?)	[96]	1个脉冲
1.064	10	2.6	[81]	10 Hz
1.079 6	0.015	250~350(?)	[96]	1个脉冲

关于这一晶体

  1985 年，BBO 晶体被 Chen 等人[8]发现后，很快就成为一种在可见和紫外区应用最为普遍的晶体。我们将简短地提及最近从这种非线性材料得到的最重要的结果。在文献[61]中，利用一块 0.4 cm 长的 BBO 晶体，倍频 Nd:YAG 的激光（$\lambda = 946$ nm）产生了 550 mW 连续波（CW）的蓝光输出。在文献[97]中，对 CW 钛宝石激光器激光用 0.8 cm 长的 BBO 晶体通过二次谐波发生过程产生了输出功率为 400 mW 的 400 nm 激光。还可以通过千赫兹频率脉冲泵浦源的 SHG 来达到更高水平输出功率的 SHG 激光。最近，Watanabe 等人[91]研究了亚 10 fs 量级钛宝石激光器的二次谐波发生，获得了重复频率为 1 kHz、1.9 mJ 的 400 nm 激光脉冲，这相当于 1.9 W 蓝光的平均功率值。在文献[43]中，通过铜蒸气激光（$P = 32$ W，$\Delta f = 5$ kHz）的 SHG 产生了 255 nm 准连续紫外激光，功率为 5.1 W。在一块 I 类相位匹配的 BBO 晶体中产生了激光二极管泵浦的 Nd:YVO$_4$ 激光器辐射的三次谐波（$\lambda = 355$ nm，$P = 0.31$ W，$\tau = 23$ ns，$\Delta f = 10$ kHz）[77]。在文献[98]中，在一块 0.7 cm 长的 BBO 晶体上实现了重复频率为几千赫兹的 Nd:YAG 激光器辐射（$\tau = 25$ ns）的四次谐波发生（FoHG）（$\lambda = 266$ nm）和五次谐波发生（FiHG）（$\lambda = 213$ nm）；平均功率分别等于 2.1 W 和 0.54 W。

参考文献

[1] A. E. Kokh, V. A. Mishchenko, V. A. Antsigin, A. M. Yurkin, N. G. Kononova, V. A. Guets, Y. K. Nizienko, A. I. Zakharenko: Growth and investigation of BBO crystals with improved characteristics for UV harmonic generation. Proc. SPIE 3610, 139 – 147(1999).

[2] Data sheet of Cleveland Crystals Inc. Available at www.clevelandcrystals.com.

[3] R. Guo, A. S. Bhalla: Pyroelectric, piezoelectric, and dielectric properties of β-BaB$_2$O$_4$ single crystal. J. Appl. Phys. 66(12), 6186 – 6188(1989).

[4] C. Chen: Chinese lab grows new nonlinear optical borate crystals. Laser Focus World 25 (11), 129 – 137(1989).

[5] D. Eimerl, L. Davis, S. Velsko, E. K. Graham, A. Zalkin: Optical, mechanical, and thermal properties of barium borate. J. Appl. Phys. 62(5), 1968 – 1983(1987).

[6] R. S. Adhav, S. R. Adhav, J. M. Pelaprat: BBO's nonlinear optical phase-matching properties. Laser Focus 23(9), 88 – 100(1987).

[7] B. H. T. Chai: Optical Crystals. In: *CRC Handbook of Laser Science and Technology*, *Supplement 2: Optical Materials*, ed. by M. J. Weber (CRC Press, Boca Raton, 1995), pp. 3 – 65.

[8] C. Chen, B. Wu, A. Jiang, G. You: A new type ultraviolet SHG crystal β-BaB$_2$O$_4$. Sci-

entia Sinica B 28(3), 235 – 243(1985).

[9] Data sheet of Fujian Castech Crystals, Inc. Available at www.castech.com.

[10] J. D. Beasley: Thermal conductivities of some novel nonlinear optical materials. Appl. Opt. 33(1), 1000 – 1003(1994).

[11] R. DeSalvo, A. A. Said, D. J. Hagan, E. W. van Stryland, M. Sheik-Bahae: Infrared to ultraviolet measurements of two-photon absorption and n_2 in wide bandgap solids. IEEE J. Quant. Electr. 32(8), 1324 – 1333(1996).

[12] R. H. French, J. W. Ling, F. S. Ohuchi, C. T. Chen: Electronic structure of β-BaB_2O_4 and LiB_3O_5 nonlinear optical crystals. Phys. Rev. B 44(16), 8496 – 8502 (1991).

[13] K. Kato: Second-harmonic generation to 2 048 Å in β-BaB_2O_4. IEEE J. Quant. Electr. QE – 22(7), 1013 – 1014(1986).

[14] L. J. Bromley, A. Guy, D. C. Hanna: Synchronously pumped optical parametric oscillation in beta barium borate. Opt. Commun. 67(4), 316 – 320(1988).

[15] H. Kouta, Y. Kuwano: Attaining 186-nm light generation in cooled β-BaB_2O_4 crystal. Opt. Lett. 24(13), 1230 – 1232(1999).

[16] A. Dragomir, J. G. McInerney, D. N. Nikogosyan: Femtosecond measurements of twophoton absorption coefficients at λ = 264 nm in glasses, crystals, and liquids. Appl. Opt. 41(21), 4365 – 4376(2002).

[17] Y. X. Fan, R. C. Eckardt, R. L. Byer, C. Chen, A. D. Jiang: Barium borate optical parametric oscillator. IEEE J. Quant. Electr. 25(6), 1196 – 1199(1989).

[18] Y. X. Fan, R. C. Eckardt, R. L. Byer, J. Nolting, R. Wallenstein: Visible BaB_2O_4 optical parametric oscillator pumped at 355 nm by a single-axial-mode pulsed source. Appl. Phys. Lett. 53(21), 2014 – 2016(1988).

[19] A. Dubietis, G. Tamošauskas, A. Varanavičius, G. Valiulis: Two-photon absorbing properties of ultraviolet phase-matchable crystals at 264 and 211 nm. Appl. Opt. 39 (15), 2437 – 2440(2000).

[20] L. I. Isaenko, A. Dragomir, J. G. McInerney, D. N. Nikogosyan: Anisotropy of twophoton absorption in BBO at 264 nm. Opt. Commun. 198(4 – 6), 433 – 438(2001).

[21] G. Veitas, A. Dubietis, G. Valiulis, D. Podenas, G. Tamošauskas: Efficient femtosecond pulse generation at 264 nm. Opt. Commun. 138(4 – 6), 333 – 336(1997).

[22] D. Zhang, Y. Kong, J. Zhang: Optical parametric properties of 532-nm-pumped betabarium-borate near the infrared absorption edge. Opt. Commun. 184(5 – 6), 485 – 491(2000).

[23] C. Chen, Y. X. Fan, R. C. Eckardt, R. L. Byer: Recent developments in barium borate. Proc. SPIE 681, 12 – 19(1987).

[24] G. C. Bhar, S. Das, U. Chatterjee: Evaluation of beta barium borate crystal for nonlinear optics. Appl. Opt. 28(2), 202 – 204(1989).

[25] M.-H. Lu, Y.-M. Liu: Infrared up-conversion with beta barium borate crystal. Opt. Commun. 84(3-4), 193-198(1991).

[26] M. Oka, L. Y. Li, W. Wiechmann, N. Eguchi, S. Kubota: All solid-state continuous-wave frequency-quadrupled Nd: YAG laser. IEEE J. Sel. Topics Quant. Electr. 1(3), 859-866(1995).

[27] H. P. Li, C. H. Kam, Y. L. Lam, W. Ji: Femtosecond Z-scan measurements of nonlinear refraction in nonlinear optical materials. Opt. Mater. 15(4), 237-242(2001).

[28] M. Sheik-Bahae, M. Ebrahimzadeh: Measurements of nonlinear refraction in the secondorder $\chi^{(2)}$ materials $KTiOPO_4$, $KNbO_3$, β-BaB_2O_4, and LiB_3O_5. Opt. Commun. 142(4-6), 294-298(1997).

[29] C. A. Ebbers: Linear electro-optic effect in β-BaB_2O_4. Appl. Phys. Lett. 52(23), 948-1949(1988).

[30] H. Nakatani, W. Bosenberg, L. K. Cheng, C. L. Tang: Linear electro-optic effect in barium metaborate. Appl. Phys. Lett. 52(16), 1288-1290(1988).

[31] M. Abarkan, J. P. Salvestrini, M. D. Fontana, M. Aillerie: Frequency and wavelength dependences of electro-optic coefficients in inorganic crystals. Appl. Phys. B 76(7), 765-769(2003).

[32] I. Shoji, H. Nakamura, K. Ohdaira, T. Kondo, R. Ito, T. Okamoto, K. Tatsuki, S. Kubota: Absolute measurement of second-order nonlinear-optical coefficients of β-BaB_2O_4 for visible to ultraviolet second-harmonic wavelengths. J. Opt. Soc. Am. B 16(4), 620-624(1999).

[33] J. E. Midwinter, J. Warner: The effects of phase matching method and of uniaxial crystal symmetry on the polar distribution of second-order non-linear optical polarization. Brit. J. Appl. Phys. 16(11), 1135-1142(1965).

[34] R. S. Klein, G. E. Kugel, A. Maillard, A. Sifi, K. Polgar: Absolute non-linear optical coefficients measurements of BBO single crystal and determination of angular acceptance by second harmonic generation. Opt. Mater. 22(2), 163-169(2003).

[35] R. C. Eckardt, H. Masuda, Y. X. Fan, R. L. Byer: Absolute and relative nonlinear optical coefficients of KDP, KD*P, BaB_2O_4, $LiIO_3$, MgO: $LiNbO_3$, and KTP measured by phasematched second-harmonic generation. IEEE J. Quant. Electr. 26(5), 922-933(1990).

[36] S. P. Velsko, M. Webb, L. Davis, C. Huang: Phase-matched harmonic generation in lithium triborate(LBO). IEEE J. Quant. Electr. 27(9), 2182-2192(1991).

[37] W. J. Alford, A. V. Smith: Wavelength variation of the second-order nonlinear coefficients of $KNbO_3$, $KTiOPO_4$, $KTiOAsO_4$, $LiNbO_3$, $LiIO_3$, β-BaB_2O_4, KH_2PO_4, and LiB_3O_5 crystals: a test of Miller wavelength scaling. J. Opt. Soc. Am. B 18(4), 524-533(2001).

[38] K. Miyazaki, H. Sakai, T. Sato: Efficient deep-ultraviolet generation by frequency doubling in β-BaB$_2$O$_4$ crystals. Opt. Lett. 11(12), 797–799(1986).

[39] H. Yamamoto, K. Toyoda, K. Matsubara, M. Watanabe, S. Urabe: Development of a tunable 209 nm continuous-wave light source using two-stage frequency doubling of a Ti: sapphire laser. Jpn. J. Appl. Phys. 41(6A), 3710–3713(2002).

[40] K. Matsubara, U. Tanaka, H. Imajo, K. Hayasaka, R. Ohmukai, M. Watanabe, S. Urabe: An all-solid-state tunable 214.5 – nm continuous-wave light source by using two-stage frequency doubling of a diode laser. Appl. Phys. B 67(1), 1–4(1998).

[41] G. Xinan, Y. Shuzhong, B. Wu: Autocorrelation measurements of mode-locked Ar$^+$ laser pulses with a novel frequency doubling crystal β-BaB$_2$O$_4$. Chin. J. Lasers 13(12), 771–773(1986) [In Chinese, English trans.: Chinese Physics-Lasers 13(12), 892–894(1986)].

[42] A. A. Isaev, D. R. Jones, C. E. Little, G. G. Petrash, C. G. Whyte, K. I. Zemskov: 1.3 W average power at 255 nm by second harmonic generation in BBO by a copper HyBrID laser. Opt. Commun. 132(3–4), 302–306(1996).

[43] N. Huot, C. Jonin, N. Sanner, E. Baubeau, E. Audouard, P. Laporte: High UV average power at 15 kHz by frequency doubling of a copper HyBrID vapor laser in β-barium borate. Opt. Commun. 211(1–6), 277–282(2002).

[44] L. K. Cheng, W. R. Bosenberg, C. L. Tang: Broadly tunable optical parametric oscillation in β-BaB$_2$O$_4$. Appl. Phys. Lett. 53(3), 175–177(1988).

[45] H. Masuda, N. Umezu, K. Kimura, S. Kubota: High-repetition-rate, 192–197 nm pulse generation in β-BaB$_2$O$_4$ by intracavity sum-frequency-mixing of a Ti: sapphire laser with a frequency-quadrupled Nd: YAG laser. In: *Advanced Solid-State Lasers*, *OSA Trends in Optics and Photonics Series*, *Vol.* 26, ed. by M. M. Fejer, H. Injeyan, U. Keller(OSA, Washington DC, 1999), pp. 63–69.

[46] W. R. Bosenberg, L. K. Cheng, C. L. Tang: Ultraviolet optical parametric oscillation in β-BaB$_2$O$_4$. Appl. Phys. Lett. 54(1), 13–15(1989).

[47] A. Fix, T. Schroder, R. Wallenstein: The optical parametric oscillators of beta barium borate and lithium triborate: new sources of powerful tunable laser radiation in the ultraviolet, visible and near infrared. Laser und Optoelektronik 23(3), 106–110(1991).

[48] D. A. V. Kliner, F. Di Teodoro, J. P. Koplow, S. W. Moore, A. V. Smith: Efficient second, third, fourth, and fifth harmonic generation of a Yb-doped fiber amplifier. Opt. Commun. 210(3–6), 393–398(2002).

[49] S. J. Rehse, S. A. Lee: Generation of 125 mW frequency stabilized continuous-wave tunable laser light at 295 nm by frequency doubling in a BBO crystal. Opt. Commun. 213(4–6), 347–350(2002).

[50] H. J. Muschenborn, W. Theiss, W. Demtroder: A tunable UV-light source for laser

spectroscopy using second harmonic generation in β-BaB$_2$O$_4$. Appl. Phys. B 50(5), 365–369(1990).

[51] M. Ebrahimzadeh, A. J. Henderson, M. H. Dunn: An excimer-pumped β-BaB$_2$O$_4$ optical parametric oscillator tunable from 354 nm to 2.370 μm. IEEE J. Quant. Electr. 26(7), 1241–1252(1990).

[52] A. J. S. McGonigle, A. A. Anrews, D. W. Coutts, G. P. Hogan, K. S. Johnston, J. D. Moorhouse, C. E. Webb: Compact 2.5 – W 10 – kHz Nd: YLF-pumped dye laser. Appl. Opt. 41(9), 1714–1717(2002).

[53] J. Y. Huang, J. Y. Zhang, Y. R. Shen, C. Chen, B. Wu: High-power, widely tunable, picosecond coherent source from optical parametric amplification in barium borate. Appl. Phys. Lett. 57(19), 1961–1963(1990).

[54] J. Y. Zhang, J. Y. Huang, Y. R. Shen, C. Chen: Optical parametric generation and amplification in barium borate and lithium triborate crystals. J. Opt. Soc. Am. B 10(9), 1758–1764(1993).

[55] H. Vanherzeele, C. Chen: Widely tunable parametric generation in beta barium borate. Appl. Opt. 27(13), 2634–2636(1988).

[56] H. Komine: Average power scaling for ultraviolet-pumped β-barium borate and lithium triborate optical parametric oscillators. J. Opt. Soc. Am. B 10(9), 1751–1757(1993).

[57] G. C. Bhar, S. Das, U. Chatterjee: Noncollinear third harmonic generation and tunable second harmonic generation in barium borate. J. Appl. Phys. 66(10), 5111–5113(1989).

[58] V. Krylov, O. Ollikainen, J. Gallus, U. Wild, A. Rebane, A. Kalintsev: Efficient noncollinear parametric amplification of weak femtosecond pulses in the visible and near-infrared spectral range. Opt. Lett. 23(2), 100–102(1998).

[59] V. Krylov, J. Gallus, U. P. Wild, A. Kalintsev, A. Rebane: Femtosecond noncollinear and collinear parametric generation and amplification in BBO crystal. Appl. Phys. B 70(2), 163–168(2000).

[60] Y.-C. Chen, X.-J. Fang, J. Li, X.-Y. Liang, H.-L. Zhang, B.-H. Feng, X.-L. Zhang, L.-A. Wu, Z.-Y. Xu: Efficient femtosecond optical parametric generator with a birefringent delay compensator. Appl. Opt. 40(15), 2579–2582(2001).

[61] T. Kellner, F. Heine, G. Huber: Efficient laser performance of Nd: YAG at 946 nm and intracavity frequency doubling with LiJO$_3$, β-BaB$_2$O$_4$, and LiB$_3$O$_5$. Appl. Phys. B65(6), 789–792(1997).

[62] G. C. Bhar, S. Das, U. Chatterjee: Noncollinear phase-matched second-harmonic generation in beta barium borate. Appl. Phys. Lett. 54(15), 1383–1384(1989).

[63] X. D. Zhu, L. Deng: Broadly tunable picosecond pulses generated in a β-BaB$_2$O$_4$ optical parametric amplifier pumped by 0.532 μm pulses. Appl. Phys. Lett. 61(13), 1490–

1492(1992).

[64] M. Watanabe, K. Hayasaka, H. Imajo, J. Umezu, S. Urabe: Generation of continuous-wave coherent radiation tunable down to 190.8 nm in β-BaB$_2$O$_4$. Appl. Phys. B 53(1), 11 - 13(1991).

[65] I. V. Tomov, T. Anderson, P. M. Rentzepis: High repetition rate picosecond laser system at 193 nm. Appl. Phys. Lett. 61(10), 1157 - 1159(1992).

[66] W. L. Glab, J. P. Hessler: Efficient generation of 200 - nm light in β-BaB$_2$O$_4$. Appl. Opt. 26(16), 3181 - 3182(1987).

[67] U. Heitmann, M. Kotteritzsch, S. Heitz, A. Hese: Efficient generation of tunable VUV laser radiation below 205 nm by SFG in BBO. Appl. Phys. B 55(5), 419 - 423 (1992).

[68] H. Kitano, H. Kawai, K. Miramitsu, S. Owa, M. Yoshimura, Y. Mori, T. Sasaki: 387 - nm generation in Gd$_x$Y$_{1-x}$Ca$_4$O(BO$_3$)$_3$ crystal and its utilization for 193 - nm light source. Jpn. J. Appl. Phys. 42(2B), L166 - L169(2003).

[69] M. Hacker, T. Feurer, R. Sauerbrey, T. Lucza, G. Szabo: Programmable femtosecond laser pulses in the ultraviolet. J. Opt. Soc. Am. B 18(6), 866 - 871(2001).

[70] S. Sayama, M. Ohtsu: Tunable UV CW generation by frequency tripling of a Ti: sapphire laser. Opt. Commun. 137(4 - 6), 295 - 298(1997).

[71] T. Fujii, H. Kumagai, K. Midorikawa, M. Obara: Development of a high-power deep-ultraviolet continuous-wave coherent light source for laser cooling of silicon atoms. Opt. Lett. 25(19), 1457 - 1459(2000).

[72] H. Kumagai, K. Midorikawa, T. Iwane, M. Obara: Efficient sum-frequency generation of continuous-wave single-frequency coherent light at 252 nm with dual wavelength enhancement. Opt. Lett. 28(20), 1969 - 1971(2003).

[73] J. F. Pinto, L. Esterowitz, T. J. Carrig: Extended wavelength coverage of a Ce^{3+}: LiCAF laser between 223 and 243 nm by sum frequency mixing in β-barium borate. Appl. Opt. 37(6), 1060 - 1061(1998).

[74] D. W. Coutts, M. D. Ainsworth, J. A. Piper: Sum frequency mixing of copper vapor laser output in KDP and β-BBO. IEEE J. Quant. Electr. 25(9), 1985 - 1987(1989).

[75] S. Lu, Y. Yuan, Y. Tang, W. Xu, C. Wu: Mixing frequency generation of 271.0 - 291.5 nm in β-BaB$_2$O$_4$. In: *Proceedings of the Topical Meeting on Laser Materials and Laser Spectroscopy*, ed. by Z. Wang, Z. Zhang (World Scientific, Singapore, 1989), pp. 77 - 79.

[76] S. Sayama, M. Ohtsu: Tunable UV CW generation at 276 nm wavelength by frequency conversion of laser diodes. Opt. Commun. 145(1 - 6), 95 - 97(1998).

[77] Y.-L. Jia, J.-L. He, H.-T. Wang, S.-N. Zhu, Y.-Y. Zhu: Single pass third-harmonic generation of 310 mW of 355 nm with an all-solid-state laser. Chin. Phys. Lett.

18(12), 1589 – 1591(2001).

[78] K. Kurokawa, M. Nakazawa: Femtosecond 1.4 – 1.6 μm infrared pulse generation at a high repetition rate by difference frequency generation. Appl. Phys. Lett. 55(1), 7 – 9 (1989).

[79] G. C. Bhar, S. Das, U. Chatterjee: Second harmonic generation from non-collinear orthogonally polarized Nd: YAG laser radiation in β-BaB$_2$O$_4$. J. Phys. D 22(4), 562 – 563(1989).

[80] S. C. Matthews, J. S. Sorce: Fourth harmonic conversion of 1.06 μm in BBO and KD*P. Proc. SPIE 1220, 137 – 147(1990).

[81] H. Kouta: Wavelength dependence of repetitive-pulse laser-induced damage threshold in β-BaB$_2$O$_4$. Appl. Opt. 38(3), 545 – 547(1999).

[82] H. Kouta, Y. Kuwano: Improvement of laser-induced damage threshold in CZ-BBO by reducing the light scattering center with annealing. In: *Advanced Solid-State Lasers, OSA Trends in Optics and Photonics Series, Vol.19*, ed. by W. R. Bosenberg, M. M. Fejer(OSA, Washington DC,1998), pp. 28 – 31.

[83] H. Komine: Optical parametric oscillation in a beta-barium borate crystal pumped by an XeCl excimer laser. Opt. Lett. 13(8), 643 – 645(1988).

[84] H. Nakatani, W. R. Bosenberg, L. K. Cheng, C. L. Tang: Laser-induced damage in beta barium metaborate. Appl. Phys. Lett. 53(26), 2587 – 2589(1988).

[85] F. Huang, L. Huang: Picosecond optical parametric generation and amplification in LiB$_3$O$_5$ and β-BaB$_2$O$_4$. Appl. Phys. Lett. 61(15), 1769 – 1771(1992).

[86] K. Kuroda, T. Omatsu, T. Shimura, M. Chihara, I. Ogura: Second harmonic generation of a copper vapor laser in barium borate. Opt. Commun. 75(1), 42 – 46(1990).

[87] D. W. Coutts, M. D. Ainsworth, J. A. Piper: Enhanced efficiency of UV second harmonic and sum frequency generation from copper vapor lasers. IEEE J. Quant. Electr. 26(9), 1555 – 1558(1990).

[88] Y. Taira: High-power continuous-wave ultraviolet generation by frequency doubling of an argon laser. Jpn. J. Appl. Phys. 31(6A), L682 – L684(1992).

[89] W. Joosen, H. J. Bakker, L. D. Noordam, H. G. Muller, H. B. van Linden van den Heuvell: Parametric generation in β-barium borate of intense femtosecond pulses near 800 nm. J. Opt. Soc. Am. B 8(10), 2087 – 2093(1991).

[90] T. R. Zhang, H. R. Choo, M. C. Downer: Phase and group velocity matching for second harmonic generation of femtosecond pulses. Appl. Opt. 29(27), 3927 – 3933 (1990).

[91] T. Kanai, X. Zhou, T. Sekikawa, S. Watanabe, T. Togashi: Generation of subterawatt sub – 10 – fs blue pulses at 1 – 5 kHz by broadband frequency doubling. Opt. Lett. 28(16), 1484 – 1486(2003).

[92] M. A. Krumbügel, J. N. Sweetser, D. N. Fittinghoff, K. W. DeLong, R. Trebino: Ultrafast optical switching by use of fully phase-matched cascaded second-order nonlinearities in a polarization-gate geometry. Opt. Lett. 22(4), 245 – 247(1997).

[93] P. Qiu, A. Penzkofer: Picosecond third-harmonic light generation in β-BaB$_2$O$_4$. Appl. Phys. B 45(4), 225 – 236(1988).

[94] M. Yoshimura, T. Kamimura, K. Murase, Y. Mori, H. Yoshida, M. Nakatsuka, T. Sasaki: Bulk laser damage in CsLiB$_6$O$_{10}$ crystal and its dependence on crystal structure. Jpn. J. Appl. Phys. 38(2A), L129 – L131 (1999).

[95] D. W. Coutts, J. A. Piper: One watt average power by second harmonic and sum frequency generation from a single medium scale copper vapor laser. IEEE J. Quant. Electr. 28(8), 1761 – 1764(1992).

[96] G. G. Gurzadyan, A. S. Oganesyan, A. V. Petrosyan, R. O. Sharkhatunyan: Growth of β-barium borate crystals and study of their nonlinear properties. Zh. Tekh. Fiz. 61(3), 152 – 154(1991)[In Russian, English trans.: Sov. Phys. -Tech. Phys. 36(3), 341 – 342 (1991)].

[97] W.-L. Zhou, Y. Mori, T. Sasaki, S. Nakai: High-efficiency intracavity continuous-wave ultraviolet generation using crystals CsLiB$_6$O$_{10}$, β-BaB$_2$O$_4$, and LiB$_3$O$_5$. Opt. Commun. 123(4 – 6), 583 – 586(1996).

[98] L. B. Chang, S. C. Wang, A. H. Kung: Efficient compact watt-level deep-ultraviolet laser generated from a multi-kHz Q-switched diode-pumped solid-state laser system. Opt. Commun. 209(4 – 6), 397 – 401(2002).

2.2 LiB$_3$O$_5$，三硼酸锂(LBO)

负单轴晶：在 $\lambda = 0.532\ 1\ \mu m$ 时，$2V_z = 109.2°$[1]

分子量：119.371

密度：2.474 g/cm^3 [2]

点群：$mm2$

晶格常数：

$a = 8.46$ Å[3]；8.49 Å[4]；在 $T = 273$ K 时，8.461 Å[5]；(8.447 3 ± 0.000 7) Å[2]

$b = 7.38$ Å[3]；7.42 Å[4]；在 $T = 273$ K 时，7.412 Å[5]；(7.378 8 ± 0.000 6) Å[2]

$c = 5.13$ Å[3]；5.17 Å[4]；在 $T = 273$ K 时，5.179 Å[5]；(5.139 5 ± 0.000 5) Å[2]

介电轴和晶体学轴的变换：$X, Y, Z \Rightarrow a, c, b$

莫氏硬度：$6^{[2]}$；$7^{[6]}$

维氏硬度[7]：

$400 \sim 450 (\parallel X)$

$650 \sim 700 (\parallel Y)$

熔点：$1\,107\,K^{[2,8]}$

线性热膨胀系数 α_t [5]

T/K	$\alpha_t \times 10^6/K^{-1}$, $\parallel X$	$\alpha_t \times 10^6/K^{-1}$, $\parallel Y$	$\alpha_t \times 10^6/K^{-1}$, $\parallel Z$
273	107.1	-95.4	33.7
323	108.2	-88.0	33.6
373	108.3	-80.9	33.2
423	107.3	-74.0	32.6
473	105.3	-67.3	31.7
523	102.3	-60.7	30.5
573	98.2	-54.4	29.1
673	87.0	-42.3	25.5
723	79.8	-36.5	23.3
773	71.6	-30.9	20.9
873	52.1	-20.3	15.3
923	40.8	-15.3	12.1
973	28.5	-10.6	8.7
1 023	15.1	-5.9	5.0
1 073	0.8	-1.5	1.1

温度范围 $298 \sim 423\,K$ 之间线性热膨胀系数 α_t 的平均值[4]：

$$\alpha_t = 66.4 \times 10^{-6}\,K^{-1}(沿 X 轴)$$

$$\alpha_t = -52.8 \times 10^{-6}\,K^{-1}(沿 Y 轴)$$

$$\alpha_t = 27.3 \times 10^{-6}\,K^{-1}(沿 Z 轴)$$

温度范围 $273 \sim 1\,073\,K$ 之间线性热膨胀系数的温度关系（T 以 K 为单位）[5]：

$$\alpha_t(\parallel X) = 1.071 \times 10^{-4} + 3.204 \times 10^{-8}(T-273) - 2.063 \times 10^{-10}(T-273)^2$$

$$\alpha_t(\parallel Y) = -9.535 \times 10^{-5} - 1.481 \times 10^{-7}(T-273) - 3.489 \times 10^{-11}(T-273)^2$$

$$\alpha_t(\parallel Z) = 3.374 \times 10^{-5} + 3.400 \times 10^{-10}(T-273) - 5.067 \times 10^{-11}(T-273)^2$$

在 $p = 0.101\,325\,MPa$ 时的比热容 c_p [2]

T/K	$c_p/(\mathrm{J\cdot kg^{-1}\cdot K^{-1}})$
298	1 060

热导率：

$\kappa = 3.5\ \mathrm{W/(m\cdot K)}$ [9]

$\kappa = 2.7\ \mathrm{W/(m\cdot K)}\ (\parallel X)$ [2]

$\kappa = 3.1\ \mathrm{W/(m\cdot K)}\ (\parallel Y)$ [2]

$\kappa = 4.5\ \mathrm{W/(m\cdot K)}\ (\parallel Z)$ [2]

室温下的直接带隙能量：$E_g = 7.75\ \mathrm{eV}$ [6]，$7.78\ \mathrm{eV}$ [10]

透明范围：

以 0.5 透过计：对 0.3 cm 长晶体 0.16 ~ 2.3 μm [2]

以"0"透过计：0.155 ~ 3.2 μm [1,11]

线性吸收系数 α [12]

$\lambda/\mu\mathrm{m}$	$\alpha/\mathrm{cm^{-1}}$	$\lambda/\mu\mathrm{m}$	$\alpha/\mathrm{cm^{-1}}$
0.35 ~ 0.36	0.003 1	1.064 2	0.000 35

双光子吸收系数 β [13]

$\lambda/\mu\mathrm{m}$	τ_p/ns	$\beta\times 10^{11}/(\mathrm{cm\cdot W^{-1}})$	备注
0.211	0.000 9	103 ± 36	$\theta = 90°, \phi = 30°$
0.264	0.000 8	15 ± 5	$\theta = 90°, \phi = 30°$

折射率的实验值

$\lambda/\mu\mathrm{m}$	n_X	n_Y	n_Z	参考文献
0.253 7	1.633 5	1.658 2	1.679 2	[1]
0.289 4	1.620 9	1.646 7	1.668 1	[1]
0.296 8	1.618 2	1.645 0	1.667 4	[1]
0.312 5	1.609 7	1.641 5	1.658 8	[1]
0.334 1	1.604 3	1.634 6	1.650 9	[1]
0.365 0	1.595 23	1.625 18	1.640 25	[12]
	1.595 4	1.625 0	1.640 7	[1]
0.400 0	1.589 95	1.619 18		[12]
0.404 7	1.590 7	1.621 6	1.635 3	[1]
0.435 8	1.585 9	1.614 8	1.629 7	[1]
0.450 0	1.584 49	1.613 01	1.627 93	[12]

续表

$\lambda/\mu m$	n_X	n_Y	n_Z	参 考 文 献
0.486 1	1.581 7	1.609 9	1.624 8	[1]
0.500 0	1.580 59	1.608 62	1.623 48	[12]
0.525 0	1.579 06	1.606 86		[12]
0.532 1	1.578 68	1.606 42	1.621 22	[12]
	1.578 5	1.606 5	1.621 2	[1]
0.546 1	1.578 0	1.605 7	1.620 6	[1]
0.550 0	1.577 72	1.605 35	1.620 14	[12]
0.578 0	1.576 5	1.603 9	1.618 7	[1]
0.589 3	1.576 0	1.603 5	1.618 3	[1]
0.600 0	1.575 41	1.602 76	1.617 53	[12]
0.632 8	1.574 2	1.601 4	1.616 3	[1]
0.656 3	1.573 4	1.600 6	1.615 4	[1]
0.700 0		1.598 93	1.613 63	[12]
0.800 0	1.569 59	1.596 15	1.610 78	[12]
0.900 0	1.567 64	1.593 86	1.608 43	[12]
1.000 0	1.565 86	1.591 87	1.606 37	[12]
1.064 2	1.564 87	1.590 72	1.605 15	[12]
	1.565 6	1.590 5	1.605 5	[1]
1.100 0	1.564 32	1.590 05	1.604 49	[12]

最佳色散关系方程(λ 以 μm 为单位,$T = 293$ K)[14]:

$$n_X^2 = 2.454\,2 + \frac{0.011\,25}{\lambda^2 - 0.011\,35} - 0.013\,88\lambda^2$$

$$n_Y^2 = 2.539\,0 + \frac{0.012\,77}{\lambda^2 - 0.011\,89} - 0.018\,49\lambda^2 + 4.302\,5 \times 10^{-5}\lambda^4$$
$$\quad - 2.913\,1 \times 10^{-5}\lambda^6$$

$$n_Z^2 = 2.586\,5 + \frac{0.013\,10}{\lambda^2 - 0.012\,23} - 0.018\,62\lambda^2 + 4.577\,8 \times 10^{-5}\lambda^4$$
$$\quad - 3.252\,6 \times 10^{-5}\lambda^6$$

在[1]、[11]、[12]、[15]、[16]、[17]、[18]、[19]、[20]、[21]、[22]、[23]、[24]中给出了其他的 Sellmeier 方程组。

折射率的温度微商:

光谱范围 $0.4 \sim 1.0$ μm 和温度范围 $293 \sim 338$ K 之间(λ 以 μm 为单位)[12]:

$$dn_X/dT = -1.8 \times 10^{-6} K^{-1}$$

$$\mathrm{d}n_Y/\mathrm{d}T = -13.6 \times 10^{-6} \mathrm{K}^{-1}$$
$$\mathrm{d}n_Z/\mathrm{d}T = -(6.3 + 2.1\lambda) \times 10^{-6} \mathrm{K}^{-1}$$

光谱范围 0.4~1.0 μm 和温度范围 293~383 K 之间(λ 以 μm 为单位)[14]:
$$\mathrm{d}n_X/\mathrm{d}T = -(3.76\lambda - 2.3) \times 10^{-6} \mathrm{K}^{-1}$$
$$\mathrm{d}n_Y/\mathrm{d}T = -(19.40 - 6.01\lambda) \times 10^{-6} \mathrm{K}^{-1}$$
$$\mathrm{d}n_Z/\mathrm{d}T = -(9.70 - 1.50\lambda) \times 10^{-6} \mathrm{K}^{-1}$$

对于 λ = 0.632 8 μm 和温度范围 293~473 K 之间(λ 以 μm 为单位,T 以 K 为单位)[25]:
$$\mathrm{d}n_X/\mathrm{d}T = [0.203\,42 - 1.969\,7 \times 10^{-2}(T-273) - 1.441\,5 \times 10^{-5}(T-273)^2] \times 10^{-6} \mathrm{K}^{-1}$$
$$\mathrm{d}n_Y/\mathrm{d}T = -[10.748 + 7.103\,4 \times 10^{-2}(T-273) + 5.738\,7 \times 10^{-5}(T-273)^2] \times 10^{-6} \mathrm{K}^{-1}$$
$$\mathrm{d}n_Z/\mathrm{d}T = -[0.859\,98 + 1.547\,6 \times 10^{-1}(T-273) - 9.467\,5 \times 10^{-4}(T-273)^2 + 2.237\,5 \times 10^{-6}(T-273)^3] \times 10^{-6} \mathrm{K}^{-1}$$

非线性折射率 γ

λ/μm	$\gamma \times 10^{15}/(\mathrm{cm}^2 \cdot \mathrm{W}^{-1})$	参考文献	备注
0.780	0.26 ± 0.03	[26]	[100]方向
	0.19 ± 0.03	[26]	[010]方向
0.850	0.19 ± 0.04	[27]	$\theta = 90°$, $\phi = 31.7°$

LBO 晶体主平面上有效二阶非线性系数的表达式(Kleinman 对称条件不成立时)[28]:

XY 面: $d_{ooe} = d_{32}\cos\phi$

YZ 面: $d_{oeo} = d_{eoo} = d_{15}\cos\theta$

XZ 面, $\theta < V_Z$: $d_{eoe} = d_{oee} = d_{24}\sin^2\theta + d_{15}\cos^2\theta$

XZ 面, $\theta > V_Z$: $d_{eeo} = d_{32}\sin^2\theta + d_{31}\cos^2\theta$

LBO 晶体主平面上有效二阶非线性系数的表达式(Kleinman 对称条件成立时, $d_{15} = d_{31}$ 和 $d_{24} = d_{32}$)[28]:

XY 面: $d_{ooe} = d_{32}\cos\phi$

YZ 面: $d_{oeo} = d_{eoo} = d_{31}\cos\theta$

XZ 面, $\theta < V_Z$: $d_{eoe} = d_{oee} = d_{32}\sin^2\theta + d_{31}\cos^2\theta$

XZ 面, $\theta > V_Z$: $d_{eeo} = d_{32}\sin^2\theta + d_{31}\cos^2\theta$

文献[28]中给出了 LBO 晶体内任意方向上有效二阶非线性系数的表达式。

二阶非线性系数[29]：

$$|d_{31}(1.064\ 2\ \mu m)| = 0.67\ pm/V$$
$$|d_{32}(1.064\ 2\ \mu m)| = 0.85\ pm/V$$
$$|d_{33}(1.064\ 2\ \mu m)| = 0.04\ pm/V$$

LBO 的二阶非线性系数 d_{15}、d_{31} 和 d_{24}、d_{32}、d_{33} 的符号是相反的[29]。

相位匹配角的实验值($T = 293\ K$)

相互作用的波长/μm	$\phi_{exp}/(°)$	$\theta_{exp}/(°)$	参 考 文 献
XY 面，$\theta = 90°$			
SHG，o + o ⇒ e			
1.908⇒0.954	23.8		[16]
1.5⇒0.75	7		[16]
1.079 6⇒0.539 8	10.6		[16]
	10.7		[1], [30]
1.064 2⇒0.532 1	11.3		[12]
	11.4		[20], [31], [32]
	11.6		[11], [16], [33]
	11.8		[34]
0.946⇒0.473	19.4		[35], [36]
	19.5		[37], [38]
0.930⇒0.465	21.3		[39]
0.896⇒0.448	23.25		[40]
0.88⇒0.44	24.53		[40]
0.850⇒0.425	27		[41]
0.84⇒0.42	27.92		[40]
0.836⇒0.418	28.3		[42]
0.80⇒0.40	31.70		[40]
0.794⇒0.397	32.3		[43]
0.786⇒0.393	33		[44]
0.78⇒0.39	33.70		[40]
0.773 5⇒0.386 75	34.4		[45]
0.75⇒0.375	37.13		[40]
	37		[46]
0.746⇒0.373	37.5		[47], [48]
0.709 4⇒0.354 7	41.8		[16]
	41.9		[49]

续表

相互作用的波长/μm	$\phi_{exp}/(°)$	$\theta_{exp}/(°)$	参 考 文 献
	42		[50]
	43.5		[51]
0.63⇒0.315	55.6		[52]
0.555⇒0.2775	86		[16]
0.554⇒0.277	90		[53]
SFG, o+o⇒e			
1.3414+0.6707⇒0.44713	20		[19]
1.0642+0.5321⇒0.35473	37		[54]
	37.1		[19]
	37.2		[11], [12]
1.053+0.5265⇒0.351	38.2		[55]
1.0642+0.35473⇒0.26605	60.7		[11]
	61		[16]
0.86+0.43⇒0.2867	61		[41]
1.3188+0.26605⇒0.22139	70.2		[11]
0.21284+2.35524⇒0.1952	50.3		[19]
0.21284+1.90007⇒0.1914	63.8		[19]
0.21284+1.58910⇒0.18774	88		[19]
YZ 面,$\phi=90°$			
SHG, o+e⇒o			
1.908⇒0.954		46.2	[16]
1.5⇒0.75		14.7	[16]
1.0796⇒0.5398		19.2	[16]
1.0642⇒0.5321		19.9	[12]
		20.5	[11]
		20.6	[56]
		21.0	[16]
SFG, o+e⇒o			
1.0641+0.53205⇒0.3547		42	[57]
		42.7	[58]
1.0642+0.5321⇒0.35473		42.2	[11]
		42.5	[19]
		43.2	[12]
XZ 面,$\phi=0°$,$\theta<V_Z$			
SHG, e+o⇒e			
1.3414⇒0.6707		3.6	[59]
		4.2	[16]

续表

相互作用的波长/μm	$\phi_{exp}/(°)$	$\theta_{exp}/(°)$	参 考 文 献
		5.0	[33]
1.318 8⇒0.659 4		5.2	[11]
1.3⇒0.65		5.4	[33]
XZ 面，$\phi = 0°$，$\theta > V_Z$			
SHG, e + e⇒o			
1.341 4⇒0.670 7		86.1	[59], [60]
		86.3	[16]
		86.6	[33]
1.318 8⇒0.659 4		86.0	[11]
1.3⇒0.65		86.1	[33]
1.24⇒0.62		86	[61]

非临界相位匹配(NCPM)温度的实验值

相互作用的波长/μm	$T/℃$	参 考 文 献
沿 X 轴		
SHG，Ⅰ类		
1.547⇒0.773 5	117	[45]
1.46⇒0.73	50	[62]
1.252⇒0.626	3.5	[63]
1.25⇒0.625	−2.9	[20], [31]
1.215⇒0.607 5	21	[20]
1.211⇒0.605 5	20	[11]
1.206⇒0.603	24	[64]
1.2⇒0.6	24.3	[20], [31]
1.15⇒0.575	61.1	[20], [31]
1.135⇒0.567 5	77.4	[34]
1.11⇒0.555	108.2	[20], [31]
1.079 6⇒0.539 8	112	[1]
1.064 2⇒0.532 1	134(?)	[65]
	148	[20], [31]
	148.5	[66], [67]
	149	[34], [68]
	149.5	[69]
	151	[56]
1.047⇒0.523 5	166.5	[70]

续表

相互作用的波长/μm	T/℃	参 考 文 献
	167	[71]
	≈170	[72]
	172	[73]
	175	[74]
	176.5	[75]
	180	[76]
1.025⇒0.512 5	190.3	[20], [31]
SFG, I 类		
1.908+1.064 2⇒0.683 2	81	[34]
1.444+1.08⇒0.617 9	23	[77]
1.135+1.064 2⇒0.549 1	112	[34]
1.547+0.773 5⇒0.515 7	141	[45]
DFG, I 类		
0.532−0.8⇒1.588	135	[78]
沿 Z 轴		
SHG, II 类		
1.342⇒0.671	35	[79]
1.3⇒0.65	46	[62]

内角带宽、温度带宽和光谱带宽的实验值

相互作用的波长/μm	T/℃	$\Delta\phi^{int}$/(°)	$\Delta\theta^{int}$/(°)	ΔT/℃	参考文献
沿 X 轴					
SHG, I 类					
1.46⇒0.73	50			6	[62]
1.252⇒0.626	3.5			9	[63]
1.206⇒0.603	24			13	[64]
1.135⇒0.567 5	77.4			4.7	[34]
1.064 2⇒0.532 1	148	3.54	2.57	3.9	[31]
	148.5			2.7	[66]
	148.5			4.2	[67]
	149	2.3	1.9	4.0	[34]
	149.5			4.1	[69]
	151	2.1	2.1	2.9	[56]
1.047⇒0.523 5	175			3.5	[74]
	176.5			3.5	[75]

续表

相互作用的波长/μm	T/℃	$\Delta\phi^{\text{int}}$/(°)	$\Delta\theta^{\text{int}}$/(°)	ΔT/℃	参考文献
SFG，I 类					
1.908 + 1.064 2⇒0.683 2	81			7.4	[34]
1.444 + 1.08⇒0.617 9	23	4.2	3.0		[77]
1.135 + 1.064 2⇒0.549 1	112			5.0	[34]
DFG，I 类					
0.532 - 0.8⇒1.588	135			3.8	[78]

内角带宽、温度带宽和光谱带宽的实验值

相互作用的波长/μm	ϕ_{pm}/(°)	θ_{pm}/(°)	$\Delta\phi^{\text{int}}$/(°)	$\Delta\theta^{\text{int}}$/(°)	ΔT/℃	$\Delta\nu$/cm^{-1}	参考文献
XY 平面，$\theta = 90°$ ($T = 293$ K)							
SHG，o + o⇒e							
1.079 6⇒0.539 8		10.7	0.31				[30]
1.064 2⇒0.532 1		10.8	0.27	2.63			[56]
		11.4	0.24	1.79			[20]
		11.6			5.8		[11]
			0.34	2.64	6.7	8.8	[80]
0.886⇒0.443		24.1			7.8	15.9	[40]
0.870⇒0.435		25.4	0.12				[81]
			0.10				[40]
0.78⇒0.39		33.7	0.08				[81]
			0.07				[40]
0.760 5⇒0.380 25		35.9			15.3	10.5	[40]
0.715⇒0.357 5		41	0.06				[81]
SFG，o + o⇒e							
1.064 2 + 0.354 7⇒0.266 1		60.7			3.8		[11]
YZ 平面，$\phi = 90°$ ($T = 293$ K)							
SHG，o + e⇒o							
1.064 2⇒0.532 1		20.6	3.20	0.77			[56]
			3.00	0.81		11.5	[80]
					6.2		[11]
SFG，o + e⇒o							
1.064 1 + 0.532 05⇒0.354 7		42	0.79	0.16	6		[57]
1.064 2 + 0.532 1⇒0.354 73		42.2		0.18			[11]
		41	3.07	0.18			[15]

LBO 中 SHG 过程逆群速失配的计算值

相互作用的波长/μm	$\phi_{pm}/(°)$	$\theta_{pm}/(°)$	$\beta/(fs \cdot mm^{-1})$
XY 平面，$\theta=90°$			
SHG, o + o ⇒ e			
1.2 ⇒ 0.6	2.36		18
1.1 ⇒ 0.55	9.37		37
1.0 ⇒ 0.5	15.74		59
0.9 ⇒ 0.45	22.94		86
0.8 ⇒ 0.4	31.69		123
0.7 ⇒ 0.35	43.38		175
0.6 ⇒ 0.3	62.63		257
YZ 平面，$\phi=90°$			
SHG, o + e ⇒ o			
1.1 ⇒ 0.55		15.98	82
1.0 ⇒ 0.5		28.96	106
0.9 ⇒ 0.45		45.36	139
0.8 ⇒ 0.4		76.88	186

激光诱导的表面损伤阈值

$\lambda/\mu m$	τ_p/ns	$I_{thr}/(GW \cdot cm^{-2})$	参 考 文 献	备 注
0.266 1	12	>0.04	[82]	
	0.07	>3	[83]	
0.308	17	>0.05	[84]	
		>0.06	[85]	
	10	>0.1	[86]	
	0.000 3	47 000	[87]	细聚焦
0.354 7	18	>0.18	[88]	10 Hz
	10	>0.04	[49]	
		>0.2	[89]	
	8	>0.1	[17]	
	7	>0.14	[90]	
	0.03	>9.4	[91]	10 Hz
		>18	[92]	10 Hz
	0.015	>2.8	[51]	
	0.018	>5	[50]	
	0.025	>6	[93]	10 Hz
0.514 5	CW	>0.000 03	[94]	
0.523 5	0.055	>1.1	[76]	500 Hz

续表

$\lambda/\mu m$	τ_p/ns	$I_{thr}/(GW \cdot cm^{-2})$	参 考 文 献	备 注
		>5	[95]	500 Hz
0.532 1	CW	>0.000 4	[69]	
	60	>0.07	[96]	900 Hz
	10	>0.22	[33]	
	0.1	>4.5	[97]	500 Hz
	0.035	>3.1	[66]	
	0.015	>4.4	[18]	
0.592	0.000 5	>50	[98]	1 kHz
0.605	0.000 2	>25	[99]	
0.616	0.000 4	31 000	[87]	细聚焦
		35 000	[100]	细聚焦
		38 000	[101]	细聚焦
0.652	0.02	>0.81	[21]	
0.7~0.9	10	>0.03	[40]	10 Hz
0.71~0.87	25	1.1~1.4	[81]	25 Hz
0.72~0.85	0.001	>8	[102]	
0.77~0.83	0.000 05	>22	[103]	80 MHz
1.064 2	CW	>0.001	[69]	
	60	>0.06	[96]	1 333 Hz
	18	>0.6	[88]	10 Hz
	9	>0.9	[104]	10 Hz
	8	>0.5	[56]	
	1.3	19	[80]	
	1.1	45	[105]	体损伤
	0.1	25	[1]	
	0.035	>4.8	[66]	
	0.025	>3.3	[93]	10 Hz
1.079 6	5	20	[106]	1~25 Hz
	0.04	30	[87]	

关于这一晶体

与 BBO 晶体相比较,LBO 的应用主要集中于近红外激光的 SHG 和可见及近红外区的 OPO,我们将会给出一些典型应用的例子。在最近的实验中,三硼酸锂被用于以下激光器的激光 CW 倍频，包括 Nd：YVO$_4$ 激光器（λ = 1 342 nm）[79]、Nd：YAlO$_3$ 激光器（λ = 1 341.4 nm）[60]、Nd：YAG 激光器（λ = 946 nm）[35]、InGaAs 激光二极管参量发生器（λ = 930 nm）[39]和钛宝石激光器（λ = 746 nm）[47]。所获得的 CW SH 输出功率范围约在 0.6 ~ 1.2 W 之间。LBO 晶体也被用于二极管泵浦高平均功率调 Q Nd：YAG 激光器（λ = 1 064 nm）的 SHG；产生了 138 W 的绿光输出功率[107]。尽管在这类应用中使用 KTP 可产生更高的 SH 功率，但使用 LBO 的优点是没有光致色变损伤（灰迹）。结果表明，输出的绿光功率不随时间降低。

在文献[108]中，以 LBO 为基础的 OPO 是由锁模 ps Nd：YLF 激光器的二次谐波激光（λ = 527 nm，P = 5.6 W，τ = 35 ps，Δf = 76 MHz）泵浦的，信号光和闲频光分别可在 750 ~ 930 nm 以及 1 220 ~ 1 770 nm 的范围内调谐。信号光的输出功率最大可达到 1.6 W。在文献[58]中，以 LBO 为基础的 OPO 由锁模 ps Nd：YVO$_4$ 激光器的三次谐波激光（λ = 355 nm，P = 9.0 W，τ = 7.5 ps，Δf = 84 MHz）所同步泵浦。信号光的调谐范围为 457 ~ 479 nm，在 462 nm 处信号光输出功率可达到 5.0 W，在 1 535 nm 处闲频光的输出功率为 1.7 W。

■ 参考文献

[1] C. Chen, Y. Wu, A. Jiang, B. Wu, G. You, R. Li, S. Lin：New nonlinear-optical crystal：LiB$_3$O$_5$. J. Opt. Soc. Am. B 6(4)，616 – 621(1989).

[2] Data sheet of Cleveland Crystals Inc. Available at www.clevelandcrystals.com.

[3] Y.-N. Xu, W. Y. Ching：Electronic structure and optical properties of LiB$_3$O$_5$. Phys. Rev. B 41(8)，5471 – 5474(1990).

[4] R. Guo, S. A. Markgraf, Y. Furukawa, M. Sato, A. S. Bhalla：Pyroelectric, dielectric, and piezoelectric properties of LiB$_3$O$_5$. J. Appl. Phys. 78(12)，7234 – 7239(1995).

[5] L. Wei, D. Guiqing, H. Qingzhen, Z. An, L. Jingkui：Anisotropic thermal expansion of LiB$_3$O$_5$. J. Phys. D 23(8)，1073 – 1075 (1990).

[6] B. H. T. Chai：Optical Crystals. In：CRC Handbook of Laser Science and Technology, Supplement 2：Optical Materials, ed. by M. J. Weber (CRC Press, Boca Raton, 1995)，pp. 3 – 65.

[7] Y. Mori, S. Nakajima, A. Miyamoto, M. Inagaki, T. Sasaki, H. Yoshida, S. Nakai：Generation of ultraviolet light using a new nonlinear optical crystal CsLiB$_6$O$_{10}$. Proc. SPIE, 2633, 299 – 307(1995).

[8] Data sheet of Fujian Castech Crystals, Inc. Available at www.castech.com.

[9] J. D. Beasley：Thermal conductivities of some novel nonlinear optical materials. Appl.

Opt. 33(1), 1000 – 1003(1994).

[10] R. H. French, J. W. Ling, F. S. Ohuchi, C. T. Chen: Electronic structure of β-BaB$_2$O$_4$ and LiB$_3$O$_5$ nonlinear optical crystals. Phys. Rev. B 44(16), 8496 – 8502 (1991).

[11] K. Kato: Tunable UV generation to 0.232 5 μm in LiB$_3$O$_5$. IEEE J. Quant. Electr. 26(7), 1173 – 1175(1990).

[12] S. P. Velsko, M. Webb, L. Davis, C. Huang: Phase-matched harmonic generation in lithium triborate(LBO). IEEE J. Quant. Electr. 27(9), 2182 – 2192(1991).

[13] A. Dubietis, G. Timošauskas, A. Varanavičius, G. Valiulis: Two-photon absorbing properties of ultraviolet phase-matchable crystals at 264 and 211 nm. Appl. Opt. 39(15), 2437 – 2440(2000).

[14] K. Kato: Temperature-tuned 90° phase-matching properties of LiB$_3$O$_5$. IEEE J. Quant. Electr. 30(12), 2950 – 2952(1994).

[15] B. Wu, N. Chen, C. Chen, D. Deng, Z. Xu: Highly efficient ultraviolet generation at 355 nm in LiB$_3$O$_5$. Opt. Lett. 14(19), 1080 – 1081(1989).

[16] S. Lin, B. Wu, F. Xie, C. Chen: Phase-matching retracing behavior: new features in LiB$_3$O$_5$. Appl. Phys. Lett. 59(13), 1541 – 1543(1991).

[17] F. Hanson, D. Dick: Blue parametric generation from temperature-tuned LiB$_3$O$_5$. Opt. Lett. 16(4), 205 – 207(1991).

[18] S. Lin, J. Y. Huang, J. Ling, C. Chen, Y. R. Shen: Optical parametric amplification in a lithium triborate crystal tunable from 0.65 to 2.5 μm. Appl. Phys. Lett. 59(22), 2805 – 2807(1991).

[19] B. Wu, F. Xie, C. Chen, D. Deng, Z. Xu: Generation of tunable coherent vacuum ultraviolet radiation in LiB$_3$O$_5$ crystal. Opt. Commun. 88(4 – 6), 451 – 454(1992).

[20] T. Ukachi, R. J. Lane, W. R. Bosenberg, C. L. Tang: Phased-matched second-harmonic generation and growth of a LiB$_3$O$_5$ crystal. J. Opt. Soc. Am. B 9(7), 1128 – 1133 (1992).

[21] H. Mao, B. Wu, C. Chen, D. Zhang, P. Wang: Broadband optical parametric amplification in LiB$_3$O$_5$. Appl. Phys. Lett. 62(16), 1866 – 1868(1993).

[22] G. C. Bhar, P. K. Datta, A. M. Rudra: Noncollinear ultraviolet generation in a lithium borate crystal. Appl. Phys. B 57(6), 431 – 434(1993).

[23] T. Schröder, K.-J. Boller, A. Fix, R. Wallenstein: Spectral properties and numerical modeling of a critically phase-matched nanosecond LiB$_3$O$_5$ optical parametric oscillator. Appl. Phys. B 58(5), 425 – 438 (1994).

[24] I. T. Bodnar, A. U. Sheleg, L. V. Losovskaya: Refractive indices of lithium triborate in a temperature range of 20 – 600 ℃. Opt. Spektrosk. 86(4), 571 – 573 (1999) [In Russian, English trans.: Opt. Spectrosc. 86(4), 640 – 642(1999)].

[25] Y. Tang, Y. Cui, M. H. Dunn: Thermal dependence of the principal refractive indices

of lithium triborate. J. Opt. Soc. Am. B 12(4), 638-643(1995).

[26] H. P. Li, C. H. Kam, Y. L. Lam, W. Ji: Femtosecond Z-scan measurements of nonlinear refraction in nonlinear optical materials. Opt. Mater. 15(4), 237-242(2001).

[27] M. Sheik-Bahae, M. Ebrahimzadeh: Measurements of nonlinear refraction in the secondorder $\chi^{(2)}$ materials $KTiOPO_4$, $KNbO_3$, β-BaB_2O_4, and LiB_3O_5. Opt. Commun. 142(4-6), 294-298(1997).

[28] V. G. Dmitriev, D. N. Nikogosyan: Effective nonlinearity coefficients for three-wave interactions in biaxial crystals of $mm2$ point group symmetry. Opt. Commun. 95(1-3), 173-182(1993).

[29] D. A. Roberts: Simplified characterization of uniaxial and biaxial nonlinear optical crystals: a plea for standardization of nomenclature and conventions. IEEE J. Quant. Electr. 28(10), 2057-2074(1992).

[30] S. Lin, Z. Sun, B. Wu, C. Chen: The nonlinear optical characteristics of a LiB_3O_5 crystal. J. Appl. Phys. 67(2), 634-638 (1990).

[31] T. Ukachi, R. J. Lane, W. R. Bosenberg, C. L. Tang: Measurements of noncritically phasematched second-harmonic generation in a LiB_3O_5 crystal. Appl. Phys. Lett. 57(10), 980-982(1990).

[32] N. Pavel, J. Saikawa, T. Taira: Diode end-pumped passively Q-switched Nd: YAG laser intra-cavity frequency doubled by LBO crystal. Opt. Commun. 195(1-4), 233-240(2001).

[33] K. Kato: Parametric oscillation in LiB_3O_5 pumped at 0.532 μm. IEEE J. Quant. Electr. 26(12), 2043-2045(1990).

[34] J. T. Lin, J. L. Montgomery, K. Kato: Temperature-tuned noncritically phase-matched frequency conversion in LiB_3O_5 crystal. Opt. Commun. 80(2), 159-165(1990).

[35] D.-H. Li, P.-X. Li, Z.-G. Zhang, S.-W. Zhang: Compact high-power blue light from a diode-pumped intracavity-doubled Nd: YAG laser. Chin. Phys. Lett. 19(11), 1632-1634(2002).

[36] P.-X. Li, D.-H. Li, Z.-G. Zhang, S.-W. Zhang: Diode-pumped compact CW frequencydoubled Nd: YAG laser in the Watt range at 473 nm. Chin. Phys. Lett. 20(7), 1064-1066(2003).

[37] T. Kellner, F. Heine, G. Huber: Efficient laser performance of Nd: YAG at 946 nm and intracavity frequency doubling with $LiJO_3$, β-BaB_2O_4, and LiB_3O_5. Appl. Phys. B 65(6), 789-792(1997).

[38] X.-C. Lin, R.-N. Li, D.-F. Cui, A.-Y. Yao, Y. Feng, Y. Bi, Z.-Y. Xu: Highly-efficient blue-light generation by intracavity frequency doubling with LiB_3O_5. Chin. Phys. Lett. 19(8), 1106-1107(2002).

[39] D. Woll, B. Beier, K.-J. Boller, R. Wallenstein, M. Hagberg, S. O'Brien: 1 W of

blue 465-nm radiation generated by frequency doubling of the output of a high-power diode laser in critically phase-matched LiB_3O_5. Opt. Lett. 24 (10), 691 – 693 (1999).

[40] D.-W. Chen, J. T. Lin: Temperature-tuned phase-matching properties of LiB_3O_5 for Ti: sapphire laser frequency doubling. IEEE J. Quant. Electr. 29 (2), 307 – 310 (1993).

[41] F. Balembois, M. Gaignet, F. Louradour, V. Couderc, A. Barthelemy, P. Georges, A. Brun: Tunable picosecond UV source at 10 kHz based on an all-solid-state diode-pumped laser system. Appl. Phys. B 65(2), 255 – 258(1997).

[42] H. Yamamoto, K. Toyoda, K. Matsubara, M. Watanabe, S. Urabe: Development of a tunable 209 nm continuous-wave light source using two-stage frequency doubling of a Ti: sapphire laser. Jpn. J. Appl. Phys. 41(6A), 3710 – 3713(2002).

[43] T. Kaing, M. Houssin: Ring cavity enhanced second harmonic generation of a diode laser using LBO crystal. Opt. Commun. 157(1 – 6), 155 – 160(1998).

[44] P. L. Ramazza, S. Ducci, A. Zavatta, M. Bellini, F. T. Arecchi: Second-harmonic generation from a picosecond Ti: Sa laser in LBO: conversion efficiency and spatial properties. Appl. Phys. B 75(1), 53 – 58(2002).

[45] H. Kitano, H. Kawai, K. Miramitsu, S. Owa, M. Yoshimura, Y. Mori, T. Sasaki: 387-nm generation in $Gd_xY_{1-x}Ca_4O(BO_3)_3$ crystal and its utilization for 193-nm light source. Jpn. J. Appl. Phys. 42(2B), L166-L169(2003).

[46] S. Sayama, M. Ohtsu: Tunable UV CW generation by frequency tripling of a Ti: sapphire laser. Opt. Commun. 137(4 – 6), 295 – 298(1997).

[47] Y. Asakawa, H. Kumagai, K. Midorikawa, M. Obara: 50% frequency doubling efficiency of 1.2-W CW Ti: sapphire laser at 746 nm. Opt. Commun. 217(1 – 6), 311 – 315 (2003).

[48] T. Fujii, H. Kumagai, K. Midorikawa, M. Obara: Development of a high-power deep-ultraviolet continuous-wave coherent light source for laser cooling of silicon atoms. Opt. Lett. 25(19), 1457 – 1459(2000).

[49] Y. Wang, Z. Xu, D. Deng, W. Zheng, B. Wu, C. Chen: Visible optical parametric oscillation in LiB_3O_5. Appl. Phys. Lett. 59(5), 531 – 533(1991).

[50] H.-J. Krause, W. Daum: Efficient parametric generation of high-power coherent picosecond pulses in lithium borate tunable from 0.405 to 2.4 μm. Appl. Phys. Lett. 60(18), 2180 – 2182(1992).

[51] J. Y. Zhang, J. Y. Huang, Y. R. Shen, C. Chen, B. Wu: Picosecond optical parametric amplification in lithium triborate. Appl. Phys. Lett. 58(3), 213 – 215(1991).

[52] W. S. Pelouch, T. Ukachi, E. S. Wachman, C. L. Tang: Evaluation of LiB_3O_5 for secondharmonic generation of femtosecond optical pulses. Appl. Phys. Lett. 57(2), 111 –

113(1990).

[53] A. Borsutzky, R. Brunger, C. Huang, R. Wallenstein: Harmonic and sum-frequency generation of pulsed laser radiation in BBO, LBO and KD*P. Appl. Phys. B 52(1), 55 −62(1991).

[54] D. A. V. Kliner, F. Di Teodoro, J. P. Koplow, S. W. Moore, A. V. Smith: Efficient second, third, fourth, and fifth harmonic generation of a Yb-doped fiber amplifier. Opt. Commun. 210(3 −6), 393 −398(2002).

[55] D. Wang, C. Grässer, R. Beigang, R. Wallenstein: The generation of tunable blue ps-light-pulses from a CW mode-locked LBO optical parametric oscillator. Opt. Commun. 138(1 −3), 87 −90(1997).

[56] V. A. Dyakov, M. K. Dzhafarov, A. A. Lukashev, A. A. Podshivalov, V. I. Pryalkin: Conversion of the frequency of laser radiation in lithium triborate LiB_3O_5 crystals. Kvant. Elektron. 18 (3), 339 − 341 (1991) [In Russian, English trans.: Sov. J. Quantum Electron. 21(3), 307 −308(1991)].

[57] J. J. McFerran, A. N. Luiten: Efficient continuous-wave ultraviolet generation in LiB_3O_5 and RbD_2AsO_4. Appl. Opt. 39(18), 3115 −3119(2000).

[58] B. Ruffing, A. Nebel, R. Wallenstein: High-power picosecond LiB_3O_5 optical parametric oscillators tunable in the blue spectral range. Appl. Phys. B 72(2), 137 −149 (2001).

[59] Q. Zheng, H. Tan, L. Zhao, L. Quan: Diode-pumped 671 nm laser frequency doubled by CPM LBO. Opt. Laser Technol. 34(4), 329 −331(2002).

[60] G. Zhang, H. Shen, R. Zeng, C. Huang, W. Lin, J. Huang: The study of 1 341.4 nm Nd:$YalO_3$ laser intracavity frequency doubling by LiB_3O_5. Opt. Commun. 183(5 −6), 461 −466(2000).

[61] X. Liu, L. Qian, F. W. Wise: Efficient generation of 50-fs red pulses by frequency doubling in LiB_3O_5. Opt. Commun. 144(4 −6), 265 −268(1997).

[62] S. French, M. Ebrahimzadeh, A. Miller: Visible picosecond pulse generation in a frequency doubled optical parametric oscillator based on LiB_3O_5. Opt. Commun. 128(1 −3), 166 −176(1996).

[63] V. Shcheslavskiy, V. Petrov, F. Noack, N. Zhavoronkov: An all-solid-state laser system for generation of 100 μJ femtosecond pulses near 625 nm at 1 kHz. Appl. Phys. B 69(2), 167 −169(1999).

[64] N. Zhavoronkov, V. Petrov, F. Noack: Powerful and tunable operation of a 1 −2-kHz repetition-rate gain-switched Cr:forsterite laser and its frequency doubling. Appl. Opt. 38(15), 3285 −3293(1999).

[65] M. Tsunekane, S. Kimura, M. Kimura, N. Taguchi, H. Inaba: Broadband tuning of a continuous-wave, doubly resonant, lithium triborate optical parametric oscillator from

791 to 1620 nm. Appl. Opt. 37(27), 6459 – 6462(1998).

[66] J. Y. Huang, Y. R. Shen, C. Chen, B. Wu: Noncritically phase-matched second-harmonic generation and optical parametric amplification in a lithium triborate crystal. Appl. Phys. Lett. 58(15), 1579 – 1581(1991).

[67] I. Gontijo: Determination of important parameters for second harmonic generation in LBO. Opt. Commun. 108(4 – 6), 324 – 328 (1994).

[68] Y. Bi, H.-B. Zhang, Z.-P. Sun, Z.-R.-G.-T. Bao, H.-Q. Li, Y.-P. Kong, X.-C. Lin, G.-L. Wang, J. Zhang, W. Hou, R.-N. Li, D.-F. Cui, Z.-Y. Xu, L.-W. Song, P. Zhang, J.-F. Cui, Z.-W. Fan: High-power blue light generation by external frequency doubling of an optical parametric oscillator. Chin. Phys. Lett. 20(11), 1957 – 1959(2003).

[69] S. T. Yang, C. C. Pohalski, E. K. Gustafson, R. L. Byer, R. S. Feigelson, R. J. Raymakers, R. K. Route: 6.5-W, 532-nm radiation by CW resonant external-cavity second-harmonic generation of an 18-W Nd: YAG laser in LiB_3O_5. Opt. Lett. 16 (19), 1493 – 1495(1991).

[70] G. J. Hall, A. I. Ferguson: LiB_3O_5 optical parametric oscillator pumped by a Q-switched frequency-doubled all-solid-state laser. Opt. Lett. 18(18), 1511 – 1513(1993).

[71] G. P. A. Malcolm, M. Ebrahimzadeh, A. I. Ferguson: Efficient frequency conversion of mode-locked diode-pumped lasers and tunable all-solid-state laser sources. IEEE J. Quant. Electr. 28(4), 1172 – 1178(1992).

[72] K. F. Wall, J. S. Smucz, B. Pati, Y. Isyanova, P. Moulton, J. G. Manni: A quasi-continuouswave deep ultraviolet laser source. IEEE J. Quant. Electr. 39(9), 1160 – 1169(2003).

[73] A. Robertson, A. I. Ferguson: Synchronously pumped all-solid-state lithium triborate optical parametric oscillator in a ring configuration. Opt. Lett. 19 (2), 117 – 119 (1994).

[74] S. D. Butterworth, M. J. McCarthy, D. C. Hanna: Widely tunable synchronously pumped optical parametric oscillator. Opt. Lett. 18(17), 1429 – 1431(1993).

[75] M. J. McCarthy, S. D. Butterworth, D. C. Hanna: High-power widely-tunable picosecond pulses from an all-solid-state synchronously-pumped optical parametric oscillator. Opt. Commun. 102(3 – 4), 297 – 303(1993).

[76] M. Ebrahimzadeh, G. J. Hall, A. I. Ferguson: Temperature-tuned noncritically phase-matched picosecond LiB_3O_5 optical parameter oscillator. Appl. Phys. Lett. 60(12), 1421 – 1423(1992).

[77] H. M. Kretschmann, F. Heine, G. Huber, T. Halldorsson: All-solid-state continuous-wave doubly resonant all-intracavity sum-frequency mixer. Opt. Lett. 22(19), 1461 – 1463(1997).

[78] J. Hong, A. D. O. Bawagan, S. Charbonneau, A. Stolow: Broadly tunable femtosecond pulse generation in the near and mid-infrared. Appl. Opt. 36(9), 1894 – 1897 (1997).

[79] A. Agnesi, A. Guandalini, G. Reali: Efficient 671-nm pump source by intracavity doubling of a diode-pumped Nd: YVO_4 laser. J. Opt. Soc. Am. B 19(5), 1078 – 1082 (2002).

[80] C. Chen: Chinese lab grows new nonlinear optical borate crystals. Laser Focus World 25 (11), 129 – 137(1989).

[81] G. A. Skripko, S. G. Bartoshevich, I. V. Mikhnyuk, I. G. Tarazevich: LiB_3O_5: a highly efficient frequency converter for Ti: sapphire lasers. Opt. Lett. 16(22), 1726 – 1728(1991).

[82] Y. Tang, Y. Cui, M. H. Dunn: Lithium triborate optical parametric oscillator pumped at 266 nm. Opt. Lett. 17(3), 192 – 194(1992).

[83] J. Izawa, K. Midorikawa, M. Obara, K. Toyoda: Picosecond ultraviolet optical parametric generation using a type-II phase-matched lithium triborate crystal for an injection seed of VUV lasers. IEEE J. Quant. Electr. 33(11), 1997 – 2001(1997).

[84] G. Robertson, A. Henderson, M. H. Dunn: Broadly tunable LiB_3O_5 optical parametric oscillator. Appl. Phys. Lett. 60(3), 271 – 273(1992).

[85] G. Robertson, A. Henderson, M. Dunn: Attainment of high efficiencies in optical parametric oscillators. Opt. Lett. 16(20), 1584 – 1586(1991).

[86] M. Ebrahimzadeh, G. Robertson, M. H. Dunn: Efficient ultraviolet LiB_3O_5 optical parametric oscillator. Opt. Lett. 16(10), 767 – 769(1991).

[87] I. M. Bayanov, V. M. Gordienko, M. S. Djidjoev, V. A. Dyakov, S. A. Magnitskii, V. I. Pryalkin, A. P. Tarasevitch: Parametric generation of high-peak-power femtosecond light pulses in LBO crystal. Proc. SPIE 1800, 2 – 17(1991).

[88] Y. Cui, M. H. Dunn, C. J. Norrie, W. Sibbett, B. D. Sinclair, Y. Tang, J. A. C. Terry: All-solid-state optical parametric oscillator for the visible. Opt. Lett. 17(9), 646 – 648 (1992).

[89] Y. Cui, D. E. Withers, C. F. Rae, C. J. Norrie, Y. Tang, B. D. Sinclair, W. Sibbett, M. H. Dunn: Widely tunable all-solid-state optical parametric oscillator for the visible and near infrared. Opt. Lett. 18(2), 122 – 124(1993).

[90] A. Fix, T. Schröder, R. Wallenstein: The optical parametric oscillators of beta barium borate and lithium triborate: new sources of powerful tunable laser radiation in the ultraviolet, visible and near infrared. Laser und Optoelektronik 23(3), 106 – 110 (1991).

[91] F. Huang, L. Huang: Picosecond optical parametric generation and amplification in LiB_3O_5 and β-BaB_2O_4. Appl. Phys. Lett. 61(15), 1769 – 1771(1992).

[92] F. Huang, L. Huang, B.-I. Yin, Y. Hua: Generation of 415.9 – 482.6 nm tunable intense picosecond single pulse in LiB$_3$O$_5$. Appl. Phys. Lett. 62(7), 672 – 674 (1993).

[93] H.-J. Krause, W. Daum: High-power source of coherent picosecond light pulses tunable from 0.41 to 12.9 μm. Appl. Phys. B 56(1), 8 – 13(1993).

[94] F. G. Colville, A. J. Henderson, M. J. Padgett, J. Zhang, M. H. Dunn: Continuous-wave parametric oscillation in lithium triborate. Opt. Lett. 18(3), 205 – 207 (1993).

[95] M. Ebrahimzadeh, G. J. Hall, A. I. Ferguson: Singly resonant, all-solid-state, mode-locked LiB$_3$O$_5$ optical parametric oscillator tunable from 652 nm to 2.65 μm. Opt. Lett. 17(9), 652 – 654(1992).

[96] F. Hanson, P. Poirier: Efficient intracavity frequency doubling of a high-repetition-rate diode-pumped Nd: YAG laser. Opt. Lett. 19(19), 1526 – 1528(1994).

[97] H. Zhou, J. Zhang, T. Chen, C. Chen, Y. R. Shen: Picosecond, narrow-band, widely tunable optical parametric oscillator using a temperature-tuned lithium borate crystal. Appl. Phys. Lett. 62(13), 1457 – 1459 (1993).

[98] G. P. Banfi, C. Solcia, P. Di Trapani, R. Danielius, A. Piskarskas, R. Righini, R. Torre: Travelling-wave parametric conversion of microjoule pulses with LBO. Opt. Commun. 118 (3 – 4), 353 – 359(1995).

[99] G. P. Banfi, R. Danielius, A. Piskarskas, P. Di Trapani, P. Foggi, R. Righini: Femtosecond traveling-wave parametric generation with lithium triborate. Opt. Lett. 18(19), 1633 – 1635(1993).

[100] S. A. Akhmanov, I. M. Bayanov, V. M. Gordienko, V. A. Dyakov, S. A. Magnitskii, V. I. Pryalkin, A. P. Tarasevitch: Parametric generation of femtosecond pulses by LBO crystal in the near IR. In: *Ultrafast Processes in Spectroscopy 1991*, *IOP Conf. Ser. No. 126*, ed. by A. Laubereau, A. Seilmeier (IOP Publishing, Bristol, 1992), pp. 67 – 70.

[101] V. M. Gordienko, S. A. Magnitskii, A. P. Tarasevitch: Injection-locked femtosecond parametric oscillators on LBO crystal; towards 10^{17} Wcm^{-2}. In: *Frontiers in Nonlinear Optics. The Sergei Akhmanov Memorial Volume*, ed. by H. Walther, N. Koroteev, M. O. Scully(IOP Publishing, Bristol, 1993), pp. 286 – 292.

[102] A. Nebel, R. Beigang: External frequency conversion of cw mode-locked Ti: Al$_2$O$_3$ laser radiation. Opt. Lett. 16(22), 1729 – 1731(1991).

[103] J. Jiang, T. Hasama: High repetition-rate femtosecond optical parametric oscillator based on LiB$_3$O$_5$. Opt. Commun. 211(1 – 6), 295 – 302(2002).

[104] F. Xie, B. Wu, G. You, C. Chen: Characterization of LiB$_3$O$_5$ crystal for second-harmonic generation. Opt. Lett. 16(16), 1237 – 1239(1991).

[105] M. Yoshimura, T. Kamimura, K. Murase, Y. Mori, H. Yoshida, M. Nakatsuka, T. Sasaki: Bulk laser damage in CsLiB$_6$O$_{10}$ crystal and its dependence on crystal structure. Jpn. J. Appl. Phys. 38(2A), L129-L131(1999).

[106] S. V. Muraviov, A. A. Babin, F. I. Feldstein, A. M. Yurkin, V. A. Kamenskii, A. Y. Malyshev, M. S. Kitai, N. M. Bityurin: Efficient conversion to the fifth harmonic of spatially multimode radiation of a repetitively pulsed Nd:YAP laser. Kvant. Elektron. 25(6), 535–536(1998)[In Russian, English trans.: Quantum Electron. 28(6), 520–521(1998)].

[107] S. Konno, T. Kojima, S. Fujikawa, K. Yasui: High-brightness 138-W green laser based on an intracavity-frequency-doubled diode-side-pumped Q-switched Nd:YAG laser. Opt. Lett. 25(2), 105–107(2000).

[108] T. W. Tukker, C. Otto, J. Greve: A narrow-bandwidth optical parametric oscillator. Opt. Commun. 154(1–3), 83–86(1998).

2.3 LiNbO$_3$, 铌酸锂(LN)

负单轴晶: $n_o > n_e$

分子量: 147.846

密度:

$T = 296$ K, 4.628 g/cm^3 [1]

$T = 300$ K, (4.620 ± 0.020) g/cm^3 (化学计量比 LN) [2]

$T = 300$ K, (4.617 ± 0.020) g/cm^3 (同成分 LN) [2]

$T = 298$ K, (4.635 ± 0.005) g/cm^3 (化学计量比 LN) [3]

$T = 298$ K, (4.648 ± 0.005) g/cm^3 (同成分 LN, Li/Nb 摩尔比 = 0.940) [3]

点群: $3m$

晶格常数:

$a = (5.148\ 29 \pm 0.000\ 02)$ Å [4], $c = (13.863\ 1 \pm 0.000\ 4)$ Å [4]

$a = 5.148\ 9$ Å (同成分 LN) [5], $c = 13.863\ 1$ Å (同成分 LN) [5]

$a = (5.150\ 2 \pm 0.000\ 5)$ Å (同成分 LN) [2], $c = (13.863\ 6 \pm 0.001\ 0)$ Å (同成分 LN) [2]

$a = (5.150\ 52 \pm 0.000\ 06)$ Å (同成分 LN, Li/Nb 摩尔比 = 0.940) [3], $c = (13.864\ 96 \pm 0.000\ 03)$ Å (同成分 LN, Li/Nb 摩尔比 = 0.940) [3]

$a = (5.148\ 3 \pm 0.000\ 5)$ Å (化学计量比 LN) [2], $c = (13.857\ 3 \pm 0.001\ 0)$ Å (化学计量比 LN) [2]

$a = (5.147\ 39 \pm 0.000\ 08)$ Å (化学计量比 LN) [3], $c = (13.856\ 14 \pm 0.000\ 09)$ Å (化学计量比 LN) [3]

莫氏硬度：5[6,7]；5~5.5[8]

维氏硬度：(630±30)kgf/mm², 压痕载荷为15~200 g[9]

在100 g水中的溶解度[7]

T/K	s/g	T/K	s/g
273	0.003 4	348	0.008 9
298	0.004 1	373	0.010 9
323	0.006 4		

熔点：1 530 K[10]；1 533 K[11]

居里温度：1 411 K(同成分LN,Li/Nb 摩尔比=0.942)[12]；1 438 K(同成分LN)[5]；1 466 K(化学计量比LN)[13]；(1 466±2)K(化学计量比LN,Li/Nb摩尔比=0.988)[12]；1 475 K(化学计量比LN)[14]

居里温度与Li浓度的关系([Li]以mol%计,46%<[Li]<50%,T_C以K为单位)[15]：

$$T_C = -473.57 + 39.064[Li]$$

线性热膨胀系数

T/K	$\alpha_t \times 10^6/K^{-1}$, $\parallel c$	$\alpha_t \times 10^6/K^{-1}$, $\perp c$	参考文献
100	1.0	1.9	[16]
200	3.8	8.5	[16]
300	4.0	15.7	[16]
	4.1	15.0	[10]
400	2.0	17.5	[16]
600	2.0	19.0	[16]

线性热膨胀系数的平均值[4]

T/K	$\alpha_t \times 10^6/K^{-1}$, $\parallel c$	$\alpha_t \times 10^6/K^{-1}$, $\perp c$
297~873	≈2	
297~1 073		16.7

在温度范围298 K<T<773 K之间$\parallel c$的热膨胀[5]：

$$L(T) = L(T_0)[1 + \alpha(T-298) + \beta(T-298)^2]$$

其中T以K为单位，$T_0 = 298$ K，$\alpha = 7.5 \times 10^{-6}$ K^{-1}，$\beta = -7.7 \times 10^{-9}$ K^{-2}。

在温度范围298 K<T<773 K间$\perp c$的热膨胀[5]：

$$L(T) = L(T_0)[1 + \alpha(T-298) + \beta(T-298)^2]$$

其中 T 以 K 为单位，$T_0 = 298$ K，$\alpha = 15.4 \times 10^{-6}$ K^{-1}，$\beta = 5.3 \times 10^{-9}$ K^{-2}。

在 $p = 0.101\,325$ MPa 时的比热容 c_p[17]

T/K	$c_p/(\mathrm{J \cdot kg^{-1} \cdot K^{-1}})$	T/K	$c_p/(\mathrm{J \cdot kg^{-1} \cdot K^{-1}})$
80	136	270	619
100	218	290	639
150	379	300	648
200	514	340	682
250	592	390	718

热导率

T/K	$\kappa/(\mathrm{W \cdot m^{-1} \cdot K^{-1}})$	参考文献	备注
300	4.4	[17]	$\parallel c$
	4.5	[17]	$\perp c$
	4.6	[18]	

室温下带隙能量（直接跃迁）：$E_g = 3.9$ eV[19]，4.0 eV[20,6]；4.3 eV[21]

室温下带隙能量（间接跃迁）：$E_g = 3.3$ eV[21]

以"0"透过计的透明范围：0.4~5.5 μm[22,23]

化学计量比 LN 的 UV 透过截止边在 0.3 μm[14]

LN 晶体中以 $\alpha = 20$ cm^{-1} 计 LN 中作为 Li 的相对浓度函数的 UV 透过截止边（$T = 295$ K）[24]

$\dfrac{[\mathrm{Li}]}{[\mathrm{Li}]+[\mathrm{Nb}]}$/%	λ/μm	$\dfrac{[\mathrm{Li}]}{[\mathrm{Li}]+[\mathrm{Nb}]}$/%	λ/μm
47.8	324	49.7	309
48.5	320	50.0	303
49.2	314		

线性吸收系数 α

λ/μm	α/cm^{-1}	参考文献	备 注
0.326	2.0	[2]	同成分 LN
0.514 5	0.025	[25]	
	0.019~0.025	[26]	$\parallel c$
	0.035~0.045	[26]	e 光，$\perp c$
0.659 4	0.002 1~0.004 4	[26]	$\parallel c$

续表

$\lambda/\mu m$	α/cm^{-1}	参考文献	备注
	0.008 5 ~ 0.009 6	[26]	e光, $\perp c$
1.064 2	0.001 9 ~ 0.002 3	[26]	$\parallel c$
	0.001 4 ~ 0.001 9	[26]	e光, $\perp c$
	0.004 2	[27]	$\parallel c$
	0.002 8	[27]	$\perp c$
	0.001 1	[27]	$\parallel c$, 最佳晶体
1.318 8	0.001 8 ~ 0.004 4	[26]	$\parallel c$
	0.001 7 ~ 0.011 0	[26]	e光, $\perp c$
4.0	0.08	[28], [29]	e光
	≈ 0.1	[30]	e光
5.0	0.94	[28], [29]	e光
5.3	≈ 3	[31]	e光

双光子吸收系数 β

$\lambda/\mu m$	τ_p/ns	$\beta \times 10^{11}/(cm \cdot W^{-1})$	参考文献	备注
0.528 8	0.007	15	[32]	
0.53	0.01	500(?)	[33]	
0.532 1	10	290	[34]	o光
	10	160	[34]	e光
	0.025	350(?)	[35]	
		25	[36]	o光
	0.022	38 ± 8	[19]	$E \parallel c$
0.694 3	30	1 000	[37]	

折射率的实验值

$\lambda/\mu m$	n_o	n_e
对富锂铌酸锂(摩尔比 Li/Nb = 0.996)在 $T = 298$ K 时由气相输运平衡法生长[38]		
0.325 0	2.636 0	2.467 0
0.454 5	2.375 1	2.260 8
0.457 9	2.371 9	2.258 4
0.465 8	2.365 8	2.253 0
0.472 7	2.360 4	2.248 9

续表

$\lambda/\mu m$	n_o	n_e
0.476 5	2.357 3	2.246 5
0.488 0	2.349 5	2.239 8
0.496 5	2.343 7	2.235 2
0.501 7	2.340 5	2.232 9
0.514 5	2.333 4	2.227 0
0.632 8	2.287 8	2.189 0
1.064 2	2.233 9	2.144 0
对从化学计量比熔体中生长的铌酸锂(摩尔比 Li/Nb≈1.0), $T=293$ K[22]		
0.42	2.408 9	2.302 5
0.45	2.378 0	2.277 2
0.50	2.341 0	2.245 7
0.55	2.313 2	2.223 7
0.60	2.296 7	2.208 2
0.70	2.271 6	2.187 4
0.80	2.257 1	2.174 5
0.90	2.244 8	2.164 1
1.00	2.237 0	2.156 7
1.20	2.226 9	2.147 8
1.40	2.218 4	2.141 7
1.60	2.211 3	2.136 1
1.80	2.204 9	2.130 6
2.00	2.197 4	2.125 0
2.20	2.190 9	2.118 3
2.40	2.185 0	2.112 9
2.60	2.177 8	2.107 1
2.80	2.170 3	2.100 9
3.00	2.162 5	2.094 5
3.20	2.154 3	2.087 1
3.40	2.145 6	2.080 4
3.60	2.136 3	2.072 5
3.80	2.126 3	2.064 2
4.00	2.115 5	2.055 3
对从同成分比熔体中生长的铌酸锂(摩尔比 Li/Nb=0.946), $T=293$ K[39]		
0.435 84	2.392 76	2.292 78
0.546 08	2.316 57	2.228 16
0.632 82	2.286 47	2.202 40

续表

$\lambda/\mu m$	n_o	n_e
1.152 3	2.227 3	2.151 5
3.391 3	2.145 1	2.082 2
对从同成分比熔体中生长的铌酸锂(摩尔比 Li/Nb = 0.946),T = 297.5 K[40]		
0.404 63	2.431 7	2.326 0
0.435 84	2.392 8	2.293 2
0.467 82	2.363 4	2.268 3
0.479 99	2.354 1	2.260 5
0.508 58	2.335 6	2.244 8
0.546 07	2.316 5	2.228 5
2.576 96	2.304 0	2.217 8
0.578 97	2.303 2	2.217 1
0.587 56	2.300 2	2.214 7
0.643 85	2.283 5	2.200 2
0.667 82	2.277 8	2.195 3
0.706 52	2.269 9	2.188 6
0.809 26	2.254 1	2.174 9
0.871 68	2.247 1	2.168 8
0.935 64	2.241 2	2.163 9
0.959 98	2.239 3	2.162 2
1.014 00	2.235 1	2.158 4
1.092 14	2.230 4	2.154 5
1.153 92	2.227 1	2.151 7
1.157 94	2.226 9	2.151 5
1.287 70	2.221 1	2.146 4
1.439 97	2.215 1	2.141 3
1.638 21	2.208 3	2.135 6
1.911 25	2.199 4	2.128 0
2.184 28	2.191 2	2.121 1
2.399 95	2.184 0	2.115 1
2.615 04	2.176 5	2.108 7
2.730 35	2.172 4	2.105 3
2.897 33	2.165 7	2.099 9
3.051 48	2.159 4	2.094 6

在 T = 298 K 时用气相输运平衡法生长的富锂铌酸锂晶体(摩尔比 Li/Nb = 0.996)折射率的温度微商[38]

$\lambda/\mu m$	$dn_o/dT \times 10^6/K^{-1}$	$dn_e/dT \times 10^6/K^{-1}$
0.325 0	87	129
0.454 5	19	62
0.632 8	5.2	43
1.064 2	1.4	39

在 $T = 293$ K 时从化学计量比熔体中生长的铌酸锂晶体(摩尔比 Li/Nb ≈ 1.0)折射率的温度微商

$\lambda/\mu m$	$dn_o/dT \times 10^6/K^{-1}$	$dn_e/dT \times 10^6/K^{-1}$	参考文献
0.45 ~ 0.70	20	76	[41]
0.632 8	8	50	[23]

Sellmeier 方程(λ 以 μm 为单位):

对于用气相输运平衡法生长的富锂铌酸锂(摩尔比 Li/Nb = 0.996, 0.325 $\mu m < \lambda < 1.064$ μm, $T = 298$ K)[38]:

$$n_o^2 = 4.912\,96 + \frac{0.116\,275}{\lambda^2 - 0.048\,398} - 0.027\,3\lambda^2$$

$$n_e^2 = 4.545\,28 + \frac{0.091\,649}{\lambda^2 - 0.046\,079} - 0.030\,3\lambda^2$$

对于从化学计量比熔体中生长的铌酸锂(摩尔比 Li/Nb ≈ 1.0, 0.4 $\mu m < \lambda < 4.0$ μm, $T = 293$ K)[42]:

$$n_o^2 = 4.913\,00 + \frac{0.118\,717}{\lambda^2 - 0.045\,932} - 0.027\,8\lambda^2$$

$$n_e^2 = 4.579\,06 + \frac{0.099\,318}{\lambda^2 - 0.042\,286} - 0.022\,4\lambda^2$$

对于同成分 LN(摩尔比 Li/Nb = 0.937, 0.4 $\mu m < \lambda < 5.0$ μm, $T = 294$ K)[43]:

$$n_o^2 = 1 + \frac{2.673\,4\lambda^2}{\lambda^2 - 0.017\,64} + \frac{1.229\,0\lambda^2}{\lambda^2 - 0.059\,14} + \frac{12.614\lambda^2}{\lambda^2 - 474.6}$$

$$n_e^2 = 1 + \frac{2.980\,4\lambda^2}{\lambda^2 - 0.020\,47} + \frac{0.598\,1\lambda^2}{\lambda^2 - 0.066\,6} + \frac{8.954\,3\lambda^2}{\lambda^2 - 416.08}$$

在[39]、[40]中给出了室温下同成分 $LiNbO_3$ 晶体的其他色散关系方程组。

Sellmeier 方程的温度关系(λ 以 μm 为单位, T 以 K 为单位):

对于用气相输运平衡法生长的富锂铌酸锂(摩尔比 Li/Nb = 0.996, 0.325 $\mu m < \lambda < 1.064$ μm)[38]:

$$n_o^2 = 4.913 + 1.6 \times 10^{-8}(T^2 - 88\,506.25)$$
$$+ \frac{0.116\,3 + 0.94 \times 10^{-8}(T^2 - 88\,506.25)}{\lambda^2 - [0.220\,1 + 3.98 \times 10^{-8}(T^2 - 88\,506.25)]^2} - 0.027\,3\lambda^2$$

$$n_e^2 = 4.546 + 2.72 \times 10^{-7}(T^2 - 88\,506.25)$$
$$+ \frac{0.091\,7 + 1.93 \times 10^{-8}(T^2 - 88\,506.25)}{\lambda^2 - [0.214\,8 + 5.3 \times 10^{-8}(T^2 - 88\,506.25)]^2} - 0.030\,3\lambda^2$$

对波长 $0.4\ \mu m < \lambda < 4.0\ \mu m$ 范围内化学计量比熔体中生长的铌酸锂(摩尔比 Li/Nb ≈ 1.0)[42]：

$$n_o^2 = 4.913\,0 + \frac{0.117\,3 + 1.65 \times 10^{-8}T^2}{\lambda^2 - (0.212 + 2.7 \times 10^{-8}T^2)^2} - 0.027\,8\lambda^2$$

$$n_e^2 = 4.556\,7 + 2.605 \times 10^{-7}T^2 + \frac{0.097\,0 + 2.70 \times 10^{-8}T^2}{\lambda^2 - (0.201 + 5.4 \times 10^{-8}T^2)^2} - 0.022\,4\lambda^2$$

对波长 $0.4\ \mu m < \lambda < 3.05\ \mu m$ 范围内同成分熔体中生长的铌酸锂(摩尔比 Li/Nb = 0.946)[44]：

$$n_o^2 = 4.904\,8 + 2.142\,9 \times 10^{-8}(T^2 - 88\,506.25)$$
$$+ \frac{0.117\,75 + 2.231\,4 \times 10^{-8}(T^2 - 88\,506.25)}{\lambda^2 - [0.218\,02 - 2.967\,1 \times 10^{-8}(T^2 - 88\,506.25)]^2} - 0.027\,153\lambda^2$$

$$n_e^2 = 4.582\,0 + 2.297\,1 \times 10^{-7}(T^2 - 88\,506.25)$$
$$+ \frac{0.099\,21 + 5.271\,6 \times 10^{-8}(T^2 - 88\,506.25)}{\lambda^2 - [0.210\,90 - 4.914\,3 \times 10^{-8}(T^2 - 88\,506.25)]^2} - 0.021\,940\lambda^2$$

对同成分 LN(摩尔比 Li/Nb = 0.937)异常光折射率 Sellmeier 方程红外校正的温度关系[28]：

$$n_e^2 = 5.355\,83 + 4.629 \times 10^{-7}(T^2 - 88\,601.475\,6)$$
$$+ \frac{0.100\,473 + 3.862 \times 10^{-8}(T^2 - 88\,601.475\,6)}{\lambda^2 - [0.206\,92 - 0.89 \times 10^{-8}(T^2 - 88\,601.475\,6)]^2}$$
$$+ \frac{100 + 2.657 \times 10^{-5}(T^2 - 88\,601.475\,6)}{\lambda^2 - (11.349\,27)^2} - 1.533\,4 \times 10^{-2}\lambda^2$$

在文献[28]、[48]中给出了波长 $0.4\ \mu m < \lambda < 1.2\ \mu m$ 及温度范围 $50\ K < T < 600\ K$ 之间不同组分(摩尔比 Li/Nb = 0.887 1)LN 色散关系的温度依赖性。

非线性折射率 γ [19]

$\lambda/\mu m$	$\gamma \times 10^{15}/(cm^2 \cdot W^{-1})$	备 注
0.532 1	8.3 ± 1.3	$k \parallel X, E \parallel Z$
1.064 2	0.91 ± 0.13	$k \parallel X, E \parallel Z$

室温下低频(远低于 LN 晶体声学共振频率,即对于"自由"晶体)测量的线性电

光系数

$\lambda/\mu m$	r_{13}^T /(pm·V^{-1})	r_{22}^T /(pm·V^{-1})	r_{33}^T /(pm·V^{-1})	r_{51}^T /(pm·V^{-1})	参考文献	备注
0.632 8	+9.6	+6.8	+30.9	+32.6	[46]	
	+9.7		+31.4		[47]	同成分 LN
	+10.0	+6.81	+32.2		[48]	
	+10.0±0.8		+31.5±1.4		[49]	同成分 LN
	+10.5±0.07		+31.4±0.2		[50]	同成分 LN
	+10.4±0.8		+38.3±1.4		[49]	化学计量比 LN
	+10.9±1.0		+34.0±2.5		[51]	
		+3.3		+32±2	[52]	
		+6.4±0.3			[53]	同成分 LN
		+6.7			[54],[55]	
		+6.8±0.4			[53]	化学计量比 LN
1.047	+8		+24.6		[47]	同成分 LN
1.152 3		+5.4			[46]	
3.391 3		+3.1			[46]	

室温下高频(远高于 LN 晶体的声学共振频率,即对于"受夹"晶体)的线性电光系数

$\lambda/\mu m$	r_{13}^S /(pm·V^{-1})	r_{22}^S /(pm·V^{-1})	r_{33}^S /(pm·V^{-1})	r_{51}^S /(pm·V^{-1})	参考文献	备注
0.632 8	7.68		28.8	18.2(?)	[56]	
	8.6	3.4	30.8	28	[57]	
		3.8±0.2			[53]	同成分 LN
		4.5±0.2			[53]	化学计量比 LN
1.152 3	6.65		27.2		[56]	
3.391 3	5.32~6.5	3.1	25.5~28	23	[56]	

在 1 kHz 测量的线性电光系数 r_{22}^T 对于 LN 中 Li 相对摩尔浓度的函数关系[58]

$\lambda/\mu m$	$\frac{[Li]}{[Li]+[Nb]}/\%$	r_{22}^T /(pm·V^{-1})	$\lambda/\mu m$	$\frac{[Li]}{[Li]+[Nb]}/\%$	r_{22}^T /(pm·V^{-1})
0.6328	48.51	6.07		49.09	1.97
	48.69	4.67		49.36	6.50
	48.90	1.51		49.95	9.89

矫顽场值：

$$\approx 21 \text{ kV/mm}(\text{同成分 LN})^{[59,60]};$$
$$\approx 4 \text{ kV/mm}(\text{化学计量比 LN})^{[61]}$$

一般情况下，有效二阶非线性系数的表达式（当 Kleinman 对称条件成立时，$d_{15} = d_{24} = d_{31} = d_{32}$）[62]：

$$d_{ooe} = d_{31}\sin(\theta+\rho) - d_{22}\cos(\theta+\rho)\sin 3\phi$$
$$d_{eoe} = d_{oee} = d_{22}\cos^2(\theta+\rho)\cos 3\phi$$

有效二阶非线性系数的简化表达式（小双折射角近似，当 Kleinman 对称条件成立时，$d_{15} = d_{24} = d_{31} = d_{32}$）[63]：

$$d_{ooe} = d_{31}\sin\theta - d_{22}\cos\theta\sin 3\phi$$
$$d_{eoe} = d_{oee} = d_{22}\cos^2\theta\cos 3\phi$$

从同成分熔体生长铌酸锂的二阶非线性系数的绝对值（摩尔比 Li/Nb = 0.946）[64]：

$$|d_{31}(0.852 \ \mu m)| = 4.8 \text{ pm/V}$$
$$|d_{33}(0.852 \ \mu m)| = 25.7 \text{ pm/V}$$
$$|d_{31}(1.064 \ \mu m)| = 4.6 \text{ pm/V}$$
$$|d_{33}(1.064 \ \mu m)| = 25.2 \text{ pm/V}$$
$$|d_{31}(1.313 \ \mu m)| = 3.2 \text{ pm/V}$$
$$|d_{33}(1.313 \ \mu m)| = 19.5 \text{ pm/V}$$

从同成分熔体中生长铌酸锂的二阶非线性系数值（摩尔比 Li/Nb = 0.946）[65,66]：

$$d_{22}(1.064 \ \mu m) = (2.10 \pm 0.21) \text{ pm/V}$$
$$d_{31}(1.064 \ \mu m) = (-4.35 \pm 0.44) \text{ pm/V}$$
$$d_{33}(1.064 \ \mu m) = (-27.2 \pm 2.7) \text{ pm/V}$$

从化学计量比熔体中生长铌酸锂的二阶非线性系数值（摩尔比 Li/Nb = 1.000）[22,66]：

$$d_{22}(1.058\ \mu m) = (2.46 \pm 0.23)\ pm/V$$
$$d_{31}(1.058\ \mu m) = (-4.64 \pm 0.66)\ pm/V$$
$$d_{33}(1.058\ \mu m) = (-41.7 \pm 7.8)\ pm/V$$

相位匹配角的实验值

相互作用的波长/μm	$\theta_{exp}/(°)$	参 考 文 献
富锂铌酸锂(摩尔比 Li/Nb = 0.996, T = 295 K)		
SHG, o + o ⇒ e		
1.064 2 ⇒ 0.532 1	67.5	[38]
化学计量比熔体(摩尔比 Li/Nb ≈ 1.0, T = 293 K)		
SHG, o + o ⇒ e		
1.118 ⇒ 0.559	71.7	[42]
1.152 3 ⇒ 0.576 15	67.6	[42]
	68	[22]
	69	[11]
SFG, o + o ⇒ e		
2.179 33 + 0.852 9 ⇒ 0.613	55	[67]
4.0 + 0.723 94 ⇒ 0.613	47.5	[67]
同成分熔体(摩尔比 Li/Nb = 0.946, T = 293 K)		
SHG, o + o ⇒ e		
1.152 3 ⇒ 0.576 15	72	[11]
2.12 ⇒ 1.06	43.8	[68]
2.128 4 ⇒ 1.064 2	44.6	[69]
	47	[70]
SFG, o + o ⇒ e		
1.951 60 + 1.064 2 ⇒ 0.688 67	52.7	[71]
2.578 87 + 1.064 2 ⇒ 0.753 33	48.1	[71]
3.222 41 + 1.064 2 ⇒ 0.800 00	46.5	[71]
4.190 39 + 1.064 2 ⇒ 0.848 67	47	[71]

注:相位匹配(PM)角的值与熔体化学计量比强烈相关。

NCPM 温度的实验值

相互作用的波长/μm	$T/℃$	参 考 文 献
富锂铌酸锂(摩尔比 Li/Nb = 0.996)		
SHG, o + o ⇒ e		
0.954 ⇒ 0.477	-62.5	[38]
1.064 2 ⇒ 0.532 1	233.7	[27], [15]
	238	[38]
1.318 8 ⇒ 0.659 4	520	[38]

续表

相互作用的波长/μm	T/℃	参 考 文 献
化学计量比熔体(摩尔比 Li/Nb≈1.0)		
SHG, o + o⇒e		
1.029⇒0.514 5	15	[72]
1.058⇒0.529	0	[73]
1.064 2⇒0.532 1	43	[74]
	72	[75]
1.084⇒0.542	97	[76]
1.118⇒0.559	153.5	[42]
1.152 3⇒0.576 15	193	[73]
	208	[42]
	211	[75]
同成分熔体(摩尔比 Li/Nb = 0.946)		
SHG, o + o⇒e		
1.029⇒0.514 5	−66	[72]
1.057 6⇒0.528 8	−14	[32]
1.064 2⇒0.532 1	−8	[77]
	6	[78]
	11.5	[74]
1.084⇒0.542	38	[79]
	42	[77]
	46	[72]
1.152 3⇒0.576 15	172	[77]
	174	[40]

注:NCPM 温度值与熔体计量比强烈相关。

内角带宽的实验值[80]

相互作用的波长/μm	$\Delta\theta^{int}$/(°)	相互作用的波长/μm	$\Delta\theta^{int}$/(°)
SHG, o + o⇒e		1.06⇒0.53	0.040

温度带宽和光谱带宽的实验值

相互作用的波长/μm	T/℃	θ_{pm}/(°)	ΔT/℃	$\Delta\nu_1$/cm^{-1}	参考文献
SHG, o+o⇒e					
1.06⇒0.53	20	68		3.2	[80]
1.064 2⇒0.532 1	−1.6	90	0.74		[81]
	51	90	0.72		[82]
	234	90	0.52		[27]
1.084⇒0.542	38	90	0.74		[72]
	46	90	0.74		[79]
1.152 3⇒0.576 15	172	90	0.66		[77]
SFG, o+o⇒e					
1.7+0.694 3⇒0.493	70	90	1.6	7.9	[83]
2.65+0.488⇒0.411 5	90	90		2.9	[84]

激光诱导的体损伤阈值

λ/μm	τ_p/ns	I_{thr}/(GW·cm^{-2})	参考文献	备注
0.53	0.007	>10	[85]	
0.532 1	0.002	>70	[86]	10 Hz
0.59~0.596	≈10	>0.35	[86]	10 Hz
0.694 3	25	0.15	[87]	1个脉冲
1.06	30	0.06	[88]	25个脉冲
		0.12	[88]	10个脉冲
		0.17	[89]	
		0.47	[88]	1个脉冲
	10~30	0.3	[90]	
	14	10~13	[91]	束腰直径100 μm
		36	[91]	束腰直径21 μm
	10	0.25	[92]	
	0.006	>10	[68]	
1.064 2	30	15~20	[93]	镀膜
	20	>0.1	[69]	
	10	0.5~2	[94]	
	7	0.84	[95]	
		0.43	[96]	100 Hz
1.56	50	0.35	[97]	300 Hz,束腰直径80 μm

激光诱导的表面损伤阈值

$\lambda/\mu m$	τ_p/ns	$I_{thr} \times 10^{-12}/(W \cdot m^{-2})$	参考文献	备 注
1.064 2	12	111	[98]	[100]方向,束腰直径30 μm
	10	5~30	[94]	
	7	8.4	[95]	

关于这一晶体

  LiNbO$_3$是被最早应用的晶体之一,特别是用于非线性频率变换应用[22,99]。在20世纪60年代末到70年代初成功地用于第一个OPO系统[100],并随后成为应用非常普遍的非线性光学材料。然而,当更多比其效率和抗损伤阈值更高的晶体(KTP、BBO和LBO)被引入后,体块LN晶体的应用被完全替代了。随后,确实神奇的是,周期性极化LN(PPLN),由于其沿光轴方向有非常高的有效非线性系数值(高达20 pm/V)可以被应用,成为20世纪90年代最为流行的非线性材料。令人啼笑皆非的是,在周期性极化材料中的准相位匹配(QPM)方法是由Bloembergen等人早在1962年就提出的[101],甚至于比双折射相位匹配的提出还要早,只是在当时缺乏极化的手段而停止了这一方法的发展。在1980年,一个中国科研组发现了周期极化LN晶体中SHG的增强效应[102],十年后,有了PPLN在SHG中的第一个应用[103-106],以后又先后报道了在DFG中的应用[107,108]以及在OPO中的应用[109]。在其时,有数以百计的工作致力于PPLN及其应用。QPM和PPLN的综述可见文献[110-112]。

  铌酸锂晶体有一些基本的缺点,即损伤阈值低并对光折变损伤敏感[113,114]。为了避免光折变效应,LN(或PPLN)元件应保持在升温状态,典型的是升温至140~230 ℃[115-118]。另一条途径就是掺MgO(见氧化镁掺杂铌酸锂晶体)。实际表明1.8 mol%掺杂的化学计量比LN的抗光折变损伤阈值要比未掺杂的化学计量比以及同成分熔化LN的阈值要高4个数量级[12]。在ZnO掺杂的LN晶体中也观察到了抗光折变损伤阈值的相似效应[119,120]。

参考文献

[1] L. G. van Uitert, J. J. Rubin, W. A. Bonner: Growth of Ba$_2$NaNb$_5$O$_{15}$ single crystals for optical applications. IEEE J. Quant. Electr. QE-4(10), 622-627(1968).

[2] D. Redfield, W. J. Burke: Optical absorption edge of LiNbO$_3$. J. Appl. Phys. 45(10), 4566-4571(1974).

[3] S. C. Abrahams, P. Marsh: Defect structure dependence on composition of lithium nio-

bate. Acta Crystallogr. B 42(1), 61 −68(1986).

[4] S. C. Abrahams, H. J. Levinstein, J. M. Reddy: Ferroelectric lithium niobate. V. Polycrystal X-ray diffraction study between 24° and 1 200°. J. Phys. Chem. Solids 27(6 −7), 1019 − 1026(1966).

[5] Y. S. Kim, R. T. Smith: Thermal expansion of lithium tantalate and lithium niobate crystals. J. Appl. Phys. 40(11), 4637 −4641(1969).

[6] B. H. T. Chai: Optical Crystals. In: *CRC Handbook of Laser Science and Technology, Supplement 2: Optical Materials*, ed. by M. J. Weber(CRC Press, Boca Raton, 1995), pp. 3 −65.

[7] Y. S. Kuzminov: *Lithium Niobate and Lithium Tantalate. Materials for Nonlinear Optics* (Nauka, Moscow, 1975) [In Russian].

[8] V. G. Dmitriev, G. G. Gurzadyan, D. N. Nikogosyan: *Handbook of Nonlinear Optical Crystals*; Third Revised Edition(Springer, Berlin, 1999).

[9] K. G. Subhadra, K. Kishan Rao, D. B. Sirdeshmukh: Systematic hardness studies on lithium niobate crystals. Bull. Mater. Sci. 23(2), 147 −150(2000).

[10] S. S. Ballard, J. S. Browder: Thermal Properties. In: *CRC Handbook of Laser Science and Technology*, Vol. IV, *Optical Materials*: Part 2, ed. by M. J. Weber (CRC Press, Boca Raton, 1987), pp. 49 −54

[11] A. M. Prokhorov, Y. S. Kuzminov: *Physics and Chemistry of Crystalline Lithium Niobate* (Adam Hilger, Bristol, 1990).

[12] K. Niwa, Y. Furukawa, S. Takekawa, K. Kitamura: Growth and characterization of MgO doped near stoichiometric $LiNbO_3$ crystals as a new nonlinear optical crystal. J. Cryst. Growth 208(1 −4), 493 −500(2000).

[13] K. Polgar, A. Peter, I. Földvari: Crystal growth and stoichiometry of $LiNbO_3$ prepared by the flux method. Opt. Mater. 19(1), 7 −11(2002).

[14] G. Ravi, R. Jayavel, S. Takekawa, M. Nakamura, K. Kitamura: Effect of niobium substitution in stoichiometric lithium tantalate(SLT) single crystals. J. Cryst. Growth 250 (1 −2), 146 −151(2003).

[15] P. F. Bordui, R. G. Norwood, D. H. Jundt, M. M. Fejer: Preparation and characterization of off-congruent lithium niobate crystals. J. Appl. Phys. 71 (2), 875 −879 (1992).

[16] *Physical Quantities. Handbook*, ed. by I. S. Grigoriev, E. Z. Meilikhov (Energoatomizdat, Moscow, 1991) [In Russian].

[17] V. V. Zhdanov a, V. P. Klyuev, V. V. Lemanov, I. A. Smirnov, V. V. Tikhonov: Thermal properties of lithium niobate crystals. Fiz. Tverd. Tela 10(6), 1725 −1728(1968). [In Russian, English trans.: Sov. Phys. -Solid State 10(6), 1360 −1362(1968)].

[18] A. A. Blistanov, V. S. Bondarenko, N. V. Perelomova, F. N. Strizhevskaya,

V. V. Tchkalova, M. P. Shaskolskaya: *Acoustic Crystals* (Nauka, Moscow, 1982). [In Russian].

[19] R. DeSalvo, A. A. Said, D. J. Hagan, E. W. van Stryland, M. Sheik-Bahae: Infrared to ultraviolet measurements of two-photon absorption and n_2 in wide bandgap solids. IEEE J. Quant. Electr. 32(8), 1324 – 1333 (1996).

[20] E. W. van Stryland, L. L. Chase: Two-Photon Absorption. Inorganic Materials. In: *CRC Handbook of Laser Science and Technology*, Supplement 2: *Optical Materials*, ed. By M. J. Weber (CRC Press, Boca Raton, 1995) pp. 299 – 328.

[21] S. Kase, K. Ohi: Optical absorption and interband Faraday rotation in $LiTaO_3$ and $LiNbO_3$. Ferroelectrics 8(1 – 2), 419 – 420 (1974).

[22] G. D. Boyd, R. C. Miller, K. Nassau, W. L. Bond, A. Savage: $LiNbO_3$: an efficient phase matchable nonlinear optical material. Appl. Phys. Lett. 5 (11), 234 – 236 (1964).

[23] G. V. Ageev, R. P. Bashuk, A. S. Bebchuk, N. S. Voidetskaya, D. A. Gromov, Y. N. Solovieva, A. V. Chesnokov: Optical and electrooptical properties of some alkali and alkaline earth niobates and tantalates. In: *Nonlinear Optics*, ed. by R. V. Khokhlov (Nauka, Novosibirsk, 1968) pp. 211 – 217 [In Russian].

[24] L. Kovacs, G. Ruschhaupt, K. Polgar, G. Corradi, M. Wöhlecke: Composition dependence of the ultraviolet absorption edge in lithium niobate. Appl. Phys. Lett. 70(21), 2801 – 2803 (1997).

[25] Y. C. See, S. Guha, J. Falk: Limits to the NEP of an intracavity $LiNbO_3$ upconverter. Appl. Opt. 19(9), 1415 – 1418 (1980).

[26] D. J. Gettemy, W. C. Harker, G. Lindholm, N. P. Barnes: Some optical properties of KTP, $LiIO_3$, and $LiNbO_3$. IEEE J. Quant. Electr. 24(11), 2231 – 2237 (1988).

[27] D. H. Jundt, M. M. Fejer, R. L. Byer, R. G. Norwood, P. F. Bordui: 69% efficient continuous-wave second-harmonic generation in lithium-rich lithium niobate. Opt. Lett. 16(23), 1856 – 1858 (1991).

[28] D. H. Jundt: Temperature-dependent Sellmeier equation for the index of refraction, n_e, in congruent lithium niobate. Opt. Lett. 22(20), 1553 – 1555 (1997).

[29] L. E. Myers, R. C. Eckardt, M. M. Fejer, R. L. Byer, W. R. Bosenberg: Multigrating quasiphase-matched optical parametric oscillator in periodically poled $LiNbO_3$. Opt. Lett. 21(8), 591 – 593 (1996).

[30] G. Hansson, D. D. Smith: Mid-infrared-wavelength generation in 2-μm pumped periodically poled lithium niobate. Appl. Opt. 37(24), 5743 – 5746 (1998).

[31] L. Lefort, K. Puech, S. D. Butterworth, G. W. Ross, P. G. R. Smith, D. C. Hanna, D. H. Jundt: Efficient, low-threshold synchronously-pumped parametric oscillation in periodically-poled lithium niobate over the 1.3 μm to 5.3 μm range. Opt. Commun.

152(1-3), 55-58(1998).

[32] A. Seilmeier, W. Kaiser: Generation of tunable picosecond light pulses covering the frequency range between 2 700 and 32 000 cm^{-1}. Appl. Phys. 23(2), 113-119(1980).

[33] D. von der Linde, A. M. Glass, K. F. Rodgers: Multiphoton photorefractive processes for optical storage in LiNbO$_3$. Appl. Phys. Lett. 25(3), 155-157(1974).

[34] N. M. Bityurin, V. I. Bredikhin, V. N. Genkin: Nonlinear optical absorption and energy structure of LiNbO$_3$ and α-LiIO$_3$ crystals. Kvant. Elektron. 5(11), 2453-2457(1978) [In Russian, English trans.: Sov. J. Quantum Electron. 8(11), 1377-1379(1978)].

[35] H. Kurz, D. von der Linde: Nonlinear optical excitation of photovoltaic LiNbO$_3$. Ferroelectrics 21(1-4), 621-622(1978).

[36] H. Li, F. Zhou, X. Zhang, W. Ji: Picosecond Z-scan study of bound electronic Kerr effect in LiNbO$_3$ crystal associated with two-photon absorption. Appl. Phys. B 64(6), 659-662(1997).

[37] V. V. Arseniev, V. S. Dneprovskii, D. N. Klyshko, A. N. Penin: Nonlinear absorption and restriction of light intensity in semiconductors. Zh. Eksp. Teor. Fiz. 56(3), 760-765(1969) [In Russian, English trans.: Sov. Phys. -JETP 29(3), 413-415(1969)].

[38] D. H. Jundt, M. M. Fejer, R. L. Byer: Optical properties of lithium-rich lithium niobate fabricated by vapor transport equilibration. IEEE J. Quant. Electr. 26(1), 135-138(1990).

[39] D. S. Smith, H. D. Riccius, R. P. Edwin: Refractive indices of lithium niobate. Opt. Commun. 17(3), 332-335(1976); Errata. Opt. Commun. 20(1), 188(1977).

[40] D. F. Nelson, R. M. Mikulyak: Refractive indices of congruently melting lithium niobate. J. Appl. Phys. 45(8), 3688-3689(1974).

[41] J. E. Midwinter: Lithium niobate: effects of composition on the refractive indices and optical second-harmonic generation. J. Appl. Phys. 39(7), 3033-3038(1968).

[42] M. V. Hobden, J. Warner: The temperature dependence of the refractive indices of pure lithium niobate. Phys. Lett. 22(3), 243-244(1966).

[43] D. E. Zelmon, D. L. Small, D. Jundt: Infrared corrected Sellmeier coefficients for congruently grown lithium niobate and 5 mol. % magnesium oxide-doped lithium niobate. J. Opt. Soc. Am. B 14(12), 3319-3322(1997).

[44] G. J. Edwards, M. Lawrence: A temperature-dependent dispersion equation for congruently grown lithium niobate. Opt. Quant. Electron. 16(4), 373-375(1984).

[45] U. Schlarb, K. Betzler: Refractive indices of lithium niobate as a function of temperature, wavelength, and composition: a generalized fit. Phys. Rev. B 48(21), 15613-15620(1993).

[46] A. Yariv, P. Yeh: *Optical Waves in Crystals* (JohnWiley & Sons, New York, 1984).

[47] A. Mendez, A. Garcia-Cabanes, E. Dieguez, J. M. Cabrera: Wavelength dependence of electro-optic coefficients in congruent and stoichiometric LiNbO$_3$. Electron. Lett. 35(6), 498−499(1999).

[48] J. D. Zook, D. Chen, G. N. Otto: Temperature dependence and model of electro-optic effect in LiNbO$_3$. Appl. Phys. Lett. 11(5), 159−161(1967).

[49] T. Fujiwara, M. Takahashi, M. Ohama, A. J. Ikushima, Y. Furukawa, K. Kitamura: Comparison of electro-optic effect between stoichiometric and congruent LiNbO$_3$. Electron. Lett. 35(6), 499−501(1999).

[50] J. A. de Toro, M. D. Serrano, A. Garcia Cabanes, J. M. Cabrera: Accurate interferometric measurement of electrooptic coefficients: application to quasi-stoichiometric LiNbO$_3$. Opt. Commun. 154(1−3), 23−27(1998).

[51] K. Onuki, N. Uchida, T. Saku: Interferometric method for measuring electro-optic coefficients in crystals. J. Opt. Soc. Am. 62(9), 1030−1032(1972).

[52] E. Bernal, G. D. Chen, T. C. Lee: Low frequency electro-optic and dielectric constants of lithium niobate. Phys. Lett. 21(3), 259−260(1966).

[53] M. Abarkan, J. P. Salvestrini, M. D. Fontana, M. Aillerie: Frequency and wavelength dependences of electro-optic coefficients in inorganic crystals. Appl. Phys. B 76(7), 765−769(2003).

[54] P. V. Lenzo, E. G. Spencer, K. Nassau: Electro-optic coefficients in single-domain ferroelectric lithium niobate. J. Opt. Soc. Am. 56(5), 633−635(1966).

[55] R. S. Weis, T. K. Gaylord: Lithium niobate: summary of physical properties and crystal structure. Appl. Phys. A 37(4), 191−203(1985).

[56] I. P. Kaminow: Tables of Linear Electrooptic Coefficients. In: *CRC Handbook of Laser Science and Technology, Vol. III, Optical Materials: Part 2*, ed. by M. J. Weber (CRC Press, Boca Raton, 1986), pp. 253−278.

[57] E. H. Turner: High-frequency electro-optic coefficients of lithium niobate. Appl. Phys. Lett. 8(11), 303−304(1966).

[58] F. Abdi, M. Aillerie, P. Bourson, M. D. Fontana, K. Polgar: Electro-optic properties in pure LiNbO$_3$ crystals from the congruent to the stoichiometric composition. J. Appl. Phys. 84(4), 2251−2254(1998).

[59] T. Hatanaka, K. Nakamura, T. Taniuchi, H. Ito, Y. Furukawa, K. Kitamara: Quasi-phase-matched optical parametric oscillation with periodically poled stoichiometric LiTaO$_3$. Opt. Lett. 25(9), 651−653(2000).

[60] J.-P. Meyn, C. Laue, R. Knappe, R. Wallenstein, M. M. Fejer: fabrication of periodically poled lithium tantalate for UV generation with diode lasers. Appl. Phys. B 73(2), 111−114(2001).

[61] K. Nakamura, T. Hatanaka, H. Ito: High output energy quasi-phase-matched optical

parametric oscillator using diffusion-bonded periodically poled and single domain LiNbO$_3$. Jpn. J. Appl. Phys. 40(4A), L337-L339(2001).

[62] I. Shoji, H. Nakamura, K. Ohdaira, T. Kondo, R. Ito, T. Okamoto, K. Tatsuki, S. Kubota: Absolute measurement of second-order nonlinear-optical coefficients of β-BaB$_2$O$_4$ for visible to ultraviolet second-harmonic wavelengths. J. Opt. Soc. Am. B 16(4), 620–624(1999).

[63] J. E. Midwinter, J. Warner: The effects of phase matching method and of uniaxial crystal symmetry on the polar distribution of second-order non-linear optical polarization. Brit. J. Appl. Phys. 16(11), 1135–1142(1965).

[64] I. Shoji, T. Kondo, A. Kitamoto, M. Shirane, R. Ito: Absolute scale of second-order nonlinear-optical coefficients. J. Opt. Soc. Am. B 14(9), 2268–2294(1997).

[65] R. C. Miller, W. A. Nordland, P. M. Bridenbaugh: Dependence of second-harmonic-generation coefficients of LiNbO$_3$ on melt composition. J. Appl. Phys. 42(11), 4145–4147 (1971).

[66] D. A. Roberts: Simplified characterization of uniaxial and biaxial nonlinear optical crystals: a plea for standardization of nomenclature and conventions. IEEE J. Quant. Electr. 28(10), 2057–2074(1992).

[67] D. S. Moore, S. C. Schmidt: Tunable subpicosecond infrared pulse generation to 4 μm. Opt. Lett. 12(7), 480–482(1987).

[68] A. Laubereau, L. Greiter, W. Kaiser: Intense tunable picosecond pulses in the infrared. Appl. Phys. Lett. 25(1), 87–89(1974).

[69] R. L. Herbst, R. N. Fleming, R. L. Byer: A 1.4–4.0-μm high-energy angle-tuned LiNbO$_3$ parametric oscillator. Appl. Phys. Lett. 25(9), 520–522(1974).

[70] Z. I. Ivanova, V. Kabelka, S. A. Magnitskii, A. Piskarskas, V. Smilgiavichyus, N. M. Rubinina, V. G. Tunkin: Parametric generation of infrared picosecond pulses in LiNbO$_3$ crystals. Kvant. Elektron. 4(11), 2469–2472(1977) [In Russian, English trans.: Sov. J. Quantum Electron. 7(11),1414–1416(1977)].

[71] K. Kato: High-efficiency high-power difference-frequency generation at 2-4 μm in LiNbO$_3$. IEEE J. Quant. Electr. QE–16(10), 1017–1018(1980).

[72] P. M. Bridenbaugh, J. R. Carruthers, J. M. Dziedzic, F. R. Nash: Spatually uniform and alterable SHG phase-matching temperatures in lithium niobate. Appl. Phys. Lett. 17(3), 104–106(1970).

[73] R. C. Miller, G. D. Boyd, A. Savage: Nonlinear optical interactions in LiNbO$_3$ without double refraction. Appl. Phys. Lett. 6(4), 77–79(1965).

[74] H. Fay, W. J. Alfred, H. M. Dess: Dependence of second-harmonic phase-matching temperature in LiNbO$_3$ crystals on melt composition. Appl. Phys. Lett. 12(3), 89–92 (1968).

[75] N. B. Angert, O. F. Butyagin, V. P. Zorenko, A. P. Kudryavtseva, V. R. Kushnir, S. R. Rustamov: Phase-matching angles and temperatures of lithium metaniobate crystals of different stoichiometries. Kvant. Elektron. No. 5, 128 – 129 (1971) [In Russian, English trans. : Sov. J. Quantum Electron. 1(5), 542 – 543 (1971)].

[76] J. C. Bergman, A. Ashkin, A. A. Ballman, J. M. Dziedzic, H. J. Levinstein, R. G. Smith: Curie temperature, birefringence, and phase-matching temperature variations in $LiNbO_3$ as a function of melt stoichiometry. Appl. Phys. Lett. 12(3), 92 – 94 (1968).

[77] R. L. Byer, J. F. Young, R. S. Feigelson: Growth of high-quality $LiNbO_3$ crystals from the congruent melt. J. Appl. Phys. 41(6), 2320 – 2325 (1970).

[78] T. R. Volk, N. M. Rubinina, A. I. Kholodnykh: Efficient laser frequency converters made of nonphotorefractive lithium niobate. Kvant. Elektron. 15(8), 1705 – 1706 (1988) [In Russian, English trans. : Sov. J. Quantum Electron. 18(8), 1061 – 1062 (1988)].

[79] F. R. Nash, G. D. Boyd, M. Sargent III, P. M. Bridenbaugh: Effect of optical inhomogeneities on phase matching in nonlinear crystals. J. Appl. Phys. 41(6), 2564 – 2576 (1970).

[80] W. F. Hagen, P. C. Magnante: Efficient second-harmonic generation with diffraction-limited and high-spectral-radiance Nd-glass lasers. J. Appl. Phys. 40(1), 219 – 224 (1969).

[81] E. O. Ammann, S. Guch, Jr. : 1.06 – 0.53 μm second harmonic generation using congruent lithium niobate. Appl. Phys. Lett. 52(17), 1374 – 1376 (1988).

[82] V. A. Dyakov, V. I. Pryalkin, A. I. Kholodnykh: Potassium niobate optical parametric oscillator pumped by the second harmonic of a garnet laser. Kvant. Elektron. 8(4), 715 – 721 (1981) [In Russian, English trans. : Sov. J. Quantum Electron. 11(4), 433 – 436 (1981)].

[83] J. E. Midwinter, J. Warner: Up-conversion of near infra-red to visible radiation in lithium meta-niobate. J. Appl. Phys. 38(2), 519 – 523 (1967).

[84] E. N. Antonov, V. G. Koloshnikov, D. N. Nikogosyan: Nonlinear frequency converter as infrared spectrometer and detector. Opt. Spektrosk. 36(4), 768 – 772 (1974) [In Russian, English trans. : Opt. Spectrosc. USSR 36(4), 446 – 448 (1974)].

[85] T. Kushida, Y. Tanaka, M. Ojima, Y. Nakazaki: Generation of widely tunable picosecond pulses by optical parametric effect. Jpn. J. Appl. Phys. 14(7), 1097 – 1098 (1975).

[86] M. Berg, C. B. Harris, T. W. Kenny, P. L. Richards: Generation of intense tunable picosecond pulses in the far-infrared. Appl. Phys. Lett. 47(3), 206 – 208 (1985).

[87] J. Falk, J. E. Murray: Single-cavity noncollinear optical parametric oscillation. Appl. Phys. Lett. 14(8), 245 – 247 (1969).

[88] G. M. Zverev, E. A. Levchuk, V. A. Pashkov, Y. D. Poryadin: Laser-radiation-induced damage to the surface of lithium niobate and tantalate single crystals. Kvant. Elektron. No. 2, 94 – 96 (1972) [In Russian, English trans. : Sov. J. Quantum Electron. 2(2),167 – 169(1972)].

[89] G. M. Zverev, S. A. Kolyadin, E. A. Levchuk, L. A. Skvortsov: Influence of the surface layer on the optical strength of lithium niobate. Kvant. Elektron. 4(9), 1882 – 1889 (1977) [In Russian, English trans. : Sov. J. Quantum Electron. 7 (9), 1071 – 1075 (1977)].

[90] S. J. Brosnan, R. L. Byer: Optical parametric oscillator threshold and linewidth studies. IEEE J. Quant. Electr. QE – 15(6), 415 – 431(1979).

[91] G. M. Zverev, E. A. Levchuk, E. K. Maldutis: Destruction of KDP, ADP, and LiNbO$_3$ crystals by powerful laser radiation. Zh. Eksp. Teor. Fiz. 57(3), 730 – 736 (1969) [In Russian, English trans. : Sov. Phys. - JETP 30(3), 400-403 (1970)].

[92] G. M. Zverev: Materials for quantum electronics (yttrium-aluminium garnet, lithium niobate). Izv. Akad. Nauk SSSR, Ser. Fiz. 44(8), 1614 – 1621 (1980) [In Russian, English trans. : Bull. Acad. Sci. USSR, Phys. Ser. 44(8), 49 – 54(1980)].

[93] M. J. Soileau: Mechanism of laser-induced failure in antireflection-coated LiNbO$_3$ crystals. Appl. Opt. 20(6), 1030 – 1033(1981).

[94] R. M. Wood, R. T. Taylor, R. L. Rouse: Laser damage in optical materials at 1.06 μm. Opt. Laser Technol. 7(3), 105 – 111(1975).

[95] M. Bass: Nd: YAG laser-irradiation-induced damage to LiNbO$_3$ and KDP. IEEE J. Quant. Electr. QE – 7(7), 350 – 359(1971).

[96] L. E. Myers, G. D. Miller, R. C. Eckardt, M. M. Fejer, R. L. Byer, W. R. Bosenberg: Quasi-phase-matched 1.064-μm-pumped optical parametric oscillator in bulk periodically poled LiNbO$_3$. Opt. Lett. 20(1), 52 – 54(1995).

[97] P. E. Britton, D. Taverner, K. Puech, D. J. Richardson, P. G. R. Smith, G. W. Ross, D. C. Hanna: Optical parametric oscillation in periodically poled lithium niobate driven by a diode-pumped Q-switched erbium fiber laser. Opt. Lett. 23(8), 582 – 584 (1998).

[98] M. Bass, H. H. Barrett: Avalanche breakdown and the probabilistic nature of laser-induced damage. IEEE J. Quant. Electr. QE – 8(3), 338 – 343(1972).

[99] A. A. Ballman: Growth of piezoelectric and ferroelectric materials by Czochralski technique. J. Am. Ceram. Soc. 48(2), 112 – 113(1965).

[100] J. A. Giordmaine, R. C. Miller: Tunable coherent parametric oscillation in LiNbO$_3$ at optical frequencies. Phys. Rev. Lett. 14(24), 973 – 976(1965).

[101] J. A. Armstrong, N. Bloembergen, J. Ducuing, P. S. Pershan: Interactions between light waves in a nonlinear dielectric. Phys. Rev. 127(6), 1918 – 1939(1962).

[102] D. Feng, N.-B. Ming, J.-F. Hong, Y.-S. Yang, J.-S. Zhu, Z. Yang, Y.-N. Wang: Enhancement of second-harmonic generation in LiNbO$_3$ crystals with periodic laminar ferroelectric domains. Appl. Phys. Lett. 37(7), 607–609(1980).

[103] D. H. Jundt, G. A. Magel, M. M. Fejer, R. L. Byer: Periodically poled LiNbO$_3$ for high-efficiency second-harmonic generation. Appl. Phys. Lett. 59(21), 2657–2659 (1991).

[104] W. K. Burns, R. W. McElhanon, L. Goldberg: Second harmonic generation in field poled, quasi-phase-matched, bulk LiNbO$_3$. IEEE Photon. Technol. Lett. 6(2), 252–254(1994).

[105] V. Pruneri, J. Webjörn, P. S. J. Russell, J. R. M. Barr, D. C. Hanna: Intracavity second harmonic generation of 0.532 μm in bulk periodically poled lithium niobate. Opt. Commun. 116(1–3), 159–162(1995).

[106] V. Pruneri, R. Koch, P. G. Kazansky, W. A. Clarkson, P. S. J. Russell, D. C. Hanna: 49 mW of CW blue light generated by first-order quasi-phase-matched frequency doubling of a diode-pumped 946-nm Nd: YAG laser. Opt. Lett. 20(23), 2375–2377 (1995).

[107] L. Goldberg, W. K. Burns, R. W. McElhanon: Difference-frequency generation of tunable mid-infrared radiation in bulk periodically poled LiNbO$_3$. Opt. Lett. 20(11), 1280–1282(1995).

[108] L. Goldberg, W. K. Burns, R. W. McElhanon: Wide acceptance bandwidth difference frequency generation in quasi-phase-matched LiNbO$_3$. Appl. Phys. Lett. 67(20), 2910–2912(1995).

[109] L. E. Myers, G. D. Miller, R. C. Eckardt, M. M. Fejer, R. L. Byer, W. R. Bosenberg: Quasiphase-matched 1.064-μm-pumped optical parametric oscillator in bulk periodically poled LiNbO$_3$. Opt. Lett. 20(1), 52–54(1995).

[110] M. M. Fejer, G. A. Magel, D. H. Jundt, R. L. Byer: Quasi-phase-matched second harmonic generation: tuning and tolerances. IEEE. J. Quant. Electr. 28(11), 2631–2654 (1992).

[111] L. E. Myers, R. C. Eckardt, M. M. Fejer, R. L. Byer, W. R. Bosenberg, J. W. Pierce: Quasiphase-matched optical parametric oscillators in bulk periodically poled LiNbO$_3$. J. Opt. Soc. Am. B 12(11), 2102–2116(1995).

[112] L. E. Myers, W. R. Bosenberg: Periodically poled lithium niobate and quasi-phase-matched optical parametric oscillators. IEEE. J. Quant. Electr. 33(10), 1663–1672 (1997).

[113] A. Ashkin, G. D. Boyd, J. M. Dziedzic, R. G. Smith, A. A. Ballman, J. J. Levinstein, K. Nassau: Optically-induced refractive index inhomogeneities in LiNbO$_3$ and LiTaO$_3$. Appl. Phys. Lett. 9(1), 72–74(1966).

[114] A. M. Glass: The photorefractive effect. Opt. Eng. 17(5), 470 – 479(1978).

[115] G. D. Miller, R. G. Batchko, W. M. Tulloch, D. R. Weise, M. M. Fejer, R. L. Byer: 42%-efficient single-pass CW second-harmonic generation in periodically poled lithium niobate. Opt. Lett. 22(24), 1834 – 1836(1997).

[116] W. R. Bosenberg, J. I. Alexander, L. E. Myers, R. W. Wallace: 2.5-W, continuous-wave, 629-nm solid-state laser source. Opt. Lett. 23(3), 207 – 209(1998).

[117] P. Schlup, S. D. Butterworth, I. T. McKinnie: Efficient single-frequency pulsed periodically poled lithium niobate optical parametric oscillator. Opt. Commun. 154(4), 191 – 195(1998).

[118] U. Bäder, J. -P. Meyn, J. Bartschke, T. Weber, A. Borsutzky, R. Wallenstein, R. G. Batchko, M. M. Fejer, R. L. Byer: Nanosecond periodically poled lithium niobate optical parametric generator pumped at 532 nm by a single-frequency passively Q-switched Nd: YAG laser. Opt. Lett. 24(22), 1608 – 1610(1999).

[119] T. R. Volk, V. I. Pryalkin, N. M. Rubinina: Optical-damage-resistant $LiNbO_3$: Zn crystal. Opt. Lett. 15(18), 996 – 998(1990).

[120] Y. Zhang, Y. H. Xu, M. H. Li, Y. Q. Zhao: Growth and properties of Zn doped lithium niobate crystal. J. Cryst. Growth 233(3), 537 – 540(2001).

2.4 $KTiOPO_4$, 磷酸钛氧钾(KTP)

正双轴晶：在 $\lambda = 0.546\ 1\ \mu m$ 时 $2V_z = 37.4°$[1]

分子量：197.949

密度：2.945 g/cm^3[2,3]; 3.023 g/cm^3[4]; 3.024 g/cm^3[5]; 3.03 g/cm^3[6]

点群：$mm2$

晶格常数：a = 12.814 Å[7]; 12.815 7 Å[6]; 在 T = 298 K 时, (12.816 4 ± 0.001 4) Å[8]; 12.822 Å[9]

b = 6.404 Å[7]; 6.402 7 Å[6]; 在 T = 298 K 时, (6.403 3 ± 0.000 6) Å[8]; 6.405 4 Å[9]

c = 10.616 Å[7]; 10.586 6 Å[6]; 在 T = 298 K 时, (10.589 7 ± 0.001 4) Å[8]; 10.589 Å[9]

介电轴和晶体学轴的变换：$X, Y, Z \Rightarrow a, b, c$

莫氏硬度：5[3]

维氏硬度：531[4]; 566[10]

诺氏硬度：702[4]

熔点(熔化时分解)：1 421 K[9]; 1 423 K[7]; 1 445 K[11]

居里温度：1 211 K[12]; 1 213 K[13]; 1 189 K(在熔体起始组分中钾浓度最

低,[K]/[P]=1)[14];1 231 K(在熔体起始组分中钾浓度最高,[K]/[P]=2)[14]

线性热膨胀系数[7]

$\alpha_t \times 10^6/\mathrm{K}^{-1}$, ∥ X	$\alpha_t \times 10^6/\mathrm{K}^{-1}$, ∥ Y	$\alpha_t \times 10^6/\mathrm{K}^{-1}$, ∥ Z
11	9	0.6

线性热膨胀系数[15]

T/K	$\alpha_t \times 10^6/\mathrm{K}^{-1}$, ∥ X	$\alpha_t \times 10^6/\mathrm{K}^{-1}$, ∥ Y	$\alpha_t \times 10^6/\mathrm{K}^{-1}$, ∥ Z
373	8.7	10.5	−0.2

在温度范围 298 K < T < 473 K 之间沿 X 轴的热膨胀系数[16]:

$$L(T) = L(T_0)[1 + \alpha(T - 298) + \beta(T - 298)^2]$$

其中 T 以 K 为单位,T_0 = 298 K,α = (6.7 ± 0.7) × 10^{-6} K^{-1},β = (11 ± 2) × 10^{-9} K^{-2}。

在 p = 0.101 325 MPa 时的比热容 c_p

T/K	$c_p/(\mathrm{J \cdot kg^{-1} \cdot K^{-1}})$	参考文献	T/K	$c_p/(\mathrm{J \cdot kg^{-1} \cdot K^{-1}})$	参考文献
298	688	[4]		729	[7]
	727	[17]			

热导率[7]

$\kappa/(\mathrm{W \cdot m^{-1} \cdot K^{-1}})$, ∥ X	$\kappa/(\mathrm{W \cdot m^{-1} \cdot K^{-1}})$, ∥ Y	$\kappa/(\mathrm{W \cdot m^{-1} \cdot K^{-1}})$, ∥ Z
2	3	3.3

室温下带隙能量:E_g = 3.54 eV[18];3.8 eV[19]

以"0"透过计的透明范围:0.35~4.5 μm[20,21],在 3.5 μm 处有正磷酸根的谐波吸收[13]

UV 透过截止波长(α = 2 cm^{-1})为 0.352 μm($\boldsymbol{E} \parallel \boldsymbol{X}$);0.359 μm($\boldsymbol{E} \parallel \boldsymbol{Y}$);0.365 μm($\boldsymbol{E} \parallel \boldsymbol{Z}$)[22]

线性吸收系数 α

$\lambda/\mu m$	α/cm^{-1}	参考文献	备 注
0.4	0.025 ~ 0.036	[23]	取决于 Pt 杂质含量
0.423	0.151 ± 0.024	[24]	熔盐法生长 PPKTP
0.43 ~ 0.78	< 0.004	[25]	氧退火 + 铈掺杂
0.473	0.021 ~ 0.067	[22]	熔盐法生长，$E \parallel X$
	0.023 ~ 0.053	[22]	熔盐法生长，$E \parallel Y$
	0.034 ~ 0.085	[22]	熔盐法生长，$E \parallel Z$
	0.037	[22]	水热法生长，$E \parallel X$
	0.049	[22]	水热法生长，$E \parallel Y$
	0.076	[22]	水热法生长，$E \parallel Z$
0.514 5	0.013	[26]	沿 a 轴
	0.027	[26]	沿 b 轴
	0.026	[26]	沿 c 轴
0.53 ~ 0.78	< 0.005	[25]	氧气氛下退火
0.532 1	0.04	[27]	沿 SHG 方向
	< 0.02	[7]	
	0.009 ~ 0.036	[22]	熔盐法生长，$E \parallel X$
	0.011 ~ 0.024	[22]	熔盐法生长，$E \parallel Y$
	0.019 ~ 0.039	[22]	熔盐法生长，$E \parallel Z$
	0.017	[22]	水热法生长，$E \parallel X$
	0.025	[22]	水热法生长，$E \parallel Y$
	0.040	[22]	水热法生长，$E \parallel Z$
0.659 4	0.006 5	[26]	沿 a 轴
	0.008 7	[26]	沿 b 轴
	0.006 5	[26]	沿 c 轴
0.846	0.018 ± 0.009	[24]	熔盐法生长 PPKTP
1.06	< 0.01	[6]	
1.064 2	< 0.006	[7]	
	0.005	[27]	沿 SHG 方向
	0.000 2	[26]	沿 a 轴
	0.000 5	[26]	沿 b 轴
	0.000 4	[26]	沿 c 轴
	0.000 3	[28]	
1.079 6	0.012	[29]	沿 SHG 方向

续表

$\lambda/\mu m$	α/cm^{-1}	参考文献	备注
1.318 8	0.001 5	[26]	沿 a 轴
	0.000 4	[26]	沿 b 轴
	0.001	[26]	沿 c 轴
3.297	0.59	[30]	

双光子吸收系数 β

$\lambda/\mu m$	τ_p/ns	$\beta \times 10^{11}/(cm \cdot W^{-1})$	参考文献	备注
0.532 1	0.022	10 ± 2	[19]	$k \parallel X, E \parallel Z$
	0.021	24 ± 4.8	[31]	[100]方向
		16 ± 3.2	[31]	[010]方向
		14 ± 2.8	[31]	[110]方向
0.6	0.0012	3.5	[32]	$\theta = 67.3°, \phi = 0°$

熔盐法生长 KTP 折射率的实验值

$\lambda/\mu m$	n_X	n_Y	n_Z	参考文献
0.404 7	1.824 9	1.841 0	1.962 9	[1]
0.435 8	1.808 2	1.822 2	1.935 9	[1]
0.491 6	1.788 3	1.800 0	1.904 4	[1]
0.534 3	1.778 0	1.788 7	1.888 8	[1]
0.539 75	1.776 4	1.786 9	1.886 3	[33]
0.541 0	1.776 7	1.787 3	1.886 9	[1]
0.546 1	1.775 6	1.786 0	1.885 0	[1]
0.577 0	1.770 3	1.780 3	1.876 9	[1]
0.579 0	1.769 9	1.779 8	1.876 4	[1]
0.585 3	1.768 9	1.778 7	1.874 9	[1]
0.589 3	1.768 4	1.778 0	1.874 0	[1]
0.623 4	1.763 7	1.773 2	1.867 2	[1]
0.632 8	1.762 2	1.771 4	1.864 9	[33]
0.641 0	1.761 7	1.770 9	1.864 1	[1]
0.693 9	1.756 5	1.765 2	1.856 4	[1]
0.694 3	1.756 4	1.765 2	1.856 4	[1]
0.705 0	1.755 5	1.764 2	1.855 0	[1]
1.064 0	1.738 1	1.745 8	1.830 2	[1]
1.064 2	1.737 9	1.745 4	1.829 7	[33]
1.079 5	1.737 5	1.745 0	1.829 1	[33]
1.341 4	1.731 4	1.738 7	1.821 1	[33]

熔盐法生长 KTP 折射率的温度微商

$\lambda/\mu m$	T/K	$dn_X/dT \times 10^6/K^{-1}$	$dn_Y/dT \times 10^6/K^{-1}$	$dn_Z/dT \times 10^6/K^{-1}$	参考文献
0.632 8	302~399	9.6±1.1	13.0±0.7	22.4±0.9	[34]
1.064 2	288~313	6.1	8.3	14.5	[35]

在 0.43~3.54 μm 范围内熔盐法生长 KTP 的最佳色散关系方程(λ 以 μm 为单位,$T = 293$ K)[36]:

$$n_X^2 = 3.291\,00 + \frac{0.041\,40}{\lambda^2 - 0.039\,78} + \frac{9.355\,22}{\lambda^2 - 31.455\,71}$$

$$n_Y^2 = 3.450\,18 + \frac{0.043\,41}{\lambda^2 - 0.045\,97} + \frac{16.988\,25}{\lambda^2 - 39.437\,99}$$

$$n_Z^2 = 4.594\,23 + \frac{0.062\,06}{\lambda^2 - 0.047\,63} + \frac{110.806\,72}{\lambda^2 - 86.121\,71}$$

在文献[1]、[21]、[33]、[37]、[38]、[39]、[40]、[41]、[42]、[43]、[44]、[45]、[46]中给出了其他色散关系方程组。

在 0.38 μm < λ < 4.5 μm 光谱范围内折射率 n_Z 红外修正的 Sellmeier 方程(λ 以 μm 为单位,$T = 293$ K)[47]:

$$n_Z^2 = 1 + \frac{1.716\,45\lambda^2}{\lambda^2 - 0.013\,346} + \frac{0.592\,4\lambda^2}{\lambda^2 - 0.065\,03} + \frac{0.322\,6\lambda^2}{\lambda^2 - 67.120\,8} - 0.011\,33\lambda^2$$

在文献[48]中给出了折射 n_Z 的另一个红外修正的 Sellmeier 方程。

$T = 293~353$ K 之间及光谱范围 0.43 μm < λ < 1.58 μm 之间熔盐法生长 KTP 晶体折射率的温度微商(λ 以 μm 为单位)[36]:

$$\frac{dn_X}{dT} = \left(\frac{0.171\,7}{\lambda^3} - \frac{0.535\,3}{\lambda^2} + \frac{0.841\,6}{\lambda} + 0.162\,7\right) \times 10^{-5} K^{-1}$$

$$\frac{dn_Y}{dT} = \left(\frac{0.199\,7}{\lambda^3} - \frac{0.406\,3}{\lambda^2} + \frac{0.515\,4}{\lambda} + 0.542\,5\right) \times 10^{-5} K^{-1}$$

对光谱范围 0.53 μm < λ < 1.57 μm(λ 以 μm 为单位)[36]:

$$\frac{dn_Z}{dT} = \left(\frac{0.922\,1}{\lambda^3} - \frac{2.922\,0}{\lambda^2} + \frac{3.667\,7}{\lambda} - 0.189\,7\right) \times 10^{-5} K^{-1}$$

对光谱范围 1.32 μm < λ < 3.53 μm(λ 以 μm 为单位)[36]:

$$\frac{dn_Z}{dT} = \left(-\frac{0.552\,3}{\lambda} + 3.392\,0 - 1.710\,1\lambda + 0.342\,4\lambda^2\right) \times 10^{-5} K^{-1}$$

非线性折射率 γ

$\lambda/\mu m$	$\gamma \times 10^{15}/(cm^2 \cdot W^{-1})$	参考文献	备注
0.532 1	2.3 ± 0.4	[19]	$\boldsymbol{k} \parallel X$, $\boldsymbol{E} \parallel Z$
0.780	1.20 ± 0.16	[49]	[100]方向
	0.94 ± 0.16	[49]	[010]方向
0.850	1.08 ± 0.20	[50]	$\theta = 90°$, $\phi = 23°$
1.064 2	1.4	[51]	XY 平面
	1.4 ± 0.28	[31]	[110]方向
	1.8 ± 0.36	[31]	[010]方向
	2.1 ± 0.42	[31]	[100]方向
	2.4 ± 0.5	[19]	$\boldsymbol{k} \parallel Z$, $\boldsymbol{E} \parallel Y$
	3.1	[18]	

室温下在低频(远低于 KTP 晶体声学共振频率,即对于"自由"晶体)测量的线性电光系数[7,52]

$\lambda/\mu m$	$r_{13}^T/(pm \cdot V^{-1})$	$r_{23}^T/(pm \cdot V^{-1})$	$r_{33}^T/(pm \cdot V^{-1})$	$r_{42}^T/(pm \cdot V^{-1})$	$r_{51}^T/(pm \cdot V^{-1})$
0.632 8	+9.5 ± 0.5	+15.7 ± 0.8	+36.3 ± 1.8	9.3 ± 0.9	7.3 ± 0.7

室温下在高频(远高于 KTP 晶体声学共振频率,即对于"受夹"晶体)测量的线性电光系数[7]

$\lambda/\mu m$	$r_{13}^S/(pm \cdot V^{-1})$	$r_{23}^S/(pm \cdot V^{-1})$	$r_{33}^S/(pm \cdot V^{-1})$	$r_{42}^S/(pm \cdot V^{-1})$	$r_{51}^S/(pm \cdot V^{-1})$
0.632 8	+8.8 ± 0.8	+13.8 ± 1.4	+35.0 ± 3.5	8.8 ± 1.8	6.9 ± 1.4

矫顽场值: ≈ 2 kV/mm[53-55]

KTP 晶体主平面中有效二阶非线性系数的表达式(Kleinman 对称条件不成立时)[56]:

XY 面

$$d_{eoe} = d_{oee} = d_{15}\sin^2\phi + d_{24}\cos^2\phi$$

YZ 面

$$d_{oeo} = d_{eoo} = d_{15}\sin\theta$$

XZ 面,$\theta < V_Z$

$$d_{ooe} = d_{32}\sin\theta$$

XZ 面,$\theta > V_Z$

$$d_{oeo} = d_{eoo} = d_{24}\sin\theta$$

KTP 晶体主平面中有效二阶非线性系数的表达式(Kleinman 对称条件成立,

$d_{15} = d_{31}$ 和 $d_{24} = d_{32}$)[56]:

XY 面
$$d_{eoe} = d_{oee} = d_{31}\sin^2\phi + d_{32}\cos^2\phi$$

YZ 面
$$d_{oeo} = d_{eoo} = d_{31}\sin\theta$$

XZ 面，$\theta < V_z$
$$d_{ooe} = d_{32}\sin\theta$$

XZ 面，$\theta > V_z$
$$d_{oeo} = d_{eoo} = d_{32}\sin\theta$$

在文献[56]中给出了 KTP 晶体内任意方向上有效二阶非线性系数的表达式。

二阶非线性系数的绝对值[57]：

$$d_{15}(0.852~\mu m) = (1.9 \pm 0.1)\text{pm/V}$$
$$d_{24}(0.852~\mu m) = (3.9 \pm 0.2)\text{pm/V}$$
$$d_{33}(0.852~\mu m) = (16.6 \pm 0.8)\text{pm/V}$$
$$d_{15}(1.064~\mu m) = (1.9 \pm 0.1)\text{pm/V}$$
$$d_{24}(1.064~\mu m) = (3.7 \pm 0.2)\text{pm/V}$$
$$d_{31}(1.064~\mu m) = (2.2 \pm 0.1)\text{pm/V}$$
$$d_{32}(1.064~\mu m) = (3.7 \pm 0.2)\text{pm/V}$$
$$d_{33}(1.064~\mu m) = (14.6 \pm 0.7)\text{pm/V}$$
$$d_{15}(1.313~\mu m) = (1.4 \pm 0.1)\text{pm/V}$$
$$d_{24}(1.313~\mu m) = (2.6 \pm 0.1)\text{pm/V}$$
$$d_{33}(1.313~\mu m) = (11.1 \pm 0.6)\text{pm/V}$$

KTP 二阶非线性系数所有的符号都是相同的[58]。

二阶非线性系数的其他可靠数值：

$$d_{24}(0.6~\mu m) = 4.2~\text{pm/V}^{[32]}$$
$$d_{15}(1.064~\mu m) = 1.8~\text{pm/V}^{[58]}$$
$$d_{24}(1.054~\mu m) = (4.1 \pm 0.4)\text{pm/V}^{[59]}$$
$$d_{24}(1.064~\mu m) = 3.4~\text{pm/V}^{[58]}$$
$$(3.9 \pm 0.3)\text{pm/V}^{[60]}$$
$$(4.2 \pm 0.2)\text{pm/V}^{[61]}$$
$$d_{33}(1.064~\mu m) = 17.4~\text{pm/V}^{[58]}$$
$$d_{15}(1.32~\mu m) = (1.2 \pm 0.1)\text{pm/V}^{[45]}$$
$$d_{24}(1.32~\mu m) = (2.4 \pm 0.2)\text{pm/V}^{[45]}$$

相位匹配角的实验值（$T = 293$ K）

相互作用的波长/μm	$\phi_{exp}/(°)$	$\theta_{exp}/(°)$	参考文献
水热法生长 KTP			
XY 面，$\theta = 90°$			
SHG，e + o ⇒ e			
1.053 ⇒ 0.526 5	34		[62]
1.062 ⇒ 0.531	25		[6]
1.064 2 ⇒ 0.532 1	24		[39]
	26		[2]，[63]，[64]
SFG，e + o ⇒ e			
1.318 8 + 0.659 4 ⇒ 0.439 6	3.8		[65]
YZ 面，$\phi = 90°$			
SFG，o + e ⇒ o			
1.318 8 + 0.659 4 ⇒ 0.439 6		65.1	[65]
1.338 + 0.669 ⇒ 0.446		63.2	[65]
XZ 面，$\phi = 0°$，$\theta > V_z$			
SFG，o + e ⇒ o			
1.318 8 + 0.659 4 ⇒ 0.439 6		87.7	[65]
1.338 + 0.669 ⇒ 0.446		79.9	[65]
1.064 2 + 1.458 1 ⇒ 0.615 2		78	[66]
1.064 2 + 1.476 2 ⇒ 0.618 4		76.6	[66]
1.064 2 + 1.591 8 ⇒ 0.637 8		75.8	[66]
熔盐法生长 KTP			
XY 面，$\theta = 90°$			
SHG，e + o ⇒ e			
1.064 1 ⇒ 0.532 05	23.5		[67]
	23.6		[68]
1.064 2 ⇒ 0.532 1	23.0		[69]
	23.2		[1]
	23.3		[70]
	24.1		[71]
	24.7		[72]
	25.0		[3]
	25.2		[20]，[39]，[73]
	25.3		[10]
YZ 面，$\phi = 90°$			
SHG，o + e ⇒ o			
1.064 2 ⇒ 0.532 1		69.0	[74]
		69.2	[39]
1.068 ⇒ 0.534		67.8	[74]

续表

相互作用的波长/μm	$\phi_{exp}/(°)$	$\theta_{exp}/(°)$	参考文献
1.182⇒0.591		57.4	[74]
1.318 8⇒0.659 4		50.0	[39]
1.5⇒0.75		44.6	[74]
XZ 面，$\phi=0°$，$\theta>V_Z$			
SHG，o+e⇒o			
1.079 6⇒0.539 8		85.3	[75]
		86.7	[33]
1.235⇒0.617 5		65	[76]
1.318 8⇒0.659 4		58.3	[39]
		58.6	[77]
		58.9	[78]
1.341 4⇒0.670 7		58.7	[33]
		58.9	[79]
1.54⇒0.77		53	[80]
1.907 68⇒0.953 84		51.1	[78]
2.05⇒1.025		50.8	[78]
2.128 4⇒1.064 2		53.7	[81]
		54	[78]
SFG，o+e⇒o			
1.318 8+0.659 4⇒0.439 6		87.1	[65]
		87.6	[39]
1.338+0.669⇒0.446		79.8	[65]
1.341 4+0.670 7⇒0.447 13		78.1	[82]
1.064 2+1.907 68⇒0.683 33		77.2	[78]
1.079 6+1.341 4⇒0.598 17		74.9	[33]
1.54+0.78⇒0.517 76		61	[83]
1.907 68+2.406 88⇒1.064 2		58.6	[78]
1.770+0.76⇒0.532 1		55	[84]
1.580 53+1.54⇒0.78		52.1	[83]
1.907 68+1.064 2⇒0.683 33		48.7	[78]

NCPM 温度的实验值及相应的温度带宽

相互作用的波长/μm	T/℃	ΔT/℃	参考文献
水热法生长 KTP			
沿 X 轴			
SFG，Ⅱ类			
$1.318\,8^Y + 0.659\,4^Z \Rightarrow 0.439\,6^Y$	47	8.5	[65]
$1.338^Y + 0.669^Z \Rightarrow 0.446^Y$	463	8.5	[65]
沿 Y 轴			
SHG，Ⅱ类			
$0.994\,3^X + 0.994\,3^Z \Rightarrow 0.497\,15^X$	20	175	[85]
SFG，Ⅱ类			
$1.064\,2^X + 0.809^Z \Rightarrow 0.459\,61^X$	20	122	[86]
熔盐法生长 KTP			
沿 X 轴			
SHG，Ⅱ类			
$1.079\,6^Y + 1.079\,6^Z \Rightarrow 0.539\,8^Y$	153(?)	20	[75]
	63	30	[87]
$1.08^Y + 1.08^Z \Rightarrow 0.54^Y$	≈20		[88]
SFG，Ⅱ类			
$1.090^Y + 1.039^Z \Rightarrow 0.532\,1^Y$	20		[89]
	20		[90]
$2.15^Z + 1.04^Y \Rightarrow 0.700\,94^Y$	20		[91]
$2.402^Z + 1.08^Y \Rightarrow 0.745^Y$	≈20		[88]
$2.75^Z + 1.16^Y \Rightarrow 0.816^Y$	≈20		[92]
$2.756^Z + 1.182^Y \Rightarrow 0.827^Y$	≈20		[93]
$3.09^Z + 1.38^Y \Rightarrow 0.953\,96^Y$	20		[91]
$3.297^Z + 1.571^Y \Rightarrow 1.047^Y$	20		[94]
$3.276^Z + 1.539^Y \Rightarrow 1.064\,2^Y$	20		[94]
$3.303^Z + 1.57^Y \Rightarrow 1.064\,2^Y$	≈20		[95]
$3.290^Z + 1.573^Y \Rightarrow 1.064\,2^Y$	≈20		[96]
$1.182^Y + 0.827^Z \Rightarrow 0.487^Y$	≈20		[93]
$1.318\,8^Y + 0.659\,4^Z \Rightarrow 0.439\,6^Y$	60.2	8.5	[65]
	53	10.1	[77]
$1.32^Y + 0.66^Z \Rightarrow 0.44^Y$	128	8.7	[77]
$1.338^Y + 0.669^Z \Rightarrow 0.446^Y$	484	8.5	[65]
沿 Y 轴			
SHG，Ⅱ类			
$0.99^X + 0.99^Z \Rightarrow 0.495^X$	20		[37]
SFG，Ⅱ类			
$1.064\,2^X + 0.806\,8^Z \Rightarrow 0.458\,9^X$	20		[37]
$1.064\,2^X + 0.808^Z \Rightarrow 0.459\,29^X$	20		[39]
$1.064\,2^X + 0.969\,1^Z \Rightarrow 0.507\,2^X$	20		[37]

注：相互作用波长的上标代表偏振方向。

内角带宽、温度带宽和光谱带宽的实验值

相互作用的波长/μm	ϕ_{pm}/(°)	θ_{pm}/(°)	$\Delta\phi^{int}$/(°)	$\Delta\theta^{int}$/(°)	ΔT/℃	$\Delta\nu$/cm^{-1}	参考文献
XY 平面,$\theta=90°(T=293\ K)$							
SHG,e + o⇒e							
1.058 2⇒0.592 1			0.43	2.01			[97]
1.062⇒0.531	25		0.49	2.23	25	4.9	[6]
1.064 2⇒0.532 1	23		0.53		20		[69]
	23				23.3		[98]
	23.2		0.58	1.82	24		[1]
	23.3		0.43		20	4.0	[70]
	25				6.2		[3]
	25.2				25		[20]
	25.2		0.42		17.5		[73]
	25.2		0.52	2.52	25.7		[10]
YZ 平面,$\phi=90°(T=293\ K)$							
SHG,o + e⇒o							
0.994 3⇒0.497 15		90	5.70	2.96	175	7.1	[85]
1.064 2⇒0.532 1		69			100		[99]
		69		0.11			[74]
2.532⇒1.266		56		0.20		30.7	[100]
SFG,Ⅱ类							
1.064 2^x + 0.809^z⇒0.459 61^x		90	6.13	2.72		17.6($\Delta\nu_2$)	[86]
XZ 平面,$\phi=0°$,$\theta>V_z$							
SHG,o + e⇒o							
1.079 6⇒0.539 8($T=293\ K$)		85.3		0.34			[75]
($T=426\ K$)		90		1.70			[75]

注:相互作用波长的上标代表偏振方向。

熔盐法生长 KTP 晶体 SHG 过程逆群速失配的计算值

相互作用的波长/μm	ϕ_{pm}/(°)	θ_{pm}/(°)	β/(fs·mm^{-1})
XY 平面,$\theta=90°$			
SHG,e + o⇒e			
1.0⇒0.5	73.18		475
1.05⇒0.525	35.03		434
YZ 平面,$\phi=90°$			
SHG,o + e⇒o			
1.0⇒0.5		83.17	490
1.1⇒0.55		64.36	361
1.2⇒0.6		56.22	329
1.3⇒0.65		51.02	228

续表

相互作用的波长/μm	$\phi_{pm}/(°)$	$\theta_{pm}/(°)$	$\beta/(fs·mm^{-1})$
1.4⇒0.7		47.46	186
1.5⇒0.75		45.02	153
1.6⇒0.8		43.40	126
1.7⇒0.85		42.44	103
1.8⇒0.9		41.99	84
1.9⇒0.95		41.98	83
2.0⇒1.0		42.35	100
XZ 平面,$\phi=0°$,$\theta>V_Z$			
SHG, o+e⇒o			
1.1⇒0.55		80.31	391
1.2⇒0.6		67.47	307
1.3⇒0.65		61.25	246
1.4⇒0.7		57.32	200
1.5⇒0.75		54.70	164
1.6⇒0.8		52.99	135
1.7⇒0.85		51.94	111
1.8⇒0.9		51.42	90
1.9⇒0.95		51.32	81
2.0⇒1.0		51.57	98

水热法生长 KTP 激光诱导的损伤阈值

$\lambda/μm$	τ_p/ns	$I_{thr}/(GW·cm^{-2})$	参考文献	备注
0.526	0.03	30	[40]	
		30	[7]	10 Hz
0.72~0.99	0.000 14	>35	[101]	76 MHz
1.064 2	125 000	0.001	[102]	
	30	0.15	[103]	
	20	>0.15	[104]	
	11	2~3	[105]	10 Hz

熔盐法生长 KTP 激光诱导的体损伤阈值

$\lambda/μm$	τ_p/ns	$I_{thr}/(GW·cm^{-2})$	参考文献	备 注
0.523 5	0.003 5	>8	[106]	50 Hz
0.526	0.03	10	[7]	10 Hz
0.532 1	400 000	0.001	[107]	在 1.064 2 μm 波存在时

续表

$\lambda/\mu m$	τ_p/ns	$I_{thr}/(GW \cdot cm^{-2})$	参考文献	备 注
	50 000	0.002 5	[107]	在 1.064 2 μm 波存在时
	220	0.051	[108]	1.2 kHz, 有基波存在
	220	0.089	[108]	1.2 kHz, PPKTP 有基波存在
	14	0.05	[73]	60 个脉冲
	8	2.0 ~ 3.2	[109]	2 Hz
	3	> 0.6	[110]	100 Hz, PPKTP 镀膜
	0.06	> 1.8	[71]	5 Hz
0.74 ~ 0.84	0.000 2	> 200	[111]	1 kHz
0.75 ~ 0.85	0.000 045	> 16	[112]	84 MHz
0.816	0.000 085	> 50	[92]	76 MHz
	0.000 09	> 1 000	[92]	250 kHz
1.053	0.1	> 7	[113]	1 kHz
1.064 2	30	> 3.3	[78]	
	25	> 0.6	[114]	250 000 个脉冲, 整体变黑
		> 0.3	[114]	3 500 000 个脉冲, 整体变黑
	20	0.15	[73]	60 个脉冲
	17	2.8 ± 0.1	[115]	1 Hz, 10 000 次, 商用 KTP
		6.2 ± 0.1	[115]	1 Hz, 10 000 次, 高纯 KTP
	16	> 0.14	[116]	1 Hz
	15	> 0.5	[96]	2.5 Hz
	11	2.4 ~ 3.5	[109]	2 Hz
	10	0.9 ~ 1.0	[69]	
	9	31	[117]	1 个脉冲
	5	> 0.9	[118]	20 Hz, PPKTP
	1	15	[1]	1 个脉冲
		> 15	[119]	
1.235	35	> 0.5	[76]	20 Hz

熔盐法生长 KTP 激光诱导的表面损伤阈值

$\lambda/\mu m$	τ_p/ns	$I_{thr}/(GW \cdot cm^{-2})$	参考文献	备注
0.529 1	18	0.08 ~ 0.1	[97]	
0.532 1	8	1.4 ~ 2.2	[109]	2 Hz
1.058 2	25	0.18 ~ 0.22	[97]	
1.064 2	11	1.5 ~ 2.2	[109]	2 Hz
	1.3	4.6	[120]	
1.618	0.022	50 ± 10	[121]	10 Hz

熔盐法生长 KTP 激光诱导的灰迹损伤阈值

$\lambda/\mu m$	τ_p/ns	$I_{thr}/(GW \cdot cm^{-2})$	参考文献	备注
0.514 5	CW	0.000 026	[122], [123]	$\boldsymbol{E} \perp \boldsymbol{Z}$, $\theta = 90°$, $\phi = 23.4°$
		0.000 130	[122], [123]	$\boldsymbol{E} \perp \boldsymbol{Z}$, $\theta = 90°$, $\phi = 23.4°$
0.532 1	75	0.015	[124]	6.3 kHz, $\theta = 90°$, $\phi = 23.1°$
		0.125	[124]	1 kHz, $\theta = 90°$, $\phi = 23.1°$
	25	0.045	[114]	10 Hz, $\theta = 90°$, $\phi = 23°$
	20	0.05 ~ 0.1	[125]	20 Hz, $\theta = 90°$, $\phi = 23°$
	10	0.08	[126]	10 Hz, $\theta = 90°$, $\phi = 23°$
	1	> 0.1	[64]	3.7 kHz, $\theta = 90°$, $\phi = 26°$
	0.026	2	[84]	10 Hz, $\theta = 55°$, $\phi = 0°$

关于这一晶体

　　KTP 是主要的基本非线性光学晶体之一；在 20 世纪 80 年代和 90 年代，KTP 晶体被广泛地应用于频率转换装置中。KTP 晶体的缺点之一是产生光色损伤的敏感性，这种损伤也称作灰色或灰迹，这是在绿色脉冲或 CW 激光照射下发生的，最一般的是在 532 或 514.5 nm 的辐射下产生(最近的综述，可见[127])。在 150 ℃ 以上灰迹不会形成[128,129]，在升高温度进行退火时迅速消除[127,130]。除了光色效应外，在 KTP 晶体中也存在光折变效应[129]。

　　近年来，激光二极管泵浦的、以 Nd:YAG 激光器为基础的、以 KTP 晶体作倍频的绿光激光源变得非常重要，因为它的效率高(电光的效率 ≈ 5% ~ 10%)而且平均输出功率也高(在 532 nm 最高达 300 W)[131-133]。KTP 倍频器的缺点是易于产生灰迹。结果是绿光输出功率会慢慢降低，即在[134]中报道的在 100 小时连续运转后从 106 W 降到 97.4 W。

1994 年[53]，引入了周期性极化的 KTP 晶体（PPKTP）。KTP（有关 QPM）具有低矫顽场（这使得大口径极化晶体的制作得以进行）、没有光折变损伤（器件可以在室温下运转）以及高得多的抗光伤阈值（与 LN 相比）等独特优点。PPKTP 的主要缺点是元件尺寸相对较短，通常是约 10 mm 或者更短，然而，PPKTP 的各类应用十分广泛。我们可以在此列举一些关于 QPM SHG[108,135-138]，QPM DFG[48,139] 和 QPM OPO[110,140-144] 等在周期性极化 KTP 晶体中的开创性工作。

参考文献

[1] T. Y. Fan, C. E. Huang, B. Q. Hu, R. C. Eckardt, Y. X. Fan, R. L. Byer, R. S. Feigelson: Second harmonic generation and accurate index of refraction measurements in fluxgrown KTiOPO$_4$. Appl. Opt. 26(12), 2390-2394(1987).

[2] Y. S. Liu, D. Dentz, R. Belt: High-average-power intracavity second-harmonic generation using KTiOPO$_4$ in an acousto-optically Q-switched Nd: YAG laser oscillator at 5 kHz. Opt. Lett. 9(3), 76-78(1984).

[3] D. N. Dovchenko, V. A. Dyakov, V. I. Pryalkin: Growth and applications of potassium titanyl phosphate crystals. Izv. Akad. Nauk SSSR, Ser. Fiz. 52(2), 225-230(1988)[In Russian, English trans.: Bull. Acad. Sci. USSR, Phys. Ser. 52(2), 13-17(1988)].

[4] J. C. Jacco: KTiOPO$_4$ (KTP)-past, present and future. Proc. SPIE 968, 93-99(1988).

[5] B. H. T. Chai: Optical Crystals. In: *CRC Handbook of Laser Science and Technology, Supplement 2: Optical Materials*, ed. by M. J. Weber(CRC Press, Boca Raton, 1995), pp. 3-65.

[6] R. F. Belt, G. Gashurov, Y. S. Liu: KTP as a harmonic generator of Nd: YAG lasers. Laser Focus 21(10), 110-124(1985).

[7] J. D. Bierlein, H. Vanherzeele: Potassium titanyl phosphate: properties and new applications. J. Opt. Soc. Am. B 6(4), 622-633(1989).

[8] I. V. Voloshina, R. G. Gerr, M. Y. Antipin, V. G. Tsirelson, N. I. Pavlova, Y. T. Struchkov, R. P. Ozerov, I. S. Rez: Electron density distribution in a nonlinear KTiOPO$_4$ crystal according to precision X-ray data. Kristallogr. 30(4), 668-676(1985)[In Russian, English trans.: Sov. Phys. -Crystallogr. 30(4), 389-393(1985)].

[9] L. K. Cheng, E. M. McCarron Ⅲ, J. Calabrese, J. D. Bierlein, A. A. Ballman: Development of the nonlinear optical crystal CsTiOAsO$_4$. I. Structural stability. J. Crystal Growth 132(1-2), 280-288(1993).

[10] Y. Kitaoka, T. Sasaki, S. Nakai, Y. Goto: New nonlinear optical crystal thienylchalcone and its harmonic generation properties. Appl. Phys. Lett. 59(1), 19-21

(1991).

[11] I. Bhaumik, S. Ganesamoorthy, R. Bhatt, R. Sundar, A. K. Karnal, V. K. Wadhawan: Novel seeding technique for growing KTiOPO$_4$ single crystals by the TSSG method. J. Cryst. Growth 243(3-4), 522-525(2002).

[12] J. D. Bierlein, H. Vanherzeele, A. A. Ballman: Linear and nonlinear optical properties of flux-grown KTiOAsO$_4$. Appl. Phys. Lett. 54(9), 783-785(1989).

[13] L. K. Cheng, J. D. Bierlein: KTP and isomorphs—recent progress in device and material development. Ferroelectrics 142(1-2), 209-228(1993).

[14] M. Roth, N. Angert, M. Tseitlin, A. Alexandrovski: On the optical quality of KTP crystals for nonlinear optical and electro-optic applications. Opt. Mater. 16(1-2), 131-136(2001).

[15] Data sheet of Molecular Technology GmbH. Available at www. mt-berlin. com.

[16] S. Emanueli, A. Arie: Temperature-dependent dispersion equations for KTiOPO$_4$ and KTiOAsO$_4$. Appl. Opt. 42(33), 6661-6665(2003).

[17] J. D. Bierlein: Potassium titanyl phosphate (KTP): properties, recent advances and new applications. Proc. SPIE 1104, 2-12(1989).

[18] M. Sheik-Bahae, D. C. Hutchings, D. J. Hagan, E. W. Van Stryland: Dispersion of bound electron nonlinear refraction in solids. IEEE J. Quant. Electr. 27(6), 1296-1309(1991).

[19] R. DeSalvo, A. A. Said, D. J. Hagan, E. W. van Stryland, M. Sheik-Bahae: Infrared to ultraviolet measurements of two-photon absorption and n_2 in wide bandgap solids. IEEE J. Quant. Electr. 32(8), 1324-1333(1996).

[20] A. L. Aleksandrovskii, S. A. Akhmanov, V. A. Dyakov, N. I. Zheludev, V. I. Pryalkin: Efficient nonlinear optical converters made of potassium titanyl phosphate. Kvant. Elektron. 12(7), 1333-1334(1985) [In Russian, English trans. : Sov. J. Quantum Electron. 15(7), 885-886(1985)].

[21] K. Kato: Parametric oscillation at 3.2 μm in KTP pumped at 1.064 μm. IEEE J. Quant. Electr. 27(5), 1137-1140(1991).

[22] G. Hansson, H. Karlsson, S. Wang, F. Laurell: Transmission measurements in KTP and isomorphic compounds. Appl. Opt. 39(27), 5058-5069(2000).

[23] A. Miyamoto, Y. Mori, T. Sasaki, S. Nakai: Improvement of optical transmission of KTiOPO$_4$ crystals by growth in nitrogen ambient. Appl. Phys. Lett. 69(8), 1032-1034(1996).

[24] F. Torabi-Goudarzi, E. Riis: Efficient CW high-power frequency doubling in periodically poled KTP. Opt. Commun. 227(4-6), 389-403(2003).

[25] P. F. Bordui, R. Blachman, R. G. Norwood: Improved optical transmission of KTiOPO$_4$ crystals through cerium-doping and oxygen annealing. Appl. Phys. Lett. 61(12),

1369 – 1371(1992).

[26] D. J. Gettemy, W. C. Harker, G. Lindholm, N. P. Barnes: Some optical properties of KTP, $LiIO_3$, and $LiNbO_3$. IEEE J. Quant. Electr. 24(11), 2231 – 2237(1988).

[27] P. E. Perkins, T. S. Fahlen: 20-W average-power KTP intracavity-doubled Nd: YAG laser. J. Opt. Soc. Am. B 4(7), 1066 – 1071(1987).

[28] T. Kojima, S. Fujikawa, K. Yasui: Stabilization of a high-power diode-side-pumped intracavity-frequency-doubled CW Nd: YAG laser by compensating for thermal lensing of a KTP crystal and Nd: YAG rods. IEEE J. Quant. Electr. 35(3), 377-380 (1999).

[29] C. H. Huang, H. Y. Shen, Z. D. Zeng, Y. P. Zhou, R. R. Zeng, G. F. Yu, A. D. Jiang, T. B. Chen: Measurement of the total absorption coefficient of a KTP crystal. Opt. Laser Technol. 22(5), 345 – 347(1990).

[30] M. S. Webb, P. F. Moulton, J. J. Kasinski, R. L. Burnham, G. Loiacono, R. Stolzenberger: High-average-power $KTiOAsO_4$ optical parametric oscillator. Opt. Lett. 23(15), 1161 – 1163(1998).

[31] R. DeSalvo, M. Sheik-Bahae, A. A. Said, D. J. Hagan, E. W. Van Stryland: Z-scan measurements of the anisotropy of nonlinear refraction and absorption in crystals. Opt. Lett. 18(3), 194 – 196(1993).

[32] T. Nishikawa, N. Uesugi: Effects of walk-off and group velocity difference on the optical parametric generation in $KTiOPO_4$ crystals. J. Appl. Phys. 77(10), 4941 – 4947 (1995).

[33] H. Y. Shen, Y. P. Zhou, W. X. Lin, Z. D. Zeng, R. R. Zeng, G. F. Yu, C. H. Huang, A. D. Jiang, S. Q. Jia, D. Z. Shen: Second harmonic generation and sum frequency mixing of dual wavelength Nd: $YAlO_3$ laser in flux grown $KTiOPO_4$ crystal. IEEE J. Quant. Electr. 28(1), 48 – 51(1992).

[34] Z. Zeng, H. Shen, H. Xu, Y. Zhou, C. Huang, D. Shen: Measurements of the refractive index and its thermal coefficient for $KTiOPO_4$. J. Synth. Cryst. 16(3), 274 – 277 (1987) [In Chinese].

[35] W. Wiechmann, S. Kubota, T. Fukui, H. Masuda: Refractive-index temperature derivatives of potassium titanyl phosphate. Opt. Lett. 18(15), 1208 – 1210(1993).

[36] K. Kato, E. Takaoka: Sellmeier and thermo-optic dispersion formulas for KTP. Appl. Opt. 41(24), 5040 – 5044(2002).

[37] K. Kato: Second-harmonic and sum-frequency generation to 4950 and 4589 Å in KTP. IEEE J. Quant. Electr. QE – 24(1), 3 – 4(1988).

[38] H. Liao, H. Shen, T. Lian, Y. Zhou, C. Huang, R. Zeng, G. Yu: Accurate values for the index of refraction and the optimum phase match parameters in a flux grown $KTiOPO_4$ crystal. Opt. Laser Technol. 20(2), 103 – 104(1988).

[39] D. W. Anthon, C. D. Crowder: Wavelength dependent phase matching in KTP. Appl. Opt. 27(13), 2650-2652(1988).

[40] H. Vanherzeele, J. D. Bierlein, F. C. Zumsteg: Index of refraction measurement and parametric generation in hydrothermally-grown KTP. Appl. Opt. 27(16), 3314-3316 (1988).

[41] V. A. Dyakov, V. V. Krasnikov, V. I. Pryalkin, M. S. Pshenichnikov, T. B. Razumikhina, V. S. Solomatin, A. I. Kholodnykh: Sellmeier equation and tuning characteristics of KTP crystal frequency converters in the 0.4-4.0 μm range. Kvant. Elektron. 15(9), 1703-1704(1988) [In Russian, English trans.: Sov. J. Quantum Electron. 18(9), 1059-1060 (1988)].

[42] G. Ghosh: Temperature dispersion in KTP for nonlinear devices. IEEE Phot. Technol. Lett. 7(1), 68-70(1995).

[43] L. Carrion, J. P. Girardeau-Montaut: Performance of a new picosecond KTP optical parametric generator and amplifier. Opt. Commun. 152(4-6), 347-350(1998).

[44] B. Boulanger, J. P. Feve, G. Marnier, C. Bonnin, P. Villeval, J. J. Zondy: Absolute measurement of quadratic nonlinearities from phase-matched second-harmonic generation in a single KTP crystal cut as a sphere. J. Opt. Soc. Am. B 14(6), 1380-1386 (1997).

[45] B. Boulanger, J. P. Feve, G. Marnier, B. Menaert: Methodology for optical studies of nonlinear crystals: application to the isomorph family $KTiOPO_4$, $KTiOAsO_4$, $RbTiOAsO_4$ and $CsTiOAsO_4$. Pure Appl. Opt. 7(2), 239-256(1998).

[46] D. Y. Zhang, H. Y. Shen, W. Liu, G. F. Zhang, W. Z. Chen, G. Zhang, R. R. Zeng, C. H. Huang, W. X. Lin, J. K. Liang: The principal refractive indices and nonlinear optical phase matched properties of Nb: KTP crystals. Opt. Mater. 15(2), 99-102 (2000).

[47] M. Katz, D. Eger, M. B. Ogon, A. Hardy: Refractive dispersion curve measurement of $KTiOPO_4$ using periodically segmented waveguides and periodically poled crystals. J. Appl. Phys. 90(1), 53-58(2001).

[48] K. Fradkin, A. Arie, A. Skliar, G. Rosenman: Tunable midinfrared source by difference frequency generation in bulk periodically poled $KTiOPO_4$. Appl. Phys. Lett. 74 (7), 914-916(1999); Errata, Appl. Phys. Lett. 74(18), 2723(1999).

[49] H. P. Li, C. H. Kam, Y. L. Lam, W. Ji: Femtosecond Z-scan measurements of nonlinear refraction in nonlinear optical materials. Opt. Mater. 15(4), 237-242(2001).

[50] M. Sheik-Bahae, M. Ebrahimzadeh: Measurements of nonlinear refraction in the second-order $\chi^{(2)}$ materials $KTiOPO_4$, $KNbO_3$, β-BaB_2O_4, and LiB_3O_5. Opt. Commun. 142 (4-6), 294-298(1997).

[51] R. Adair, L. L. Chase, S. A. Payne: Nonlinear refractive index of optical crystals.

Phys. Rev. B 39(5), 3337 – 3350(1989).

[52] L. K. Cheng, L. T. Cheng, J. Galperin, P. A. Morris Hotsenpiller, J. D. Bierlein: Crystal growth and characterization of KTiOPO$_4$ isomorphs from the self-fluxes. J. Cryst. Growth 137(1 – 2), 107 – 115(1994).

[53] Q. Chen, W. P. Risk: Periodic poling of KTiOPO$_4$ using an applied electric field. Electron. Lett. 30(18), 1516 – 1517(1994).

[54] H. Karlsson, F. Laurell: Electric field poling of flux grown KTiOPO$_4$. Appl. Phys. Lett. 71(24), 3474 – 3476(1997).

[55] J. Hellström, V. Pasiskevicius, F. Laurell, H. Karlsson: Efficient nanosecond optical parametric oscillators based on periodically poled KTP emitting in the 1.8 – 2.5-μm spectral range. Opt. Lett. 24(17), 1233 – 1235(1999).

[56] V. G. Dmitriev, D. N. Nikogosyan: Effective nonlinearity coefficients for three-wave interactions in biaxial crystals of mm2 point group symmetry. Opt. Commun. 95(1 – 3), 173 – 182(1993).

[57] I. Shoji, T. Kondo, A. Kitamoto, M. Shirane, R. Ito: Absolute scale of second-order nonlinear-optical coefficients. J. Opt. Soc. Am. B 14(9), 2268 – 2294(1997).

[58] A. Anema, T. Rasing: Relative signs of the nonlinear coefficients of potassium titanyl phosphate. Appl. Opt. 36(24), 5902 – 5904(1997).

[59] E. C. Cheung, K. Koch, G. T. Moore, J. M. Liu: Measurements of second-order nonlinear optical coefficients from the spectral brightness of parametric fluorescence. Opt. Lett. 19(3), 168 – 170(1994).

[60] W. J. Alford, A. V. Smith: Wavelength variation of the second-order nonlinear coefficients of KNbO$_3$, KTiOPO$_4$, KTiOAsO$_4$, LiNbO$_3$, LiIO$_3$, β-BaB$_2$O$_4$, KH$_2$PO$_4$, and LiB$_3$O$_5$ crystals: a test of Miller wavelength scaling. J. Opt. Soc. Am. B 18(4), 524 – 533(2001).

[61] T. Nishikawa, N. Uesugi: Effects of walk-off and group velocity difference on the optical parametric generation in KTiOPO$_4$ crystals. J. Appl. Phys. 77(10), 4941 – 4947 (1995).

[62] H. Vanherzeele: Optimization of a CW mode-locked frequency-doubled Nd: YLF laser. Appl. Opt. 27(17), 3608 – 3615(1988).

[63] P. E. Perkins, T. S. Fahlen: Half watt average power at 25 kHz from fourth harmonic of Nd: YAG. IEEE J. Quant. Electr. QE – 21(10), 1636 – 1638(1985).

[64] D. A. V. Kliner, F. Di Teodoro, J. P. Koplow, S. W. Moore, A. V. Smith: Efficient second, third, fourth, and fifth harmonic generation of a Yb-doped fiber amplifier. Opt. Commun. 210(3 – 6), 393 – 398(2002).

[65] R. A. Stolzenberger, C. C. Hsu, N. Peyghambarian, J. J. E. Reid, R. A. Morgan: Type II sum frequency generation in flux and hydrothermally grown KTP at 1.319 and

1.338 μm. IEEE Photon. Technol. Lett. 1(12), 446 – 448(1989).

[66] K. Kurokawa, M. Nakazawa: Femtosecond 1.4 – 1.6 μm infrared pulse generation at a high repetition rate by difference frequency generation. Appl. Phys. Lett. 55(1), 7 – 9 (1989).

[67] H. Wang, Y. Ma, Z. Zhai, J. Gao, C. Xie, K. Peng: Tunable continuous-wave doubly resonant optical parametric oscillator by use of a semimonolithic KTP crystal. Appl. Opt. 41(6), 1124 – 1127(2002).

[68] Y. -L. Jia, J. -L. He, H. -T. Wang, S. -N. Zhu, Y. -Y. Zhu: Single pass third-harmonic generation of 310 mW of 355 nm with an all-solid-state laser. Chin. Phys. Lett. 18 (12), 1589 – 1591(2001).

[69] O. I. Lavrovskaya, N. I. Pavlova, A. V. Tarasov: Second harmonic generation of light from an YAG: Nd^{3+} laser in an optically biaxial crystal $KTiOPO_4$. Kristallogr. 31 (6), 1145 – 1151(1986) [In Russian, English trans. : Sov. Phys. -Crystallogr. 31(6), 678 – 682(1986)].

[70] R. A. Stolzenberger: Nonlinear optical properties of flux grown $KTiOPO_4$. Appl. Opt. 27(18), 3883 – 3886(1988).

[71] L. J. Bromley, A. Guy, D. C. Hanna: Synchronously pumped optical parametric oscillation in KTP. Opt. Commun. 70(4), 350 – 354(1989).

[72] Y. Huo, F. Chen, S. He, D. Shen, X. Ma, Y. Lu: Realizing intracavity frequency doubling and Q switching by KTP. Proc. SPIE 3556, 31 – 36(1998).

[73] Y. A. Galaichuk, V. A. Dyakov, N. I. Likholit, V. S. Ovechko, R. A. Petrenko, T. V. Rozhdestvenskaya, V. L. Strizhevskii, A. I. Khilchevskii, Y. N. Yashkir: $KTiOPO_4$ as an optical-frequency converter. Izv. Akad. Nauk SSSR, Ser. Fiz. 52 (3), 560 – 563 (1988) [In Russian, English trans. : Bull. Akad. Sci. USSR, Phys. Ser. 52(3), 131 – 133(1988)].

[74] T. Nishikawa, N. Uesugi, H. Ito: Angle tuning characteristics of second-harmonic generation in $KTiOPO_4$. Appl. Phys. Lett. 55(19), 1943 – 1945(1989).

[75] V. M. Garmash, G. A. Ermakov, N. I. Pavlova, A. V. Tarasov: Efficient second-harmonic generation in potassium titanyl-phosphate crystals with noncritical matching. Pisma Zh. Tekh. Fiz. 12 (20), 1222 – 1225 (1986) [In Russian, English trans. : Sov. Tech. Phys. Lett. 12(10),505 – 506(1986)].

[76] I. T. McKinnie, A. L. Oien: Tunable red-yellow laser based on second harmonic generation of Cr: forsterite in KTP. Opt. Commun. 141(3 – 4), 157 – 161(1997).

[77] J. P. Feve, B. Boulanger, X. Cabirol, B. Menaert, G. Marnier, C. Bonnin, P. Villeval: Non-critically phase-matched cascaded THG at 440 nm in $KTiOP_{1-y}As_yO_4$ crystals. Opt. Commun. 115(3 – 4), 323 – 326(1995).

[78] R. Burnham, R. A. Stolzenberger, A. Pinto: Infrared optical parametric oscillator in po-

tassium titanyl phosphate. IEEE Photon. Technol. Lett. 1(1), 27−28(1989).

[79] Q. Zheng, H. Tan, L. Zhao, L. Quan: Diode-pumped 671 nm laser frequency doubled by CPM LBO. Opt. Laser Technol. 34(4), 329−331(2002).

[80] W. Wang, K. Nakagawa, Y. Toda, M. Ohtsu: 1.5 μm diode laser-based nonlinear frequency conversions by using potassium titanyl phosphate. Appl. Phys. Lett. 61(16), 1886−1888(1992).

[81] J. T. Lin, J. L. Montgomery: Generation of tunable mid-IR (1.8−2.4 μm) laser radiation from optical parametric oscillation in KTP. Opt. Commun. 75(3−4), 315−320 (1990).

[82] W. X. Lin, H. Y. Shen, Y. P. Zhou, R. R. Zeng, G. F. Yu, C. H. Huang, Z. D. Zeng, W. J. Zhang: Tripling the harmonic generation of a 1 341.4 nm Nd:YAP laser in $LiIO_3$ and KTP crystals to get 447.1 nm blue coherent radiation. Opt. Commun. 82(3−4), 333−336(1991).

[83] W. Wang, M. Ohtsu: Frequency-tunable sum- and difference-frequency generation by using two diode lasers and a KTP crystal. Opt. Commun. 102(3−4), 304−308 (1993).

[84] L. Carrion, J.-P. Girardeau-Montaut: Gray-track damage in potassium titanyl-phosphate under a picosecond regime at 532 nm. Appl. Phys. Lett. 77(8), 1074−1076(2000).

[85] W. P. Risk, R. N. Payne, W. Lenth, C. Harder, H. Meier: Noncritically phase-matched frequency doubling using 994 nm dye and diode laser radiation in $KTiOPO_4$. Appl. Phys. Lett. 55(12), 1179−1181(1989).

[86] J.-C. Baumert, F. M. Schellenberg, W. Lenth, W. P. Risk, G. C. Bjorklund: Generation of blue CW coherent radiation by sum frequency mixing in $KTiOPO_4$. Appl. Phys. Lett. 51 (26), 2192−2194(1987).

[87] Z. Y. Ou, S. F. Pereira, E. S. Polzik, H. J. Kimble: 85% efficiency for CW frequency doubling from 1.08 to 0.54 μm. Opt. Lett. 17(9), 640−642(1992).

[88] T. Kartaloǧlu, K. G. Köprülü, O. Aytür: Phase-matched self-doubling optical parametric oscillator. Opt. Lett. 22(5), 280−282(1997).

[89] S. T. Yang, R. C. Eckardt, R. L. Byer: Power and spectral characteristics of continuouswave parametric oscillators: the doubly to singly resonant transition. J. Opt. Soc. Am. B 10(9), 1684−1695(1993).

[90] S. T. Yang, R. C. Eckardt, R. L. Byer: Continuous-wave singly resonant optical parametric oscillator pumped by a single-frequency resonantly doubled Nd:YAG laser. Opt. Lett. 18(12), 971−973(1993).

[91] K. Kato, M. Masutani: Widely tunable 90° phase-matched KTP parametric oscillator. Opt. Lett. 17(3), 178−179(1992).

[92] G. R. Holtom, R. A. Crowell, X. S. Xie: High-repetition-rate femtosecond optical parametric-amplifier system near 3 μm. J. Opt. Soc. Am. B 12(9), 1723 – 1731(1995).

[93] K. G. Köprülü, T. Kartaloglu, Y. Dikmelik, O. Aytür: Single-crystal sum-frequency-generating optical parametric oscillator. J. Opt. Soc. Am. B 16(9), 1546 – 1552 (1999).

[94] J. A. C. Terry, Y. Cui, Y. Yang, W. Sibbett, M. H. Dunn: Low-threshold operation of an all-solid-state KTP optical parametric oscillator. J. Opt. Soc. Am. B 11(5), 758 – 769(1994).

[95] R. Dabu, C. Fenic, A. Stratan: Intracavity pumped nanosecond optical parametric oscillator emitting in the eye-safe range. Appl. Opt. 40(24), 4334 – 4340(2001).

[96] V. L. Naumov, A. M. Onishchenko, A. S. Podstavkin, A. V. Shestakov: High-efficiency parametric converter based on KTP crystals. Kvant. Elektron. 30(7), 632 – 634(2000) [In Russian, English trans.: Quantum Electron. 30(7), 632 – 634(2000)].

[97] G. I. Dyakonov, V. A. Maslov, V. A. Mikhailov, S. K. Pak, V. N. Semenenko, I. A. Shcherbakov: Highly-efficient YSGG: Cr: Nd laser with frequency doubling in a KTP crystal. Kvant. Elektron. 16(8), 1601 – 1603(1989) [In Russian, English trans.: Sov. J. Quantum Electron. 19(8), 1031 – 1032(1989)].

[98] S. G. Grechin, V. G. Dmitriev, V. A. Dyakov, V. I. Pryalkin: Temperature-independent phase matching for second-harmonic generation in a KTP crystal. Kvant. Elektron. 26(1), 77 – 81(1999) [In Russian, English trans.: Quantum Electron. 29(1), 77 – 81(1999)].

[99] K. Kato: Temperature insensitive SHG at 0.532 1 μm in KTP. IEEE J. Quant. Electr. 28, 1974 – 1976(1992).

[100] J. -J. Zondy, M. Abed, A. Clairon: Type II frequency doubling of $\lambda = 1.30$ μm and $\lambda = 2.53$ μm in flux-grown potassium titanyl phosphate. J. Opt. Soc. Am. B 11(10), 2004 – 2015(1994).

[101] Y. Wang, V. Petrov, Y. J. Ding, Y. Zheng, J. B. Khurgin, W. P. Risk: Ultrafast generation of blue light by efficient second-harmonic generation in periodically-poled bulk and waveguide potassium titanyl phosphate. Appl. Phys. Lett. 73(7), 873 – 875 (1998).

[102] S. E. Moody, J. M. Eggleston, J. F. Seamans: Long-pulse second-harmonic generation in KTP. IEEE J. Quant. Electr. QE – 23(3), 335 – 340(1987).

[103] T. A. Driscoll, H. J. Hoffman, R. E. Stone, P. E. Perkins: Efficient second-harmonic generation in KTP crystals. J. Opt. Soc. Am. B 3(5), 683 – 686(1986).

[104] F. C. Zumsteg, J. D. Bierlein, T. E. Gier: $K_xRb_{1-x}TiOPO_4$: a new nonlinear optical material. J. Appl. Phys. 47(11), 4980 – 4985(1976).

[105] F. Ahmed: Laser damage threshold of $KTiOPO_4$. Appl. Opt. 28(1), 119 – 122

(1989).

[106] C. Rauscher, T. Roth, R. Laenen, A. Laubereau: Tunable femtosecond-pulse generation by an optical parametric oscillator in the saturation regime. Opt. Lett. 20(19), 2003-2005(1995).

[107] S. Favre, T. C. Sidler, R. -P. Salathe: High-power long-pulse second harmonic generation and optical damage with free-running Nd:YAG laser. IEEE J. Quant. Electr. 39(6), 733-740(2003).

[108] V. Pasiskevicius, S. Wang, J. A. Tellefsen, F. Laurell, H. Karlsson: Periodically poled flux grown KTP for Nd:YAG frequency doubling. Appl. Opt. 37(30), 7116-7119(1998).

[109] P. Yankov, D. Schumov, A. Nenov, A. Monev: Laser damage tests of large flux-grown $KTiOPO_4$ crystals. Opt. Lett. 18(21), 1771-1773(1993).

[110] M. Oba, M. Kato, Y. Maruyama: Optical parametric oscillator with periodically poled $KTiOPO_4$ pumped by 100 Hz Nd:YAG green laser. Jpn. J. Appl. Phys. 41 (8A), L881-L883(2002).

[111] V. Petrov, F. Noack, R. Stolzenberger: Seeded femtosecond optical parametric amplification in the mid-infrared spectral region above 3 μm. Appl. Opt. 36(6), 1164-1172(1997).

[112] J. Jiang, T. Hasama: Femtosecond optical parametric oscillator with a repetition rate of 504 MHz. Jpn. J. Appl. Phys. 41(3A), 1365-1368(2002).

[113] K. Finsterbusch, R. Urschel, H. Zacharias: Tunable, high-power, narrow-band picosecond IR radiation by optical parametric amplification in KTP. Appl. Phys. B 74(4-5), 319-322(2002).

[114] J. C. Jacco, D. R. Rockafellow, E. A. Teppo: Bulk-darkening threshold of flux-grown $KTiOPO_4$. Opt. Lett. 16(17), 1307-1309(1991).

[115] X. B. Hu, H. Liu, J. Y Wang, H. J. Zhang, H. D. Jiang, S. S. Jiang, Q. Li, Y. L. Tian, Y. Y. Huang, W. X. Huang, W. He: Comparative study of $KTiOPO_4$ crystals. Opt. Mater. 23(1-2), 369-372(2003).

[116] J. Sorce, K. Palombo, S. Matthews, E. Gregor: Phase conjugate laser producing 1 J at 532 nm with 80% second-harmonic-generation efficiency. In: *OSA Proceedings on Advanced Solid-State Lasers*, Vol. 13, ed. by L. L. Chase, A. A. Pinto(OSA,Washington DC,1992), pp. 366-368.

[117] R. J. Bolt, M. van der Mooren: Single shot bulk damage threshold and conversion efficiency measurements on flux grown $KTiOPO_4$ (KTP). Opt. Commun. 100(1-4), 399-410(1993).

[118] J. Hellström, G. Karlsson, V. Pasiskevicius, F. Laurell: Optical parametric amplification in periodically poled $KTiOPO_4$ seeded by an Er-Yb: glass microchip laser.

Opt. Lett. 26(6), 352 – 354(2001).

[119] D. Eimerl, S. Velsko, L. Davis, F. Wang, G. Loiacono, G. Kennedy: Deuterated L-arginine phosphate: a new efficient nonlinear crystal. IEEE J. Quant. Electr. 25 (2), 179 – 193(1989).

[120] C. Chen: Chinese lab grows new nonlinear optical borate crystals. Laser Focus World 25(11), 129 – 137(1989).

[121] J. P. Feve, B. Boulanger, Y. Guillien: Efficient energy conversion for cubic third-harmonic generation that is phase-matched in $KTiOPO_4$. Opt. Lett. 25(18), 1373 – 1375 (2000).

[122] X. Mu, Y. J. Ding, J. Wang, Y. Li, J. Wei, J. B. Khurgin: Investigation of damage mechanisms for $KTiOPO_4$ crystals under irradiation of a CW argon laser. Proc. SPIE 3610, 9 – 14(1999).

[123] X. B. Hu, J. Y. Wang, H. J. Zhang, H. D. Jiang, H. Liu, X. D. Mu, Y. J. Ding: Dependence of photochromic damage on polarization in $KTiOPO_4$ crystals. J. Cryst. Growth 247(1 – 2), 137 – 140(2003).

[124] J. P. Feve, B. Boulanger, G. Marnier, H. Albrecht: Repetition rate dependence of graytracking in $KTiOPO_4$ during second-harmonic generation at 532 nm. Appl. Phys. Lett. 70(3), 277 – 279(1997).

[125] G. M. Loiacono, D. N. Loiacono, T. McGee, M. Babb: Laser damage formation in $KTiOPO_4$ and $KTiOAsO_4$: grey tracks. J. Appl. Phys. 72(7), 2705 – 2712 (1992).

[126] B. Boulanger, M. M. Fejer, R. Blachman, P. F. Bordui: Study of $KTiOPO_4$ graytracking at 1064, 532, and 355 nm. Appl. Phys. Lett. 65(19), 2401 – 2403 (1994).

[127] B. Boulanger, I. Rousseau, J. P. Feve, M. Maglione, M. Menaert, G. Marnier: Optical studies of laser-induced gray-tracking in KTP. IEEE J. Quant. Electr. 35(3), 281 – 286(1999).

[128] N. B. Angert, V. M. Garmash, N. I. Pavlova, A. V. Tarasov: Influence of color centers on the optical properties of KTP crystals and on the efficiency of the laser radiation frequency conversion in these crystals. Kvant. Elektron. 18(4), 470 – 472(1991)[In Russian, English trans.: Sov. J. Quantum Electron. 21(4), 426 – 428(1991)].

[129] J. K. Tyminski: Photorefractive damage in KTP used as second-harmonic generator. J. Appl. Phys. 70(10), 5570 – 5576(1991).

[130] R. Blachman, P. F. Bordui, M. M. Fejer: Laser-induced photochromic damage in potassium titanyl phosphate. Appl. Phys. Lett. 64(11), 1318 – 1320(1994).

[131] B. J. Le Garrec, G. J. Raze, P. Y. Thro, M. Gilbert: High-average-power diode-array-

pumped frequency-doubled YAG laser. Opt. Lett. 21(24), 1990 – 1992(1996).

[132] R. J. S. Pierre, D. W. Mordaunt, H. Injeyan, J. G. Berg, R. C. Hilyard, M. E. Weber, M. G. Wickham, G. M. Harpole, R. Senn: Diode array pumped kilowatt laser. IEEE J. Sel. Topics Quant. Electr. 3(1), 53 – 58(1997).

[133] E. C. Honea, C. A. Ebbers, R. J. Beach, J. A. Speth, J. A. Skidmore, M. A. Emmanuel, S. A. Payne: Analysis of an intracavity-doubled diode-pumped Q-switched Nd: YAG laser producing more than 100 W of power at 0.532 µm. Opt. Lett. 23(15), 1203 – 1205(1998).

[134] Y. Hirano, N. Pavel, S. Yamamoto, Y. Koyata, T. Tajime: 100-W, 100-h external green generation with Nd: YAG rod master-oscillator power-amplifier system. Opt. Commun. 184(1 – 4), 231 – 236(2000).

[135] A. Arie, G. Rosenman, V. Mahal, A. Skliar, M. Oron, M. Katz, D. Eger: Green and ultraviolet quasi-phase-matched second harmonic generation in bulk periodically-poled $KTiOPO_4$. Opt. Commun. 142(4 – 6), 265 – 268(1997).

[136] A. Englander, R. Lavi, M. Katz, M. Oron, D. Eger, E. Lebiush, G. Rosenman, A. Skliar: Highly efficient doubling of a high-repetition-rate diode-pumped laser with bulk periodically poled KTP. Opt. Lett. 22(21), 1598 – 1599(1997).

[137] A. Arie, G. Rosenman, A. Korenfeld, A. Skliar, M. Oron, M. Katz, D. Eger: Efficient resonant frequency doubling of a CW Nd: YAG laser in bulk periodically poled $KTiOPO_4$. Opt. Lett. 23(1), 28 – 30(1998).

[138] S. Wang, V. Pasiskevicius, F. Laurell, H. Karlsson: Ultraviolet generation by first order frequency doubling in periodically poled $KTiOPO_4$. Opt. Lett. 23(24), 1883 – 1885(1998).

[139] G. M. Gibson, G. A. Turnbull, M. Ebrahimzadeh, M. H. Dunn, H. Karlsson, G. Arvidsson, F. Laurell: Temperature-tuned difference-frequency mixing in periodically poled $KTiOPO_4$. Appl. Phys. B 67(5), 675 – 677(1998).

[140] T. Kartaloglu, K. G. Köprülü, O. Aytür, M. Sundheimer, W. P. Risk: Femtosecond optical parametric oscillator based on periodically poled $KTiOPO_4$. Opt. Lett. 23(1), 61 – 63(1998).

[141] A. Garashi, A. Arie, A. Skliar, G. Rosenman: Continuous-wave optical parametric oscillator based on periodically poled $KTiOPO_4$. Opt. Lett. 23(22), 1739 – 1741 (1998).

[142] D. R. Weise, U. Strößner, A. Peters, J. Mlynek, S. Schiller, A. Arie, A. Skliar, G. Rosenman: Continuous-wave 532-nm-pumped singly resonant optical parametric oscillator with periodically poled $KTiOPO_4$. Opt. Commun. 184(1 – 4), 329 – 333 (2000).

[143] J. Hellström, V. Pasiskevicius, H. Karlsson, F. Laurell: High-power optical paramet-

ric oscillation in large-aperture periodically poled KTiOPO$_4$. Opt. Lett. 25(3), 174 – 176(2000).

[144] G. W. Baxter, P. Schlup, I. T. McKinnie, J. Hellström, F. Laurell: Single-mode near-infrared optical parametric oscillator amplifier based on periodically poled KTiOPO$_4$. Appl. Opt. 40(36), 6659 – 6662(2001).

第3章
主要的红外材料

本章包括了最重要的红外非线性材料,即硫镓银(AGS)、硒镓银(AGSe)、磷锗锌(ZGP)和硒化镓(GaSe)。

3.1　AgGaS$_2$,硫镓银(AGS)

负单轴晶:$n_o > n_e$(在 $\lambda < 0.497$ μm 时,$n_e > n_o$)

分子量:241.723

密度:4.58 g/cm^3 [1];4.7 g/cm^3 [2];4.702 g/cm^3 [3]

点群:$\bar{4}2m$

晶格常数:

$a = 5.742$ Å [4];5.755 Å [5];5.757 Å [2,6]

$c = 10.26$ Å [4];10.28 Å [5];10.304 Å [2];10.305 Å [6]

$a = (5.757\ 22 \pm 0.000\ 03)$ Å,$T = 298$ K [1]

$c = (10.303\ 6 \pm 0.000\ 2)$ Å,$T = 298$ K [1]

莫氏硬度:3~3.5

熔点：$(1\,235\pm2)\,\text{K}^{[1]}$；$(1\,238\pm2)\,\text{K}^{[7]}$；$1\,269\,\text{K}^{[8]}$；$1\,270\,\text{K}^{[9]}$；$1\,271\,\text{K}^{[10]}$；$1\,323\,\text{K}^{[11]}$

线性热膨胀系数 α_t 的平均值[7]

T/K	$\alpha_t \times 10^6/\text{K}^{-1}$, $\parallel c$	$\alpha_t \times 10^6/\text{K}^{-1}$, $\perp c$
298 ~ 523	−13.2	12.7
298 ~ 773	−15.2	17.3
298 ~ 973	−16.7	20.1

$p = 0.101\,325\,\text{MPa}$ 时的比热容 c_p[12]

T/K	$c_p/(\text{J}\cdot\text{kg}^{-1}\cdot\text{K}^{-1})$
292	404

热导率 κ[13]

T/K	$\kappa/(\text{W}\cdot\text{m}^{-1}\cdot\text{K}^{-1})$, $\parallel c$	$\kappa/(\text{W}\cdot\text{m}^{-1}\cdot\text{K}^{-1})$, $\perp c$
293	1.4	1.5

室温下带隙能量：$E_g = 2.62\,\text{eV}^{[14]}$，$2.75\,\text{eV}^{[1]}$；$2.76\,\text{eV}^{[9]}$；$2.655\,\text{eV}(\boldsymbol{E}\perp c)^{[15]}$；$2.572\,\text{eV}(\boldsymbol{E}\parallel c)^{[15]}$

透明范围：

以 $\alpha = 1\,\text{cm}^{-1}$ 计：$0.48 \sim 11.4\,\mu\text{m}^{[8]}$

以 $\alpha = 3\,\text{cm}^{-1}$ 计：$0.50 \sim 13.2\,\mu\text{m}^{[14]}$

以"0"透过计：$0.47 \sim 13\,\mu\text{m}^{[16]}$

线性吸收系数 α

$\lambda/\mu\text{m}$	α/cm^{-1}	参考文献	备注
0.5 ~ 13	<0.1	[17]	
0.6 ~ 0.65	0.04	[18]	
0.6 ~ 12	<0.09	[16]	
0.633	0.02 ~ 0.05	[9]	典型晶体
	0.015 ~ 0.017	[9]	最佳晶体
0.633	0.05	[19]	
0.7 ~ 9	0.01	[20]	
0.8 ~ 9	0.01 ~ 0.02	[9]	典型晶体
	0.015	[9]	最佳晶体
0.845	0.01	[12]	

续表

$\lambda/\mu m$	α/cm^{-1}	参考文献	备注
0.9~8.5	<0.9	[11]	
1.064 2	0.01	[19]	
	0.01~0.02	[21]	典型晶体
	0.001~0.009	[9]	典型晶体
	0.000 5~0.005	[21]	最佳晶体
	<0.000 5	[9]	最佳晶体
1.15	0.02~0.07	[22]	
1.26	0.26	[23]	e 光
1.8	<0.1	[15]	e 光
1.9	0.05~0.15	[9]	典型晶体，e 光
	0.03	[9]	最佳晶体，e 光
2.1	<0.02	[15]	e 光
2.15	0.08~0.25	[9]	典型晶体，e 光
	0.05	[9]	最佳晶体，e 光
2.8	0.012~0.024	[9]	典型晶体
	0.009	[9]	最佳晶体
4~8.5	<0.04	[18]	
4.64	0.03	[24]	
9.27	0.19	[24]	
9.55	<0.1	[25]	
10.2	0.43	[23]	o 光
10.6	0.6	[9]	最佳晶体

双光子吸收系数 β

$\lambda/\mu m$	τ_p/ns	$\beta \times 10^{11}/(cm \cdot W^{-1})$	参考文献	备注
0.8	0.000 2	350	[26]	o 光
0.8~0.87	0.000 2	18	[27]	

折射率的实验值[6]

$\lambda/\mu m$	n_o	n_e	$\lambda/\mu m$	n_o	n_e	$\lambda/\mu m$	n_o	n_e
0.490	2.714 8	2.728 7	0.625	2.557 7	2.511 6	0.850	2.480 2	2.427 9
0.500	2.691 6	2.686 7	0.650	2.543 7	2.496 1	0.900	2.471 6	2.419 2
0.525	2.650 3	2.623 9	0.675	2.531 0	2.482 4	0.950	2.464 4	2.411 8
0.550	2.619 0	2.583 4	0.700	2.520 5	2.470 6	1.000	2.458 2	2.405 3
0.575	2.594 4	2.553 7	0.750	2.504 9	2.454 0	1.100	2.448 6	2.395 4
0.600	2.574 8	2.530 3	0.800	2.490 9	2.439 5	1.200	2.441 4	2.388 1

续表

$\lambda/\mu m$	n_o	n_e	$\lambda/\mu m$	n_o	n_e	$\lambda/\mu m$	n_o	n_e
1.300	2.435 9	2.381 9	3.200	2.406 8	2.353 4	7.500	2.378 7	2.325 2
1.400	2.431 5	2.378 1	3.400	2.406 2	2.352 2	8.000	2.375 7	2.321 9
1.500	2.428 0	2.374 5	3.600	2.404 6	2.351 1	8.500	2.369 9	2.316 3
1.600	2.425 2	2.371 6	3.800	2.402 4	2.349 1	9.000	2.366 3	2.312 1
1.800	2.420 6	2.367 0	4.000	2.402 4	2.348 8	9.500	2.360 6	2.306 4
2.000	2.416 4	2.363 7	4.500	2.400 3	2.346 1	10.00	2.354 8	2.301 2
2.200	2.414 2	2.368 4	5.000	2.395 5	2.341 9	10.50	2.348 6	2.294 8
2.400	2.411 9	2.358 3	5.500	2.393 8	2.340 1	11.00	2.341 7	2.288 0
2.600	2.410 2	2.356 7	6.000	2.390 8	2.336 9	11.50	2.332 9	2.278 9
2.800	2.409 4	2.355 9	6.500	2.387 4	2.333 4	12.00	2.326 6	2.271 6
3.000	2.408 0	2.354 5	7.000	2.382 7	2.329 1	12.50	2.317 7	

折射率的实验值[28]

$\lambda/\mu m$	n_o	n_e	$\lambda/\mu m$	n_o	n_e
0.632 8	2.547 6	2.503 9	5.295 5	2.394 5	2.340 5
1.064 2	2.451 3	2.398 2	9.271 4	2.362 7	2.307 4
1.152 3	2.444 3	2.391 1	10.591 0	2.347 6	2.291 9
3.391 3	2.405 5	2.351 9	10.632 1	2.347 1	2.291 4

旋光性[16,29]：$\rho = 522$ (°)/mm，在各向同性点($n_o = n_e, \lambda = 0.497\ 3\ \mu m$)

最佳色散关系方程(λ 以 μm 为单位, $0.58\ \mu m < \lambda < 10.59\ \mu m, T = 293\ K$)[30]：

$$n_o^2 = 5.797\ 5 + \frac{0.231\ 1}{\lambda^2 - 0.068\ 8} - 0.002\ 57\lambda^2$$

$$n_e^2 = 5.543\ 6 + \frac{0.223\ 0}{\lambda^2 - 0.094\ 6} - 0.002\ 61\lambda^2$$

较宽波长范围的色散关系($0.54\ \mu m < \lambda < 12.9\ \mu m$)[24,28]：

$$n_o^2 = 5.794\ 19 + \frac{0.231\ 14}{\lambda^2 - 0.068\ 82} - 2.453\ 4 \times 10^{-3}\lambda^2 + 3.181\ 4 \times 10^{-7}\lambda^4 - 9.705\ 1 \times 10^{-9}\lambda^6$$

$$n_e^2 = 5.541\ 20 + \frac{0.220\ 41}{\lambda^2 - 0.098\ 24} - 2.524\ 0 \times 10^{-3}\lambda^2 + 3.621\ 4 \times 10^{-7}\lambda^4 - 8.360\ 5 \times 10^{-9}\lambda^6$$

在文献[23]、[31]、[32]、[33]、[34]、[35]、[36]中给出了其他的 Sellmeier 方程组。

光谱范围 $0.56 \sim 10.59\ \mu m$ 以及温度范围 $293 \sim 473\ K$ 之间折射率的温度微商(λ 以 μm 为单位)[30]：

$$\frac{dn_o}{dT} = \left(\frac{0.3180}{\lambda^3} + \frac{2.8968}{\lambda^2} - \frac{0.8685}{\lambda} + 15.2679\right) \times 10^{-5} \text{ K}^{-1}$$

$$\frac{dn_e}{dT} = \left(\frac{6.1742}{\lambda^3} - \frac{12.0868}{\lambda^2} + \frac{8.2485}{\lambda} + 14.4365\right) \times 10^{-5} \text{ K}^{-1}$$

在文献[31]、[36]、[37]中给出了其他热光色散公式。

室温下低频(远低于$AgGaS_2$晶体的声学共振频率,即对"自由"晶体)下测量的线性电光系数[38]

$\lambda/\mu m$	$r_{41}^T/(pm \cdot V^{-1})$	$r_{63}^T/(pm \cdot V^{-1})$
0.6328	4.0±0.2	3.0±0.1

一般情况下有效非线性系数的表达式(Kleinman对称条件成立,$d_{14} = d_{25} = d_{36}$)[39]:

$$d_{ooe} = -d_{36}\sin(\theta+\rho)\sin 2\phi$$
$$d_{eoe} = d_{oee} = 2d_{36}\sin(\theta+\rho)\cos(\theta+\rho)\cos 2\phi$$

有效二阶非线性系数的简化表达式(小双折射角近似,Kleinman对称条件成立,$d_{14} = d_{25} = d_{36}$)[40]:

$$d_{ooe} = -d_{36}\sin\theta\sin 2\phi$$
$$d_{eoe} = d_{oee} = d_{36}\sin 2\theta\cos 2\phi$$

二阶非线性系数的值:

$$d_{36}(1.054 \text{ μm}) = (23.6 \pm 2.4) \text{ pm/V}^{[41]}$$

$$d_{36}(2.53 \text{ μm}) = (13.7 \pm 2.2) \text{ pm/V}^{[23]}$$

$$d_{36}(10.6 \text{ μm}) = 0.134 \times d_{36}(GaAs) \pm 15\% = (11.1 \pm 1.7) \text{ pm/V}^{[6,42]}$$

$$d_{36}(10.6 \text{ μm}) = 0.15 \times d_{36}(GaAs) \pm 20\% = (12.5 \pm 2.5) \text{ pm/V}^{[42,43]}$$

下面给出的$AgGaS_2$其他二阶非线性系数的值与最近LIS、$LiInSe_2$及$HgGaS_4$晶体测量的相对值不相符:

$$d_{36}(1.2 \text{ μm}) = (31 \pm 5) \text{ pm/V}^{[44]}$$

$$d_{36}(9.2714 \text{ μm}) = 0.84 \pm 0.10 \times d_{36}(AgGaSe_2) = (34.8 \pm 4.0) \text{ pm/V}^{[24,45]}$$

$$d_{36}(10.591 \text{ μm}) = 0.84 \pm 0.10 \times d_{36}(AgGaSe_2) = (32.0 \pm 4.0) \text{ pm/V}^{[24,45]}$$

相位匹配角的实验值($T = 293$ K)

相互作用的波长/μm	$\theta_{exp}/(°)$	参 考 文 献
SHG, o+o⇒e		
2.0970⇒1.0485	56.1	[24]
2.1284⇒1.0642	54.7	[24]
3.3913⇒1.69565	34.1	[24]

续表

相互作用的波长/μm	$\theta_{\exp}/(°)$	参考文献
SHG, o+o⇒e		
	33	[16]
5.295 5⇒2.647 75	32.7	[28]
9.271 4⇒4.635 7	54.2	[28]
10.571 0⇒5.285 5	68.2	[28]
10.591 0⇒5.295 5	68.5	[28]
10.6⇒5.3	67	[43]
	67.5	[10]
	68	[16]
	70.8	[6]
10.632 1⇒5.316 05	69.1	[28]
11.10⇒5.55	78.5	[28]
SFG, o+o⇒e		
11.538+1.172 33⇒1.064 2	34.7	[35]
10.591 0+5.295 5⇒3.530 3	43.4	[28]
9.9+1.192 37⇒1.064 2	35.9	[46]
8.7+1.212 52⇒1.064 2	37	[47]
6.24+1.283 01⇒1.064 2	41.1	[48]
5.89+1.298 88⇒1.064 2	42.1	[46]
4.8+1.367 35⇒1.064 2	44	[47]
4.0+1.449 96⇒1.064 2	47.7	[48]
3.09+1.623 25⇒1.064 2	51	[34]
2.53+1.836 83⇒1.064 2	53.4	[34]
6.85+1.064 2⇒0.921 10	42	[49]
4.43+1.064 2⇒0.858 07	55	[49]
6.6+0.775 93⇒0.694 3	60	[50]
4.8+0.811 71⇒0.694 3	75.5	[50]
11.663 29+0.617⇒0.586	64	[51]
10.124 78+0.622⇒0.586	70	[51]
SFG, e+o⇒e		
10.9+1.179 34⇒1.064 2	38.3	[52]
8.8+1.210 60⇒1.064 2	40.3	[52]
7.0+1.255 00⇒1.064 2	43.6	[52]
5.2+1.338 03⇒1.064 2	50.6	[52]
10.6+1.064 2⇒0.967 11	39.8	[53]
9.6+1.064 2⇒0.958 00	41.5	[53]
10.6+0.694 3⇒0.651 62	55	[54]

$T = 293$ K 时内角宽度和温度带宽的实验值

相互作用的波长/μm	$\theta_{pm}/(°)$	$\Delta\theta^{int}/(°)$	$\Delta\nu_1/cm^{-1}$	参考文献
SHG, o + o ⇒ e				
10.6 ⇒ 5.3	67.5	0.41		[16]
SFG, o + o ⇒ e				
10.53 + 0.589 ⇒ 0.565 89	90	2.34		[33]
10.619 + 0.634 ⇒ 0.598	90		1.73	[18]
10.6 + 0.598 ⇒ 0.566	90		1.5	[55]
10.6 + 0.596 8 ⇒ 0.565	90		1.44	[56]
DFG, e - o ⇒ o				
0.6943 - 0.817 7 ⇒ 4.6	82.7	0.42		[50]
0.871 63 - 1.064 2 ⇒ 4.817	52		5.9	[49]
1.064 2 - 1.283 ⇒ 6.24	41.1		9.8	[48]
DFG, e - o ⇒ e				
0.76 - 0.830 1 ⇒ 9.0	46.5	0.4	11($\Delta\nu_3$)	[57]
0.80 - 0.870 7 ⇒ 9.85			2.9	[58]
0.80 - 0.871 5 ⇒ 9.75			3.4	[27]
0.80 - 0.885 3 ⇒ 8.3			3.1	[58]

相位匹配角的温度变化[53]

相互作用的波长/μm	$T/°C$	$\theta_{pm}/(°)$	$d\theta_{pm}/dT/[(°)\cdot K^{-1}]$
SFG, e + o ⇒ e			
10.6 + 1.064 2 ⇒ 0.967 1	20	39.8	0.03

非临界 SFG 的温度调谐[44]

相互作用的波长/μm	$d\lambda_1/dT/(nm\cdot K^{-1})$
SHG, o + o ⇒ e	
7.8 + 0.65 ⇒ 0.6	≈ 4

温度带宽的实验值[30]

相互作用的波长/μm	$\theta_{exp}/(°)$	$\Delta T/°C$	相互作用的波长/μm	$\theta_{exp}/(°)$	$\Delta T/°C$
SHG, o + o ⇒ e			SHG, o + o ⇒ e		
10.591 ⇒ 5.295 5	68.5	139	9.271 4 ⇒ 4.635 7	55.0	118
10.246 6 ⇒ 5.123 3	64.3	135	5.295 5 ⇒ 2.647 75	33.2	59
9.552 5 ⇒ 4.776 25	57.4	123	3.530 3 ⇒ 1.765 15	33.7	22

非临界和频发生(SFG)过程的温度带宽实验值

相互作用的波长/μm	T/℃	ΔT/℃	参 考 文 献
SFG, o + o ⇒ e			
10.591 + 0.598 3 ⇒ 0.566 32		2.5	[37]
3.262 7 + 1.064 2 ⇒ 0.802 5	192	6.4	[30]

激光诱导的表面损伤阈值

λ/μm	τ_p/ns	$I_{thr}/(GW \cdot cm^{-2})$	参 考 文 献	备 注
0.59	500	0.02	[51]	10 个脉冲
0.598	3	0.015	[55]	
0.625	500	0.025 ~ 0.036	[51]	10 个脉冲
0.694 3	30	0.000 6	[54]	1 Hz, 1 000 个脉冲
	10	0.01	[50]	100 个脉冲
		0.02	[31]	
0.75	50	0.025	[20]	
0.8	30	0.01	[57]	10 Hz
0.8 ~ 0.87	0.000 2	>60	[27]	
1.06	35	0.02 ~ 0.025	[31]	
1.064 2	100	0.002	[59]	10 Hz, 3 000 个脉冲
	20	0.01	[34]	10 Hz
	17.5	>0.012	[60]	1 000 个脉冲
	15	0.02	[10]	
	12	0.035	[52]	10 Hz
	11	0.03 ~ 0.05	[61]	10 Hz, 50 个脉冲
	10	0.03	[21]	10 Hz
	8	0.034 ~ 0.06	[61]	单脉冲
	0.025	>0.7	[35]	10 Hz
	0.023	>0.075	[62]	10 Hz
	0.021	>2	[49]	
	0.020	3	[46]	
	0.002	>1	[63]	
2	6	0.017	[64]	未镀膜晶体
		0.035	[64]	镀膜晶体
2.079	180	0.001 4	[24]	5 Hz, 未镀膜晶体
9.27	50	>0.044	[24]	1 Hz, 未镀膜晶体
9.55	30	0.18	[25]	SHG 方向
10.6	220	0.025	[60]	1 000 个脉冲
	150	0.01	[33]	
		0.02	[65]	

激光诱导的体损伤阈值[21]

$\lambda/\mu m$	τ_p/ns	$I_{thr}/(GW \cdot cm^{-2})$
1.064	10	>0.5

关于这一晶体

最近几年，AGS 被广泛地用于 DFG 和 OPO；下面，我们将只是列出最佳的技术成就。在文献[66]中，由钛宝石激光器($\lambda = 800$ nm)50 fs、150 kHz 脉冲泵浦 BBO 晶体 OPA 的信号光和闲散光在 0.1 cm 长的 II 类相位匹配硫镓银晶体中混频，获得可调谐波范围为 2.4～12 μm。在文献[64]中，由 Nd:YAG 激光器(8 ns、30 Hz)泵浦 LN 晶体 OPO 的信号光和闲频光的波长用于 DFG。调谐范围为 2.4～12 μm，在接近 7.5 μm 处最大的差频脉冲能量为 95 μJ。对于以 AGS 为基础的 OPO，在接近 6 μm 处最高的闲频光脉冲所达到的最高能量是 400 μJ[59]。在同一工作中，所报道最宽的闲频光调谐范围是 3.9～11.3 μm。文献[59]的作者们应用了一台纳秒的 Nd:YAG 激光器(1.064 μm、30 ns、10 Hz)和 2 cm 长 II 类相位匹配的 AGS 晶体。在文献[67]中，采用一台调 Q 锁模 Nd:YAG 激光器($\lambda = 1.064$ μm，$P = 8$ W)泵浦一块 3 cm 长 I 类相位匹配的 $AgGaS_2$ 晶体，闲频光波长在 4.06 μm 处的最高平均功率达到 600 mW 左右。同时，1.44 μm 的信号光的平均功率达到 1.5 W。

参考文献

[1] *Physical-Chemical Properties of Semiconductors. Handbook.* (Nauka, Moscow, 1979) [In Russian].

[2] D. M. Bercha, Y. V. Voroshilov, V. Y. Slivka, I. D. Turyanitsa: *Complex Chalcogenides and Chalcohalogenides*, ed. by D. V. Chepur (Vishcha Shkola, Lvov, 1983) [In Russian].

[3] B. H. T. Chai: Optical Crystals. In: *CRC Handbook of Laser Science and Technology, Supplement 2: Optical Materials*, ed. by M. J. Weber (CRC Press, Boca Raton, 1995) pp. 3-65.

[4] V. I. Gavrilenko, A. M. Grekhov, D. B. Korbutyak, V. G. Litovchenko: *Optical Properties of Semiconductors. Handbook.* (Naukova Dumka, Kiev, 1987) [In Russian].

[5] *Chemist's Handbook*, Vol. I, ed. by B. P. Nikolskii (Goskhimizdat, Leningrad, 1962) [In Russian].

[6] G. D. Boyd, H. Kasper, J. H. McFee: Linear and nonlinear optical properties of $AgGaS_2$, $CuGaS_2$ and $CuInS_2$, and the theory of the wedge technique for the measurement of nonlinear coefficients. IEEE J. Quant. Electr. QE-7(12), 563-573(1971).

[7] P. Korczak, C. B. Staff: Liquid encapsulated Czochralski growth of silver thiogallate. J.

Cryst. Growth 24 – 25, 386 – 389(1974).

[8] R. S. Feigelson, R. K. Route: Recent developments in the growth of chalcopyrite crystals for nonlinear infrared applications. Opt. Eng. 26(2), 113 – 119(1987).

[9] G. C. Catella, D. Burlage: Crystal growth and optical properties of $AgGaS_2$ and $AgGaSe_2$ MRS Bulletin 23(7), 28 – 36(1998).

[10] D. S. Chemla, P. J. Kupecek, D. S. Robertson, R. C. Smith: Silver thiogallate, a new material with potential for infrared devices. Opt. Commun. 3(1), 29 – 31(1971).

[11] H. Matthes, R. Viehmann, N. Marschall: Improved optical quality of $AgGaS_2$. Appl. Phys. Lett. 26(5), 237 – 239(1975).

[12] A. Douillet, J.-J. Zondy, A. Yelisseyev, S. Lobanov, L. Isaenko: Stability and frequency tuning of thermally loaded continuous-wave $AgGaS_2$ optical parametric oscillators. J. Opt. Soc. Am. B 16(9), 1481 – 1498(1999).

[13] J. D. Beasley: Thermal conductivities of some novel nonlinear optical materials. Appl. Opt. 33 (6), 1000 – 1003(1994).

[14] G. C. Bhar, R. C. Smith: Optical properties of II-IV-V_2 and I-III-VI_2 crystals with particular reference to transmission limits. Phys. Stat. Solidi A 13 (1), 157 – 168 (1972).

[15] Data sheet of Cleveland Crystals Inc. Available at www. clevelandcrystals. com.

[16] V. V. Badikov, O. N. Pivovarov, Y. V. Skokov, O. V. Skrebneva, N. K. Trotsenko: Some optical properties of silver thiogallate single crystals. Kvant. Elektron. 2(3), 618 – 621(1975)[In Russian,English trans. :Sov. J. Quantum Electron. 5(3),350 – 351(1975)].

[17] E . S. Voronin, V. S. Solomatin, N. I. Cherepov, V. V. Shuvalov, V. V. Badikov, O. N. Pivovarov: Conversion of infrared radiation in an $AgGaS_2$ crystal. Kvant. Elektron. 2(5), 1090 – 1092(1975)[In Russian,English trans. :Sov. J. Quantum Electron. 5(5),597 – 598(1975)].

[18] P. Canarelli, Z. Benko, R. Curl, F. K. Tittel: Continuous-wave infrared laser spectrometer based on difference frequency generation in $AgGaS_2$ for high-resolution spectroscopy. J. Opt. Soc. Am. B 9(2), 197 – 202(1992).

[19] A. H. Hielscher, C. E. Miller, D. C. Bayard, U. Simon, K. P. Smolka, R. F. Curl, F. K. Tittel: Optimisation of a midinfrared high-resolution difference-frequency laser spectrometer. J. Opt. Soc. Am. B 9(11), 1962 – 1967(1992).

[20] A. O. Okorogu, S. B. Mirov, W. Lee, D. I. Crouthamel, N. Jenkins, A. Y. Dergachev, K. L. Vodopyanov, V. V. Badikov: Tunable middle infrared downconversion in GaSe and $AgGaS_2$. Opt. Commun. 155(4 – 6), 307 – 312 (1998).

[21] S. Chandra, T. H. Allik, G. Catella, J. A. Hutchinson: Tunable output around 8 μm from a single step $AgGaS_2$ OPO pumped at 1. 064 μm. In: *Advanced Solid-State Lasers, OSA Trends in Optics and Photonics Series, Vol. 19*, ed. by W. R. Bosenberg,

M. M. Fejer(OSA, Washington DC, 1998), pp. 282 – 284.

[22] V. V. Badikov, P. S. Blinov, A. A. Kosterev, V. S. Letokhov, A. L. Malinovskii, E. A. Ryabov: Efficient parametric generators of picosecond mid-infrared pulses based on $AgGaS_2$ crystals. Kvant. Elektron. 24(6), 537 – 540(1997) [In Russian, English trans.: Quantum Electron. 27(6), 523 – 526(1997)].

[23] J.-J. Zondy, D. Touahri, O. Acef: Absolute value of the d_{36} nonlinear coefficient of $AgGaS_2$: prospect for a low-threshold doubly resonant oscillator-based 3:1 frequency divider. J. Opt. Soc. Am. B 14(10), 2481 – 2497(1997).

[24] A. Harasaki, K. Kato: New data on the nonlinear optical constant, phase-matching, and optical damage of $AgGaS_2$. Jpn. J. Appl. Phys. 36(2), 700 – 703(1997).

[25] Y. M. Andreev, V. V. Badikov, V. G. Voevodin, L. G. Geiko, P. P. Geiko, M. V. Ivashchenko, A. I. Karapuzikov, I. V. Sherstov: Radiation resistance of nonlinear crystals at a wavelength of 9.55 μm. Kvant. Elektron. 31(12), 1075 – 1078(2001) [In Russian, English trans.: Quantum Electron. 31(12), 1075 – 1078(2001)].

[26] F. Rotermund, V. Petrov, F. Noack, L. Isaenko, A. Yelisseyev, S. Lobanov: Optical parametric generation of femtosecond pulses up to 9 μm with $LiInS_2$ pumped at 800 nm. Appl. Phys. Lett. 78(18), 2623 – 2625(2001).

[27] J. Song, J. F. Xia, Z. Zhang, D. Strickland: Mid-infrared pulses generated from the mixing output of an amplified, dual-wavelength Ti: sapphire system. Opt. Lett. 27(3), 200 – 202(2002).

[28] K. Kato, H. Shirahata: Nonlinear IR generation in $AgGaS_2$. Jpn. J. Appl. Phys. 35(9A), 4645 – 4648(1996).

[29] V. V. Badikov, I. N. Matveev, S. M. Pshenichnikov, O. V. Skrebneva, N. K. Trotsenko, N. D. Ustinov: Dispersion of birefringence and the optical activity of $AgGa(S_{1-x}Se_x)_2$ crystals. Kristallogr. 26(3), 537 – 539(1981) [In Russian, English trans.: Sov. Phys. -Crystallogr. 26(3), 304 – 305(1981)].

[30] E. Takaoka, K. Kato: Thermo-optic dispersion formula for $AgGaS_2$. Appl. Opt. 38(21), 4577 – 4580(1999).

[31] G. C. Bhar, R. C. Smith: Silver thiogallate ($AgGaS_2$)—Part II: linear optical properties. IEEE J. Quant. Electr. QE – 10(7), 546 – 550(1974).

[32] G. C. Bhar: Refractive index interpolation in phase-matching. Appl. Opt. 15(2), 305 – 307(1976).

[33] T. Itabe, J. L. Bufton: Phase-matching measurements for 10-μm upconversion in $AgGaS_2$. Appl. Opt. 23(18), 3044 – 3047(1984).

[34] Y. X. Fan, R. C. Eckardt, R. L. Byer, R. K. Route, R. S. Feigelson: $AgGaS_2$ infrared parametric oscillator. Appl. Phys. Lett. 45(4), 313 – 315(1984).

[35] H.-J. Krause, W. Daum: High-power source of coherent picosecond light pulses tunable

from 0.41 to 12.9 μm. Appl. Phys. B 56(1), 8-13(1993).

[36] J.-J. Zondy, D. Touahri: Updated thermo-optic coefficients of AgGaS$_2$ from temperature-tuned noncritical $3\omega - \omega \to 2\omega$ infrared parametric amplification. J. Opt. Soc. Am. B 14(6), 1331-1338(1997).

[37] G.C. Bhar, D.K. Ghosh, P.S. Ghosh, D. Schmitt: Temperature effects in AgGaS$_2$ nonlinear devices. Appl. Opt. 22(16), 2492-2494(1983).

[38] V.M. Cound, P.H. Davies, K.F. Hulme, D. Robertson: The electrooptic coefficients of silver thiogallate(AgGaS$_2$). J. Phys. C3(4), L83-L84(1970).

[39] R.C. Eckardt, H. Masuda, Y.X. Fan, R.L. Byer: Absolute and relative nonlinear optical coefficients of KDP, KD*P, BaB$_2$O$_4$, LiIO$_3$, MgO: LiNbO$_3$, and KTP measured by phasematched second-harmonic generation. IEEE J. Quant. Electr. 26(5), 922-933(1990).

[40] J.E. Midwinter, J. Warner: The effects of phase matching method and of uniaxial crystal symmetry on the polar distribution of second-order non-linear optical polarization. Brit. J. Appl. Phys. 16(11), 1135-1142(1965).

[41] E.C. Cheung, K. Koch, G.T. Moore, J.M. Liu: Measurements of second-order nonlinear optical coefficients from the spectral brightness of parametric fluorescence. Opt. Lett. 19(3), 168-170(1994).

[42] D.A. Roberts: Simplified characterization of uniaxial and biaxial nonlinear optical crystals: a plea for standardization of nomenclature and conventions. IEEE J. Quant. Electr. 28(10), 2057-2074(1992).

[43] P.J. Kupeček, C.A. Schwartz, D.S. Chemla: Silver thiogallate (AgGaS$_2$)—Part I: nonlinear optical properties. IEEE J. Quant. Electr. QE-10(7), 540-545(1974).

[44] P. Canarelli, Z. Benko, A.H. Hielscher, R.F. Curl, F.K. Tittel: Measurements of nonlinear coefficient and phase matching characteristics of AgGaS$_2$. IEEE J. Quant. Electr. 28(1), 52-55(1992).

[45] K. Kato: Second-harmonic and sum-frequency generation in ZnGeP$_2$. Appl. Opt. 36(12), 2506-2510(1997).

[46] T. Elsaesser, A. Seilmeier, W. Kaiser, P. Koidl, G. Brandt: Parametric generation of tunable picosecond pulses in the medium infrared using AgGaS$_2$ crystals. Appl. Phys. Lett. 44(4), 383-385(1984).

[47] H.J. Bakker, J.T.M. Kennis, H.J. Kop, A. Lagendijk: Generation of intense picosecond pulses tunable between 1.2 and 8.7 μm. Opt. Commun. 86(1), 58-64(1991).

[48] T. Elsaesser, H. Lobentanzer, A. Seilmeier: Generation of tunable picosecond pulses in the medium infrared by down-conversion in AgGaS$_2$. Opt. Commun. 52(5), 355-359(1985).

[49] A.G. Yodh, H.W.K. Tom, G.D. Aumiller, R.S. Miranda: Generation of tunable mid-

infrared picosecond pulses at 76 MHz. J. Opt. Soc. Am. B 8(8), 1663 – 1667(1991).

[50] D. C. Hanna, V. V. Rampal, R. C. Smith: Tunable infrared down-conversion in silver thiogallate. Opt. Commun. 8(2), 151 – 153(1973).

[51] D. C. Hanna, V. V. Rampal, R. C. Smith: Tunable medium infrared generation in silver thiogallate ($AgGaS_2$) by down-conversion of flash-pumped dye-laser radiation. IEEE J. Quant. Electr. QE – 10(4), 461 – 462(1974).

[52] K. Kato: High-power difference-frequency generation at 5 – 11 μm in $AgGaS_2$. IEEE J. Quant. Electr. QE – 20(7), 698 – 699(1984).

[53] G. C. Bhar, S. Das, U. Chatterjee, R. S. Feigelson, R. K. Route: Synchronous and noncollinear infrared upconversion in $AgGaS_2$. Appl. Phys. Lett. 54(16), 1489 – 1491 (1989).

[54] S. A. Andreev, I. N. Matveev, I. P. Nekrasov, S. M. Pshenichnikov, N. P. Sopina: Parametric conversion of infrared radiation in an $AgGaS_2$ crystal. Kvant. Elektron. 4(3), 657 – 659(1977) [In Russian, English trans. : Sov. J. Quantum Electron. 7(3) ,366 – 368(1977)].

[55] W. Jantz, P. Koidl: Efficient up-conversion of 10.6-μm radiation into the green spectral range. Appl. Phys. Lett. 31(2), 99 – 101(1977).

[56] A. P. Gorchakov, A. A. Popesku, V. S. Solomatin: Nonlinear spectroscopy based on a silver thiogallate crystal. Kvant. Elektron. 5(2), 413 – 415(1978) [In Russian, English trans. : Sov. J. Quantum Electron. 8(2) ,236 – 237(1978)].

[57] N. Saito, K. Akagawa, S. Wada, H. Tashiro: Difference-frequency generation by mixing dual-wavelength pulses emitting from an electronically tuned Ti: sapphire laser. Appl. Phys. B 69(2), 93 – 97(1999).

[58] J. F. Xia, J. Song, D. Strickland: Development of a dual-wavelength Ti: sapphire multipass amplifier and its application to intense mid-infrared generation. Opt. Commun. 206(1 – 3), 149 – 157(2002).

[59] K. L. Vodopyanov, J. P. Maffetone, I. Zwieback, W. Ruderman: $AgGaS_2$ optical parametric oscillator continuously tunable from 3.9 to 11.3 μm. Appl. Phys. Lett. 75(9), 1204 – 1206(1999).

[60] D. C. Hanna, B. Luther-Davies, H. N. Rutt, R. C. Smith, C. R. Stanley: Q-switched laser damage of infrared nonlinear materials. IEEE J. Quant. Electr. QE – 8(3), 317 – 324(1972).

[61] P. B. Phua, R. F. Wu, T. C. Chong, B. X. Xu: Nanosecond $AgGaS_2$ optical parametric oscillator with more than 4 micron output. Jpn. J. Appl. Phys. 36(12B), L1661-L1664 (1997).

[62] K. G. Spears, X. Zhu, X. Yang, L. Wang: Picosecond infrared generation from Nd: YAG and a visible short-cavity dye laser. Opt. Commun. 66(2 – 3), 167 – 171(1988).

[63] T. Dahinten, U. Plödereder, A. Seilmeier, K. L. Vodopyanov, K. R. Allakhverdiev,

Z. A. Ibragimov: Infrared pulses of 1 picosecond duration tunable between 4 μm and 18 μm. IEEE J. Quant. Electr. 29(7), 2245 - 2250(1993).

[64] S. Haidar, K. Nakamura, E. Niwa, K. Masumoto, H. Ito: Mid-infrared(5 - 12-μm) and limited(5.5 - 8.5-μm) single-knob tuning generated by difference frequency mixing in single-crystal $AgGaS_2$. Appl. Opt. 38(9), 1798 - 1801(1999).

[65] H. Kildal, G. W. Iseler: Laser-induced surface damage of infrared nonlinear materials. Appl. Opt. 15(12), 3062 - 3065(1976).

[66] B. Golubovic, M. K. Reed: All-solid-state generation of 100-kHz tunable mid-infrared 50-fs pulses in type Ⅰ and type Ⅱ $AgGaS_2$. Opt. Lett. 23(22), 1760 - 1762(1998).

[67] K. J. McEwan: High-power synchronously pumped $AgGaS_2$ optical parametric oscillator. Opt. Lett. 23(9), 667 - 669(1998).

3.2 $AgGaSe_2$，硒镓银(AGSe)

负单轴晶：$n_o > n_e$(在 $\lambda < 0.804$ μm 时，$n_e > n_o$)

分子量：335.511

密度：5.70 g/cm³[1,2]；5.71 g/cm³[3]；5.76 g/cm³[4]

点群：$\bar{4}2m$

晶格常数：

$a = 5.9220$ Å[5]；(5.99202 ± 0.00018) Å[6]

$c = 10.8803$ Å[5]；(10.88626 ± 0.00030) Å[6]

莫氏硬度：3 ~ 3.5

熔点：1 123 K[3]；1 124 K[2]；1 129 K[7]；1 133 K[8]

线性热膨胀系数 α_t 的平均值

T/K	$\alpha_t \times 10^6/K^{-1}$, ∥ c	$\alpha_t \times 10^6/K^{-1}$, ⊥ c	参 考 文 献
298 ~ 423	- 8.1	19.8	[9]
	- 6.4	23.4	[4]
298 ~ 573	- 9.6	24.6	[9]
	- 15.7	16.3	[10]
423 ~ 773	- 12.6	28.0	[9]
423 ~ 873	- 16.0	18.0	[4]

$p = 0.101\ 325$ MPa 时的比热容 c_p[11]

T/K	$c_p/(J \cdot kg^{-1} \cdot K^{-1})$	T/K	$c_p/(J \cdot kg^{-1} \cdot K^{-1})$
300	297	500	318
400	311		

热导率 $\kappa^{[12]}$

T/K	$\kappa/(W \cdot m^{-1} \cdot K^{-1})$, $\parallel c$	$\kappa/(W \cdot m^{-1} \cdot K^{-1})$, $\perp c$
293	1.0	1.1

室温下带隙能量：$E_g = 1.65\ eV^{[3]}$；$1.72\ eV^{*[13]}$；$1.8\ eV^{[14]}$；$1.803\ eV^{[15]}$；$1.83\ eV^{[2]}$；$1.713\ eV(E \perp c)^{[16]}$；$1.689\ eV(E \parallel c)^{[16]}$

透明范围：

以 $\alpha = 1\ cm^{-1}$ 透过计：$0.76 \sim 17\ \mu m^{[7]}$

以 $\alpha = 3\ cm^{-1}$ 透过计：$0.78 \sim 18\ \mu m^{[13]}$

以 "0" 透过计：$0.71 \sim 19\ \mu m^{[8,17]}$

线性吸收系数 α

$\lambda/\mu m$	α/cm^{-1}	参考文献	备注
1	<0.02	[18]	
1~11	0.01~0.18	[2]	典型值
1.06	0.018	[19]	e 光
1.064	0.012~0.2	[2]	典型值
	0.006	[2]	最佳值
1.3	0.002	[20]	o 光，OPO 方向
	0.002	[20]	e 光，OPO 方向
1.45~1.6	<0.015	[19]	o 光，e 光
1.8	<0.02	[16]	e 光
1.9	0.012~0.2	[2]	典型值，e 光
	0.01	[2]	最佳值，e 光
2.0	0.012	[21]	o 光，OPO 方向
	0.030	[21]	e 光，OPO 方向
	0.004	[22]	
2~5	<0.05	[23]	
2.05	0.015~0.058	[24]	
	<0.01	[25]	

* 译者注：原书为 "1.72"，应为 "1.72 eV"。

续表

$\lambda/\mu m$	α/cm^{-1}	参 考 文 献	备 注
2.1	0.06~0.07	[17]	
	<0.05	[26]	
	0.012~0.072	[27]	
2.15	0.03~0.08	[2]	典型值,e 光
	0.013 5	[19]	e 光
	0.01	[2]	最佳值,e 光
2.2	0.002~0.004	[20]	o 光,OPO 方向
	0.02~0.05	[20]	e 光,OPO 方向
2.8	0.008~0.012	[2]	典型值
	0.006	[2]	最佳值
4.65	0.05	[28]	SHG 方向
4.775	<0.02	[29]	
5~11	<0.02	[18]	
9.2~10.8	0.02	[30]	o 光
9.3	0.05	[28]	SHG 方向
9.5	0.03	[31]	
9.55	<0.1	[32]	SHG 方向
	<0.02	[29]	
10.6	0.089	[33]	
	<0.02	[16]	
	0.01~0.06	[34]	
	0.01~0.018	[2]	典型值
	0.007	[2]	最佳值
	0.002	[8]	

双光子吸收系数 β

$\lambda/\mu m$	τ_p/ns	$\beta\times 10^{11}/(cm\cdot W^{-1})$	参 考 文 献	备 注
1.06	~10	140(?)	[15]	$\perp c$,e 光
1.08	0.04	2 500	[35]	$\parallel c$
1.319	70	3 600	[36]	o 光,Eksma 样品
		1 800	[36]	e 光,Eksma 样品
		1 800	[36]	o 光,Cleveland 晶体样品
		600	[36]	e 光,Cleveland 晶体样品
1.338	70	3 000	[36]	o 光,Eksma 样品
		1 300	[36]	e 光,Eksma 样品
1.395	15	3 700	[36]	o 光,Eksma 样品
1.540	15	800	[36]	o 光,Eksma 样品
1.590	15	200	[36]	o 光,Eksma 样品
		80	[36]	e 光,Eksma 样品

折射率的实验值[5]

$\lambda/\mu m$	n_o	n_e	$\lambda/\mu m$	n_o	n_e
0.725	2.845 2	2.893 2	3.800	2.620 3	2.587 6
0.750	2.819 1	2.841 5	4.000	2.618 9	2.586 3
0.800	2.784 9	2.786 6	4.500	2.616 6	2.584 0
0.850	2.759 8	2.752 2	5.000	2.614 4	2.581 9
0.900	2.740 6	2.727 5	5.500	2.612 8	2.580 0
0.950	2.725 2	2.708 5	6.000	2.611 3	2.578 4
1.000	2.713 2	2.693 4	6.500	2.609 4	2.576 5
1.100	2.694 2	2.671 2	7.000	2.607 0	2.574 3
1.200	2.680 6	2.655 4	7.500	2.604 9	2.572 3
1.300	2.670 5	2.643 8	8.000	2.603 2	2.570 4
1.400	2.662 4	2.634 7	8.500	2.600 9	2.568 1
1.600	2.651 6	2.622 4	9.000	2.598 8	2.565 9
1.800	2.643 2	2.613 1	9.500	2.596 4	2.563 5
2.000	2.637 6	2.607 1	10.00	2.593 9	2.560 8
2.200	2.633 6	2.602 7	10.50	2.591 7	2.558 5
2.400	2.630 4	2.599 2	11.00	2.589 0	2.555 5
2.600	2.628 6	2.596 8	11.50	2.586 8	2.553 6
2.800	2.626 1	2.594 3	12.00	2.583 7	2.550 5
3.000	2.624 5	2.592 5	12.50	2.580 5	2.547 3
3.200	2.623 1	2.591 2	13.00	2.577 1	2.543 9
3.400	2.622 1	2.589 9	13.50	2.573 1	2.540 4
3.600	2.621 3	2.588 9			

旋光性[37]：$\rho = 7(°)/mm$，在各向同性点（$n_o = n_e, \lambda = 0.804\ \mu m$）

折射率的温度微商

$\lambda/\mu m$	T/K	$dn_o/dT \times 10^6/K^{-1}$	$dn_e/dT \times 10^6/K^{-1}$	参 考 文 献
2.05			57 ± 9	[24]
3.391 3	308	45	76	[17]

非线性折射率 γ[38]

$\lambda/\mu m$	$\gamma \times 10^{15}/(cm^2 \cdot W^{-1})$
1.55	35

最佳 Sellmeier 方程组（λ 以 μm 为单位, $T = 293$ K）[19,39]：

$$n_o^2 = 6.8507 + \frac{0.4297}{\lambda^2 - 0.1584} - 0.00125\lambda^2$$

$$n_e^2 = 6.6792 + \frac{0.4598}{\lambda^2 - 0.2122} - 0.00126\lambda^2$$

在文献[40]、[41]中给出了对 11~16 μm 波长范围内修正的色散关系。在文献[10]、[18]、[42]、[43]、[44]中给出了其他的色散关系方程组。

在 2.05~10.59 μm 光谱范围内和 293~393 K 温度范围之间折射率的温度微商 (λ 以 μm 为单位)[45]:

$$dn_o/dT = (0.046\lambda + 7.514) \times 10^{-5} \text{ K}^{-1}$$

$$dn_e/dT = (0.061\lambda + 7.984) \times 10^{-5} \text{ K}^{-1}$$

室温下低频(远低于 $AgGaSe_2$ 晶体的声学共振频率,即对"自由"晶体)测量的线性电光系数[46]:

λ/μm	r_{41}^T/(pm·V^{-1})	r_{63}^T/(pm·V^{-1})
1.15	4.5	3.9

普遍情况下有效二阶非线性系数的表达式(Kleinman 对称条件成立, $d_{14} = d_{25} = d_{36}$)[47]:

$$d_{ooe} = -d_{36}\sin(\theta + \rho)\sin 2\phi$$

$$d_{eoe} = d_{oee} = 2d_{36}\sin(\theta + \rho)\cos(\theta + \rho)\cos 2\phi$$

有效二阶非线性系数的简化表达式(小双折射角近似,Kleinman 对称条件成立,$d_{14} = d_{25} = d_{36}$)[48]:

$$d_{ooe} = -d_{36}\sin\theta\sin 2\phi$$

$$d_{eoe} = d_{oee} = d_{36}\sin 2\theta\cos 2\phi$$

二阶非线性系数的绝对值:

$$d_{36}(9.2714 \text{ μm}) = (41.4 \pm 2.0) \text{ pm/V}^{[49,39]}$$

$$d_{36}(10.591 \text{ μm}) = (39.5 \pm 1.9) \text{ pm/V}^{[49,39]}$$

相位匹配角的实验值($T = 293$ K)

相互作用的波长/μm	θ_{exp}/(°)	参 考 文 献
SHG, o + o ⇒ e		
10.63⇒5.315	55.9	[42]
10.6114⇒5.3057	55.6	[39]
10.6⇒5.3	57.5	[8]
10.591⇒5.2955	55.5	[39]
10.55⇒5.275	55.3	[42]

续表

相互作用的波长/μm	θ_{exp}/(°)	参 考 文 献
SHG, o + o ⇒ e		
10.3 ⇒ 5.15	53.7	[42]
10.21 ⇒ 5.105*	53.1	[42]
9.66 ⇒ 4.83	~49	[50]
9.64 ⇒ 4.82	50.0	[44]
9.552 5 ⇒ 4.776 25	49.6	[39]
9.55 ⇒ 4.775	48.8	[44]
9.503 9 ⇒ 4.751 95	49.3	[39]
9.31 ⇒ 4.655	48.3	[44]
9.282 4 ⇒ 4.641 2	48.3	[39]
9.271 4 ⇒ 4.635 7	48.2	[39]
9.200 7 ⇒ 4.600 35	47.9	[39]
6 ⇒ 3	42.2	[42]
5.295 5 ⇒ 2.647 75	41.3	[39]
5.2 ⇒ 2.6	40.3	[42]
4.635 7 ⇒ 2.317 85	44.6	[39]
4.1 ⇒ 2.05	49.7	[25]
	50.0	[39]
3.391 3 ⇒ 1.695 65	65.8	[39]
SFG, o + o ⇒ e		
12.15 + 10.63 ⇒ 5.67	61	[42]
10.63 + 5.33 ⇒ 3.55	42.7	[42]
5.515 + 3.391 3 ⇒ 2.1	≈48	[17]
4.84 + 3.55 ⇒ 2.047 9	49.2	[25]
5.13 + 2.685 ⇒ 1.763	61.3	[51]
6.00 + 2.586 ⇒ 1.807	56	[51]
7.43 + 2.484 ⇒ 1.862	49.5	[51]
9.93 + 2.384 ⇒ 1.923	45.8	[51]
6.95 + 1.66 ⇒ 1.34	≈78	[25]
7.4 + 1.604 ⇒ 1.318	80	[8]
8.8 + 1.550 ⇒ 1.318	70	[8]
12.3 + 1.476 ⇒ 1.318	60	[8]

* 译者注：原书为"10.21⇒5.15"，应为"10.21⇒5.105"。

内角带宽和温度带宽的实验值

相互作用波长/μm	$\Delta\theta^{int}/(°)$	$\Delta\nu_1/cm^{-1}$	参 考 文 献
SHG, o + o ⇒ e			
9.3 ⇒ 4.65	0.85		[28]
10.25 ⇒ 5.125	0.85		[34]
SFG, o + o ⇒ e			
5.515 + 3.391 3 ⇒ 2.1	0.54		[17]
10.6 + 1.318 ⇒ 1.172 2	1.2		[52]
DFG, e − o ⇒ o			
1.289 9 − 1.571 5 ⇒ 7.2		14.8	[53]

温度带宽的实验值[45]

相互作用的波长/μm	$\theta_{exp}/(°)$	$\Delta T/℃$
SHG, o + o ⇒ e		
10.591 ⇒ 5.295 5	55.5	350
5.295 5 ⇒ 2.647 8	41.3	230
SHG, e + o ⇒ e		
5.295 5 ⇒ 2.647 8	72.2	260
SFG, o + o ⇒ e		
10.591 + 5.295 5 ⇒ 3.530 3	42.4	390
10.591 + 3.530 3 ⇒ 2.647 8	41.3	220
SFG, e + o ⇒ e		
10.591 + 5.295 5 ⇒ 3.530 3	56.6	550
10.591 + 3.530 3 ⇒ 2.647 8	50.4	260

激光诱导的表面损伤阈值

$\lambda/\mu m$	τ_p/ns	$I_{thr}/(GW \cdot cm^{-2})$	参 考 文 献	备 注
1.064 2	35	0.011	[42]	1 000 个脉冲
		0.03	[42]	单脉冲
	23	0.013 ~ 0.04	[25]	
1.57	6	> 0.02	[54]	5 kHz
2.0	30	0.008 3	[21]	5 kHz, 未镀膜晶体
		< 0.013	[21]	5 kHz, 镀膜晶体
	20 ~ 30	0.02 ~ 0.03	[22]	
2.05	55	0.006 ~ 0.01	[26]	
	50	0.025	[25]	
2.097	180	0.007	[39]	5 Hz
2.1	180	0.009 4	[27]	未镀膜晶体
		0.017	[27]	镀膜晶体

续表

$\lambda/\mu m$	τ_p/ns	$I_{thr}/(GW \cdot cm^{-2})$	参 考 文 献	备 注
	50	0.013	[17]	
2.79	40	0.025	[23]	
9.2 ~ 10.8	CW	0.000 04	[30]	
9.27	50	0.05	[39]	1 Hz
9.3	50	0.03	[28]	1 Hz
9.5	30	0.033	[31]	
9.55	30	0.15	[32]	SHG 方向
10.2	CW	0.000 01 ~ 0.000 04	[55]	未镀膜晶体
		0.000 06	[55]	镀膜晶体
10.25	75	0.012	[34]	10 个脉冲
10.6	150	0.01 ~ 0.02	[56]	

激光诱导的体损伤阈值[30]

$\lambda/\mu m$	τ_p/ns	$I_{thr}/(GW \cdot cm^{-2})$
9.2 ~ 10.8	CW	0.000 1

关于这一晶体

与 AGS 晶体相比，AGSe 在红外区有更宽的透明范围，可达 17 μm，而 AGS 则为 11.4 μm（以 1 cm^{-1} 水平计）。所以，硒镓银晶体不仅能广泛用于 DFG 和 OPO，而且还可用于 CO_2 激光器的 SHG。在文献[30]中，在一块 1.9 cm 长的 AGSe 晶体中，产生了 4.6 ~ 5.4 μm 范围约 6 mW 的 CW 二次谐波输出。在文献[29]中，采用了 2 ns 的脉冲 CO_2 激光（λ = 9.55 μm），在一块 4 cm 长的晶体中产生的二次谐波脉冲能量达到 100 mJ。通过用 AGSe 晶体 OPO 的闲频光和信号光的混频可以实现差频输出[57,58]。在文献[57]中，采用了以锁模钛宝石激光器泵浦以 KTP 为基础的 OPO，所获得的调谐范围是 8 ~ 18 μm。在文献[58]中，所应用的是 $LiNbO_3$ 为基础的 OPO，差频波长可在 5 ~ 18 μm 范围内调谐。以 AGSe* 为基础的 OPO 可以用 KTP 晶体 OPO 的 1.57 μm 输出来泵浦[54]，也可以用 CTA 晶体 OPO 的 1.55 μm 输出来泵浦[38]。在前面一种情况下，调谐范围为 5 ~ 18 μm，所获得的红外脉冲能量最高达 1.2 mJ。在后一个实验中，调谐范围为 4 ~ 8 μm，平均红外功率在 4.55 μm 处达 35 mW。

---

* 译者注：原书为"ASGe"，应为"AGSe"。

参考文献

[1] B. H. T. Chai: Optical Crystals. In: *CRC Handbook of Laser Science and Technology, Supplement 2: Optical Materials*, ed. by M. J. Weber (CRC Press, Boca Raton, 1995), pp. 3 – 65.

[2] G. C. Catella, D. Burlage: Crystal growth and optical properties of $AgGaS_2$ and $AgGaSe_2$. MRS Bulletin 23(7), 28 – 36(1998).

[3] *Physical-Chemical Properties of Semiconductors. Handbook.* (Nauka, Moscow, 1979) [In Russian].

[4] D. Eimerl, J. Marion, E. K. Graham, H. A. McKinstry, S. Haussühl: Elastic components and thermal fracture of $AgGaSe_2$ and d-LAP. IEEE J. Ouant. Electr. 27(1), 142-145(1991).

[5] G. D. Boyd, H. M. Kasper, J. H. McFee, F. G. Storz: Linear and nonlinear optical properties of some ternary selenides. IEEE J. Quant. Electr. QE – 8(12), 900 – 908(1972).

[6] V. I. Gavrilenko, A. M. Grekhov, D. B. Korbutyak, V. G. Litovchenko: *Optical Properties of Semiconductors. Handbook.* (Naukova Dumka, Kiev, 1987) [In Russian].

[7] R. S. Feigelson, R. K. Route: Recent developments in the growth of chalcopyrite crystals for nonlinear infrared applications. Opt. Eng. 26(2), 113 – 119(1987).

[8] R. L. Byer, M. M. Choy, R. L. Herbst, D. S. Chemla, R. S. Feigelson: Second harmonic generation and infrared mixing in $AgGaSe_2$. Appl. Phys. Lett. 24(2), 65 – 68(1974).

[9] G. W. Iseler: Thermal expansion and seeded Bridgman growth of $AgGaSe_2$. J. Cryst. Growth 41(1), 146 – 150(1977).

[10] P. G. Schunemann, S. D. Setzler, T. M. Pollak: Phase-matched crystal growth of $AgGaSe_2$ and $AgGa_{1-x}In_xSe_2$. J. Cryst. Growth 211(1 – 4), 257 – 264(2000).

[11] H. Neumann, G. Kühn, W. Möller: High-temperature specific heat of $AgInS_2$ and $AgGaSe_2$. Cryst. Res. Technol. 20(9), 1225 – 1229(1985).

[12] J. D. Beasley: Thermal conductivities of some novel nonlinear optical materials. Appl. Opt. 33(6), 1000 – 1003(1994).

[13] G. C. Bhar, R. C. Smith: Optical properties of II-IV-V_2 and I-III-VI_2 crystals with particular reference to transmission limits. Phys. Stat. Solidi A 13(1), 157 – 168(1972).

[14] E. W. van Stryland, L. L. Chase: Two-Photon Absorption. Inorganic Materials. In: *CRC Handbook of Laser Science and Technology, Supplement 2: Optical Materials*, ed. By M. J. Weber (CRC Press, Boca Raton, 1995) pp. 299 – 328.

[15] A. Miller, G. S. Ash: Two-photon absorption and short pulse stimulated recombination in $AgGaSe_2$. Opt. Commun. 33(3), 297 – 300(1980).

[16] Data sheet of Cleveland Crystals Inc. Available at www.clevelandcrystals.com.

[17] N. P. Barnes, D. J. Gettemy, J. R. Hietanen, R. A. Iannini: Parametric amplification in

AgGaSe$_2$. Appl. Opt. 28(23), 5162 – 5168(1989).

[18] V. V. Badikov, V. B. Laptev, V. L. Panyutin, E. A. Ryabov, G. S. Shevyrdyaeva, O. B. Scherbina: Growth and optical properties of nonlinear silver selenogallate single crystals. Kvant. Elektron. 19(8), 782 – 784 (1992) [In Russian, English trans.: Sov. J. Quantum Electron. 22(8), 722 – 724 (1992)].

[19] H. Komine, J. M. Fukumoto, W. H. Long, Jr., E. A. Stappaerts: Noncritically phase matched mid-infrared generation in AgGaSe$_2$. IEEE J. Sel. Topics Quant. Electr. 1(1), 44 – 49(1995).

[20] G. C. Catella, L. R. Shiozawa, J. R. Hietanen, R. C. Eckardt, R. K. Route, R. S. Feigelson, D. G. Cooper, C. L. Marquardt: Mid-IR absorption in AgGaSe$_2$ optical parametric oscillator crystal. Appl. Opt. 32(21), 3948 – 3951(1993).

[21] P. A. Budni, M. G. Knights, E. P. Chicklis, K. L. Schepler: Kilohertz AgGaSe$_2$ optical parametric oscillator pumped at 2 μm. Opt. Lett. 18(13), 1068 – 1070(1993).

[22] U. Simon, Z. Benko, M. W. Sigrist, R. F. Curl, F. K. Tittel: Design consideration of an infrared spectrometer based on difference-frequency generation in AgGaSe$_2$. Appl. Opt. 32(33), 6650 – 6655(1993).

[23] J. Kirton: A 2.54 μm-pumped type II AgGaSe$_2$ mid-IR optical parametric oscillator. Opt. Commun. 115(1 – 2), 93 – 98(1995).

[24] C. L. Marquardt, D. G. Cooper, P. A. Budni, M. G. Knights, K. L. Schepler, R. DeDomenico, G. C. Catella: Thermal lensing in silver gallium selenide parametric oscillator crystal. Appl. Opt. 33(15), 3192 – 3197(1994).

[25] R. C. Eckardt, Y. X. Fan, R. L. Byer, C. L. Marquardt, M. E. Storm, L. Esterowitz: Broadly tunable infrared parametric oscillator using AgGaSe$_2$. Appl. Phys. Lett. 49(11), 608 – 610(1986).

[26] P. G. Schunemann, K. L. Schepler, P. A. Budni: Nonlinear frequency conversion performance of AgGaSe$_2$, ZnGeP$_2$, and CdGeAs$_2$. MRS Bulletin 23(7), 45 – 49(1998).

[27] B. C. Ziegler, K. L. Schepler: Transmission and damage-threshold measurements in AgGaSe$_2$ at 2.1 μm. Appl. Opt. 30(34), 5077 – 5080(1991).

[28] Y. M. Andreev, V. V. Butuzov, G. A. Verozubova, A. I. Gribenyukov, S. V. Davydov, V. P. Zakharov: Generation of the second harmonic of pulsed CO_2-laser radiation in AgGaSe$_2$ and ZnGeP$_2$ single crystals. Laser Phys. 5(5), 1014 – 1019(1995).

[29] H. P. Chou, R. C. Slater, Y. Wang: High-energy, fourth-harmonic generation using CO_2 lasers. Appl. Phys. B 66(5), 555 – 559(1998).

[30] S. Y. Tochitsky, V. O. Petukhov, V. A. Gorobets, V. V. Churakov, V. N. Yakimovich: Efficient continuous-wave frequency doubling of a tunable CO_2 laser in AgGaSe$_2$. Appl. Opt. 36(9), 1882 – 1888(1997).

[31] D. A. Russell, R. Ebert: Efficient generation and heterodyne detection of 4.75-μm light

with second-harmonic generation. Appl. Opt. 32(33), 6638 – 6644(1993).

[32] Y. M. Andreev, V. V. Badikov, V. G. Voevodin, L. G. Geiko, P. P. Geiko, M. V. Ivashchenko, A. I. Karapuzikov, I. V. Sherstov: Radiation resistance of nonlinear crystals at a wavelength of 9.55 μm. Kvant. Elektron. 31(12), 1075 – 1078(2001) [In Russian, English trans. : Quantum Electron. 31(12), 1075 – 1078(2001)].

[33] N. P. Barnes, R. C. Eckardt, D. J. Gettemy, L. B. Edgett: Absorption coefficients and the temperature variation of the refractive index difference of nonlinear optical crystals. IEEE J. Quant. Electr. QE – 15(10), 1074 – 1076(1979).

[34] R. C. Eckardt, Y. X. Fan, R. L. Byer, R. K. Route, R. S. Feigelson, J. van der Laan: Efficient second harmonic generation of 10-μm radiation in $AgGaSe_2$. Appl. Phys. Lett. 47(8), 786 – 788(1985).

[35] A. A. Bugaev, G. K. Averkieva, V. D. Prochukhan: Two-photon absorption and nonstationary energy transfer in the ternary semiconductor $AgGaSe_2$. Fiz. Tverd. Tela 37(8), 2495 – 2502(1995) [In Russian, English trans. : Phys. Solid State 37(8), 1367 – 1370(1995)].

[36] S. Pearl, S. Fastig, Y. Ehrlich, R. Lavi: Limited efficiency of a silver selenogallate optical parametric oscillator caused by two-photon absorption. Appl. Opt. 40(15), 2490 – 2492(2001).

[37] V. V. Badikov, I. N. Matveev, S. M. Pshenichnikov, O. V. Skrebneva, N. K. Trotsenko, N. D. Ustinov: Dispersion of birefringence and the optical activity of $AgGa(S_{1-x}Se_x)_2$ crystals. Kristallogr. 26(3), 537 – 539(1981) [In Russian, English trans. : Sov. Phys. -Crystallogr. 26(3), 304 – 305(1981)].

[38] S. Marzenell, R. Beigang, R. Wallenstein: Synchronously pumped femtosecond optical oscillator based on $AgGaSe_2$ tunable from 2 μm to 8 μm. Appl. Phys. B 69(5 – 6), 423 – 428(1999).

[39] A. Harasaki, K. Kato: New data on the nonlinear optical constant, phase-matching, and optical damage of $AgGaS_2$. Jpn. J. Appl. Phys. 36(2), 700 – 703(1997).

[40] H. W. Wang, M. H. Lu: The refractive index of extraordinary wave for $AgGaSe_2$ crystal in 11 – 16 μm range. Opt. Commun. 192(3 – 6), 357 – 363(2001).

[41] H. W. Wang, M. H. Lu: A two-stage up-convertor made of $AgGaSe_2$ and β-BBO crystals. Appl. Phys. B 70(1), 15 – 21(2001).

[42] H. Kildal, J. C. Mikkelsen: The nonlinear optical coefficient, phasematching and optical damage in chalcopyrite $AgGaSe_2$. Opt. Commun. 9(3), 315 – 318(1973).

[43] G. C. Bhar: Refractive index interpolation in phase-matching. Appl. Opt. 15(2), 305 – 307(1976).

[44] Y. M. Andreev, I. S. Baturin, P. P. Geiko, A. I. Gusamov: Frequency doubling of CO_2-laser radiation in new nonlinear crystal $AgGa_xIn_{1-x}Se_2$. Kvant. Elektron. 29(1), 66 –

[45] E. Tanaka, K. Kato: Thermo-optic dispersion formula of $AgGaSe_2$ and its practical applications. Appl. Opt. 37(3), 561–564(1998).

[46] H. Horinaka, H. Sonomura, T. Miyauchi: Linear electro-optic effect of $AgGaSe_2$. Jpn. J. Appl. Phys. 21(10), 1485–1488(1982).

[47] R. C. Eckardt, H. Masuda, Y. X. Fan, R. L. Byer: Absolute and relative nonlinear optical coefficients of KDP, KD^*P, BaB_2O_4, $LiIO_3$, $MgO:LiNbO_3$, and KTP measured by phase-matched second-harmonic generation. IEEE J. Quant. Electr. 26(5), 922–933(1990).

[48] J. E. Midwinter, J. Warner: The effects of phase matching method and of uniaxial crystal symmetry on the polar distribution of second-order non-linear optical polarization. Brit. J. Appl. Phys. 16(11), 1135–1142(1965).

[49] K. Kato: Second-harmonic and sum-frequency generation in $ZnGeP_2$. Appl. Opt. 36(12), 2506–2510(1997).

[50] G. C. Bhar, S. Das, U. Chatterjee, P. K. Datta, Y. N. Andreev: Noncritical second harmonic generation of CO_2 laser radiation in mixed chalcopyrite crystal. Appl. Phys. Lett. 63(10), 1316–1318(1993).

[51] A. Bianchi, M. Garbi: Down-conversion in the 4–18 μm range with GaSe and $AgGaSe_2$ nonlinear crystals. Opt. Commun. 30(1), 122–124(1979).

[52] G. C. Bhar, S. Das, R. K. Route, R. S. Feigelson: Synchronous pulsed infrared detection in $AgGaSe_2$ crystal using 1.318 μm pump. Appl. Phys. B 65(4–5), 471–473(1997).

[53] B. Sumpf, D. Rehle, T. Kelz, H. -D. Kronfeldt: A tunable diode-laser spectrometer for the MIR region near 7.2 μm applying difference-frequency in $AgGaSe_2$. Appl. Phys. B 67(3), 369–373(1998).

[54] S. Chandra, T. H. Alik, G. Catella, R. Utano, J. A. Hutchinson: Continuously tunable, 6-14 μm silver-gallium selenide optical parametric oscillator pumped at 1.57 μm. Appl. Phys. Lett. 71(5), 584–586(1997).

[55] J.-J. Zondy: Experimental investigation of single and twin $AgGaSe_2$ crystals for CW 10.2 μm SHG. Opt. Commun. 119(3–4), 320–326(1995).

[56] H. Kildal, G. W. Iseler: Laser-induced surface damage of infrared nonlinear materials. Appl. Opt. 15(12), 3062–3065(1976).

[57] J. M. Fraser, D. Wang, A. Hache, G. R. Allan, H. M. van Driel: Generation of highrepetition-rate femtosecond pulses from 8 to 18μm. Appl. Opt. 36(21), 5044–5047(1997).

[58] K. S. Abedin, S. Haidar, Y. Konno, C. Takyu, H. Ito: Difference frequency generation of 5-18μm in a $AgGaSe_2$ crystal. Appl. Opt. 37(9), 1642–1646(1998).

3.3 $ZnGeP_2$，磷锗锌(ZGP)

正单轴晶：$n_e > n_o$

分子量：199.928

密度：4.12 g/cm³[1]；4.162 g/cm³[2]；4.175 g/cm³[3]

点群：$\bar{4}2m$

晶格常数：

a = 5.465 Å[4]；5.465 Å[1]；5.466 Å[5]

c = 10.708 Å[4]；10.717 Å[1]；10.722 Å[5]

莫氏硬度：5.5

诺氏（或维氏）硬度：1 020，压痕负荷为 50 g[6]

熔点：1 293 K[1]；1 298 K[7]；(1 300 ± 3) K[1,8]；1 298 ~ 1 301 K[9]；1 313 K[2]

线性热膨胀系数的平均值[6]

ΔT/K	$\alpha_t \times 10^6$/K^{-1}, $\parallel c$	$\alpha_t \times 10^6$/K^{-1}, $\perp c$
293 ~ 573	15.9	17.5
573 ~ 873	8.08	9.1

p = 0.101 325 MPa 时的比热容：

$$c_p = 392 \text{ J/(kg·K)}^{[2]}$$

$$c_p = 464 \text{ J/(kg·K)}^{[3]}$$

热导率 κ[10]

T/K	κ/(W·m^{-1}K^{-1}), $\parallel c$	κ/(W·m^{-1}·K^{-1}), $\perp c$
293	36	35

T = 300 K 时的带隙能量：E_g = 2.0 eV[1]；2.1 eV[1]

以 "0" 透过计的透明范围：0.74 ~ 12 μm[11,12]

线性吸收系数 α

λ/μm	α/cm^{-1}	参考文献	备注
1.064	1.52	[13]	
	1.06	[14]	最佳晶体
1.9	0.8 ~ 0.95	[15]	
2.0	0.15	[16]	o 光，最佳晶体
	0.16	[17]	
2.05	0.35	[18]	
	0.26	[19]	o 光
	0.23	[20]	
	0.2	[8]	o 光，退火后

续表

$\lambda/\mu m$	α/cm^{-1}	参 考 文 献	备 注
	<0.1	[20]	最佳晶体
	0.09	[14]	最佳晶体
	0.02~0.04	[21]	退火和辐射过程后
2.08	0.62	[22]	o 光,平均值
	1.20	[22]	e 光,平均值
2.15	0.6	[23]	
	0.09~0.25	[24]	典型晶体,o 光
	0.03	[24]	最佳晶体,o 光
2.39	0.55	[25]	
2.5	0.11	[8]	o 光,原生晶体
2.5~8	<0.1	[26]	
2.5~8.3	<0.2	[27]	
2.5~8.5	<0.1	[28]	
2.73	0.03	[20]	
2.75	0.3	[29]	
2.79	0.06	[30]	
2.8	0.01	[24]	最佳晶体,o 光
2.8~8.3	<0.1	[31]	
3~8	0.005~0.15	[32]	
	<0.1	[33]	
	<0.01	[14]	
3.15	0.17	[29]	
3.5~3.9	0.41	[34]	o 光,SFG 方向
3.5	0.4	[35]	
3.8	0.1~0.18	[15]	
3.9~4.5	0.10	[29]	
4~8.5	<0.05	[36]	
4.5~8	0.03	[37]	最佳样品
4.65	0.4	[38]	
	0.1~0.2	[39]	
	0.01~0.05	[40]	SHG 方向
4.78	<0.055	[25]	
	0.16	[41]	
5.3~6.1	0.32	[34]	e 光,SFG 方向
5.5~6.3	0.10	[29]	
7.8	0.15	[29]	
8.24	0.02	[20]	
8.3	0.45	[29]	

续表

$\lambda/\mu m$	α/cm^{-1}	参 考 文 献	备 注
8.3~9.5	<0.3	[27]	
9	0.9	[29]	
9.2	0.51	[19]	
9.28	0.4	[26]	
9.3	0.8	[38]	
	0.7	[40]	SHG 方向
	0.4~0.5	[39]	
	0.48	[42]	e 光
9.5	0.39	[42]	e 光
9.55	0.26	[43]	SHG 方向
	0.56	[41]	
9.6	0.33	[19]	
9.7	0.33	[42]	e 光
10.0	0.45	[36]	
10.3	0.42	[44]	
10.4	0.6	[23]	
10.6	0.9	[35]	
	0.83	[34]	e 光,SFG 方向
	0.65	[19]	
10.7	0.88	[42]	e 光
11.1	1.2	[36]	

双光子吸收系数 $\beta^{[45]}$

$\lambda/\mu m$	τ_p/ns	$\beta \times 10^{11}/(cm \cdot W^{-1})$
1.3	0.00013	25

折射率的实验值[11]

$\lambda/\mu m$	n_o	n_e	$\lambda/\mu m$	n_o	n_e
0.64	3.5052	3.5802	0.90	3.2830	3.3336
0.66	3.4756	3.5467	0.95	3.2638	3.3124
0.68	3.4477	3.5160	1.00	3.2478	3.2954
0.70	3.4233	3.4885	1.10	3.2232	3.2688
0.75	3.3730	3.4324	1.20	3.2054	3.2493
0.80	3.3357	3.3915	1.30	3.1924	3.2346
0.85	3.3063	3.3593	1.40	3.1820	3.2244

续表

$\lambda/\mu m$	n_o	n_e	$\lambda/\mu m$	n_o	n_e
1.60	3.1666	3.2077	5.00	3.1149	3.1533
1.80	3.1562	3.1965	5.50	3.1131	3.1518
2.00	3.1490	3.1889	6.00	3.1101	3.1480
2.20	3.1433	3.1829	6.50	3.1057	3.1445
2.40	3.1388	3.1780	7.00	3.1040	3.1420
2.60	3.1357	3.1745	7.50	3.0994	3.1378
2.80	3.1327	3.1717	8.00	3.0961	3.1350
3.00	3.1304	3.1693	8.50	3.0919	3.1311
3.20	3.1284	3.1671	9.00	3.0880	3.1272
3.40	3.1263	3.1647	9.50	3.0836	3.1231
3.60	3.1257	3.1632	10.00	3.0788	3.1183
3.80	3.1237	3.1616	10.50	3.0738	3.1137
4.00	3.1223	3.1608	11.00	3.0689	3.1087
4.20	3.1209	3.1595	11.50	3.0623	3.1008
4.50	3.1186	3.1561	12.00	3.0552	3.0949
4.70	3.1174	3.1549			

折射率的温度微商[11]

$\lambda/\mu m$	$dn_o/dT \times 10^6/K^{-1}$	$dn_e/dT \times 10^6/K^{-1}$	$\lambda/\mu m$	$dn_o/dT \times 10^6/K^{-1}$	$dn_e/dT \times 10^6/K^{-1}$
0.64	359.4	375.8	2.40	141.4	154.9
0.66	312.3	373.4	2.60	151.3	168.0
0.68	295.2	325.3	2.80	154.8	160.5
0.70	286.3	318.2	3.00	132.6	139.6
0.75	262.2	282.6	3.20	149.4	162.8
0.80	246.9	264.3	3.40	144.0	154.6
0.85	241.2	253.9	3.60	155.8	162.9
0.90	223.4	246.1	3.80	145.8	165.3
0.95	213.2	242.6	4.00	142.6	150.2
1.00	211.8	230.1	4.20	135.7	151.4
1.10	201.1	220.8	4.50	153.1	166.0
1.20	186.3	205.1	4.70	155.1	167.1
1.30	168.4	201.2	5.00	150.5	164.3
1.40	153.4	165.5	5.50	144.9	154.2
1.60	151.0	167.5	6.00	145.8	163.0
1.80	132.0	144.0	6.50	156.0	161.3
2.00	141.9	152.9	7.00	128.5	150.1
2.20	146.0	152.8	7.50	181.5	185.9

续表

$\lambda/\mu m$	$dn_o/dT \times 10^6/K^{-1}$	$dn_e/dT \times 10^6/K^{-1}$	$\lambda/\mu m$	$dn_o/dT \times 10^6/K^{-1}$	$dn_e/dT \times 10^6/K^{-1}$
8.00	161.0	174.3	10.50	154.0	168.4
8.50	151.6	173.7	11.00	152.5	163.4
9.00	155.6	175.0	11.50	147.4	183.2
9.50	162.7	171.1	12.00	142.4	165.9
10.00	165.3	184.1			

最佳色散关系方程(λ 以 μm 为单位,1.5 $\mu m < \lambda < 10.59$ μm,$T = 293$ K)[46]:

$$n_o^2 = 11.6413 + \frac{0.69363}{\lambda^2 - 0.21967} + \frac{1586.06}{\lambda^2 - 832.75}$$

$$n_e^2 = 12.1438 + \frac{0.75255}{\lambda^2 - 0.21913} + \frac{2061.68}{\lambda^2 - 951.07}$$

在文献[18]、[36]、[47]、[48]、[49]、[50]、[51]、[52]中给出了其他室温下的色散关系方程组;文献[53]中给出了 $T = 93$ K, 173 K, 373 K, 473 K, 673 K 温度下的方程;文献[54]中给出了 $T = 343$ K 时的方程。

从室温加热到 $T[K]$ 及光谱范围 $1.5 \sim 10.25$ μm 之间折射率的温度微商[46]:

$$dn_o/dT = (11.4188/\lambda^3 - 12.8971/\lambda^2 + 7.2947/\lambda + 14.2082) \times 10^{-5}$$
$$\times [1 + 3.36 \times 10^{-3}(T - 293)]$$

$$dn_e/dT = (10.3798/\lambda^3 - 10.1785/\lambda^2 + 6.3877/\lambda + 15.6688) \times 10^{-5}$$
$$\times [1 + 3.28 \times 10^{-3}(T - 293)]$$

室温下高频(远高于 $ZnGeP_2$ 晶体的声学共振频率,即对"受夹"晶体)下测量的线性电光系数[55]

$\lambda/\mu m$	$r_{41}^S/(pm \cdot V^{-1})$	$r_{63}^S/(pm \cdot V^{-1})$
3.3913	1.6	-0.8

普遍情况下有效二阶非线性系数的表达式(Kleinman 对称条件成立,$d_{14} = d_{25} = d_{36}$)[56,57]:

$$d_{eeo} = 2d_{36}\sin(\theta + \rho)\cos(\theta + \rho)\cos 2\phi$$

$$d_{oeo} = d_{eoo} = -d_{36}\sin(\theta + \rho)\sin 2\phi$$

有效二阶非线性系数的简化表达式(小双折射角近似,Kleinman 对称条件成立,$d_{14} = d_{25} = d_{36}$)[57]:

$$d_{eeo} = d_{36}\sin 2\theta \cos 2\phi$$

$$d_{oeo} = d_{eoo} = -d_{36}\sin \theta \sin 2\phi$$

二阶非线性系数值:

$d_{36}(5.295\ 5\ \mu m) = (1.70 \pm 0.17) \times d_{36}(AgGaSe_2) = (70 \pm 7)\ pm/V^{[49]}$

$d_{36}(9.6\ \mu m) = (75 \pm 8)\ pm/V^{[41]}$

$d_{36}(10.6\ \mu m) = 0.83 \times d_{36}(GaAs) \pm 15\% = (68.9 \pm 10.3)\ pm/V^{[11,58]}$

相位匹配角的实验值($T = 293$ K)

相互作用的波长/μm	$\theta_{exp}/(°)$	参 考 文 献
SHG, e + e⇒o		
3.927 8⇒1.963 9	57.8 ± 0.3	[15], [49]
4.34⇒2.17	55.8 ± 0.2	[23]
4.64⇒2.32	47.5	[59]
4.775⇒2.387 5	49.2	[25]
5.295 5⇒2.647 75	46.8	[49]
9.2⇒4.6	63.8	[60]
9.305 4⇒4.652 7	61.3	[28]
	61.3	[48]
	62.7 ~ 64.4	[39]
	63	[49]
	64	[38]
9.5⇒4.75	62.1	[28]
	62.1	[48]
	66.8	[60]
9.552 4⇒4.776 2	65.3	[49]
9.603 6⇒4.801 8	64.9	[39]
	65.8	[49]
10.2⇒5.1	72	[28]
10.303 5⇒5.151 75	74.3	[44]
	74.5	[49]
10.551 4⇒5.275 7	79.2	[49]
10.591 0⇒5.295 5	80.1	[49]
SFG, e + e⇒o		
10.668 + 4.34⇒3.085	54.3 ± 0.2	[23]
10.591 0 + 5.295 5⇒3.530 33	52.1	[49]
10.591 0 + 3.530 33⇒2.647 75	48.4	[49]
9.74 + 4.203 9⇒2.936 5	49.6	[12]
5.295 5 + 3.530 33⇒2.118 2	51.7	[49]
SFG, o + e⇒o		
6.74 + 5.203 6⇒2.936 5	76	[61]
6.45 + 5.390 8⇒2.936 5	79.2	[27]
6.25 + 5.538 9⇒2.936 5	84.0	[27]

续表

相互作用的波长/μm	$\theta_{exp}/(°)$	参 考 文 献
SFG, o+e⇒o		
6.15+5.619 9⇒2.936 5	85.5	[27]
6.29+5.017 3⇒2.791	76	[31]
6.19+5.082 8⇒2.791	77.6	[31]
6.06+5.173 9⇒2.791	80.5	[31]
6.015+5.207⇒2.791	84	[62]
5.95+5.256 9⇒2.791	83.4	[31]
5.90+5.296 5⇒2.791	87	[31]
10.591 0+1.064 2⇒0.967 03	84	[35]

内角带宽的实验值

相互作用的波长/μm	$\Delta\theta^{int}/(°)$	参 考 文 献
SHG, e+e⇒o		
3.8⇒1.9	1.33	[15]
4.34⇒2.17	1.05	[23]
5.3⇒2.65	0.69	[59]
7.8⇒3.9	0.5	[29]
9.3⇒4.65	0.74~0.80	[39]
	0.83	[40]
	1.15	[38]
9.55⇒4.775	0.89	[41]
9.6⇒4.8	0.8	[39]
10.2⇒5.1	1.35	[28]
10.3⇒5.15	1.20	[44]
SFG, e+e⇒o		
10.668+4.34⇒3.085	1.23	[23]
SFG, o+e⇒o		
10.6+1.064⇒0.967	0.55	[35]

光谱带宽的实验值

相互作用的波长/μm	$\Delta\nu/cm^{-1}$	参 考 文 献
SHG, e+e⇒o		
4.34⇒2.17	7.9	[23]
10.2⇒5.1	4.9	[28]

温度带宽的实验值[49]

相互作用的波长/μm	θ_{exp}/(°)	T/℃	参 考 文 献
SHG, e + e ⇒ o			
10.591 0 ⇒ 5.295 5	80.1	44	[49]
10.303 5 ⇒ 5.151 75	74.5	45	[49]
10.2 ⇒ 5.1	72	50	[28]
9.603 6 ⇒ 4.801 8	65.8	48	[49]
SFG, o + e ⇒ o			
10.591 0 + 1.064 2 ⇒ 0.967 03	84	81.9	[35]

相匹配角的温度变化

相互作用的波长/μm	dθ_{pm}/dT/[(°)·K^{-1}]	参 考 文 献
SHG, e + e ⇒ o		
9.2 ⇒ 4.6	0.014	[60]
10.3 ⇒ 5.15	0.072	[28]
10.6 ⇒ 5.3	0.107	[28]
SFG, o + e ⇒ o		
10.6 + 1.064 2 ⇒ 0.967 1	0.007	[35]

激光诱导的表面损伤阈值

λ/μm	τ_p/ns	I_{thr}/(GW·cm^{-2})	参 考 文 献	备 注
1.064 2	30	>0.003	[63]	12.5 Hz
	10	0.003	[11]	
1.3	0.000 13	>150	[45]	1 kHz
1.66	0.000 13	>100	[45]	1 kHz
2.05	30	0.013~0.016	[19]	5 kHz
	10	>0.074	[64]	10 kHz
2.79	50	>0.014	[30]	10 Hz
		0.018	[30]	10 Hz
	0.15	30	[31]	
	0.1	35	[16], [62]	1 Hz
2.8	70	0.056	[22]	1 Hz, 未镀膜样品
		0.08	[22]	1 Hz, 镀膜样品
2.94	0.11	30	[61]	
		30	[12]	
5.3~6.1	CW	>0.000 01	[59]	
		0.000 25	[34]	
7.8	5 000	10	[29]	

续表

$\lambda/\mu m$	τ_p/ns	$I_{thr}/(GW \cdot cm^{-2})$	参考文献	备注
9.2~10.8	CW	>0.000 08	[42]	
9.28	2	1.25	[26]	
9.3	100	0.012	[38]	100 Hz
	50	>0.06	[40]	1 Hz
9.3~10.6	125	0.025	[44]	20 Hz
		0.03~0.04	[44]	2 Hz
9.55	220	0.078	[41],[65]	
	30	0.14	[43]	SHG 方向
10.2~10.8	CW	>0.000 001	[28]	
	100 000~10 000 000	0.06	[28]	1 500 Hz
10.6	CW	>0.000 000 01	[63]	
		0.000 2	[34]	

关于这一晶体

近年来由于 Schunemann[14,19] 和 Gribenyukov[8,21] 对 ZPG 晶体的大力研究,降低了这种晶体在 2.05 μm 处的 IR 吸收(降到 0.02~0.04 cm^{-1})。这就使 Vodopyanov 可以大大地改进以 ZGP 为基础的单共振 OPO 系统的特性,这一系统由 Er,Cr:YSGG 激光器($\lambda = 2.8$ μm)泵浦。以 I 类相位匹配为基础的 OPO 的调谐范围为 3.8~12.4 μm,而对于 II 类相位匹配调谐范围可达到 4~10 μm[66]。在 6~8 μm 范围内 IR 脉冲能量在频率为 25 Hz 时达到 300 μJ[67]。在文献[68]中,全固态激光二极管泵浦的 Nd:YAG 激光器泵浦以 PPLN 为基础的 OPO,使用它(OPO)的闲频光($\lambda = 2.3~3.7$ μm)来泵浦单共振 ZGP 为基础的 OPO(重复频率为 1~10 kHz),输出脉冲具有 20 μJ 以上的能量,可以在 6~8 μm 范围内调谐。

参考文献

[1] *Physical-Chemical Properties of Semiconductors. Handbook.* (Nauka,Moscow,1979)[In Russian].

[2] Data sheet of Inrad, Inc. Available at www.inrad.com.

[3] J. E. Tucker, C. L. Marquardt, S. R. Bowman, B. J. Feldman:Transient thermal lens in a ZnGeP$_2$ crystal. Appl. Opt. 34(15),2678-2682(1995).

[4] V. I. Gavrilenko, A. M. Grekhov, D. B. Korbutyak, V. G. Litovchenko:*Optical Properties of Semiconductors. Handbook.* (Naukova Dumka,Kiev,1987)[In Russian].

[5] G. D. Boyd, E. Buehler, F. G. Storz, J. H. Wernick:Linear and nonlinear optical prop-

erties of ternary $A^{II} B^{IV} C_2^V$ chalcopyrite semiconductors. IEEE J. Quant. Electr. QE – 8 (4), 419 – 426(1972).

[6] I. I. Kozhina, A. S. Borshchevskii: High-temperature x-ray investigations of $A^{II} B^{IV} C_2^V$ compounds. Vestnik LGU No. 22, 87 – 92(1971) [In Russian].

[7] H. M. Hobgood, T. Henningsen, R. N. Thomas, R. H. Hopkins, M. C. Ohmer, W. C. Mitchel, D. W. Fischer, S. M. Hegde, F. K. Hopkins: $ZnGeP_2$ grown by the liquid encapsulated Czochralski method. J. Appl. Phys. 73(8), 4030 – 4036(1993).

[8] Y. M. Andreev, G. A. Verozubova, A. I. Gribenyukov, V. V. Korotkova: $ZnGeP_2$ crystals for infrared laser radiation frequency conversion. J. Korean Phys. Soc. 33(3), 356 – 361(1998).

[9] G. A. Verozubova, A. I. Gribenyukov, V. V. Korotkova, M. P. Ruzaikin: $ZnGeP_2$ synthesis and growth from melt. Mater. Sci. Eng. B 48(3), 191 – 197(1997).

[10] J. D. Beasley: Thermal conductivities of some novel nonlinear optical materials. Appl. Opt. 33(6), 1000 – 1003(1994).

[11] G. D. Boyd, E. Buehler, F. G. Storz: Linear and nonlinear optical properties of $ZnGeP_2$ and CdSe. Appl. Phys. Lett. 18(7), 301 – 304(1971).

[12] K. L. Vodopyanov: Parametric generation of tunable infrared radiation in $ZnGeP_2$ and GaSe pumped at 3 μm. J. Opt. Soc. Am. B 10(9), 1723 – 1729(1993).

[13] W. Shi, Y. J. Ding: Continuously tunable and coherent terahertz radiation by means of phase-matched difference-frequency generation in zinc germanium phosphide. Appl. Phys. Lett. 83(5), 848 – 851(2003).

[14] P. G. Schunemann, T. M. Pollak: Ultralow gradient HGF-grown $ZnGeP_2$ and $CdGeAs_2$ and their optical properties. MRS Bulletin 23(7), 23 – 27(1998).

[15] Y. M. Andreev, S. D. Velikanov, A. S. Elutin, A. F. Zapolskii, D. V. Konkin, S. N. Mikshin, S. V. Smirnov, Y. N. Frolov, V. V. Shchurov: Second harmonic generation from DF laser radiation in $ZnGeP_2$. Kvant. Elektron. 19(11), 1110(1992) [In Russian, English trans.: Quantum Electron. 22(11), 1035(1992)].

[16] Y. M. Andreev, G. C. Bhar, A. I. Gribenyukov, G. A. Verozubova, K. L. Vodopyanov: Nonlinear tunable parametric luminescence in $ZnGeP_2$ crystals. Proc. SPIE 3403, 336 – 340(1997).

[17] R. F. Wu, K. S. Lai, E. Lau, H. F. Wong, W. J. Xie, Y. L. Lim, K. W. Lim, L. Chia: Multi-watt ZGP OPO based on diffusion bonded walkoff compensated KTP OPO and Nd: YALO laser. In: *Advanced Solid-State Lasers*, *OSA Trends in Optics and Photonics Series*, *Vol. 68*, ed. by M. E. Fermann, L. R. Marshall (OSA, Washington DC, 2002), pp. 194 – 197.

[18] N. P. Barnes, K. E. Murray, M. G. Jani, P. G. Schunemann, T. M. Pollak: $ZnGeP_2$ parametric oscillator. J. Opt. Soc. Am. B 15(1), 232 – 238(1998).

[19] P. G. Schunemann, K. L. Schepler, P. A. Budni: Nonlinear frequency conversion performance of AgGaSe$_2$, ZnGeP$_2$, and CdGeAs$_2$. MRS Bulletin 23(7), 45–49(1998).

[20] P. A. Ketteridge, P. A. Budni, P. G. Schunemann, M. L. Lemons, T. M. Pollak, E. P. Chicklis: Tunable all solid state average power ZGP OPO at 2.7 and 8.5 microns. In: *Advanced Solid-State Lasers*, *OSA Trends in Optics and Photonics Series*, *Vol. 19*, ed. by W. R. Bosenberg, M. M. Fejer(OSA, Washington DC, 1998), pp. 233–235.

[21] A. I. Gribenyukov: Nonlinear optical ZnGeP$_2$ crystals: the history of technology. Atmos. Oceanic Opt. 15(1), 61–68(2002).

[22] R. D. Peterson, K. L. Schepler, J. L. Brown, P. G. Schunemann: Damage properties of ZnGeP$_2$ at 2 μm. J. Opt. Soc. Am. B 12(11), 2142–2146(1995).

[23] Y. M. Andreev, V. G. Voevodin, P. P. Geiko, A. I. Gribenyukov, V. V. Zuev, A. S. Solodukhin, S. A. Trushin, V. V. Churakov, S. F. Shubin: Transformation of the frequencies of nontraditional(4.3 and 10.4 μm) CO$_2$ laser radiation bands in ZnGeP$_2$. Kvant. Elektron. 14(11), 2137–2138(1987)[In Russian, English trans. : Sov. J. Quantum Electron. 17(11),1362–1363(1987)].

[24] G. C. Catella, D. Burlage: Crystal growth and optical properties of AgGaS$_2$ and AgGaSe$_2$. MRS Bulletin 23(7), 28–36(1998).

[25] H. P. Chou, R. C. Slater, Y. Wang: High-energy, fourth-harmonic generation using CO$_2$ lasers. Appl. Phys. B 66(5), 555–559(1998).

[26] Y. M. Andreev, V. Y. Baranov, V. G. Voevodin, P. P. Geiko, A. I. Gribenyukov, S. V. Izyumov, S. M. Kozochkin, V. D. Pismenny, Y. A. Satov, A. P. Streltsov: Efficient generation of the second harmonic of a nanosecond CO$_2$ laser radiation pulse. Kvant. Elektron. 14(11), 2252–2254(1987)[In Russian, English trans. : Sov. J. Quantum Electron. 17(11),1435–1436(1987)].

[27] K. L. Vodopyanov, V. G. Voevodin, A. I. Gribenyukov, L. A. Kulevskii: Picosecond parametric superluminescence in the ZnGeP$_2$ crystal. Izv. Akad. Nauk SSSR, Ser. Fiz. 49(3), 569–572(1985)[In Russian, English trans. : Bull. Acad. Sci. USSR, Phys. Ser. 49(3),146–149(1985)].

[28] Y. M. Andreev, V. G. Voevodin, A. I. Gribenyukov, O. Y. Zyryanov, I. I. Ippolitov, A. N. Morozov, A. V. Sosnin, G. S. Khmelnitskii: Efficient generation of the second harmonic of tunable CO$_2$ laser radiation in ZnGeP$_2$. Kvant. Elektron. 11(8), 1511–1512(1984)[In Russian, English trans. : Sov. J. Quantum Electron. 14(8),1021–1022(1984)].

[29] J. M. Auerhammer, A. F. G. van der Meer, P. W. van Amersfoort, Q. H. F. Vrehen, E. R. Eliel: Efficient frequency doubling of ps-pulses from a free-electron laser in ZnGeP$_2$. Opt. Commun. 118(1–2), 85–89(1995).

[30] T. H. Allik, S. Chandra, D. M. Rines, P. G. Schunemann, J. A. Hutchinson, R. Utano: Tuna-

ble 7 – 12-μm optical parametric oscillator using a Cr, Er: YSGG laser to pump CdSe and ZnGeP$_2$ crystals. Opt. Lett. 22(9), 597 – 599(1997).

[31] K. L. Vodopyanov, V. G. Voevodin, A. I. Gribenyukov, L. A. Kulevskii: High-efficiency picosecond parametric superradiance emitted by a ZnGeP$_2$ crystal in 5 – 6.3 μm range. Kvant. Elektron. 14(9), 1815 – 1819(1987) [In Russian, English trans. : Sov. J. Quantum Electron. 17(9),1159 – 1161(1987)].

[32] G. A. Verozubova, A. I. Gribenyukov, V. V. Korotkova, O. Semchinova, D. Uffmann: Synthesis and growth of ZnGeP$_2$ crystals for nonlinear optical applications. J. Cryst. Growth 213(3 – 4), 334 – 339(2000).

[33] Y. M. Andreev, A. D. Belykh, V. G. Voevodin, P. P. Geiko, A. I. Gribenyukov, V. A. Gurashvili, S. V. Izyumov: Doubling of the emission of CO lasers with the efficiency of 3%. Kvant. Elektron. 14(4), 782 – 783(1987) [In Russian, English trans. : Sov. J. Quantum Electron. 17(4),490 – 491(1987)].

[34] Y. M. Andreev, V. G. Voevodin, A. I. Gribenyukov, V. P. Novikov: Mixing of frequencies of CO$_2$ and CO lasers in ZnGeP$_2$ crystals. Kvant. Elektron. 14(6), 1177 – 1179 (1987) [In Russian, English trans. : Sov. J. Quantum Electron. 17(6), 748 – 749 (1987)].

[35] G. D. Boyd, W. B. Gandrud, E. Buehler: Phase-matched upconversion of 10.6-μm radiation in ZnGeP$_2$. Appl. Phys. Lett. 18(10), 446 – 448(1971).

[36] S. V. Zakharov, A. E. Negin, P. G. Filippov, E. F. Zhilis: Sellmeier equation and conversion of the radiation of a repetitively pulsed tunable TEA CO$_2$ laser into the second harmonic in a ZnGeP$_2$ crystal. Kvantovaya Elektron. 28(3), 251 – 255(1999) [English trans. : Quantum Electron. 29(9),806 – 810(1999)].

[37] V. E. Zuev, M. V. Kabanov, Y. M. Andreev, V. G. Voevodin, P. P. Geiko, A. I. Gribenyukov, V. V. Zuev: Applications of efficient parametric IR-laser frequency converters. Izv. Akad. Nauk SSSR, Ser. Fiz. 52(6), 1142 – 1148(1988) [In Russian, English trans. : Bull. Acad. Sci. USSR, Phys. Ser. 52(6),87 – 92(1988)].

[38] A. A. Barykin, S. V. Davydov, V. P. Dorokhov, V. P. Zakharov, V. V. Butuzov: Generation of the second harmonic of CO$_2$ laser pulses in a ZnGeP$_2$ crystal. Kvant. Elektron. 20(8), 794 – 800(1993) [In Russian, English trans. : Quantum Electron. 23, 688 – 693(1993)].

[39] Y. M. Andreev, A. N. Bykanov, A. I. Gribenyukov, V. V. Zuev, V. D. Karyshev, A. V. Kisletsov, I. O. Kovalev, V. I. Konov, G. P. Kuzmin, A. A. Nesterenko, A. E. Osorgin, Y. M. Starodumov, N. I. Chapliev: Conversion of pulsed laser radiation from the 9.3 – 9.6 μm range to the second harmonic in ZnGeP$_2$ crystals. Kvant. Elektron. 17(4), 476 – 480(1990) [In Russian, English trans. : Sov. J. Quantum Electron. 20(4),410 – 414(1990)].

[40] Y. M. Andreev, V. V. Butuzov, G. A. Verozubova, A. I. Gribenyukov, S. V. Davydov,

V. P. Zakharov: Generation of the second harmonic of pulsed CO_2-laser radiation in $AgGaSe_2$ and $ZnGeP_2$ single crystals. Laser Phys. 5(5), 1014–1019(1995).

[41] P. D. Mason, D. J. Jackson, E. K. Gorton: CO_2 laser frequency doubling in $ZnGeP_2$. Opt. Commun. 110(1–2), 163–166(1994).

[42] S. Y. Tochitsky, V. O. Petukhov, V. A. Gorobets, V. V. Churakov, V. N. Yakimovich: Efficient continuous-wave frequency doubling of a tunable CO_2 laser in $AgGaSe_2$. Appl. Opt. 36(9), 1882–1888(1997).

[43] Y. M. Andreev, V. V. Badikov, V. G. Voevodin, L. G. Geiko, P. P. Geiko, M. V. Ivashchenko, A. I. Karapuzikov, I. V. Sherstov: Radiation resistance of nonlinear crystals at a wavelength of 9.55 μm. Kvant. Elektron. 31(12), 1075–1078(2001) [In Russian, English trans.: Quantum Electron. 31(12),1075–1078(2001)].

[44] G. B. Abdullaev, K. R. Allakhverdiev, M. E. Karasev, V. I. Konov, L. A. Kulevskii, N. B. Mustafaev, P. P. Pashinin, A. M. Prokhorov, Y. M. Starodumov, N. I. Chapliev: Efficient generation of the second harmonic of CO_2 laser radiation in a GaSe crystal. Kvant. Elektron. 16(4), 757–763(1989) [In Russian, English trans.: Sov. J. Quantum Electron. 19(4),494–498(1989)].

[45] V. Petrov, F. Rotermund, F. Noack, P. Schunemann: Femtosecond parametric generation in $ZnGeP_2$. Opt. Lett. 24(6), 414–416(1999).

[46] K. Kato, E. Takaoka, N. Umemura: New Sellmeier and thermo-optic dispersion formulas for $ZnGeP_2$. In: *Conference on Lasers and Electrooptics CLEO/QELS 2003, Technical Digest*(OSA, Washington DC,2003), paper CTuM17.

[47] G. C. Bhar: Refractive index interpolation in phase-matching. Appl. Opt. 15(2), 305–307(1976).

[48] G. C. Bhar, L. K. Samanta, D. K. Ghosh, S. Das: Tunable parametric $ZnGeP_2$ crystal oscillator. Kvant. Elektron. 14(7), 1361–1363(1987) [In Russian, English trans.: Sov. J. Quantum Electron. 17(7),860–861(1987)].

[49] K. Kato: Second-harmonic and sum-frequency generation in $ZnGeP_2$. Appl. Opt. 36(12), 2506–2510(1997).

[50] G. Ghosh: Sellmeier coefficients for the birefringence and refractive indices of $ZnGeP_2$ nonlinear crystal at different temperatures. Appl. Opt. 37(7), 1205–1212(1998).

[51] D. E. Zelmon, E. A. Hanning, P. G. Schunemann: Refractive-index measurements and Sellmeier coefficients for zinc germanium phosphide from 2 to 9 μm with implications for phase matching in optical frequency-conversion devices. J. Opt. Soc. Am. B 18(9), 1307–1310(2001).

[52] S. Das, G. C. Bhar, S. Gangopadhyay, C. Ghosh: Linear and nonlinear optical properties of $ZnGeP_2$ crystal for infrared laser device applications: revisited. Appl. Opt. 42(21), 4335–4340(2003).

[53] G. C. Bhar, G. C. Ghosh: Temperature dependent phase-matched nonlinear optical devices using CdSe and $ZnGeP_2$. IEEE J. Quant. Electr. QE–16(8), 838–843(1980).

[54] G. C. Bhar, G. Ghosh: Temperature-dependent Sellmeier coefficients and coherence lengths for some chalcopyrite crystals. J. Opt. Soc. Am. 69(5), 730–733(1979).

[55] E. H. Turner, E. Buehler, H. Kasper: Electro-optic behavior and dielectric constants of $ZnGeP_2$ and $CuGaS_2$. Phys. Rev. B 9(2), 558–561(1974).

[56] R. C. Eckardt, H. Masuda, Y. X. Fan, R. L. Byer: Absolute and relative nonlinear optical coefficients of KDP, KD^*P, BaB_2O_4, $LiIO_3$, MgO: $LiNbO_3$, and KTP measured by phase-matched second-harmonic generation. IEEE J. Quant. Electr. 26(5), 922–933(1990).

[57] J. E. Midwinter, J. Warner: The effects of phase matching method and of uniaxial crystal symmetry on the polar distribution of second-order non-linear optical polarization. Brit. J. Appl. Phys. 16(11), 1135–1142(1965).

[58] D. A. Roberts: Simplified characterization of uniaxial and biaxial nonlinear optical crystals: a plea for standardization of nomenclature and conventions. IEEE J. Quant. Electr. 28(10), 2057–2074(1992).

[59] Y. M. Andreev, T. V. Vedernikova, A. A. Betin, V. G. Voevodin, A. I. Gribenyukov, O. Y. Zyryanov, I. I. Ippolitov, V. I. Masychev, O. V. Mitropolskii, V. P. Novikov, M. A. Novikov, A. V. Sosnin: Conversion of CO_2 and CO laser radiation in a $ZnGeP_2$ crystal to the 2.3–3.1 μm spectral range. Kvant. Elektron. 12(7), 1535–1537 (1985) [In Russian, English trans.: Sov. J. Quantum Electron. 15(7), 1014–1015 (1985)].

[60] G. C. Bhar, S. Das, U. Chatterjee, K. L. Vodopyanov: Temperature-tunable secondharmonic generation in zinc germanium diphosphide. Appl. Phys. Lett. 54(4), 313–314 (1989).

[61] K. L. Vodopyanov, L. A. Kulevskii, V. G. Voevodin, A. I. Gribenyukov, K. R. Allakhverdiev, T. A. Kerimov: High efficiency middle IR parametric superradiance in $ZnGeP_2$ and GaSe crystals pumped by an erbium laser. Opt. Commun. 83(5–6), 322–326(1991).

[62] K. L. Vodopyanov, Y. A. Andreev, G. C. Bhar: Parametric superluminescence in a $ZnGeP_2$ crystal with temperature tuning and pumping by an erbium laser. Kvant. Elektron. 20(9), 879–881(1993) [In Russian, English trans.: Quantum Electron. 23(9), 763–765(1993)].

[63] N. P. Andreeva, S. A. Andreev, I. N. Matveev, S. M. Pshenichnikov, N. D. Ustinov: Parametric conversion of infrared radiation in zinc germanium diphosphide. Kvant. Elektron. 6(2), 357–359(1979) [In Russian, English trans.: Sov. J. Quantum Electron. 9(2), 208–210(1979)].

[64] P. A. Budni, L. A. Pomeranz, M. L. Lemons, P. G. Schunemann, T. M. Pollak,

E. P. Chicklis: 10 W mid-IR holmium pumped ZnGeP$_2$ OPO. In: *Advanced Solid-State Lasers*, *OSA Trends in Optics and Photonics Series*, *Vol.19*, ed. by W. R. Bosenberg, M. M. Fejer(OSA,Washington DC 1998), pp. 226 – 229.

[65] P. D. Mason, D. J. Jackson, E. K. Gorton: CO$_2$ laser frequency doubling in ZnGeP$_2$. Erratum. Opt. Commun. 114(5 – 6), 529(1995).

[66] K. L. Vodopyanov, F. Ganikhanov, J. P. Maffetone, I. Zwieback, W. Ruderman: ZnGeP$_2$ optical parametric oscillator with 3.8 – 12.4-μm tunability. Opt. Lett. 25(11), 841 – 843(2000).

[67] M. W. Todd, R. A. Provencal, T. G. Owano, B. A. Paldus, A. Kachanov, K. L. Vodopyanov, M. Hunter, S. L. Coy, J. I. Steinfeld, J. T. Arnold: Application of mid-infrared cavity-ringdown spectroscopy to trace explosives vapor detection using a broadly-tunable(6 – 8 μm)optical parametric oscillator. Appl. Phys. B 75(2 – 3), 367 – 376(2002).

[68] K. L. Vodopyanov, P. G. Schunemann: Broadly tunable noncritically phase-matched ZnGeP$_2$ optical parametric oscillator with a 2-μJ pump threshold. Opt. Lett. 28(6), 441 – 443(2003).

3.4　GaSe，硒化镓

负单轴晶：$n_o > n_e$

分子量：148.683

密度：5.03 g/cm^3 [1]

点群：$\bar{6}2m$

晶格常数[2]：
　　$a = 3.755$ Å
　　$c = 15.94$ Å

莫氏硬度：≈0

熔点：1 211 K[2]；1 233 K[2]

线性热膨胀系数[1]

T/K	$\alpha_t \times 10^6/K^{-1}$, ∥ c	$\alpha_t \times 10^6/K^{-1}$, ⊥ c
300	9.15	10.85

热导率 κ [3]

T/K	$\kappa/(W \cdot m^{-1} \cdot K^{-1})$, ∥ c	$\kappa/(W \cdot m^{-1} \cdot K^{-1})$, ⊥ c
293	16.2	2.0

室温下带隙能量：$E_g = 2.0$ eV[17]；2.09 eV[2]

以"0"透过计的透明范围：$0.62 \sim 20$ μm[4]

线性吸收系数 α

λ/μm	α/cm^{-1}	参 考 文 献	λ/μm	α/cm^{-1}	参 考 文 献
0.65~18	<1	[5]	1.9	0.1	[8]
0.7	<0.3	[6]	2	<0.1	[10]
0.7~0.8	0.3	[7]	9.3~10.6	<0.05	[11]
1.06	0.45	[8]	9.55	<0.1	[12]
	<0.25	[9]	10	<0.1	[10]
	<0.1	[10]	10.6	0.081	[13]
1.5~12	<0.03	[7]			

283~343 K 范围内 $\lambda = 0.6328$ μm 处线性吸收系数的温度关系（T 以 K 为单位）[14]：

$$\alpha(T) = 7.39\exp[0.0558 \times (T - 273)]$$

双光子吸收系数 β

λ/μm	τ_p/ns	$\beta \times 10^{11}$/(cm·W^{-1})	参 考 文 献	备 注
0.700	0.070	600	[15]	∥ c
	0.0002	216	[16]	∥ c
0.725	0.0002	190	[16]	∥ c
0.750	0.0002	78	[16]	∥ c
0.775	0.0002	50	[16]	∥ c
0.800	0.0002	56	[16]	∥ c
0.825	0.0002	45	[16]	∥ c
0.850	0.0002	43	[16]	∥ c
0.875	0.0002	48	[16]	∥ c
0.900	0.0002	68	[16]	∥ c
1.06	20	11 000	[17]	

折射率的实验值[5]

λ/μm	n_o	n_e	λ/μm	n_o	n_e
0.6328	2.97	2.74	3.3913	2.81	2.46
1.1523	2.90	2.54			

在 $T = 75 \sim 300$ K 间寻常光折射率的温度微商[18]

$\lambda/\mu m$	$dn_o/dT \times 10^6/K^{-1}$	$\lambda/\mu m$	$dn_o/dT \times 10^6/K^{-1}$
0.6	182.7	1.0	117.3
0.8	134.6	2.0	95.4

在 298~373 K 范围内 $\lambda = 0.6328\ \mu m$ 处 n_o 的温度关系(T 以 K 为单位)[14]：

$$n_o = 2.93323 + 2.55921 \times 10^{-4}(T-273) - 3.26264 \times 10^{-6}(T-273)^2$$
$$+ 8.06267 \times 10^{-8}(T-273)^3 - 5.20204 \times 10^{-10}(T-273)^4$$

最佳色散关系方程(λ 以 μm 为单位,$T = 293$ K)[19]：

$$n_o^2 = 7.4437 + \frac{0.3757}{\lambda^2 - 0.1260} - 0.00154\lambda^2$$

$$n_e^2 = 5.7608 + \frac{0.2908}{\lambda^2 - 0.1628} - 0.00131\lambda^2$$

在文献[5]、[9]、[20]中给出了其他色散关系方程组。

对 293~393 K 的温度范围和 0.9~14 μm 的光谱范围折射率的温度微商(λ 以 μm 为单位)[19]：

$$\frac{dn_o}{dT} = \left(\frac{0.69}{\lambda^3} + \frac{3.43}{\lambda^2} - \frac{2.03}{\lambda} + 9.65\right) \times 10^{-5}\ K^{-1}$$

$$\frac{dn_e}{dT} = \left(\frac{16.75}{\lambda^3} + \frac{41.31}{\lambda^2} - \frac{7.51}{\lambda} + 7.32\right) \times 10^{-5}\ K^{-1}$$

$T = 298$ K 时的 Verdet 常数[21]

$\lambda/\mu m$	$V/[(°) \cdot T^{-1} \cdot m^{-1}]$	$\lambda/\mu m$	$V/[(°) \cdot T^{-1} \cdot m^{-1}]$
0.6265	21 420	0.6356	12 330
0.6275	19 170	0.6381	11 830
0.6287	17 420	0.6420	10 830
0.6306	15 170	0.6459	10 250
0.6328	13 420	0.6494	9 920

有效二阶非线性系数的表达式[22]：

$$d_{ooe} = d_{22}\cos\theta\sin 3\phi$$
$$d_{eoe} = d_{oee} = d_{22}\cos^2\theta\cos 3\phi$$

二阶非线性系数：

$$|d_{22}(10.6\ \mu m)| = 3 \times |d_{31}(CdSe)| \pm 20\% = (54 \pm 11)\ pm/V^{[5,23]}$$

相位匹配角的实验值($T = 293$ K)

相互作用的波长/μm	$\theta_{exp}/(°)$	参 考 文 献
SHG, $o+o \Rightarrow e$		
2.36⇒1.18	18.7	[5]
5.30⇒2.65	10.2	[5]
9.30⇒4.65	12.8	[11]
9.60⇒4.80	13.2	[11]
10.3⇒5.15	14.0	[11]
10.6⇒5.3	12.7	[5]
	14.4	[11]
SFG, $o+o \Rightarrow e$		
17.4+3.532 7⇒2.936 5	13	[4], [24]
11.6+3.931 8⇒2.936 5	10	[4], [24]
10.8+2.361 1⇒1.937 5	10.7	[8]
7.4+2.485 9⇒1.860 8	11.2	[8]
5+2.703 9⇒1.754 9	12.4	[8]
10.1+1.189 5⇒1.064 2	13.3	[25]
7.15+1.250 3⇒1.064 2	15	[25]
19.1+1.114 4⇒1.053	11.5	[26]
12+1.154 3⇒1.053	12	[26]
5.8+1.286 6⇒1.053	15.7	[26]
10.6+1.064 2⇒0.967 11	13.6	[9]
4.9+1.064 2⇒0.874 3	18.8	[9]
17.17+0.723 5⇒0.694 3	15.2	[6]
9.99+0.746 2⇒0.694 3	18.3	[6]
SFG, $e+o \Rightarrow e$		
15.5+1.142 7⇒1.064 2	12.4	[25]
12.0+1.167 8⇒1.064 2	13.3	[25]
9.4+1.200 1⇒1.064 2	14.4	[25]
7.4+1.243 0⇒1.064 2	16.4	[25]
10.6+1.064 2⇒0.967 11	14.4	[9]
18.28+0.721 7⇒0.694 3	15.2	[6]
11.10+0.740 6⇒0.694 3	18.6	[6]

内角带宽的实验值

相互作用的波长/μm	$\Delta\theta^{int}/(°)$	参 考 文 献
SHG, $o+o \Rightarrow e$		
10.3⇒5.15	0.146	[11]
SFG, $o+o \Rightarrow e$		
7+2.51⇒1.847 5	0.086	[8]
12.5+0.735 1⇒0.694 3	0.021	[6]

温度带宽的实验值[19]

相互作用的波长/μm	T/℃	相互作用的波长/μm	T/℃
SHG, o+o⇒e		SFG, e+o⇒e	
10.591⇒5.295 5	172	5.295 5+3.530 3⇒2.647 8	14
5.295 5⇒2.647 8	218	SFG, o+e⇒e	
3.530 3⇒1.765 15	15	5.295 5+3.530 3⇒2.647 8	10
SFG, o+o⇒e			
10.591+3.530 3⇒2.647 8	228		

激光诱导的表面损伤阈值

λ/μm	τ_p/ns	I_{thr}/(GW·cm^{-2})	参考文献	备注
0.683	6	0.082	[15]	100个脉冲
0.694 3	30	0.02	[6]	
0.7	0.07	7	[15]	100个脉冲
0.75	50	0.02	[7]	
0.8	60	0.008	[15]	100个脉冲
1.053	0.002	>1.0	[26]	1 Hz
1.06	20	>0.01	[17]	
1.064 2	10	0.03	[9]	
2.36	40	>0.005	[5]	
2.80	0.1	>4	[27]	3 Hz
2.94	0.11	30	[4],[24]	1 Hz
9.55	30	0.12	[12]	∥c
10.6	CW	>0.000 5	[28]	
	125	0.03	[11]	2~20 Hz

关于这一晶体

GaSe 以其层状结构而出名[29]。然而,这是用于中红外 OPG 或 DFG 发生,最佳的中红外非线性光学晶体之一。最近,Vodopyanov[30,31]报道了双回路通光以 GaSe 为基础的 OPG,由 100 ps、3 mJ 输出脉冲的 Er,Cr:YSGG 激光器泵浦,运转波长为 2.8 μm。OPG 所达到的调谐范围为 3.3~19 μm。在过去五年所做的 DFG 实验中,连续可调的中红外范围持续地从 9~18 μm[32]增加到 7~20 μm[33],最后达到 2.7~28.7 μm[34]。

■ **参考文献**

[1] *Physical-Chemical Properties of Semiconductors. Handbook.* (Nauka, Moscow, 1979) [In

Russian].

[2] *Physical Quantities. Handbook*, ed. by I. S. Grigoriev and E. Z. Meilikhov(Energoatomizdat,Moscow,1991)[In Russian].

[3] G. D Guseinov, A. I. Rasulov: Heat conductivity of GaSe monocrystals. Phys. Stat. Solidi 18(2), 911 –922(1966).

[4] K. L. Vodopyanov, L. A. Kulevskii, V. G. Voevodin, A. I. Gribenyukov, K. R. Allakhverdiev, T. A. Kerimov: High efficiency middle IR parametric superradiance in $ZnGeP_2$ and GaSe crystals pumped by an erbium laser. Opt. Commun. 83(5 –6), 322 –326(1991).

[5] G. B. Abdullaev, L. A. Kulevskii, A. M. Prokhorov, A. D. Saveliev, E. Y. Salaev, V. V. Smirnov: GaSe, a new effective material for nonlinear optics. Pisma Zh. Eksp. Teor. Fiz. 16 (3), 130 –133(1972)[In Russian,English trans.: JETP Lett.16(3),90 –92(1972)].

[6] G. B. Abdullaev, L. A. Kulevskii, P. V. Nikles, A. M. Prokhorov, A. D. Saveliev, E. Y. Salaev, V. V. Smirnov: Difference frequency generation in a GaSe crystal with continuous tuning in the 560 – 1 050 cm^{-1} range. Kvant. Elektron. 3(1), 163 – 167 (1976) [In Russian,English trans.: Sov. J. Quantum Electron. 6(1),88 –90(1976)].

[7] A. O. Okorogu, S. B. Mirov, W. Lee, D. I. Crouthamel, N. Jenkins, A. Y. Dergachev, K. L. Vodopyanov, V. V. Badikov: Tunable middle infrared downconversion in GaSe and $AgGaS_2$. Opt. Commun. 155(4 –6), 307 –312(1998).

[8] A. Bianchi, A. Ferrario, M. Musci: 4 – 12 μm tunable down-conversion in GaSe from a $LiNbO_3$ parametric oscillator. Opt. Commun. 25(2), 256 –258(1978).

[9] G. B. Abdullaev, K. R. Allakhverdiev, L. A. Kulevskii, A. M. Prokhorov, E. Y. Salaev, A. D. Saveliev, V. V. Smirnov: Parametric conversion of infrared radiation in a GaSe crystal. Kvant. Elektron. 2(6), 1228 – 1233 (1975) [In Russian, English trans.: Sov. J. Quantum Electron. 5(6),665 –668(1975)].

[10] A. Bianchi, M. Garbi: Down-conversion in the 4 – 18 μm range with GaSe and $AgGaSe_2$ nonlinear crystals. Opt. Commun. 30(1), 122 – 124(1979).

[11] G. B. Abdullaev, K. R. Allakhverdiev, M. E. Karasev, V. I. Konov, L. A. Kulevskii, N. B. Mustafaev, P. P. Pashinin, A. M. Prokhorov, Y. M. Starodumov, N. I. Chapliev: Efficient generation of the second harmonic of CO_2 laser radiation in a GaSe crystal. Kvant. Elektron. 16(4), 757 –763(1989) [In Russian,English trans.: Sov. J. Quantum Electron. 19(4),494 –498(1989)].

[12] Y. M. Andreev, V. V. Badikov, V. G. Voevodin, L. G. Geiko, P. P. Geiko, M. V. Ivashchenko, A. I. Karapuzikov, I. V. Sherstov: Radiation resistance of nonlinear crystals at a wavelength of 9.55 μm. Kvant. Elektron. 31(12), 1075 –1078(2001)[In Russian,English trans.: Quantum Electron. 31(12),1075 –1078(2001)].

[13] N. P. Barnes, R. C. Eckardt, D. J. Gettemy, L. B. Edgett: Absorption coefficients and the temperature variation of the refractive index difference of nonlinear optical crystals.

IEEE J. Quant. Electr. QE – 15(10), 1074 – 1076(1979).

[14] M. A. Hernandez, M. V. Andres, A. Segura, V. Munoz: Temperature dependence of refractive index and absorption coefficient of GaSe at 633 nm. Opt. Commun. 118(3 – 4), 335 – 337(1995).

[15] K. L. Vodopyanov, S. B. Mirov, V. G. Voevodin, P. G. Schunemann: Two-photon absorption in GaSe and CdGeAs$_2$. Opt. Commun. 155(1 – 3), 47 – 50(1998).

[16] I. B. Zotova, Y. J. Ding: Spectral measurements of two-photon absorption coefficients for CdSe and GaSe crystals. Appl. Opt. 40(36), 6654 – 6658(2001).

[17] F. Adduci, I. M. Catalano, A. Cingolani, A. Minafra: Direct and indirect two-photon processes in layered semiconductors. Phys. Rev. B 15(2), 926 – 931(1977).

[18] G. Antonioli, D. Bianchi, P. Franzosi: Temperature variation of refractive index in GaSe. Appl. Opt. 18(22), 3847 – 3850(1979).

[19] E. Takaoka, K. Kato: Temperature phase-matching properties for harmonic generation in GaSe. Jpn. J. Appl. Phys. 38(5A), 2755 – 2759(1999).

[20] K. L. Vodopyanov, L. A. Kulevskii: New dispersion relationships for GaSe in the 0.65 – 18 μm spectral region. Opt. Commun. 118(3 – 4), 375 – 378(1995).

[21] A. Balbin Villaverde, D. A. Donatti: GaSe Faraday rotation near the absorption edge. J. Chem. Phys. 72(10), 5341 – 5342(1980).

[22] J. E. Midwinter, J. Warner: The effects of phase matching method and of uniaxial crystal symmetry on the polar distribution of second-order non-linear optical polarization. Brit. J. Appl. Phys. 16(11), 1135 – 1142(1965).

[23] D. A. Roberts: Simplified characterization of uniaxial and biaxial nonlinear optical crystals: a plea for standardization of nomenclature and conventions. IEEE J. Quant. Electr. 28 (10), 2057 – 2074(1992).

[24] K. L. Vodopyanov: Parametric generation of tunable infrared radiation in ZnGeP$_2$ and GaSe pumped at 3 μm. J. Opt. Soc. Am. B 10(9), 1723 – 1729(1993).

[25] Y. A. Gusev, A. V. Kirpichnikov, S. N. Konoplin, S. I. Marennikov, P. V. Nikles, Y. N. Polivanov, A. M. Prokhorov, A. D. Saveliev, R. S. Sayakhov, V. V. Smirnov, V. P. Chebotaev: Tunable mid-IR difference frequency generator. Pisma Zh. Tekh. Fiz. 6 (19 – 20), 1262 – 1265 (1980) [In Russian, English trans. : Sov. Tech. Phys. Lett. 6 (10),541 – 542(1980)].

[26] T. Dahinten, U. Plödereder, A. Seilmeier, K. L. Vodopyanov, K. R. Allakhverdiev, Z. A. Ibragimov: Infrared pulses of 1 picosecond duration tunable between 4 μm and 18 μm. IEEE J. Quant. Electr. 29(7), 2245 – 2250(1993).

[27] K. L. Vodopyanov, V. G. Voevodin: 2.8 μm laser pumped type I and type II travelling-wave optical parametric generator in GaSe. Opt. Commun. 114(3 – 4), 333 – 335(1995).

[28] J.-J. Zondy: Experimental investigation of single and twin AgGaSe$_2$ crystals for CW

10.2 μm SHG. Opt. Commun. 119(3-4), 320-326(1995).

[29] L. Kador, D. Haarer, K. R. Allakhverdiev, E. Y. Salaev: Phase-matched second-harmonic generation at 789.5 nm in a GaSe crystal. Appl. Phys. Lett. 69(6), 731-733 (1996).

[30] K. L. Vodopyanov, V. Chazapis: Extra-wide tuning range optical parametric oscillator. Opt. Commun. 135(13), 98-102(1997).

[31] K. L. Vodopyanov: Mid-infrared optical parametric generator with extra-wide (2.7-28.7 μm) tunability: applications for spectroscopy of two-dimensional electrons in quantum wells. J. Opt. Soc. Am. B 16(9), 1579-1586(1999).

[32] R. A. Kaindl, D. C. Smith, M. Joschko, M. P. Hasselbeck, M. Woerner, T. Elsaesser: Femtosecond infrared pulses tunable from 9 to 18 μm at an 88-MHz repetition rate. Opt. Lett. 23(11), 861-863(1998).

[33] R. A. Kaindl, F. Eickemeyer, M. Woerner, T. Elsaesser: Broadband phase-matched difference frequency mixing of femtosecond pulses in GaSe: experiment and theory. 75(8), 1060-1062(1999).

[34] W. Shi, Y. J. Ding, X. Mu, N. Fernelius: Tunable and coherent nanosecond radiation in the range of 2.7-28.7 μm based on difference-frequency generation in gallium selenide. Appl. Phys. Lett. 80(21), 3889-3891(2002).

第 4 章

常用晶体

这一章涉及其他常用的非线性光学晶体,诸如磷酸二氢钾(KDP)以及它最普遍的同构体磷酸二氢铵(ADP)及氘化磷酸二氢钾(DKDP);最新发展的硼酸锂铯(CLBO)、铌酸锂的同构体氧化镁掺杂铌酸锂(MgLN)、KTP 的同构体砷酸钛氧钾(KTA);最后还有铌酸钾(KN)。

4.1 KH_2PO_4,磷酸二氢钾(KDP)

负单轴晶:$n_o > n_e$

分子量:136.086

密度:2.332 5 g/cm³[1];2.338 g/cm³[2];在 293 K 时,2.338 3 g/cm³[3]

点群:$\bar{4}2m$($mm2$,$T < 122$ K[4])

晶格常数:

点群 $\bar{4}2m$:

　　　　$a = 7.448$ Å[5];7.452 Å[6];7.453 Å[4];在 $T = 296$ K[7]时,

(7.4529 ± 0.0002) Å

$c = 6.977$ Å[5], 6.959 Å[6]; 6.959 Å[4]; 在 $T = 296$ K[7] 时，(6.9751 ± 0.0006) Å

点群 $mm2$：

$a = 10.44$ Å[8]

$b = 10.53$ Å[8]

$c = 6.90$ Å[8]

从不同商业来源获得 KDP 晶体晶格常数的变化[9]

公司	a/Å	c/Å	公司	a/Å	c/Å
Inrad	7.460	6.965	Cleveland 晶体#2	7.439	6.962
Cleveland 晶体#1	7.451	6.950			

莫氏硬度：1.5[10]；2.5[11,12]

维氏硬度[13]：

沿 a 方向：122 ± 17

沿 c 方向：183 ± 12

在 100 g 水中的溶解度[3]

T/K	s/g
298	33

熔点：525 K[3]；526 K[14,15]

居里温度：122 K[16]；122.6 K[17]；123 K[7,17-19]

线性热膨胀系数[4]

T/K	$\alpha_t \times 10^6$/K^{-1}, $\parallel c$	$\alpha_t \times 10^6$/K^{-1}, $\perp c$
200	39	22
250	41	24.6
270	41.6	25.6
280	41.9	26.0
290	42.1	26.4
300	42.4	26.8

线性热膨胀系数的平均值

T/K	$\alpha_t \times 10^6$/K^{-1}, $\parallel c$	$\alpha_t \times 10^6$/K^{-1}, $\perp c$	参考文献
123~293	42	20	[4]
123~298	39.2	22	[19]

续表

T/K	$\alpha_t \times 10^6/K^{-1}$, $\parallel c$	$\alpha_t \times 10^6/K^{-1}$, $\perp c$	参 考 文 献
223~323	44.0	24.9	[19]
223~373		26.6	[4]
233~363	44.6		[4]

$p = 0.101\ 325$ MPa 时的比热容 c_p

T/K	$c_p/(J \cdot kg^{-1} \cdot K^{-1})$	参 考 文 献	T/K	$c_p/(J \cdot kg^{-1} \cdot K^{-1})$	参 考 文 献
80	341	[14]	298	857	[14]
150	552	[14]	306	879	[15]
250	764	[14]			

热导率[3]

T/K	$\kappa/(W \cdot m^{-1} \cdot K^{-1})$, $\parallel c$	$\kappa/(W \cdot m^{-1} \cdot K^{-1})$, $\perp c$
302	1.21	
319		1.34
428	1.30	1.76

室温下带隙能量：$E_g = 6.95$ eV[20]；7.0 eV[21,10]

以 $\alpha = 1$ cm^{-1} 计的透明范围：$0.176 \sim 1.4$ μm[22,7]

对 0.2 cm 长晶体以 0.5 透过计的透明范围：$0.176 \sim 1.55$ μm[1]

线性吸收系数 α

$\lambda/\mu m$	α/cm^{-1}	参 考 文 献	备 注
0.212	0.2	[23]	
0.257 25	0.01~0.2	[24]	e光, $\perp c$
	0.009	[25]	
	0.007	[26]	e光, $\perp c$
0.263	0.03	[27]	
0.351 3	0.003	[28]	e光, $\perp c$
0.514 5	0.000 1	[25]	
	0.000 05	[24]	o光
0.526 5	0.01	[29]	o光
0.694 3	0.008	[30]	
0.78	0.024	[4]	
0.89	0.015	[4]	

续表

$\lambda/\mu m$	α/cm^{-1}	参考文献	备注
0.94	0.01	[31]	
1.053	0.05	[29]	o 光
	0.03	[27]	
1.054	0.058	[28]	o 光
	0.02	[28]	e 光，$\perp c$
1.06	0.03	[4]	
1.064 2	0.03	[32]	
	0.058	[33]	o 光
	0.006	[33]	e 光
1.22	0.1	[34]	o 光
1.315 2	0.3	[35]	
1.32	0.1	[34]	e 光，$\perp c$

双光子吸收系数 β

$\lambda/\mu m$	τ_p/ns	$\beta \times 10^{11}/(cm \cdot W^{-1})$	参考文献	备注
0.211	0.000 9	136 ± 48	[36]	$\theta = 41°$，$\phi = 45°$
0.216	0.015	60 ± 5	[37]	
0.263	0.6	60	[27]	
0.263 5	0.5	50	[38]	
0.264	0.000 8	26 ± 9	[36]	$\theta = 41°$，$\phi = 45°$
0.266 1	0.015	27 ± 8	[20]	$\theta = 41°$，$\phi = 45°$
	0.6	40 ~ 80	[39]	
0.270	0.015	28 ± 3	[40]	
0.354 7	0.017	0.59 ± 0.21	[20]	e 光，$\perp c$

$T = 298$ K 时折射率的实验值[41]

$\lambda/\mu m$	n_o	n_e	$\lambda/\mu m$	n_o	n_e
0.213 856 0	1.601 77	1.546 15	0.303 578 1		1.496 67
0.228 801 8	1.585 46		0.312 566 3	1.541 17	1.494 34
0.244 690 5	1.572 28		0.313 154 5	1.540 98	1.494 19
0.246 406 8	1.571 05		0.334 147 8		1.489 54
0.253 651 9	1.566 31	1.515 86	0.365 014 6	1.529 32	1.484 32
0.280 086 9	1.552 63	1.504 16	0.365 483 3	1.529 23	1.484 23
0.298 062 8	1.546 18	1.498 24	0.366 287 8	1.529 09	1.484 09
0.302 149 9	1.544 33	1.497 08	0.390 641 0		1.480 89

续表

$\lambda/\mu m$	n_o	n_e	$\lambda/\mu m$	n_o	n_e
0.404 656 1	1.523 41	1.479 27	0.632 816 0	1.507 37	1.466 85
0.407 781 1	1.523 01	1.478 98	1.013 975 0	1.495 35	1.460 41
0.435 835 0	1.519 90	1.476 40	1.128 704 0	1.492 05	1.459 17
0.491 603 6		1.472 54	1.152 276 0	1.491 35	1.458 93
0.546 074 0	1.511 52	1.469 82	1.357 070 0	1.484 55	
0.576 958 0	1.509 87		1.523 100 0		1.455 21
0.579 065 4	1.509 77	1.468 56	1.529 525 0		1.455 12

折射率的温度微商[42]

$\lambda/\mu m$	$dn_o/dT \times 10^5/K^{-1}$	$dn_e/dT \times 10^5/K^{-1}$
0.405	−3.27	−3.15
0.436	−3.27	−2.88
0.546	−3.28	−2.90
0.578	−3.25	−2.87
0.633	−3.94	−2.54

从室温降至 $T[K]$ 时折射率的温度关系:

对于光谱范围 $0.365 \sim 0.690 \mu m$[42]:

$$n_o(T) = n_o(298) + 0.402 \times 10^{-4} \{[n_o(298)]^2 - 1.432\}(298 - T)$$

$$n_e(T) = n_e(298) + 0.221 \times 10^{-4} \{[n_e(298)]^2 - 1.105\}(298 - T)$$

对于光谱范围 $0.436 \sim 0.589 \mu m$[43]:

$$n_o(T) = n_o(300) + 10^{-4}(143.3 - 0.618T + 4.81 \times 10^{-4} T^2)$$

$$n_e(T) = n_e(300) + 10^{-4}(153.3 - 0.969T + 1.57 \times 10^{-3} T^2)$$

最佳色散关系方程组(λ 以 μm 为单位,$T = 293$ K)[41]:

$$n_o^2 = 2.259\,276 + \frac{13.005\,22 \lambda^2}{\lambda^2 - 400} + \frac{0.010\,089\,56}{\lambda^2 - (77.264\,08)^{-1}}$$

$$n_e^2 = 2.132\,668 + \frac{3.227\,992\,4 \lambda^2}{\lambda^2 - 400} + \frac{0.008\,637\,494}{\lambda^2 - (81.426\,31)^{-1}}$$

在文献[9]、[44]、[45]、[46]、[47]中给出了其他的色散关系方程组。

Sellmeier 方程的温度关系(λ 以 μm 为单位,T 以 K 为单位)[45]:

$$n_o^2 = 1.448\,96 + 3.185 \times 10^{-5} T + \frac{(0.841\,81 - 1.411\,4 \times 10^{-4} T) \lambda^2}{\lambda^2 - (0.012\,8 - 2.13 \times 10^{-7} T)}$$

$$+ \frac{(0.907\,93 + 5.75 \times 10^{-7} T) \lambda^2}{\lambda^2 - 30}$$

$$n_e^2 = 1.42961 - 1.152 \times 10^{-5}T + \frac{(0.72722 - 6.139 \times 10^{-5}T)\lambda^2}{\lambda^2 - (0.01213 + 3.104 \times 10^{-7}T)}$$
$$+ \frac{(0.22543 - 1.98 \times 10^{-7}T)\lambda^2}{\lambda^2 - 30}$$

非线性折射率 γ

$\lambda/\mu m$	$\gamma \times 10^{15}/(cm^2 \cdot W^{-1})$	参考文献	备注
0.5321	0.25 ± 0.08	[48]	$\theta = 78°$
	0.28 ± 0.08	[48]	$\theta = 41°$
	0.28 ± 0.08	[49]	
1.0642	0.20	[50]	o 光
	0.22	[50]	e 光
	0.26 ± 0.08	[48]	$\theta = 90°$
	0.28	[51]	
	0.29 ± 0.09	[52]	
	0.44 ± 0.13	[48]	$\theta = 78°$
	0.46 ± 0.14	[48]	$\theta = 59°$
	1.0(?)	[53]	

室温下低频(远低于 KDP 晶体声学共振频率,即对"自由"晶体)下测量的线性电光系数

$\lambda/\mu m$	$r_{41}^T/(pm \cdot V^{-1})$	$r_{63}^T/(pm \cdot V^{-1})$	参考文献	备注
0.20		-10.7	[54]	$T = 283$ K
0.25		-10.5	[54]	$T = 283$ K
0.500		-9.2	[55]	
0.5461	-8.77 ± 0.14		[56]	
		-10.3	[57]	
0.556	-8.6 ± 0.2		[58]	$T = 295$ K
		-10.5 ± 0.2	[58]	$T = 295$ K
0.6328	-8		[59]	
	-8.6 ± 0.2		[60]	$T = 295$ K
		-9.4 ± 0.4	[61]	
		-9.9 ± 0.2	[60]	$T = 295$ K
		-11	[59]	
0.700		-9.4	[55]	
3.3913		-9.7	[57]	

室温下高频(远高于 KDP 晶体的声学共振频率,即对"受夹"晶体)下测量线性电光系数

$\lambda/\mu m$	$r_{63}^S/(pm \cdot V^{-1})$	参 考 文 献	$\lambda/\mu m$	$r_{63}^S/(pm \cdot V^{-1})$	参 考 文 献
0.546 1	-8.5±2.4	[62]	0.632 8	-8.8±0.5	[64]
	-9.7	[63]			

纵向调制下的半波延迟电压

$\lambda/\mu m$	$V_{\lambda/2}/kV$	参 考 文 献	$\lambda/\mu m$	$V_{\lambda/2}/kV$	参 考 文 献
0.435 8	6.04±0.06	[56]		7.65±0.08	[56]
0.546 1	7.5	[63]	0.578	8.17±0.08	[56]

Verdet 常数 ($\parallel c$)

$\lambda/\mu m$	T/K	$V/[(°) \cdot T^{-1} \cdot m^{-1}]$	参 考 文 献
0.632 8	293	221±5	[2]
		213	[65]
	298	207	[66]

注：文献[65]中的测量是在室温下进行的。

计算的 Verdet 常数 ($\parallel c$)[67]

$\lambda/\mu m$	$V/[(°) \cdot T^{-1} \cdot m^{-1}]$	$\lambda/\mu m$	$V/[(°) \cdot T^{-1} \cdot m^{-1}]$
0.193	3 875	0.308	1 030
0.222	2 487	0.351	758
0.248	1 800		

ns SRS 的线束位移：$\Delta \nu = 915$ cm^{-1} [68]

普遍情况下有效二阶非线性系数的表达式 (Kleinman 对称条件成立，$d_{14} = d_{25} = d_{36}$)[69]：

$$d_{ooe} = -d_{36}\sin(\theta+\rho)\sin 2\phi$$

$$d_{eoe} = d_{oee} = 2d_{36}\sin(\theta+\rho)\cos(\theta+\rho)\cos 2\phi$$

有效二阶非线性系数的简化表达式 (小双折射角近似，Kleinman 对称条件成立，$d_{14} = d_{25} = d_{36}$)[70]：

$$d_{ooe} = -d_{36}\sin\theta\sin 2\phi$$

$$d_{eoe} = d_{oee} = d_{36}\sin 2\theta\cos 2\phi$$

二阶非线性系数的绝对值

$$d_{36}(1.319\ \mu m) = (0.31 \pm 0.02)\ pm/V^{[71]}$$

$$d_{36}(1.064\ 2\ \mu m) = 0.38\ pm/V^{[69]};\ 0.39\ pm/V^{[72]};$$

$$(0.39 \pm 0.03)\ pm/V^{[73]};\ (0.40 \pm 0.02)\ pm/V^{[71]}$$

相位匹配角的实验值 ($T = 293$ K)

相互作用的波长/μm	$\theta_{exp}/(°)$	参 考 文 献
SHG, o + o ⇒ e		
0.517 ⇒ 0.258 5	90	[44]
0.657 6 ⇒ 0.328 8	53.6	[35]
0.694 3 ⇒ 0.347 15	50.4	[74]
0.870 7 ⇒ 0.435 35	42.4	[75]
1.06 ⇒ 0.53	41	[76], [77]
1.315 2 ⇒ 0.657 6	44.3	[35]
SFG, o + o ⇒ e		
1.415 + 0.220 27 ⇒ 0.190 6	88.7	[78]
1.364 8 + 0.694 3 ⇒ 0.460 19	40.9	[75]
1.315 2 + 0.657 6 ⇒ 0.438 4	42.2	[35]
1.064 2 + 0.270 7 ⇒ 0.215 81	87.6	[79]
1.064 2 + 0.532 1 ⇒ 0.354 73	47.3	[80]
1.06 + 0.53 ⇒ 0.353 33	47.5	[77]
0.657 6 + 0.438 4 ⇒ 0.263 04	74	[81]
SHG, e + o ⇒ e		
1.315 2 ⇒ 0.657 6	61.4	[35]
1.06 ⇒ 0.53	59	[77]
SFG, e + o ⇒ e		
1.064 2 + 0.532 1 ⇒ 0.354 73	58.3	[80]
1.06 + 0.53 ⇒ 0.353 33	59.3	[77]

NCPM 温度的实验值

相互作用的波长/μm	T/℃	参 考 文 献
SHG, o + o ⇒ e		
0.514 5 ⇒ 0.257 25	-13.7	[26]
	-11	[24]
0.517 ⇒ 0.258 5	20	[44]
0.532 1 ⇒ 0.266 05	177	[82], [83]
SFG, o + o ⇒ e		
1.06 + 0.265 ⇒ 0.212	-70	[23]
1.064 2 + 0.266 05 ⇒ 0.212 84	-40	[84]
	-35	[85]
1.079 6 + 0.269 9 ⇒ 0.215 92	60	[86]

内角带宽和温度带宽的实验值

相互作用的波长/μm	T/℃	θ_{pm}/(°)	$\Delta\theta^{int}$/(°)	ΔT/℃	参考文献
SHG, o + o ⇒ e					
1.152 3 ⇒ 0.576 15	20	41	0.074		[87]
1.064 2 ⇒ 0.532 1	20	41	0.070		[69]
	25			23 ± 1	[88]
1.064 ⇒ 0.532	20	41	0.069		[89]
1.06 ⇒ 0.53	20	41	0.063		[76]
			0.065 ± 0.003		[90]
1.054 ⇒ 0.527	25	41	0.060		[91]
0.532 1 ⇒ 0.266 05	25			1.7 ± 0.1	[88]
	177	90		1.9	[82]
	177	90		2	[83]
0.53 ⇒ 0.265	20	77	0.059		[92]
	20	77	0.066	1.2(?)	[93]
SFG, o + o ⇒ e					
1.064 2 + 0.532 1 ⇒ 0.354 73	25			5.5 ± 0.2	[88]
1.054 + 0.527 ⇒ 0.351 33	25	48	0.046		[91]
1.079 6 + 0.269 9 ⇒ 0.215 92	60	90		1.3	[86]
SHG, e + o ⇒ e					
1.064 2 ⇒ 0.532 1	25			18.3 ± 1.7	[88]
1.06 ⇒ 0.53	20	59	0.129		[92]
			0.133 ± 0.002		[90]
1.054 ⇒ 0.527	25	59	0.126		[91]
SFG, e + o ⇒ e					
1.064 2 + 0.532 1 ⇒ 0.354 73	25			5.2 ± 0.2	[88]
1.06 + 0.53 ⇒ 0.353 33	20	59	0.062	2.2(?)	[93]
1.054 + 0.527 ⇒ 0.351 33	25	59	0.059		[91]

光谱带宽的实验值

相互作用的波长/μm	T/℃	θ_{pm}/(°)	$\Delta\nu$/cm^{-1}	参 考 文 献
SHG, o + o ⇒ e				
1.06 ⇒ 0.53	20	41	178	[76]
0.53 ⇒ 0.265	20	77	4.7	[92]
SHG, e + o ⇒ e				
1.06 ⇒ 0.53	20	59	101.5	[92]

相位匹配角的温度变化

相互作用的波长/μm	T/℃	θ_{pm}/(°)	$d\theta_{pm}/dT/[(°)\cdot K^{-1}]$	参考文献
SHG, o + o ⇒ e				
1.064 2 ⇒ 0.532 1	25		0.002 8	[88]
1.06 ⇒ 0.53		41	0.003 65 ± 0.000 03	
1.054 ⇒ 0.527	25	41	0.004 6	[91]
0.532 1 ⇒ 0.266 05	25		0.038 2	[88]
0.526 5 ⇒ 0.263 25		80	0.060 2	[27]
SFG, o + o ⇒ e				
1.064 2 + 0.532 1 ⇒ 0.354 73	25		0.007 3	[88]
1.054 + 0.527 ⇒ 0.351 33	25	48	0.004 6	[91]
SHG, e + o ⇒ e				
1.064 2 ⇒ 0.532 1	25	59	0.006 9 ± 0.000 3	[94]
	25		0.006 9	[88]
1.06 ⇒ 0.53	20	59	0.005 7	[92]
			0.009 7 ± 0.000 3	[90]
1.054 ⇒ 0.527	25	59	0.008 5	[91]
	20	59	0.006 9	[28]
SFG, e + o ⇒ e				
1.064 2 + 0.532 1 ⇒ 0.354 73	25	58	0.010 6 ± 0.000 3	[94]
	25		0.011 7	[88]
1.054 + 0.527 ⇒ 0.3513 3	25	59	0.015 2	[91]
	20	59	0.007 5	[28]

非临界 SHG 的温度调谐[44]

相互作用的波长/μm	$d\lambda_1/dT/(nm\cdot K^{-1})$
SHG, o + o ⇒ e	
0.517 ⇒ 0.258 5	0.048

非临界 SHG 过程的双折射温度变化

相互作用的波长/μm	$d(n_2^e - n_1^o)/dT \times 10^{-5}/K^{-1}$	参 考 文 献
0.514 5 ⇒ 0.257 25	1.745	[95]
0.532 1 ⇒ 0.266 05	1.2	[82]

SHG 过程的电光调谐敏感性

相互作用的波长/μm	$d\theta_{pm}/dE/[(°)\cdot cm\cdot kV^{-1}]$
SHG, o + o ⇒ e	
1.06 ⇒ 0.53	0.002 93 ± 0.000 02

续表

相互作用的波长/μm	$\mathrm{d}\theta_{pm}/\mathrm{d}E/[(°) \cdot \mathrm{cm} \cdot \mathrm{kV}^{-1}]$
SHG, e+o⇒e	
1.06⇒0.53	≈0

相位匹配角及"走离"角的计算值

相互作用的波长/μm	$\theta_{pm}/(°)$	$\rho_1/(°)$	$\rho_3/(°)$
SHG, o+o⇒e			
0.532 1⇒0.266 05	76.60		0.808
0.578 2⇒0.289 1	64.03		1.391
0.632 8⇒0.316 4	56.15		1.611
0.659 4⇒0.329 7	53.43		1.657
0.694 3⇒0.347 15	50.55		1.687
1.064 2⇒0.532 1	41.21		1.603
1.318 8⇒0.659 4	44.70		1.549
SFG, o+o⇒e			
0.578 2+0.510 5⇒0.271 12	72.46		1.025
1.064 2+0.532 1⇒0.354 73	47.28		1.712
1.318 8+0.659 4⇒0.439 6	42.05		1.657
SHG, e+o⇒e			
1.064 2⇒0.532 1	58.98	1.149	1.404
1.318 8⇒0.659 4	61.85	0.922	1.269
SFG, e+o⇒e			
1.064 2+0.532 1⇒0.354 73	58.23	1.166	1.521
1.318 8+0.659 4⇒0.439 6	49.42	1.104	1.634

KDP 中 SHG 过程逆群速失配的计算值

相互作用的波长/μm	$\theta_{pm}/(°)$	$\beta/(\mathrm{fs} \cdot \mathrm{mm}^{-1})$
SHG, o+o⇒e		
1.2⇒0.6	42.45	42
1.1⇒0.55	41.38	17
1.0⇒0.5	41.22	9
0.9⇒0.45	42.24	40
0.8⇒0.4	44.91	77
0.7⇒0.35	50.14	128
0.6⇒0.3	60.40	208
SHG, e+o⇒e		
1.2⇒0.6	59.54	89

续表

相互作用的波长/μm	$\theta_{pm}/(°)$	$\beta/(fs \cdot mm^{-1})$
1.1⇒0.55	58.87	67
1.0⇒0.5	59.75	89
0.9⇒0.45	62.97	118
0.8⇒0.4	70.71	158

激光诱导的体损伤阈值

$\lambda/\mu m$	τ_p/ns	$I_{thr}/(GW \cdot cm^{-2})$	参考文献	备注
0.266 1	8	2.3	[96]	
	0.75	7	[13]	$\boldsymbol{k} \parallel c, \boldsymbol{E} \parallel c$
0.354 7	0.85	5.1~6.2	[13]	
	0.017	5 000~24 000	[49]	细聚焦
0.355	7.6	5.1	[97]	$\parallel c$，大块原生晶体
		2.9	[97]	$\theta=58°$，大块原生晶体
0.52	330	0.2	[98]	
0.526 5	20	3	[29]	
	0.6	9	[29]	
0.527	0.5	>14	[99]	
0.53	10	34~57	[100]	波束腰直径 21 μm
	0.2	17	[101]	
	0.005	1 000	[102]	
0.532 1	1.0	6~20	[13]	与辐射方向和光的偏振有关
	0.6	>8	[39]	
	0.03	30	[103]	
	0.021	2 200	[49]	细聚焦
0.596	330	0.24	[98]	
	20	3	[98]	
0.694 3	20	>0.4	[104]	
	5~25	0.10~0.14	[30]	
1.053	25	4	[29]	
	1.1	10.6~20.9	[13]	与辐射方向和光的偏振有关
	1	18	[29]	
	1	15~20	[105]	KDP 溶液的 UV 辐射
	1	20	[106]	
1.054	3	>3.3	[107]	
	1	>5.1	[107]	
	0.14	>7	[108]	

续表

$\lambda/\mu m$	τ_p/ns	$I_{thr}/(GW \cdot cm^{-2})$	参考文献	备注
1.06	60	0.2	[109]	
	30	17~34	[110]	细聚焦
	14	2.5~5	[100]	波束腰直径 100 μm
		17~35	[100]	波束腰直径 21 μm
	12~25	>0.25	[76]	
	0.5	>3	[111]	
	0.2	23	[101]	
	0.003	7 000~10 000	[110]	细聚焦
1.064 2	20	0.3~0.6	[112]	
	10	6.4~18.5	[113]	
	7	2.7	[114]	
	1.3	8	[115]	
	1.1	16	[116]	
	1	3~7	[112]	
		5	[117]	
	0.1	7	[118]	
		>100	[51]	
	0.03	2 000	[53]	细聚焦
1.079 6	5	16	[86]	

激光诱导的表面损伤阈值

$\lambda/\mu m$	τ_p/ns	$I_{thr}/(GW \cdot cm^{-2})$	参考文献	备注
0.248 4	20	0.45	[119]	波束腰直径 1.5 mm
0.266 1	0.7	8.6	[119]	
0.694 3	5~25	1~5	[30]	
1.064 2	12	14.4	[120]	$\parallel c$,波束腰直径 30 μm
	10	3.9~18.5	[113]	

关于这一晶体

KDP(与其同构晶体 DKDP 和 ADP)是最老的非线性光学材料之一[8,121]。20 世纪 60 年代它被广泛用于所进行的首批实验中,直至今天仍被应用,特别是用于惯性约束聚变[122]中。在这一实验中,需要高的抗光损伤阈值和大口径的 SHG 和 THG 晶体。为适应这些目标,发展了 KDP 快速生长的方法,最高速率可达 50 mm/天，晶体尺寸可达 90 cm[123-125]。

参考文献

[1] Data sheet of Cleveland Crystals Inc. Available at www.clevelandcrystals.com.

[2] S. Haussühl, W. Effgen: Faraday effect in cubic crystals. Additivity rule and phase transitions. Z. Kristallogr. 183(1-4), 153-174(1988).

[3] E. M. Voronkova, B. N. Grechushnikov, G. I. Distler, I. P. Petrov: *Optical Materials for Infrared Technique* (Nauka, Moscow, 1965) [In Russian].

[4] A. A. Blistanov, V. S. Bondarenko, N. V. Perelomova, F. N. Strizhevskaya, V. V. Tchkalova, M. P. Shaskolskaya: *Acoustic Crystals* (Nauka, Moscow, 1982) [In Russian].

[5] *Handbook of Optical Constants of Solids II*, ed. by E. D. Palik (Academic Press, Boston, 1991).

[6] S. Haussühl: Elastische und thermoelastische Eigenschaften von KH_2PO_4, KH_2AsO_4, $NH_4H_2PO_4$, $NH_4H_2AsO_4$ und RbH_2PO_4. Z. Kristallogr. 120(6), 401-414(1964) [In German].

[7] T. R. Sliker, S. R. Burlage: Some dielectric and optical properties of KD_2PO_4. J. Appl. Phys. 34(7), 1837-1840(1963).

[8] F. Jona, G. Shirane: *Ferroelectric Crystals* (Pergamon Press, Oxford, 1962).

[9] K. W. Kirby, L. G. DeShazer: Refractive indices of 14 nonlinear crystals isomorphic to KH_2PO_4. J. Opt. Soc. Am. B 4(7), 1072-1078(1987).

[10] B. H. T. Chai: Optical Crystals. In: *CRC Handbook of Laser Science and Technology, Supplement 2: Optical Materials*, ed. by M. J. Weber (CRC Press, Boca Raton, 1995) pp. 3-65.

[11] V. G. Dmitriev, G. G. Gurzadyan, D. N. Nikogosyan: *Handbook of Nonlinear Optical Crystals, Third Revised Edition* (Springer, Berlin, 1999).

[12] D. Yuan, D. Xu, M. Liu, F. Qi, W. Yu, W. Hou, Y. Bing, S. Sun, M. Jiang: Structure and properties of a complex crystal for laser diode frequency doubling: cadmium mercury thiocyanate. Appl. Phys. Lett. 70(5), 544-546(1997).

[13] H. Yoshida, T. Jitsuno, H. Fujita, M. Nakatsuka, M. Yoshimura, T. Sasaki, K. Yoshida: Investigation of bulk laser damage in KDP crystal as a function of laser irradiation direction, polarization and wavelength. Appl. Phys. B, 70(2), 195-201(2000).

[14] *Physical Quantities. Handbook*, ed. by I. S. Grigoriev, E. Z. Meilikhov (Energoatomizdat, Moscow, 1991) [In Russian].

[15] S. S. Ballard, J. S. Browder: Thermal Properties. In: *CRC Handbook of Laser Science and Technology, Vol. IV, Optical Materials: Part 2*, ed. by M. J. Weber (CRC Press, Boca Raton, 1987) pp. 49-54.

[16] A. S. Sonin, A. S. Vasilevskaya: *Electrooptic Crystals* (Atomizdat, Moscow, 1971) [In Russian].

[17] I. P. Kaminow: Tables of Linear Electrooptic Coefficients. In: *CRC Handbook of Laser Science and Technology*, Vol. III, *Optical Materials*: Part 2, ed. by M. J. Weber (CRC Press, Boca Raton, 1986) pp. 253–278.

[18] I. P. Kaminow, G. O. Harding: Complex dielectric constant of KH_2PO_4 at 9.2 Gc/sec. Phys. Rev. 129(4), 1562–1566(1963).

[19] W. R. Cook, Jr.: Thermal expansion of crystals with KH_2PO_4 structure. J. Appl. Phys. 38(4), 1637–1642(1967).

[20] P. Liu, W. L. Smith, H. Lotem, J. H. Bechtel, N. Bloembergen, R. S. Adhav: Absolute two-photon absorption coefficients at 355 and 266 nm. Phys. Rev. B 17(12), 4620–4632(1978).

[21] E. W. van Stryland, L. L. Chase: Two-Photon Absorption. Inorganic Materials. In: *CRC Handbook of Laser Science and Technology*, Supplement 2: *Optical Materials*, ed. By M. J. Weber (CRC Press, Boca Raton, 1995) pp. 299–328.

[22] W. L. Smith: KDP and ADP transmission in the vacuum ultraviolet. Appl. Opt. 16(7), 798(1977).

[23] A. G. Akmanov, S. A. Akhmanov, B. V. Zhdanov, A. I. Kovrigin, N. K. Podsotskaya, R. V. Khokhlov: Generation of coherent radiation at $\lambda = 2120$ Å by cascade frequency conversion. Pisma Zh. Eksp. Teor. Fiz. 10(6), 244–249(1969) [In Russian, English trans.: JETP Lett. 10(6), 154–156(1969)].

[24] M. W. Dowley, E. B. Hodges: Studies of high-power CW and quasi-CW parametric UV generation by ADP and KDP in argon-ion laser cavity. IEEE J. Quant. Electr. QE-4(10), 552–558(1968).

[25] P. Huber: High power in the near ultraviolet using efficient SHG. Opt. Commun. 15(2), 196–200(1975).

[26] E. F. Labuda, A. M. Johnson: Continuous second-harmonic generation of $\lambda = 2572$ Å using Ar^{2+} laser. IEEE J. Quant. Electr. QE-3(4), 164–167(1967).

[27] D. Bruneau, A. M. Tournade, E. Fabre: Fourth harmonic generation of a large-aperture Nd: glass laser. Appl. Opt. 24(22), 3740–3745(1985).

[28] P. J. Wegner, M. A. Henesian, D. R. Speck, C. Bibeau, R. B. Ehrlich, C. W. Laumann, J. K. Lawson, T. L. Weiland: Harmonic conversion of large-aperture 1.05-μm laser beams for inertion-confinement fusion research. Appl. Opt. 31(30), 6414–6426(1992).

[29] A. Yokotani, T. Sasaki, K. Yoshida, S. Nakai: Extremely high damage threshold of a new nonlinear crystal L-arginine phosphate and its deuterium compound. Appl. Phys. Lett. 55(26), 2692–2693(1989).

[30] T. M. Christmas, J. M. Ley: Laser-induced damage in XDP materials. Electron. Lett.

7(18), 544−546(1971).

[31] E. N. Volkova, V. V. Fadeev: Linear absorption coefficient of some nonlinear optical crystals. In: *Nonlinear Optics*, ed. by R. V. Khokhlov (Nauka, Novosibirsk, 1968) pp. 185−187 [In Russian].

[32] C. Chen, Y. X. Fan, R. C. Eckardt, R. L. Byer: Recent developments in barium borate. Proc. SPIE 681, 12−19(1987).

[33] C. A. Ebbers, J. Happe, N. Nielsen, S. P. Velsko: Optical absorption at 1.06 μm in highly deuterated potassium dihydrogen phosphate. Appl. Opt. 31(12), 1960−1964 (1992).

[34] G. Dikchyus, E. Zhilinskas, A. Piskarskas, V. Sirutkaitis: Statistical properties and stabilization of a picosecond phosphate-glass laser with 2 Hz repetition frequency. Kvant. Elektron. 6(8), 1610−1619(1979) [In Russian, English trans.: Sov. J. Quantum Electron. 9(8), 950−955(1979)].

[35] E. E. Fill: Generation of higher harmonics of iodine laser radiation. Opt. Commun. 33 (3), 321−322(1980).

[36] A. Dubietis, G. Tamošauskas, A. Varanavičius, G. Valiulis: Two-photon absorbing properties of ultraviolet phase-matchable crystals at 264 and 211 nm. Appl. Opt. 39 (15), 2437−2440(2000).

[37] G. G. Gurzadyan, R. K. Ispiryan: Two-photon absorption peculiarities of potassium dihydrogen phosphate crystal at 216 nm. Appl. Phys. Lett. 59(6), 630−631(1991).

[38] I. A. Begishev, R. A. Ganeev, A. A. Gulamov, E. A. Erofeev, S. R. Kamalov, T. Usmanov, A. D. Khadzhaev: The neodymium laser fifth harmonic generation and two-photon absorption in KDP and ADP crystals. Kvant. Elektron. 15(2), 353−361(1988) [In Russian, English trans.: Sov. J. Quantum Electron. 18(2), 224−228(1988)].

[39] G. J. Linford, B. C. Johnson, J. S. Hildum, W. E. Martin, K. Snyder, R. D. Boyd, W. L. Smith, C. L. Vercimak, D. Eimerl, J. T. Hunt: Large aperture harmonic conversion experiments at Lawrence Livermore National Laboratory. Appl. Opt. 21(20), 3633−3643 (1982).

[40] G. G. Gurzadyan, R. K. Ispiryan: Two-photon absorption in potassium dihydrophosphate, potassium pentaborate and quartz crystals at 270 and 216 nm. Int. J. Nonl. Opt. Phys. 1(3), 533−540(1992).

[41] F. Zernike, Jr.: Refractive indices of ammonium dihydrogen phosphate and potassium dihydrogen phosphate between 2 000 Å and 1.5 μm. J. Opt. Soc. Am. 54 (10), 1215−1220(1964).

[42] R. A. Philips: Temperature variations of the index of refraction of ADP, KDP, and deuterated KDP. J. Opt. Soc. Am. 56(5), 629−632(1966).

[43] M. Yamazaki, T. Ogawa: Temperature dependences of the refractive indices of

$NH_4H_2PO_4$, KH_2PO_4, and partially deuterated KH_2PO_4. J. Opt. Soc. Am. 56(10), 1407–1408(1966).

[44] N. P. Barnes, D. J. Gettemy, R. S. Adhav: Variations of the refractive index with temperature and the tuning rate for KDP isomorphs. J. Opt. Soc. Am. 72(7), 895–898 (1982).

[45] G. C. Ghosh, G. C. Bhar: Temperature dispersion in ADP, KDP, and KD*P for nonlinear devices. IEEE J. Quant. Electr. QE-18(2), 143–145(1982).

[46] D. Eimerl: Electro-optic, linear and nonlinear optical properties of KDP and its isomorphs. Ferroelectrics 72(1–4), 95–139(1987).

[47] D. A. Roberts: Dispersion equations for nonlinear optical crystals: KDP, $AgGaSe_2$, and $AgGaS_2$. Appl. Opt. 35(24), 4677–4688(1966).

[48] R. A. Ganeev, I. A. Kulagin, A. I. Ryasnyansky, R. I. Tugushev, T. Usmanov: Characterization of nonlinear optical parameters of KDP, $LiNbO_3$ and BBO crystals. Opt. Commun. 229(1–6), 403–412(2004).

[49] W. L. Smith, J. H. Bechtel, N. Bloembergen: Picosecond laser-induced breakdown at 5 321 and 3 547 Å: observation of frequency-dependent behavior. Phys. Rev. B 15(8), 4039–4055(1977).

[50] R. Adair, L. L. Chase, S. A. Payne: Nonlinear refractive index of optical crystals. Phys. Rev. B 39(5), 3337–3350(1989).

[51] D. Milam, M. Weber: Time-resolved interferometric measurements of the nonlinear refractive index in laser materials. Opt. Commun. 18(1), 172–173(1976).

[52] D. Milam, M. J. Weber: Measurement of nonlinear refractive-index coefficients using time-resolved interferometry: application to optical materials for high-power neodymium lasers. J. Appl. Phys. 47(6), 2497–2501(1976).

[53] W. L. Smith, J. H. Bechtel, N. Bloembergen: Dielectric-breakdown threshold and nonlinear-refractive-index measurements with picosecond laser pulses. Phys. Rev. B 12(2), 706–714(1975).

[54] R. Onaka, H. Ito: Pockels effect of KDP and ADP in the ultraviolet region. J. Phys. Soc. Japan 41(4), 1303–1309(1976).

[55] O. G. Vlokh: Dispersion of electro-optic coefficient r_{63} in ADP and KDP crystals. Kristallogr. 7(4), 632–633 (1962) [In Russian, English trans.: Sov. Phys. -Crystallogr. 7(4), 509–511(1962)].

[56] J. H. Ott, T. R. Sliker: Linear electro-optic effect in KH_2PO_4 and its isomorphs. J. Opt. Soc. Am. 54(12), 1442–1444(1964).

[57] A. Yariv, P. Yeh: *Optical Waves in Crystals* (John Wiley & Sons, New York, 1984).

[58] R. O'B. Carpenter: The electro-optic effect in uniaxial crystals of the dihydrogen phosphate type. III. Measurements of coefficients. J. Opt. Soc. Am. 40(4), 225–229

(1950).

[59] G. W. C. Kaye, T. H. Laby: *Tables of Physical and Chemical Constants* (Longman Group Ltd., London, 1995).

[60] E. N. Volkova, I. A. Velichko: Electrooptical properties of potassium dihydrogen phosphate crystals having different degrees of deuteration. Kristallogr. 18(2), 409 – 410 (1973) [In Russian, English trans.: Sov. Phys. -Crystallogr. 18(2), 256 – 257 (1973)].

[61] K. Onuki, N. Uchida, T. Saku: Interferometric method for measuring electro-optic coefficients in crystals. J. Opt. Soc. Am. 62(9), 1030 – 1032(1972).

[62] Y. V. Pisarevskii, G. A. Tregubov, Y. V. Shaldin: The electro-optical properties of $NH_4H_2PO_4$, KH_2PO_4 and $N_4(CH_2)_6$ crystals in UHF fields. Fiz. Tverd. Tela 7(2), 661-663(1965) [In Russian, English trans.: Sov. Phys. -Solid State 7(2), 530 – 531 (1965)].

[63] S. Musikant: *Optical Materials. An Introduction to Selection and Application* (Marcel Dekker, Inc., New York, 1985).

[64] R. D. Rosner, E. H. Turner, I. P. Kaminow: Clamped electrooptic coefficients of KDP and quartz. Appl. Opt. 6(4), 778(1967).

[65] E. Munin, A. Balbin Villaverde: Magneto-optical rotatory dispersion of some non-linear crystals. J. Phys.: Condens. Matter 3(27), 5099 – 5106(1991).

[66] M. Koralewski: Dispersion of the Faraday rotation in KDP-type crystals by pulse high magnetic field. Phys. Stat. Solidi A 65(1), K49-K53(1981).

[67] J. L. Dexter, J. Landry, D. G. Cooper, J. Reintjes: Ultraviolet optical isolators utilizing KDP-isomorphs. Opt. Commun. 80(2), 115 – 118(1990).

[68] M. K. Srivastava, R. W. Crow: Raman susceptibility measurements and stimulated Raman effect in KDP. Opt. Commun. 8(1), 82 – 84(1973).

[69] R. C. Eckardt, H. Masuda, Y. X. Fan, R. L. Byer: Absolute and relative nonlinear optical coefficients of KDP, KDA^*P, BaB_2O_4, $LiIO_3$, MgO: $LiNbO_3$, and KTP measured by phase-matched second-harmonic generation. IEEE J. Quant. Electr. 26(5), 922 – 933(1990).

[70] J. E. Midwinter, J. Warner: The effects of phase matching method and of uniaxial crystal symmetry on the polar distribution of second-order non-linear optical polarization. Brit. J. Appl. Phys. 16(11), 1135 – 1142(1965).

[71] W. J. Alford, A. V. Smith: Wavelength variation of the second-order nonlinear coefficients of $KNbO_3$, $KTiOPO_4$, $KTiOAsO_4$, $LiNbO_3$, $LiIO_3$, β-BaB_2O_4, KH_2PO_4, and LiB_3O_5 crystals: a test of Miller wavelength scaling. J. Opt. Soc. Am. B 18(4), 524 – 533(2001).

[72] I. Shoji, T. Kondo, A. Kitamoto, M. Shirane, R. Ito: Absolute scale of second-order

nonlinear-optical coefficients. J. Opt. Soc. Am. B 14(9), 2268 – 2294(1997).

[73] R. J. Gehr, A. V. Smith: Separated-beam nonphase-matched second-harmonic method of characterizing nonlinear optical coefficients. J. Opt. Soc. Am. B 15 (8), 2298 – 2307 (1998).

[74] V. S. Suvorov, A. S. Sonin: Nonlinear optical materials. Kristallogr. 11(5), 832 – 848 (1966) [In Russian, English trans. : Sov. Phys. -Crystallogr. 11 (5), 711 – 723 (1966)].

[75] F. M. Johnson, J. A. Duardo: Infrared detection by parametric up-conversion. Laser Focus 3(6), 31 – 37(1967).

[76] W. F. Hagen, P. C. Magnante: Efficient second-harmonic generation with diffraction-limited and high-spectral-radiance Nd-glass lasers. J. Appl. Phys. 40(1), 219 – 224 (1969).

[77] A. P. Sukhorukov, I. V. Tomov: Tripling of optical frequencies. II. Experimental investigation of a cascade tripler. Opt. Spektrosk. 28(6), 1211 – 1213(1970) [In Russian, English trans. : Opt. Spectrosc. USSR 28(6), 651 – 653(1970)].

[78] Y. Takagi, M. Sumitani, N. Nakashima, K. Yoshihara: Efficient generation of picosecond coherent tunable radiation between 190 and 212 nm by sum frequency mixing from Raman and optical parametric radiations. IEEE J. Quant. Electr. QE – 21(3), 193 – 195(1985).

[79] G. A. Massey, J. C. Johnson: Wavelength-tunable optical mixing experiments between 208 nm and 259 nm. IEEE J. Quant. Electr. QE – 12(11), 721 – 727(1976).

[80] M. Okada, S. Ieiri: Efficiency in the optical mixing between waves at 1.06 μm and 0.53 μm. Jpn. J. Appl. Phys. 10(6), 808(1971).

[81] E. Fill, J. Wildenauer: Generation of the fifth and sixth harmonics of iodine laser pulses. Opt. Commun. 47(6), 412 – 413(1983).

[82] V. I. Bredikhin, V. N. Genkin, S. P. Kuznetsov, M. A. Novikov: 90° phase-matching in $KD_{2x}H_{2(1-x)}PO_4$ crystals upon doubling of the second harmonic of a Nd laser. Pisma Zh. Tekh. Phys. 3(9), 407 – 409 (1977) [In Russian, English trans. : Sov. Tech. Phys. Lett. 3(5), 165 – 166(1977)].

[83] V. I. Bredikhin, G. L. Galushkina, V. N. Genkin, S. P. Kuznetsov: 90° phase-matching upon frequency doubling in $Rb_xK_{(1-x)}H_2PO_4$ crystals. Pisma Zh. Tekh. Phys. 5 (7-8), 505 – 508 (1979) [In Russian, English trans. : Sov. Tech. Phys. Lett. 5 (4), 207 – 208(1977)].

[84] M. D. Jones, G. A. Massey: Milliwatt-level 213 nm source based on a repetitively Q-switched, CW-pumped Nd: YAG laser. IEEE J. Quant. Electr. QE – 15 (4), 204 – 206(1979).

[85] G. A. Massey, M. D. Jones, J. C. Johnson: Generation of pulse bursts at 212.8 nm by

intracavity modulation of an Nd: YAG laser. IEEE J. Quant. Electr. QE – 14(7), 527 – 532(1978).

[86] S. V. Muraviov, A. A. Babin, F. I. Feldstein, A. M. Yurkin, V. A. Kamenskii, A. Y. Malyshev, M. S. Kitai, N. M. Bityurin: Efficient conversion to the fifth harmonic of spatially multimode radiation of a repetitively pulsed Nd: YAP laser. Kvant. Elektron. 25(6), 535 – 536(1998) [In Russian, English trans. : Quantum Electron. 28(6), 520 – 521(1998)].

[87] A. Ashkin, G. D. Boyd, J. M. Dziedzic: Observation of continuous optical harmonic generation with gas lasers. Phys. Rev. Lett. 11(1), 14 – 17(1963).

[88] M. Webb: Temperature sensitivity of KDP for phase-matched frequency conversion of 1 μm laser light. IEEE J. Quant. Electr. 30(8), 1934 – 1942(1994).

[89] U. Deserno, S. Haussühl: Phase-matchable optical nonlinearity in strontium formate and strontium formate dihydrate. IEEE J. Quant. Electr. QE – 9(6), 598 – 601(1973).

[90] M. J. Chu, S. S. Lee: Thermo-optic and electro-optic tuning sensitivities of the second-harmonic generation in KH_2PO_4 crystal measured by diverging beam technique. J. Appl. Phys. 57(7), 2647 – 2649(1985).

[91] R. S. Craxton, S. D. Jacobs, J. E. Rizzo, R. Boni: Basic properties of KDP crystal related to the frequency conversion of 1 μm laser radiation. IEEE J. Quant. Electr. QE – 17(9), 1782 – 1786(1981).

[92] R. B. Andreev, V. D. Volosov, A. G. Kalintsev: Spectral, angular, and temperature characteristics of HIO_3, $LiIO_3$, CDA, DKDP, KDP and ADP non-linear crystals in second and fourth-harmonic generation. Opt. Spektrosk. 37(2), 294 – 299(1974) [In Russian, English trans. : Opt. Spectrosc. USSR 37(2), 169 – 171(1974)].

[93] R. B. Andreev, V. D. Volosov, V. N. Krylov: Temperature stabilization of ADP and KDP crystals in cascade UV generation. Zh. Tekh. Fiz. 47(9), 1977 – 1978(1977) [In Russian, English trans. : Sov. Phys. -Tech. Phys. 22(9), 1146(1977)].

[94] A. Yokotani, T. Sasaki, T. Yamanaka, C. Yamanaka: Temperature dependence of phasematching angle of second and third harmonic generation in type-II KDP crystal. Jpn. J. Appl. Phys. 25(1), 161 – 162(1986).

[95] M. W. Dowley: Parametric fluorescence in ADP and KDP excited by 2 573 Å CW pump. Opto-electron. 1(4), 179 – 181(1969).

[96] R. M. Wood: *Laser Damage in Optical Materials* (Adam Hilger, Bristol, 1986).

[97] M. Runkel, A. K. Burnham: Differences in bulk damage probability distributions between tripler and z-cuts of KDP and DKDP at 355 nm. Proc. SPIE 4347, 408 – 419 (2001).

[98] L. Armstrong, S. E. Neister, R. Adhav: Measuring CFP dye laser damage thresholds on UV doubling crystals. Laser Focus 18(12), 49 – 53(1982).

[99] B. F. Bareika, I. A. Begishev, S. A. Burdulis, A. A. Gulamov, E. A. Erofeev, A. S.

Piskarskas, V. A. Sirutkaitis, T. Usmanov: Highly efficient parametric generation during pumping with high-power subnanosecond pulses. Pisma Zh. Tekh. Phys. 12(2), 186 – 189(1986) [In Russian, English trans. : Sov. Tech. Phys. Lett. 12(2), 78 – 79(1986)].

[100] G. M. Zverev, E. A. Levchuk, E. K. Maldutis: Destruction of KDP, ADP, and LiNbO$_3$ crystals by powerful laser radiation. Zh. Eksp. Teor. Fiz. 57(3), 730 – 736 (1969) [In Russian, English trans. : Sov. Phys. -JETP 30(3), 400 – 403(1970)].

[101] V. D. Volosov, V. N. Krylov, V. A. Serebryakov, D. V. Sokolov: High-efficiency emission of the second and fourth harmonics of high power picosecond pulses. Pisma Zh. Eksp. Teor. Fiz. 19(1), 38 – 41(1974) [In Russian, English trans. : JETP Lett. 19(1), 23 – 25(1974)].

[102] K. P. Burneika, M. V. Ignatavichyus, V. I. Kabelka, A. S. Piskarskas, A. Y. Stabinis: Parametric generation of ultrashort pulses of tunable-frequency radiation. Pisma Zh. Eksp. Teor. Fiz. 16(7), 365 – 367(1972) [In Russian, English trans. : JETP Lett. 16(7), 257 – 258(1972)].

[103] V. Kabelka, A. Kutka, A. Piskarskas, V. Smilgiavichyus, Y. Yasevichyute: Parametric generation of picosecond light pulses with an energy conversion greater than 50%. Kvant. Elektron. 6(8), 1735 – 1739(1979) [In Russian, English trans. : Sov. J. Quantum Electron. 9(8), 1022 – 1024(1979)].

[104] V. D. Volosov, Y. E. Kamach, E. N. Kozlovsky, V. M. Ovchinnikov: The efficient generation of second harmonic of ruby laser radiation. Opt. Mekh. Promyshl. 36 (10), 3 – 4(1969) [In Russian, English trans. : Sov. J. Opt. Technol. 36(5), 656 – 657(1969)].

[105] A. Yokotani, T. Sasaki, K. Yoshida, T. Yamanaka, C. Yamanaka: Improvement of the bulk laser damage threshold of potassium dihydrogen phosphate crystals by ultraviolet irradiation. Appl. Phys. Lett. 48(16), 1030 – 1032(1986).

[106] Y. Nishida, A. Yokotani, T. Sasaki, K. Yoshida, T. Yamanaka, C. Yamanaka: Improvement of the bulk laser damage threshold of potassium dihydrogen phosphate crystal by reducing the organic impurities in growth solution. Appl. Phys. Lett. 52(6), 420 – 421(1988).

[107] C. E. Barker, B. M. van Wonterghem, J. M. Auerbach, R. J. Foley, J. R. Murray, J. H. Campbell, J. A. Caird, D. R. Speck, B. Woods: Design and performance of the Beamlet laser third harmonic frequency converter. Proc. SPIE 2633, 398 – 404(1995).

[108] W. Seka, S. D. Jacobs, J. E. Rizzo, R. Boni, R. S. Craxton: Demonstration of high efficiency third harmonic conversion of high power Nd-glass laser radiation. Opt. Commun. 34(3), 469 – 473(1980).

[109] V. D. Volosov, E. V. Nilov: Effect of the spatial structure of a laser beam on the generation of the second harmonic in ADP and KDP crystals. Opt. Spektrosk. 21(6), 715 – 719

(1966) [In Russian, English trans. : Opt. Spectrosc. USSR 21(6) ,392 – 394(1966)].

[110] R. Y. Orlov, I. B. Skidan, L. S. Telegin: Investigation of breakdown in dielectrics produced by ultrashort laser pulses. Zh. Eksp. Teor. Fiz. 61(2), 784 – 790(1971) [In Russian, English trans. : Sov. Phys. -JETP 34(2) ,418 – 421(1972)].

[111] S. A. Akhmanov, I. A. Begishev, A. A. Gulamov, E. A. Erofeev, B. V. Zhdanov, V. I. Kuznetsov, L. N. Rashkovich, T. V. Usmanov: Highly-efficient parametric frequency conversion of light in large-aperture crystals grown by a fast method. Kvant. Elektron. 11(9), 1701 – 1702 (1984) [In Russian, English trans. : Sov. J. Quantum Electron. 14(9) ,1145 – 1146(1984)].

[112] J. E. Swain, S. E. Stokowski, D. Milam, G. C. Kennedy: The effect of baking and pulsed laser irradiation on the bulk laser damage threshold of potassium dihydrogen phosphate crystals. Appl. Phys. Lett. 41(1), 12 – 14(1982).

[113] R. M. Wood, R. T. Taylor, R. L. Rouse: Laser damage in optical materials at 1.06 μm. Opt. Laser Technol. 7(3), 105 – 111(1975).

[114] M. Bass: Nd: YAGlaser-irradiation-induced damage to $LiNbO_3$ and KDP. IEEE J. Quant. Electr. QE – 7(7), 350 – 359(1971).

[115] C. Chen: Chinese lab grows new nonlinear optical borate crystals. Laser Focus World 25(11), 129 – 137(1989).

[116] M. Yoshimura, T. Kamimura, K. Murase, Y. Mori, H. Yoshida, M. Nakatsuka, T. Sasaki: Bulk laser damage in $CsLiB_6O_{10}$ crystal and its dependence on crystal structure. Jpn. J. Appl. Phys. 38(2A), L129-L131(1999).

[117] D. Eimerl, S. Velsko, L. Davis, F. Wang, G. Loiacono, G. Kennedy: Deuterated L-arginine phosphate: a new efficient nonlinear crystal. IEEE J. Quant. Electr. 25(2), 179 – 193(1989).

[118] C. Chen, Y. Wu, A. Jiang, B. Wu, G. You, R. Li, S. Lin: New nonlinear-optical crystal: LiB_3O_5. J. Opt. Soc. Am. B 6(4), 616 – 621(1989).

[119] F. Rainer, W. H. Lowdermilk, D. Milam: Bulk and surface damage thresholds of crystals and glasses at 248 nm. Opt. Eng. 22(4), 431 – 434(1983).

[120] M. Bass, H. H. Barrett: Avalanche breakdown and the probabilistic nature of laser-induced damage. IEEE J. Quant. Electr. QE – 8(3), 338 – 343(1972).

[121] L. N. Rashkovich: *KDP-family Single Crystals*(Adam Hilger, Bristol, 1991).

[122] J. J. DeYoreo, A. K. Burnham, P. K. Whitman: Developing KH_2PO_4 and KD_2PO_4 crystals for the world's most powerful laser. Int. Mater. Rev. 47(3), 113 – 152(2002).

[123] N. Zaitseva, L. Carman, I. Smolsky, R. Torres, M. Yan: The effect of impurities and supersaturation on the rapid growth of KDP crystals. J. Cryst. Growth 204(4), 512 – 524(1999).

[124] N. Zaitseva, J. Atherton, L. Carman, M. Runkel, R. Ryon, I. Smolsky, H. Spears, R. Torres, M. Yan: Rapid growth of KDP and DKDP crystals: the connection between growth quality conditions and crystal quality. Nonl. Opt. 23(3-4), 269-284(2000).

[125] N. Zaitseva, L. Carman: Rapid growth of KDP-type crystals. Progr. Cryst. Growth Character. Mater. 43(1), 1-118(2001).

4.2 $NH_4H_2PO_4$, 磷酸二氢铵(ADP)

负单轴晶: $n_o > n_e$

分子量: 115.026

密度: 1.798 g/cm^3 [1]; 1.799 g/cm^3 [2]; 在 293 K[3] 时, 1.803 g/cm^3

点群: $\bar{4}2m(222, T < 125$ K[4]$)$

晶格常数($\bar{4}2m$):

$a = 7.495$ Å[5]; 7.510 Å[6]; 7.50 Å[4]; 在 $T = 293$K[7] 时, (7.4991 ± 0.0004)Å

$c = 7.548$ Å[5], 7.564 Å[6]; 7.58 Å[4]; 在 $T = 293$ K[7] 时, (7.5493 ± 0.0012)Å

莫氏硬度: 1[8]; 2[9]

在 100 g 水中的溶解度

T/K	s/g	参 考 文 献	T/K	s/g	参 考 文 献
273	22.7	[3]	373	173.2	[3]
293	36.8	[8]			

熔点: 463 K[10]

居里温度: 147 K[7]; 148 K[11,12]

线性热膨胀系数[4]

T/K	$\alpha_t \times 10^6$/K^{-1}, $\parallel c$	$\alpha_t \times 10^6$/K^{-1}, $\perp c$
293	4	37

线性热膨胀系数平均值[7]

T/K	$\alpha_t \times 10^6$/K^{-1}, $\parallel c$	$\alpha_t \times 10^6$/K^{-1}, $\perp c$
148~298	10.7	27.2
223~323	4.2	32.0

$p = 0.101325$ MPa 时的比热容 c_p [13]

T/K	c_p/(J·kg^{-1}·K^{-1})	T/K	c_p/(J·kg^{-1}·K^{-1})
80	405	298	1 236
250	1 088		

热导率[3]

T/K	κ/(W·m^{-1}·K^{-1}), $\parallel c$	κ/(W·m^{-1}·K^{-1}), $\perp c$
315	0.71	1.26
340	0.71	1.34

室温下带隙能量：$E_g = 6.8$ eV[14]；6.81 eV[15]

以 $\alpha = 1$ cm^{-1} 计的透明范围：0.184 ~ 1.3 μm[11,16]

对 0.2 cm 长晶体以 0.5 透过计的透明范围：0.184 ~ 1.5 μm[2]

线性吸收系数 α

λ/μm	α/cm^{-1}	参考文献	备注
0.257 25	0.01	[17]	
	0.002	[18]	e 光，$\perp c$
0.265	0.07	[19]	e 光，$\perp c$
0.266 1	0.035	[20]	
0.514 5	0.000 5	[17]	
	0.000 05	[18]	o 光，$\perp c$
0.694 3	0.032	[21]	
0.79	0.03	[4]	
0.89	0.038	[4]	
1.027	0.086	[22]	
1.06	0.1	[4]	
1.083	0.208	[22]	
1.144	0.150	[22]	

双光子吸收系数 β

λ/μm	τ_p/ns	$\beta \times 10^{11}$/(cm·W^{-1})	参考文献	备注
0.263 5	0.5	35	[23]	
0.266 1	0.030	6 ± 1	[24]	
		11 ± 3	[20]	
	0.015	24 ± 7	[15]	$\theta = 42°$，$\phi = 45°$
0.308	0.120	23 ± 5	[25]	
0.354 7	0.017	0.68 ± 0.24	[15]	e 光，$\perp c$

$T = 298$ K 时折射率的实验值[26,27]

$\lambda/\mu m$	n_o	n_e	$\lambda/\mu m$	n_o	n_e
0.213 856 0	1.625 98	1.567 38	0.404 656 1	1.539 69	1.491 59
0.228 801 8	1.607 85	1.551 38	0.407 781 1	1.539 25	1.491 23
0.253 651 9	1.586 88	1.532 89	0.435 835 0	1.535 78	1.488 31
0.296 727 8	1.564 62	1.513 39	0.491 603 6		1.483 90
0.302 149 9	1.562 70	1.511 63	0.546 074 0	1.526 62	1.480 79
0.312 566 3	1.559 17	1.508 53	0.576 959 0	1.524 78	1.479 39
0.313 154 5	1.558 97	1.508 32	0.579 065 4	1.524 66	1.479 30
0.334 147 8	1.553 00	1.503 13	0.632 816 0	1.521 95	1.477 27
0.365 014 6	1.546 15	1.497 20	1.013 975 0	1.508 35	1.468 95
0.365 483 3	1.546 08	1.497 12	1.128 704 0	1.504 46	1.467 04
0.366 287 8	1.545 92	1.496 98	1.152 276 0	1.503 64	1.466 66
0.390 641 0	1.541 74				

折射率的温度微商[28]

$\lambda/\mu m$	$dn_o/dT \times 10^5/K^{-1}$	$dn_e/dT \times 10^5/K^{-1}$	$\lambda/\mu m$	$dn_o/dT \times 10^5/K^{-1}$	$dn_e/dT \times 10^5/K^{-1}$
0.405	−4.78	≈0	0.578	−4.60	≈0
0.436	−4.94	≈0	0.633	−5.08	≈0
0.546	−5.23	≈0			

从室温冷却到 $T[K]$ 时折射率的温度关系:

对于光谱范围 $0.365 \sim 0.690 \mu m$[29]:

$$n_o(T) = n_o(298) + 0.713 \times 10^{-2}\{[n_o(298)]^2 - 3.029\,7[n_o(298)] + 2.300\,4\} \times (298 - T)$$

$$n_e(T) = n_e(298) + 0.675 \times 10^{-6}[n_e(298)]^2(298 - T)$$

对于光谱范围 $0.436 \sim 0.589 \mu m$[30]:

$$n_o(T) = n_o(300) + 10^{-4}(141.8 - 0.322T - 5.02 \times 10^{-4}T^2)$$

$$n_e(T) = n_e(300) - 10^{-4}(2.5 - 0.017\,63T + 2.901 \times 10^{-5}T^2)$$

最佳色散关系方程组(λ 以 μm 为单位, $T = 293$ K)[26,27]:

$$n_o^2 = 2.302\,842 + \frac{15.102\,464\lambda^2}{\lambda^2 - 400} + \frac{0.011\,125\,165}{\lambda^2 - (75.450\,861)^{-1}}$$

$$n_e^2 = 2.163\,510 + \frac{5.919\,896\lambda^2}{\lambda^2 - 400} + \frac{0.009\,616\,676}{\lambda^2 - (76.987\,51)^{-1}}$$

文献[28]、[31]、[32]、[33]中给出了其他色散关系。

Sellmeier 方程的温度关系（λ 以 μm 为单位，T 以 K 为单位）：[31]

$$n_o^2 = (1.6996 - 8.7835 \times 10^{-4} T) + \frac{(0.64955 + 7.2007 \times 10^{-4} T)\lambda^2}{\lambda^2 - (0.01723 - 1.40526 \times 10^{-5} T)}$$

$$+ \frac{(1.10624 - 1.179 \times 10^{-4} T)\lambda^2}{\lambda^2 - 30}$$

$$n_e^2 = (1.42036 - 1.089 \times 10^{-5} T) + \frac{(0.74453 + 5.14 \times 10^{-6} T)\lambda^2}{\lambda^2 - (0.013 - 2.471 \times 10^{-7} T)}$$

$$+ \frac{(0.42033 - 9.99 \times 10^{-7} T)\lambda^2}{\lambda^2 - 30}$$

室温下低频（远低于 ADP 晶体声学共振频率，即对"自由"晶体）下测量的线性电光系数

$\lambda/\mu m$	$r_{41}^T/(pm \cdot V^{-1})$	$r_{63}^T/(pm \cdot V^{-1})$	参 考 文 献	备 注
0.20		-8.5	[34]	$T = 286$ K
0.25		-8.6	[34]	$T = 286$ K
0.30		-8.2	[34]	$T = 286$ K
0.488	-22.9 ± 0.2	-8.07 ± 0.1	[35]	$T = 298$ K
0.500		-7.8	[36]	
0.5461		-8.56	[37]	
	-23.76		[37]	
	-24.5 ± 0.4		[38]	
0.556		-8.47 ± 0.17	[39]	$T = 295$ K
	-20.8 ± 0.3(?)		[39]	$T = 295$ K
0.61 ~ 1.15		-8.55 ± 0.15	[40]	
0.6328		-7.83	[37]	
		-7.9	[36]	
	-22.2 ± 0.2	-8.14 ± 0.1	[35]	$T = 298$ K
	-23.1 ± 0.3		[41]	$T = 293$ K
	-23.41		[37]	

室温下高频（远高于 ADP 晶体的声学共振频率，即对"受夹"晶体）下测量的线性电光系数

$\lambda/\mu m$	$r_{63}^S/(pm \cdot V^{-1})$	参 考 文 献	$\lambda/\mu m$	$r_{63}^S/(pm \cdot V^{-1})$	参 考 文 献
0.5461	-4.1 ± 0.4(?)	[42]		-5.5	[44]
	-5.1 ± 1.5	[43]			

纵向调制下的半波电压

$\lambda/\mu m$	$V_{\lambda/2}/kV$	参 考 文 献	$\lambda/\mu m$	$V_{\lambda/2}/kV$	参 考 文 献
0.488	8.4	[35]	0.6	10.0	[39]
0.5	8.27	[39]	0.632 8	11.0	[35]
0.546 1	9.0	[44]			

Verdet 常数($\parallel c$)

$\lambda/\mu m$	T/K	$V/[(°) \cdot T^{-1} \cdot m^{-1}]$	参 考 文 献
0.632 8	293	251 ± 5	[1]
	298	230	[45]

计算的 Verdet 常数($\parallel c$)[46]

$\lambda/\mu m$	$V/[(°) \cdot T^{-1} \cdot m^{-1}]$	$\lambda/\mu m$	$V/[(°) \cdot T^{-1} \cdot m^{-1}]$
0.193	4 023	0.308	1 061
0.222	2 573	0.351	781
0.248	1 858		

普遍情况下有效二阶非线性系数的表达式(Kleinman 对称条件成立时,$d_{14} = d_{25} = d_{36}$)[47]:

$$d_{ooe} = -d_{36}\sin(\theta+\rho)\sin 2\phi$$
$$d_{eoe} = d_{oee} = 2d_{36}\sin(\theta+\rho)\cos(\theta+\rho)\cos 2\phi$$

有效二阶非线性系数的简化表达式(小双折射角近似,Kleinman 对称条件成立时,$d_{14} = d_{25} = d_{36}$)[48]:

$$d_{ooe} = -d_{36}\sin\theta\sin 2\phi$$
$$d_{eoe} = d_{oee} = d_{36}\sin 2\theta\cos 2\phi$$

二阶非线性系数的绝对值:

$$d_{36}(0.632\ 8\ \mu m) = (0.55 \pm 0.02)\ pm/V^{[49]}$$
$$d_{36}(1.064\ 2\ \mu m) = (0.46 \pm 0.03)\ pm/V^{[49]};\ 0.47\ pm/V^{[50]}$$

相位匹配角的实验值($T = 293$ K)

相互作用的波长/μm	$\theta_{exp}/(°)$	参 考 文 献
SHG, o + o⇒e		
0.524⇒0.262	90	[28]
0.530⇒0.265	81.7	[51]
0.694 3⇒0.347 15	51.9	[52]
0.703 5⇒0.351 75	50.5	[53]
1.06⇒0.53	41.9	[52]
	42	[54]

续表

相互作用的波长/μm	$\theta_{exp}/(°)$	参 考 文 献
SHG, o + o ⇒ e		
1.064 2 + 0.532 1 ⇒ 0.354 73	46.9	[55]
1.064 2 + 0.281 0 ⇒ 0.222 30	90	[56]
0.812 19 + 0.347 15 ⇒ 0.243 20	90	[57]
SHG, e + o ⇒ e		
1.064 2 + 0.532 1 ⇒ 0.354 73	60.2	[55]

NCPM 温度的实验值

相互作用的波长/μm	$T/℃$	参 考 文 献	备 注
SHG, o + o ⇒ e			
0.492 0 ⇒ 0.246 0	-116	[58]	
0.496 5 ⇒ 0.248 25	-93.2	[59]	
0.501 7 ⇒ 0.250 85	-68.4	[59]	
0.514 5 ⇒ 0.257 25	-11.7	[60]	
	-10.2	[59]	
	-9.2	[18]	
0.524 ⇒ 0.262	20	[28]	
0.525 34 ⇒ 0.262 67	30	[61]	
0.53 ⇒ 0.265	43	[19]	
	47	[51]	
	48	[62]	
	49.6	[63]	
0.532 1 ⇒ 0.266 05	47.1	[64]	
	49.5	[65]	
	50	[66]	
	51.2	[67]	0.1 ~ 1 Hz
	44.6	[67]	20 Hz
	51 ~ 52	[68]	
0.539 8 ⇒ 0.269 9	79	[69]	1 ~ 25 Hz
0.548 ⇒ 0.274	100	[61]	
0.557 ⇒ 0.278 5	120	[70]	
SFG, o + o ⇒ e			
1.064 2 + 0.266 05 ⇒ 0.212 84	-55	[71]	

内角带宽和温度带宽的实验值

相互作用的波长/μm	$T/$℃	$\theta_{pm}/(°)$	$\Delta\theta^{int}/(°)$	$\Delta T/$℃	参 考 文 献
SHG, o+o⇒e					
1.06⇒0.53	20	42	0.057		[54]
0.532 1⇒0.266 05	49.5	90		0.60	[65]
	51	90	1.086	0.53	[67]
0.53⇒0.265	20	82	0.118		[72]
	20	82	0.088		[73]
	20	82	0.089	0.63	[51]

光谱带宽的实验值

相互作用的波长/μm	$T/$℃	$\theta_{pm}/(°)$	$\Delta\nu/$cm^{-1}	参 考 文 献
SHG, o+o⇒e				
1.06⇒0.53	20	42	178	[54]
0.53⇒0.265	20	82	4.9	[73]

相位匹配角的温度变化[51]

相互作用的波长/μm	$T/$℃	$\theta_{pm}/(°)$	$d\theta_{pm}/dT/[(°)\cdot K^{-1}]$
SHG, o+o⇒e			
0.53⇒0.265	20	82	0.14
	47	90	1.10

非临界 SHG 的温度调谐[28]

相互作用的波长/μm	$d\lambda_1/dT/(nm\cdot K^{-1})$
SHG, o+o⇒e	
0.524⇒0.262	0.306

非临界 SFG 的温度调谐[74]

相互作用的波长/μm	$d\lambda_3/dT/(nm\cdot K^{-1})$
SHG, o+o⇒e	
0.694 3+0.399 61⇒0.253 63	0.171

非临界 SHG 过程的双折射温度变化

$(0.514\ 5\ \mu m \Rightarrow 0.257\ 25\ \mu m, o+o\Rightarrow e): d(n_2^e-n_1^o)/dT = 5.65\times 10^{-5} K^{-1}$ [60]

相位匹配角和"走离"角的计算值

相互作用的波长/μm	$\theta_{pm}/(°)$	$\rho_1/(°)$	$\rho_3/(°)$
SHG, o+o⇒e			
0.532 1⇒0.266 05	80.15		0.639
0.578 2⇒0.289 1	65.28		1.427
0.632 8⇒0.316 4	56.91		1.703
0.659 4⇒0.329 7	54.07		1.762
0.694 3⇒0.347 15	51.09		1.803
1.064 2⇒0.532 1	41.74		1.746
1.318 8⇒0.659 4	45.45		1.694
SFG, o+o⇒e			
0.578 2+0.510 5⇒0.271 12	74.84		0.955
1.064 2+0.532 1⇒0.354 73	47.82		1.836
1.318 8+0.659 4⇒0.439 6	42.56		1.794
SHG, e+o⇒e			
1.064 2⇒0.532 1	61.39	1.230	1.449
1.318 8⇒0.659 4	65.63	0.968	1.250
SFG, e+o⇒e			
1.064 2+0.532 1⇒0.354 73	59.85	1.272	1.582
1.318 8+0.659 4⇒0.439 6	50.86	1.274	1.748

ADP 中 SHG 过程逆群速失配的计算值

相互作用的波长/μm	$\theta_{pm}/(°)$	$\beta/(\mathrm{fs \cdot mm^{-1}})$	相互作用的波长/μm	$\theta_{pm}/(°)$	$\beta/(\mathrm{fs \cdot mm^{-1}})$
SHG, o+o⇒e			0.6⇒0.3	61.39	233
1.2⇒0.6	43.10	49	SHG, e+o⇒e		
1.1⇒0.55	41.94	21	1.2⇒0.6	62.50	105
1.0⇒0.5	41.71	8	1.1⇒0.55	61.39	78
0.9⇒0.45	42.68	42	1.0⇒0.5	62.02	95
0.8⇒0.4	45.34	85	0.9⇒0.45	65.24	127
0.7⇒0.35	50.67	142	0.8⇒0.4	73.80	173

激光诱导的体损伤阈值

$\lambda/\mu m$	τ_p/ns	$I_{thr}/(\mathrm{GW \cdot cm^{-2}})$	参考文献	备注
0.265	30	>1	[19]	
0.266 1	0.03	>10	[75]	
0.53	10	120~220	[76]	波束腰直径 21 μm
	0.5	>13	[77]	
0.532 1	3	>0.75	[68]	30 Hz

续表

$\lambda/\mu m$	τ_p/ns	$I_{thr}/(GW \cdot cm^{-2})$	参考文献	备注
	0.6	>8	[78]	
	0.03	>8	[24]	
0.539 8	5	6	[69]	SHG 方向
0.6	330	1.8	[79]	
0.694 3	5~25	0.15~0.24	[21]	
1.06	60	0.5	[80]	
	14	10~18	[76]	波束腰直径 100 μm
		70~130	[76]	波束腰直径 21 μm
1.064 2	10	>4.5	[81]	

激光诱导的表面损伤阈值

$\lambda/\mu m$	τ_p/ns	$I_{thr}/(GW \cdot cm^{-2})$	参考文献	备注
0.248 4	20	0.2	[82]	波束腰直径 1.5 mm
0.694 3	5~25	1~5.4	[21]	
1.064 2	12	6.4	[83]	
10	2.2 到 >4.5		[81]	$\parallel c$,波束腰直径 30 μm

关于这一晶体

ADP 是 KDP 的同构晶体,其二阶非线性系数略高于 KDP 晶体。

参考文献

[1] S. Haussühl, W. Effgen: Faraday effect in cubic crystals. Additivity rule and phase transitions. Z. Kristallogr. 183(1-4), 153-174(1988).

[2] Data sheet of Cleveland Crystals Inc. Available at www.clevelandcrystals.com.

[3] E. M. Voronkova, B. N. Grechushnikov, G. I. Distler, I. P. Petrov: *Optical Materials for Infrared Technique*(Nauka, Moscow, 1965) [In Russian].

[4] A. A. Blistanov, V. S. Bondarenko, N. V. Perelomova, F. N. Strizhevskaya, V. V. Tchkalova, M. P. Shaskolskaya: *Acoustic Crystals* (Nauka, Moscow, 1982) [In Russian].

[5] S. Haussühl: Elastische und thermoelastische Eigenschaften von KH_2PO_4, KH_2AsO_4, $NH_4H_2PO_4$, $NH_4H_2AsO_4$ und RbH_2PO_4. Z. Kristallogr. 120(6), 401-414(1964) [In German].

[6] *Handbook of Optical Constants of Solids II*, ed. by E. D. Palik (Academic Press, Boston, 1991).

[7] W. R. Cook, Jr.: Thermal expansion of crystals with KH_2PO_4 structure. J. Appl. Phys.

38(4), 1637 – 1642(1967).

[8] B. H. T. Chai: Optical Crystals. In: *CRC Handbook of Laser Science and Technology*, *Supplement 2: Optical Materials*, ed. by M. J. Weber(CRC Press, Boca Raton, 1995) pp. 3 – 65.

[9] V. G. Dmitriev, G. G. Gurzadyan, D. N. Nikogosyan: *Handbook of Nonlinear Optical Crystals*, Third Revised Edition(Springer, Berlin, 1999).

[10] S. S. Ballard, J. S. Browder: Thermal Properties. In: *CRC Handbook of Laser Science and Technology*, Vol. IV, *Optical Materials: Part 2*, ed. by M. J. Weber (CRC Press, Boca Raton, 1987), pp. 49 – 54.

[11] A. S. Sonin, A. S. Vasilevskaya: *Electrooptic Crystals* (Atomizdat, Moscow, 1971) [In Russian].

[12] I. P. Kaminow: Tables of Linear Electrooptic Coefficients. In: *CRC Handbook of Laser Science and Technology*, Vol. III, *Optical Materials: Part 2*, ed. by M. J. Weber (CRC Press, Boca Raton, 1986), pp. 253 – 278.

[13] *Physical Quantities. Handbook*, ed. by I. S. Grigoriev and E. Z. Meilikhov(Energoatomizdat, Moscow, 1991)[In Russian].

[14] E. W. van Stryland, L. L. Chase: Two-Photon Absorption. Inorganic Materials. In: *CRC Handbook of Laser Science and Technology*, Supplement 2: *Optical Materials*, ed. By M. J. Weber(CRC Press, Boca Raton, 1995), pp. 299 – 328.

[15] P. Liu, W. L. Smith, H. Lotem, J. H. Bechtel, N. Bloembergen, R. S. Adhav: Absolute two-photon absorption coefficients at 355 and 266 nm. Phys. Rev. B 17(12), 4620 – 4632(1978).

[16] W. L. Smith: KDP and ADP transmission in the vacuum ultraviolet. Appl. Opt. 16 (7), 798(1977).

[17] P. Huber: High power in the near ultraviolet using efficient SHG. Opt. Commun. 15 (2), 196 – 200(1975).

[18] M. W. Dowley, E. B. Hodges: Studies of high-power CW and quasi-CW parametric UV generation by ADP and KDP in argon-ion laser cavity. IEEE J. Quant. Electr. QE – 4 (10), 552 – 558(1968).

[19] B. V. Zhdanov, V. V. Kalitin, A. I. Kovrigin, S. M. Pershin: Parametric light generator tunable from 3 980 to 7 920 Å. Pisma Zh. Tekh. Phys. 1(18), 847 – 851 (1975)[In Russian, English trans. : Sov. Tech. Phys. Lett. 1(9), 368 – 369(1975)].

[20] J. Reintjes, R. C. Eckardt: Two-photon absorption in ADP and KD*P at 266.1 nm. IEEE J. Quant. Electr. QE – 13(9), 791 – 793(1977).

[21] T. M. Christmas, J. M. Ley: Laser-induced damage in XDP materials. Electron. Lett. 7 (18), 544 – 546(1971).

[22] E. N. Volkova, V. V. Fadeev: Linear absorption coefficient of some nonlinear optical

crystals. In: *Nonlinear Optics*, ed. by R. V. Khokhlov (Nauka, Novosibirsk, 1968), pp. 185 – 187 [In Russian].

[23] I. A. Begishev, R. A. Ganeev, A. A. Gulamov, E. A. Erofeev, S. R. Kamalov, T. Usmanov, A. D. Khadzhaev: The neodymium laser fifth harmonic generation and two-photon absorption in KDP and ADP crystals. Kvant. Elektron. 15(2), 353 – 361(1988) [In Russian, English trans.: Sov. J. Quantum Electron. 18(2), 224 – 228(1988)].

[24] J. Reintjes, R. C. Eckardt: Efficient harmonic generation from 532 to 266 nm in ADP and KD*P. Appl. Phys. Lett. 30(2), 91 – 93(1977).

[25] Y. P. Kim, M. H. R. Hutchinson: Intensity-induced nonlinear effects in UV window materials. Appl. Phys. B 49(5), 469 – 478(1989).

[26] F. Zernike, Jr. : Refractive indices of ammonium dihydrogen phosphate and potassium dihydrogen phosphate between 2 000 Å and 1.5 μm. J. Opt. Soc. Am. 54 (10), 1215 – 1220(1964).

[27] F. Zernike, Jr. : Refractive indices of ammonium dihydrogen phosphate and potassium dihydrogen phosphate between 2 000 Å and 1.5 μm. Erratum. J. Opt. Soc. Am. 55 (2), 210 – 211(1965).

[28] N. P. Barnes, D. J. Gettemy, R. S. Adhav: Variations of the refractive index with temperature and the tuning rate for KDP isomorphs. J. Opt. Soc. Am. 72(7), 895 – 898(1982).

[29] R. A. Philips: Temperature variations of the index of refraction of ADP, KDP, and deuterated KDP. J. Opt. Soc. Am. 56(5), 629 – 632(1966).

[30] M. Yamazaki, T. Ogawa: Temperature dependences of the refractive indices of $NH_4H_2PO_4$, KH_2PO_4, and partially deuterated KH_2PO_4. J. Opt. Soc. Am. 56 (10), 1407 – 1408(1966).

[31] G. C. Ghosh, G. C. Bhar: Temperature dispersion in ADP, KDP, and KD*P for nonlinear devices. IEEE J. Quant. Electr. QE – 18(2), 143 – 145(1982).

[32] D. Eimerl: Electro-optic, linear and nonlinear optical properties of KDP and its isomorphs. Ferroelectrics 72(1 – 4), 95 – 139(1987).

[33] K. W. Kirby, L. G. DeShazer: Refractive indices of 14 nonlinear crystals isomorphic to KH_2PO_4. J. Opt. Soc. Am. B 4(7), 1072 – 1078(1987).

[34] R. Onaka, H. Ito: Pockels effect of KDP and ADP in the ultraviolet region. J. Phys. Soc. Japan 41(4), 1303 – 1309(1976).

[35] Z. Li, X. Huang, D. Wu, K. Xiong: Large crystal growth and measurement of electrooptical coefficients of ADP. J. Cryst. Growth 222(3), 524 – 527(2001).

[36] O. G. Vlokh: Dispersion of electro-optic coefficient r_{63} in ADP and KDP crystals. Kristallogr. 7(4), 632 – 633(1962) [In Russian, English trans. : Sov. Phys. -Crystallogr. 7 (4), 509 – 511(1962)].

[37] A. Yariv, P. Yeh: *Optical Waves in Crystals* (John Wiley & Sons, NewYork, 1984).

[38] J. H. Ott, T. R. Sliker: Linear electro-optic effect in KH_2PO_4 and its isomorphs. J. Opt. Soc. Am. 54(12), 1442–1444(1964).

[39] R. O'B. Carpenter: The electro-optic effect in uniaxial crystals of the dihydrogen phosphate type. III. Measurements of coefficients. J. Opt. Soc. Am. 40(4), 225–229(1950).

[40] H. Koetser: Measurement of r_{63} for ADP up to electric breakdown. Electron. Lett. 3(2), 54–55(1967).

[41] J. M. Ley: Low-voltage light-amplitude modulation. Electron. Lett. 2(1), 12–13(1966).

[42] L. Silverstein, M. Sucher: Determination of the Pockels electro-optic coefficient in ADP at 5.5 GHz. Electron. Lett. 2(12), 437–438(1966).

[43] Y. V. Pisarevskii, G. A. Tregubov, Y. V. Shaldin: The electro-optical properties of $NH_4H_2PO_4$, KH_2PO_4 and $N_4(CH_2)_6$ crystals in UHF fields. Fiz. Tverd. Tela 7(2), 661-663(1965) [In Russian, English trans. : Sov. Phys. -Solid State 7(2), 530–531(1965)].

[44] S. Musikant: *Optical Materials. An Introduction to Selection and Application* (Marcel Dekker, Inc., NewYork, 1985).

[45] M. Koralewski: Dispersion of the Faraday rotation in KDP-type crystals by pulse high magnetic field. Phys. Stat. Solidi A 65(1), K49-K53(1981).

[46] J. L. Dexter, J. Landry, D. G. Cooper, J. Reintjes: Ultraviolet optical isolators utilizing KDP-isomorphs. Opt. Commun. 80(2), 115–118(1990).

[47] R. C. Eckardt, H. Masuda, Y. X. Fan, R. L. Byer: Absolute and relative nonlinear optical coefficients of KDP, KD*P, BaB_2O_4, $LiIO_3$, MgO: $LiNbO_3$, and KTP measured by phase-matched second-harmonic generation. IEEE J. Quant. Electr. 26(5), 922–933(1990).

[48] J. E. Midwinter, J. Warner: The effects of phase matching method and of uniaxial crystal symmetry on the polar distribution of second-order non-linear optical polarization. Brit. J. Appl. Phys. 16(11), 1135–1142(1965).

[49] K. Hagimoto, A. Mito: Determination of the second-order susceptibility of ammonium dihydrogen phosphate and α-quartz at 633 and 1 064 nm. Appl. Opt. 34(36), 8276–8282(1995).

[50] D. A. Roberts: Simplified characterization of uniaxial and biaxial nonlinear optical crystals: a plea for standardization of nomenclature and conventions. IEEE J. Quant. Electr. 28(10), 2057–2074(1992).

[51] R. B. Andreev, V. D. Volosov, V. N. Krylov: Temperature stabilization of ADP and KDP crystals in cascade UV generation. Zh. Tekh. Fiz. 47(9), 1977–1978(1977) [In Russian, English trans. : Sov. Phys. -Tech. Phys. 22(9), 1146(1977)].

[52] V. S. Suvorov, A. S. Sonin: Nonlinear optical materials. Kristallogr. 11(5), 832 – 848(1966) [In Russian, English trans.: Sov. Phys. -Crystallogr. 11(5), 711 – 723 (1966)].

[53] F. Wondrazek, A. Seilmeier, W. Kaiser: Picosecond light pulses tunable from the violet to near infrared. Appl. Phys. B 32(1), 39 – 42(1983).

[54] W. F. Hagen, P. C. Magnante: Efficient second-harmonic generation with diffraction limited and high-spectral-radiance Nd-glass lasers. J. Appl. Phys. 40(1), 219-224 (1969).

[55] M. Okada, S. Ieiri: Efficiency in the optical mixing between waves at 1.06 μm and 0.53 μm. Jpn. J. Appl. Phys. 10(6), 808(1971).

[56] G. A. Massey, J. C. Johnson: Wavelength-tunable optical mixing experiments between 208 and 259 nm. IEEE J. Quant. Electr. QE – 12(11), 721 – 727(1976).

[57] R. E. Stickel, Jr., F. B. Dunning: Generation of tunable coherent radiation below 250 nm at MW power levels. Appl. Opt. 17(9), 1313 – 1314(1978).

[58] R. K. Jain, T. K. Gustafson: Efficient generation of continuously tunable coherent radiation in the 2 460 – 2 650 Å spectral range. IEEE J. Quant. Electr. QE – 12(9), 555-556(1976).

[59] R. K. Jain, T. K. Gustafson: Second-harmonic generation of several argon-ion laser lines. IEEE J. Quant. Electr. QE – 9(8), 859 – 861(1973).

[60] M. W. Dowley: Parametric fluorescence in ADP and KDP excited by 2 573 Å CW pump. Opto-electron. 1(4), 179 – 181(1969).

[61] R. W. Wallace: Generation of tunable UV from 2 610 to 3 150 Å. Opt. Commun. 4 (4), 316 – 318(1971).

[62] G. V. Venkin, L. L. Kulyuk, D. I. Maleev: Investigation of stimulated Raman scattering in gases excited by fourth harmonic of neodymium laser radiation. Kvant. Elektron. 2 (11), 2475 – 2480(1975) [In Russian, English trans.: Sov. J. Quantum Electron. 5 (11),1348- 1351(1975)].

[63] B. G. Huth, Y. C. Kiang: 90° phase matching for second-harmonic conversion to the ultraviolet. J. Appl. Phys. 40(12), 4976 – 4977(1969).

[64] G. A. Massey, M. D. Jones, J. C. Johnson: Generation of pulse bursts at 212.8 nm by intracavity modulation of an Nd: YAG laser. IEEE J. Quant. Electr. QE – 14(7), 527 – 532(1978).

[65] D. P. Schinke: Generation of ultraviolet light using the Nd: YAG laser. IEEE J. Quant. Electr. QE – 8(2), 86 – 87(1972).

[66] A. H. Kung: Generation of tunable picosecond VUV radiation. Appl. Phys. Lett. 25 (11), 653 – 655(1974).

[67] K. Kato: Conversion of high power Nd: YAG laser radiation to the UV at 2661 Å. Opt.

Commun. 13(4), 361 – 362(1975).

[68] J. M. Yarborough, G. A. Massey: Efficient high-gain parametric generation in ADP continuously tunable across the visible spectrum. Appl. Phys. Lett. 18(10), 438 – 440 (1971).

[69] S. V. Muraviov, A. A. Babin, F. I. Feldstein, A. M. Yurkin, V. A. Kamenskii, A. Y. Malyshev, M. S. Kitai, N. M. Bityurin: Efficient conversion to the fifth harmonic of spatially multimode radiation of a repetitively pulsed Nd: YAP laser. Kvant. Elektron. 25(6), 535 – 536 (1998) [In Russian, English trans. : Quantum Electron. 28(6), 520 – 521 (1998)].

[70] R. S. Adhav: Materials for optical harmonic generation. Laser Focus 19(6), 73 – 78 (1983).

[71] G. A. Massey: Efficient upconversion of long-wavelength UV light into the 200-235 nm band. Appl. Phys. Lett. 24(8), 371 – 373(1974).

[72] V. D. Volosov, V. N. Krylov, V. A. Serebryakov, D. V. Sokolov: High-efficiency emission of the second and fourth harmonics of high power picosecond pulses. Pisma Zh. Eksp. Teor. Fiz. 19(1), 38 – 41(1974) [In Russian, English trans. : JETP Lett. 19(1), 23 – 25(1974)].

[73] R. B. Andreev, V. D. Volosov, A. G. Kalintsev: Spectral, angular, and temperature characteristics of HIO_3, $LiIO_3$, CDA, DKDP, KDP and ADP non-linear crystals in second-and fourth-harmonic generation. Opt. Spektrosk. 37(2), 294 – 299(1974) [In Russian, English trans. : Opt. Spectrosc. USSR 37(2), 169 – 171(1974)].

[74] T. Sato: Continuously tunable ultraviolet radiation at 2 535 Å. J. Appl. Phys. 44(5), 2257 – 2259(1973).

[75] T. A. Rabson, H. J. Ruiz, P. L. Shah, F. K. Tittel: Efficient second harmonic generation of picosecond laser pulses. Appl. Phys. Lett. 20(8), 282 – 284(1972).

[76] G. M. Zverev, E. A. Levchuk, E. K. Maldutis: Destruction of KDP, ADP, and $LiNbO_3$ crystals by powerful laser radiation. Zh. Eksp. Teor. Fiz. 57(3), 730 – 736 (1969) [In Russian, English trans. : Sov. Phys. -JETP 30(3), 400 – 403(1970)].

[77] S. A. Akhmanov, I. A. Begishev, A. A. Gulamov, E. A. Erofeev, B. V. Zhdanov, V. I. Kuznetsov, L. N. Rashkovich, T. V. Usmanov: Highly-efficient parametric frequency conversion of light in large-aperture crystals grown by a fast method. Kvant. Elektron. 11 (9), 1701 – 1702(1984) [In Russian, English trans. : Sov. J. Quantum Electron. 14 (9), 1145 – 1146(1984)].

[78] G. J. Linford, B. C. Johnson, J. S. Hildum, W. E. Martin, K. Snyder, R. D. Boyd, W. L. Smith, C. L. Vercimak, D. Eimerl, J. T. Hunt: Large aperture harmonic conversion experiments at Lawrence Livermore National Laboratory. Appl. Opt. 21(20), 3633 – 3643(1982).

[79] L. Armstrong, S. E. Neister, R. Adhav: Measuring CFP dye laser damage thresholds on UV doubling crystals. Laser Focus 18(12), 49–53(1982).

[80] V. D. Volosov, E. V. Nilov: Effect of the spatial structure of a laser beam on the generation of the second harmonic in ADP and KDP crystals. Opt. Spektrosk. 21(6), 715-719(1966) [In Russian, English trans.: Opt. Spectrosc. USSR 21(6), 392–394 (1966)].

[81] R. M. Wood, R. T. Taylor, R. L. Rouse: Laser damage in optical materials at 1.06 μm. Opt. Laser Technol. 7(3), 105–111(1975).

[82] F. Rainer, W. H. Lowdermilk, D. Milam: Bulk and surface damage thresholds of crystals and glasses at 248 nm. Opt. Eng. 22(4), 431–434(1983).

[83] M. Bass, H. H. Barrett: Avalanche breakdown and the probabilistic nature of laser-induced damage. IEEE J. Quant. Electr. QE–8(3), 338–343(1972).

4.3 KD_2PO_4，氘化磷酸二氢钾(DKDP)

负单轴晶：$n_o > n_e$

分子量：138.098

密度：2.355 g/cm³[1]；2.355 5 g/cm³[2]

点群：$\bar{4}2m$

晶格常数[3,4]：

$a = (7.469\ 7 \pm 0.000\ 3)$ Å，$T = 298$ K

$c = (6.976\ 6 \pm 0.000\ 5)$ Å，$T = 298$ K

莫氏硬度：1.5[5]；2.5[1]

居里温度：(222 ± 1) K[3]；222 K[4]；222 K，99.8% 氘化度[6]；216.3 K，98% 氘化度[6]

线性热膨胀系数[7]

$\alpha_t \times 10^6/K^{-1}$，$\parallel c$	$\alpha_t \times 10^6/K^{-1}$，$\perp c$
44	24.9

线性热膨胀系数平均值[4]

T/K	$\alpha_t \times 10^6/K^{-1}$，$\parallel c$	$\alpha_t \times 10^6/K^{-1}$，$\perp c$
223~298	39.5	19.4
223~323	40.7	20.1

热导率[8]

$\kappa/(\mathrm{W}\cdot\mathrm{m}^{-1}\cdot\mathrm{K}^{-1})$, ∥$c$	$\kappa/(\mathrm{W}\cdot\mathrm{m}^{-1}\cdot\mathrm{K}^{-1})$, ⊥$c$
1.86	2.09

室温下带隙能量：$E_g = 7.0$ eV[9]

以 $\alpha = 1$ cm^{-1} 计的透明范围：≈0.2 μm 到 ≈1.8 μm[3,10]

对 0.2 cm 长晶体以 0.5 透过计的透明范围：<0.2 μm 到 2.15 μm[2]

线性吸收系数 α

λ/μm	α/cm^{-1}	参考文献	备注
0.266 1	0.035	[11]	
0.532 1	0.004~0.005	[12]	98%~99%氘化度
	<0.001	[7]	
0.694 3	<0.004	[13]	80%~95%氘化度
0.82~1.21	<0.015	[14]	
0.94	0.005	[14]	
1.064 2	0.004~0.005	[12]	98%~99%氘化度
	0.012	[7]	o 光
	0.001 9	[15]	o 光，98%氘化度
	0.001 3	[15]	o 光，99.5%氘化度
	<0.001	[7]	e 光
	0.000 4	[15]	e 光，98%氘化度
	0.000 3	[15]	e 光，99.5%氘化度
1.315	0.025	[16]	
1.57	0.1	[17]	o 光，95%氘化度
1.74	0.1	[17]	e 光，95%氘化度

双光子吸收系数 β

λ/μm	τ_p/ns	$\beta\times 10^{11}$/(cm·W^{-1})	参考文献	备注
0.266 1	0.030	2±1	[18]	
		2.7±0.7	[11]	
0.354 7	0.017	0.54±0.19	[19]	e 光，⊥c

$T = 298$ K 时折射率的实验值[20]

λ/μm	n_o	n_e	λ/μm	n_o	n_e
0.404 7	1.518 9	1.477 6	0.435 8	1.515 5	1.474 7
0.407 8	1.518 5	1.477 2	0.491 6	1.511 1	1.471 0

续表

$\lambda/\mu m$	n_o	n_e	$\lambda/\mu m$	n_o	n_e
0.546 1	1.507 9	1.468 3	0.623 4	1.504 4	1.465 6
0.577 9	1.506 3	1.467 0	0.690 7	1.502 2	1.463 9

折射率的温度微商[21]

$\lambda/\mu m$	$dn_o/dT \times 10^5/K^{-1}$	$dn_e/dT \times 10^5/K^{-1}$	$\lambda/\mu m$	$dn_o/dT \times 10^5/K^{-1}$	$dn_e/dT \times 10^5/K^{-1}$
0.405	-3.00	-1.86	0.578	-3.00	-2.52
0.436	-3.37	-2.13	0.633	-3.16	-2.03
0.546	-2.99	-1.95			

从室温冷却至 $T[K]$ 时折射率的温度关系:

对于光谱范围 0.365~0.690 μm[20]:

$$n_o(T) = n_o(298) + 0.228 \times 10^{-4} \{[n_o(298)]^2 - 1.047\}(298 - T)$$

$$n_e(T) = n_e(298) + 0.955 \times 10^{-5} [n_e(298)]^2 (298 - T)$$

对于光谱范围 0.436~0.589 μm[22]:

$$n_o(T) = n_o(300) + 10^{-4}(85.2 - 0.069\,5T - 7.25 \times 10^{-4}T^2)$$

$$n_e(T) = n_e(300) - 10^{-4}(21.8 - 0.445T + 1.24 \times 10^{-3}T^2)$$

最佳色散关系方程组(λ 以 μm 为单位, $T = 293$ K)[23]:

$$n_o^2 = 2.240\,921 + \frac{2.246\,956\lambda^2}{\lambda^2 - (11.265\,91)^2} + \frac{0.009\,676}{\lambda^2 - (0.124\,981)^2}$$

$$n_e^2 = 2.126\,019 + \frac{0.784\,404\lambda^2}{\lambda^2 - (11.108\,71)^2} + \frac{0.008\,578}{\lambda^2 - (0.109\,505)^2}$$

文献[8]、[21]、[24]中给出了其他色散关系方程组。

Sellmeier方程的温度关系(λ 以 μm 为单位, T 以 K 为单位)[24]:

$$n_o^2 = (1.559\,34 + 3.393\,5 \times 10^{-4}T) + \frac{(0.710\,98 - 4.165\,5 \times 10^{-4})\lambda^2}{\lambda^2 - (0.014\,07 + 6.490\,4 \times 10^{-6}T)}$$

$$+ \frac{(0.676\,71 + 4.828\,1 \times 10^{-5}T)\lambda^2}{\lambda^2 - 30}$$

$$n_e^2 = (1.686\,47 + 3.43 \times 10^{-6}T) + \frac{(0.466\,29 - 6.26 \times 10^{-5}T)\lambda^2}{\lambda^2 - (0.016\,63 + 1.362\,6 \times 10^{-6}T)}$$

$$+ \frac{(0.596\,14 + 2.41 \times 10^{-7}T)\lambda^2}{\lambda^2 - 30}$$

室温下低频(远低于DKDP晶体声学共振频率,即对"自由"晶体)下测量的线性电光系数

$\lambda/\mu m$	$r_{41}^T/(pm \cdot V^{-1})$	$r_{63}^T/(pm \cdot V^{-1})$	参 考 文 献	备 注
0.546 1	-8.8 ± 0.4		[25]	83%~92%氘化度
	-8.8		[26]	
		-26.4 ± 0.7	[3]	$T = 295$ K
		-26.8	[26]	
0.500		-25.6 ± 1.3	[27]	90%氘化度
0.632 8	-10.7 ± 0.3		[6]	98%氘化度,$T = 295$ K
		-23.8 ± 0.6	[28]	
		-24.1	[29]	
		-25.8 ± 0.2	[6]	98%氘化度,$T = 295$ K
		-26.4 ± 0.7	[6]	99.8%氘化度,$T = 295$ K

室温下高频(远高于 DKDP 晶体的声学共振频率,即对"受夹"晶体)下测量的线性电光系数

$\lambda/\mu m$	$r_{63}^S/(pm \cdot V^{-1})$	参 考 文 献	备 注
0.500	-24.0 ± 1.2	[27]	90%氘化度
0.632 8	-24.1	[26]	

纵向调制下的半波电压

$\lambda/\mu m$	$V_{\lambda/2}/kV$	参 考 文 献
0.5	2.7	[3]
0.546 1	2.98	[30]

Verdet 常数,$T = 298$ K($\parallel c$)[31]

$\lambda/\mu m$	$V/[(°) \cdot T^{-1} \cdot m^{-1}]$	备 注
0.632 8	237	80%氘化度
	241	85%氘化度
	247	95%氘化度

计算的 Verdet 常数($\parallel c$)[32]

$\lambda/\mu m$	$V/[(°) \cdot T^{-1} \cdot m^{-1}]$	$\lambda/\mu m$	$V/[(°) \cdot T^{-1} \cdot m^{-1}]$
0.193	4 271	0.308	1 185
0.222	2 795	0.351	877
0.248	2 043		

普遍情况下有效二阶非线性系数的表达式(Kleinman 对称条件成立时,$d_{14} = d_{25} = d_{36}$)[33]:

$$d_{ooe} = -d_{36}\sin(\theta+\rho)\sin 2\phi$$

$$d_{eoe} = d_{oee} = 2d_{36}\sin(\theta+\rho)\cos(\theta+\rho)\cos 2\phi$$

有效二阶非线性系数的简化表达式(小双折射角近似,Kleinman 对称条件成立时,$d_{14} = d_{25} = d_{36}$)[34]:

$$d_{ooe} = -d_{36}\sin\theta\sin 2\phi$$

$$d_{eoe} = d_{oee} = d_{36}\sin 2\theta\cos 2\phi$$

二阶非线性系数的绝对值[33,35]:

$$d_{36}(1.064\ 2\ \mu m) = 0.37\ pm/V$$

相位匹配角的实验值($T = 293$ K)

相互作用的波长/μm	$\theta_{exp}/(°)$	参考文献	相互作用的波长/μm	$\theta_{exp}/(°)$	参考文献
SHG, o+o⇒e			1.062⇒0.531	37.1	[39]
0.530⇒0.265	90	[36]	SHG, e+o⇒e		
0.532 1⇒0.266 05	88	[37]	1.315 2⇒0.657 6	51.3	[40]
0.694 3⇒0.347 15	52	[38]			

NCPM 温度的实验值

相互作用的波长/μm	T/℃	参考文献	备注
SHG, o+o⇒e			
0.528⇒0.264	-30	[36]	
0.532 1⇒0.266 05	42	[41]	99%氘化度
	45	[42]	95%氘化度
	46	[43]	99%氘化度
	49.8	[44]	>95%氘化度
	60.8	[45]	90%氘化度
0.536⇒0.268	100	[36]	

内角带宽和温度带宽的实验值

相互作用的波长/μm	T/℃	$\theta_{pm}/(°)$	$\Delta\theta^{int}/(°)$	ΔT/℃	参考文献
SHG, o+o⇒e					
1.064 2⇒0.532 1	20	37	0.081		[33]
0.532 1⇒0.266 05	60.8	90		1.8	[45]
	45	90		1.9	[42]
0.53⇒0.265	25	85.4	0.099		[46]

续表

相互作用的波长/μm	T/℃	θ_{pm}/(°)	$\Delta\theta^{int}$/(°)	ΔT/℃	参考文献
SHG, e+o⇒e					
1.064 2⇒0.532 1	20	54	0.131		[47]
	20		0.126		[48]
1.06⇒0.53	20	60	0.143		[49]

光谱带宽的实验值[49]

相互作用的波长/μm	T/℃	θ_{pm}/(°)	$\Delta\nu$/cm^{-1}
SHG, e+o⇒e			
1.06⇒0.53	20	60	74.8

相位匹配角的温度变化[49]

相互作用的波长/μm	T/℃	θ_{pm}/(°)	$d\theta_{pm}/dT$/[(°)·K^{-1}]
SHG, e+o⇒e			
1.06⇒0.53	20	60	0.006 3

非临界SHG的温度调谐[21]

相互作用的波长/μm	$d\lambda_1/dT$/(nm·K^{-1})
SHG, o+o⇒e	
0.519⇒0.259 5	0.068

相位匹配角及"走离"角的计算值

相互作用的波长/μm	θ_{pm}/(°)	ρ_1/(°)	ρ_3/(°)
SHG, o+o⇒e			
0.532 1⇒0.266 05	86.20		0.225
0.578 2⇒0.289 1	66.87		1.197
0.632 8⇒0.316 4	57.53		1.467
0.659 4⇒0.329 7	54.31		1.522
0.694 3⇒0.347 15	50.86		1.558
1.064 2⇒0.532 1	36.60		1.450
1.318 8⇒0.659 4	36.36		1.412
SFG, o+o⇒e			
0.578 2+0.510 5⇒0.271 12	77.88		0.695
1.064 2+0.532 1⇒0.354 73	46.82		1.587

续表

相互作用的波长/μm	$\theta_{pm}/(°)$	$\rho_1/(°)$	$\rho_3/(°)$
1.318 8 + 0.659 4⇒0.439 6	39.18		1.515
SHG, e+o⇒e			
1.064 2⇒0.532 1	53.47	1.286	1.427
1.318 8⇒0.659 4	51.70	1.222	1.420
SFG, e+o⇒e			
1.064 2 + 0.532 1⇒0.354 73	59.38	1.174	1.378
1.318 8 + 0.659 4⇒0.439 6	47.70	1.254	1.527

DKDP 中 SHG 过程逆群速失配的计算值

相互作用的波长/μm	$\theta_{pm}/(°)$	$\beta/(fs\cdot mm^{-1})$	相互作用的波长/μm	$\theta_{pm}/(°)$	$\beta/(fs\cdot mm^{-1})$
SHG, o+o⇒e			0.6⇒0.3	62.54	218
1.2⇒0.6	35.94	<1	SHG, e+o⇒e		
1.1⇒0.55	36.28	18	1.2⇒0.6	51.62	55
1.0⇒0.5	37.47	38	1.1⇒0.55	52.73	71
0.9⇒0.45	39.79	63	1.0⇒0.5	55.37	92
0.8⇒0.4	43.75	96	0.9⇒0.45	60.41	120
0.7⇒0.35	50.37	143	0.8⇒0.4	70.43	159

激光诱导的体损伤阈值

$\lambda/\mu m$	τ_p/ns	$I_{thr}/(GW\cdot cm^{-2})$	参考文献	备注
0.266 1	0.03	>10	[11]	
0.351	3	>2.9	[50]	80%氘化度
0.355	7.6	3.5	[51]	‖c, 小容器生长块体晶体
		2.2	[52]	‖c, 常规生长, 150 个脉冲
		1.9	[51]	‖c, 大容器生长块体晶体
		1.8	[52]	‖c, 快速生长, 150 个脉冲
		1.9	[51]	$\theta=58°$, 小容器生长块体晶体
		1.4	[52]	$\theta=58°$, 常规生长, 150 个脉冲
		0.9	[51]	$\theta=58°$, 大容器生长块体晶体
		0.9	[52]	$\theta=58°$, 快速生长, 150 个脉冲
0.527	1.7	>0.5	[53]	$\theta=37°$, 89%氘化度
0.532 1	30	>0.05	[54]	
	8	17	[55]	
	0.6	>8	[56]	
	0.03	>8	[18]	

续表

$\lambda/\mu m$	τ_p/ns	$I_{thr}/(GW \cdot cm^{-2})$	参考文献	备 注
0.6	330	0.3	[57]	
0.6943	5~25	0.16~0.26	[13]	
1.062	0.007	>1	[39]	
1.0642	40	>0.25	[54]	
	18	>0.1	[12]	
	14	8	[55]	
	10	1.5~18	[58]	
	1	6	[48]	
	0.25	>3	[12]	
1.315	1	1.5	[40]	

激光诱导的表面损伤阈值

$\lambda/\mu m$	τ_p/ns	$I_{thr}/(GW \cdot cm^{-2})$	参考文献	备 注
0.2484	20	0.15	[59]	波束腰直径 1.5 mm
0.6943	5~25	1.2~5.8	[13]	95%氘化度
1.0642	10	0.7~4.3	[58]	

关于这一晶体

DKDP 是 KDP 的同构晶体,由于氘化,在 IR 区具有更高的透过率。

参考文献

[1] V. G. Dmitriev, G. G. Gurzadyan, D. N. Nikogosyan: *Handbook of Nonlinear Optical Crystals*, *Third Revised Edition* (Springer, Berlin, 1999).

[2] Data sheet of Cleveland Crystals Inc. Available at www.clevelandcrystals.com.

[3] T. R. Sliker, S. R. Burlage: Some dielectric and optical properties of KD_2PO_4. J. Appl. Phys. 34(7), 1837-1840(1963).

[4] W. R. Cook, Jr.: Thermal expansion of crystals with KH_2PO_4 structure. J. Appl. Phys. 38(4), 1637-1642(1967).

[5] B. H. T. Chai: Optical Crystals. In: *CRC Handbook of Laser Science and Technology*, *Supplement 2: Optical Materials*, ed. by M. J. Weber (CRC Press, Boca Raton, 1995), pp. 3-65.

[6] E. N. Volkova, I. A. Velichko: Electrooptical properties of potassium dihydrogen phosphate crystals having different degrees of deuteration. Kristallogr. 18(2), 409-410 (1973) [In Russian, English trans.: Sov. Phys. - Crystallogr. 18(2), 256-257(1973)].

[7] D. Eimerl: High average power harmonic generation. IEEE J. Quant. Electr. QE – 23 (5), 575 – 592(1987).

[8] D. Eimerl: Electro-optic, linear and nonlinear optical properties of KDP and its isomorphs. Ferroelectrics 72(1 – 4), 95 – 139(1987).

[9] E. W. van Stryland, L. L. Chase: Two-Photon Absorption. Inorganic Materials. In: *CRC Handbook of Laser Science and Technology*, Supplement 2: *Optical Materials*, ed. by M. J. Weber(CRC Press, Boca Raton, 1995) pp. 299 – 328.

[10] A. S. Sonin, A. S. Vasilevskaya: *Electrooptic Crystals* (Atomizdat, Moscow, 1971) [In Russian].

[11] J. Reintjes, R. C. Eckardt: Two-photon absorption in ADP and KD*P at 266.1 nm. IEEE J. Quant. Electr. QE – 13(9), 791 – 793(1977).

[12] J. P. Machewirth, R. Webb, D. Anafi: High power harmonics produced with high efficiency in KD*P. Laser Focus: 12(5), 104 – 107(1976).

[13] T. M. Christmas, J. M. Ley: Laser-induced damage in XDP materials. Electron. Lett. 7 (18), 544 – 546(1971).

[14] E. N. Volkova, V. V. Fadeev: Linear absorption coefficient of some nonlinear optical crystals. In: *Nonlinear Optics*, ed. by R. V. Khokhlov(Nauka, Novosibirsk, 1968), pp. 185 – 187[In Russian].

[15] C. A. Ebbers, J. Happe, N. Nielsen, S. P. Velsko: Optical absorption at 1.06 μm in highly deuterated potassium dihydrogen phosphate. Appl. Opt. 31(12), 1960 – 1964 (1992).

[16] G. Brederlow, E. Fill, K. J. Witte: *The High-Power Iodine Laser* (Springer, Berlin, 1983).

[17] G. Dikchyus, E. Zhilinskas, A. Piskarskas, V. Sirutkaitis: Statistical properties and stabilization of a picosecond phosphate-glass laser with 2 Hz repetition frequency. Kvant. Elektron. 6(8), 1610 – 1619 (1979) [In Russian, English trans.: Sov. J. Quantum Electron. 9(8), 950 – 955(1979)].

[18] J. Reintjes, R. C. Eckardt: Efficient harmonic generation from 532 to 266 nm in ADP and KD*P. Appl. Phys. Lett. 30(2), 91 – 93(1977).

[19] P. Liu, W. L. Smith, H. Lotem, J. H. Bechtel, N. Bloembergen, R. S. Adhav: Absolute two photon absorption coefficients at 355 and 266 nm. Phys. Rev. B 17(12), 4620 – 4632(1978).

[20] R. A. Philips: Temperature variations of the index of refraction of ADP, KDP, and deuterated KDP. J. Opt. Soc. Am. 56(5), 629 – 632(1966).

[21] N. P. Barnes, D. J. Gettemy, R. S. Adhav: Variations of the refractive index with temperature and the tuning rate for KDP isomorphs. J. Opt. Soc. Am. 72(7), 895 – 898 (1982).

[22] M. Yamazaki, T. Ogawa: Temperature dependences of the refractive indices of $NH_4H_2PO_4$, KH_2PO_4, and partially deuterated KH_2PO_4. J. Opt. Soc. Am. 56(10), 1407-1408(1966).

[23] K. W. Kirby, L. G. DeShazer: Refractive indices of 14 nonlinear crystals isomorphic to KH_2PO_4. J. Opt. Soc. Am. B 4(7), 1072–1078(1987).

[24] G. C. Ghosh, G. C. Bhar: Temperature dispersion in ADP, KDP, and KD^*P for nonlinear devices. IEEE J. Quant. Electr. QE-18(2), 143–145(1982).

[25] J. H. Ott, T. R. Sliker: Linear electro-optic effect in KH_2PO_4 and its isomorphs. J. Opt. Soc. Am. 54(12), 1442–1444(1964).

[26] A. Yariv, P. Yeh: *Optical Waves in Crystals*(John Wiley & Sons, NewYork, 1984).

[27] T. M. Christmas, C. G. Wildey: Precise pulse-transmission mode control of a ruby laser. Electron. Lett. 6(22), 152–153(1970).

[28] K. Onuki, N. Uchida, T. Saku: Interferometric method for measuring electro-optic coefficients in crystals. J. Opt. Soc. Am. 62(9), 1030–1032(1972).

[29] G. W. C. Kaye, T. H. Laby: *Tables of Physical and Chemical Constants*(Longman Group Ltd., London, 1995).

[30] S. Musikant: *Optical Materials. An Introduction to Selection and Application* (Marcel Dekker, Inc., NewYork, 1985).

[31] M. Koralewski: Dispersion of the Faraday rotation in KDP-type crystals by pulse high magnetic field. Phys. Stat. Solidi A 65(1), K49-K53(1981).

[32] J. L. Dexter, J. Landry, D. G. Cooper, J. Reintjes: Ultraviolet optical isolators utilizing KDP-isomorphs. Opt. Commun. 80(2), 115–118(1990).

[33] R. C. Eckardt, H. Masuda, Y. X. Fan, R. L. Byer: Absolute and relative nonlinear optical coefficients of KDP, KD^*P, BaB_2O_4, $LiIO_3$, MgO: $LiNbO_3$ and KTP measured by phasematched second-harmonic generation. IEEE J. Quant. Electr. 26(5), 922–933(1990).

[34] J. E. Midwinter, J. Warner: The effects of phase matching method and of uniaxial crystal symmetry on the polar distribution of second-order non-linear optical polarization. Brit. J. Appl. Phys. 16(11), 1135–1142(1965).

[35] D. A. Roberts: Simplified characterization of uniaxial and biaxial nonlinear optical crystals: a plea for standardization of nomenclature and conventions. IEEE J. Quant. Electr. 28(10), 2057–2074(1992).

[36] R. S. Adhav: Materials for optical harmonic generation. Laser Focus 19(6), 73–78 (1983).

[37] D. A. V. Kliner, F. Di Teodoro, J. P. Koplow, S. W. Moore, A. V. Smith: Efficient second, third, fourth, and fifth harmonic generation of a Yb-doped fiber amplifier. Opt. Commun. 210(3–6), 393–398(2002).

[38] V. S. Suvorov, A. S. Sonin: Nonlinear optical materials. Kristallogr. 11(5), 832 – 848(1966) [In Russian, English trans.: Sov. Phys. -Crystallogr. 11(5), 711 – 723 (1966)].

[39] T. A. Rabson, H. J. Ruiz, P. L. Shah, F. K. Tittel: Efficient second harmonic generation of picosecond laser pulses. Appl. Phys. Lett. 20(8), 282 – 284(1972).

[40] E. E. Fill: Generation of higher harmonics of iodine laser radiation. Opt. Commun. 33 (3), 321 – 322(1980).

[41] M. D. Jones, G. A. Massey: Milliwatt-level 213 nm source based on a repetitively Q-switched, CW-pumped Nd: YAG laser. IEEE J. Quant. Electr. QE-15(4), 204 – 206(1979).

[42] V. I. Bredikhin, V. N. Genkin, S. P. Kuznetsov, M. A. Novikov: 90° phase-matching in $KD_{2x}H_{2(1-x)}PO_4$ crystals upon doubling of the second harmonic of a Nd laser. Pisma Zh. Tekh. Phys. 3(9), 407 – 409(1977) [In Russian, English trans.: Sov. Tech. Phys. Lett. 3(5), 165 – 166(1977)].

[43] G. A. Massey, M. D. Jones, J. C. Johnson: Generation of pulse bursts at 212.8 nm by intracavity modulation of an Nd: YAG laser. IEEE J. Quant. Electr. QE-14(7), 527 – 532(1978).

[44] P. E. Perkins, T. S. Fahlen: Half watt average power at 25 kHz from fourth harmonic of Nd: YAG. IEEE J. Quant. Electr. QE-21(10), 1636 – 1638(1985).

[45] Y. S. Liu, W. B. Jones, J. P. Chernoch: High-efficiency high-power coherent UV generation at 266 nm in 90° phase-matched deuterated KDP. Appl. Phys. Lett. 29(1), 32 – 34(1976).

[46] S. C. Matthews, J. S. Sorce: Fourth harmonic conversion of 1.06 μm in BBO and KD*P. Proc. SPIE 1220, 137 – 147(1990).

[47] R. M. Kogan, T. G. Crow: A 1 J high-brightness frequency-doubled Nd: YAG laser. Appl. Opt. 17(6), 927 – 930(1978).

[48] R. S. Adhav, S. R. Adhav, J. M. Pelaprat: BBO's nonlinear optical phase-matching properties. Laser Focus 23(9), 88 – 100(1987).

[49] R. B. Andreev, V. D. Volosov, A. G. Kalintsev: Spectral, angular, and temperature characteristics of HIO_3, $LiIO_3$, CDA, DKDP, KDP and ADP non-linear crystals in second-and fourth-harmonic generation. Opt. Spektrosk. 37(2), 294 – 299(1974) [In Russian, English trans.: Opt. Spectrosc. USSR 37(2), 169 – 171(1974)].

[50] C. E. Barker, B. M. van Wonterghem, J. M. Auerbach, R. J. Foley, J. R. Murray, J. H. Campbell, J. A. Caird, D. R. Speck, B. Woods: Design and performance of the Beamlet laser third harmonic frequency converter. Proc. SPIE 2633, 398 – 404(1995).

[51] M. Runkel, A. K. Burnham: Differences in bulk damage probability distributions between tripler and z-cuts of KDP and DKDP at 355 nm. Proc. SPIE 4347, 408 – 419

(2001).

[52] A. K. Burnham, M. Runkel, M. D. Feit, A. M. Rubenchik, R. L. Floyd, T. A. Land, W. J. Siekhaus, R. A. Hawley-Fedder: Laser-induced damage in deuterated potassium dihydrogen phosphate. Appl. Opt. 42(27), 5483 – 5495(2003).

[53] G. Freidman, N. Andreev, V. Bespalov, V. Bredikhin, V. Ginzburg, E. Katin, E. Khazanov, A. Korytin, V. Lozhkarev, O. Palashov, A. Poteomkin, A. Sergeev, I. Yakovlev: Multicascade broadband optical parametric chirped pulse amplifier based on KD*P crystals. Proc. SPIE 4972, 90 – 101(2003).

[54] Y. S. Liu, W. B. Jones, J. P. Chernoch: High-efficiency high-power coherent UV generation at 266 nm in 90° phase-matched deuterated KDP. Appl. Phys. Lett. 29(1), 32 – 34(1976).

[55] H. Nakatani, W. R. Bosenberg, L. K. Cheng, C. L. Tang: Laser-induced damage in beta barium metaborate. Appl. Phys. Lett. 53(12), 2587 – 2589(1988).

[56] G. J. Linford, B. C. Johnson, J. S. Hildum, W. E. Martin, K. Snyder, R. D. Boyd, W. L. Smith, C. L. Vercimak, D. Eimerl, J. T. Hunt: Large aperture harmonic conversion experiments at Lawrence Livermore National Laboratory. Appl. Opt. 21(20), 3633 – 3643(1982).

[57] L. Armstrong, S. E. Neister, R. Adhav: Measuring CFP dye laser damage thresholds on UV doubling crystals. Laser Focus 18(12), 49 – 53(1982).

[58] R. M. Wood, R. T. Taylor, R. L. Rouse: Laser damage in optical materials at 1.06 μm. Opt. Laser Technol. 6(3), 105 – 111(1975).

[59] F. Rainer, W. H. Lowdermilk, D. Milam: Bulk and surface damage thresholds of crystals and glasses at 248 nm. Opt. Eng. 22(4), 431 – 434(1983).

4.4 $CsLiB_6O_{10}$，硼酸锂铯(CLBO)

负单轴晶：$n_o > n_e$

分子量：364.706

密度：2.461 g/cm³(计算值)[1]；2.472 g/cm³(计算值)[2]

点群：$\bar{4}2m$

晶格常数[3]：

$a = (10.494 \pm 0.001)$ Å

$c = (8.939 \pm 0.002)$ Å

维氏硬度：

140 ~ 170(沿[001]方向)[3]

230~260(沿[100]方向)[3]
270(对具有高的体激光损伤阈值晶体)[4]
莫氏硬度：5.5[5]
熔点：1 118 K[6]；1 121 K[7]
带隙能量：6.9 eV[8]
以"0"透过计的透明范围：0.18~2.75 μm[9]
双光子吸收系数 β [10]

$\lambda/\mu m$	τ_p/ns	$\beta \times 10^{11}/(cm \cdot W^{-1})$
0.2	0.000 14	120 ± 20

在 $T = 293$ K 时折射率的实验值[11]

$\lambda/\mu m$	n_o	n_e	$\lambda/\mu m$	n_o	n_e
0.420	1.505 8	1.451 7	0.610	1.493 5	1.441 4
0.450	1.503 0	1.449 3	0.632 8	1.492 5	1.440 9
0.480	1.500 6	1.447 4	0.670	1.491 5	1.439 8
0.500	1.499 1	1.446 2	0.700	1.490 7	1.439 2
0.532	1.497 1	1.444 5	0.720	1.490 2	1.438 7
0.560	1.495 7	1.443 4	1.064	1.483 8	1.434 0
0.590	1.494 3	1.442 2			

光谱范围 0.212 8~1.338 2 μm 和温度范围 293~373 K 间折射率的温度微商 (以 10^{-6} K^{-1} 为单位)[12]：

$$\frac{dn_o}{dT} = -12.48 - \frac{0.328}{\lambda}$$

$$\frac{dn_e}{dT} = -8.36 + \frac{0.047}{\lambda} - \frac{0.039}{\lambda^2} + \frac{0.014}{\lambda^3}$$

在文献[13]中给出了折射率温度微商的其他表达式。

最佳色散关系方程组（$T = 293$ K，λ 以 μm 为单位，0.191 4 μm < λ < 2.09 μm）[12]：

$$n_o^2 = 2.210\ 4 + \frac{0.010\ 18}{\lambda^2 - 0.014\ 24} - 0.012\ 58\lambda^2$$

$$n_e^2 = 2.058\ 8 + \frac{0.008\ 38}{\lambda^2 - 0.013\ 63} - 0.006\ 07\lambda^2$$

在文献[7]、[9]、[11]、[13]、[14]、[15]中给出了其他色散关系方程组。
普遍情况下有效二阶非线性系数的表达式（Kleinman 对称条件成立时，$d_{14} =$

$d_{25} = d_{36})^{[16]}$:

$$d_{ooe} = -d_{36}\sin(\theta+\rho)\sin 2\phi$$
$$d_{eoe} = d_{oee} = 2d_{36}\sin(\theta+\rho)\cos(\theta+\rho)\cos 2\phi$$

有效二阶非线性系数的简化表达式(小双折射角近似,Kleinman 对称条件成立时,$d_{14} = d_{25} = d_{36})^{[17]}$:

$$d_{ooe} = -d_{36}\sin\theta\sin 2\phi$$
$$d_{eoe} = d_{oee} = d_{36}\sin 2\theta\cos 2\phi$$

二阶非线性系数的绝对值[18]:

$$d_{36}(0.532\ \mu m) = 0.92\ pm/V$$
$$d_{14}(0.852\ \mu m) = 0.69\ pm/V$$
$$d_{36}(0.852\ \mu m) = 0.83\ pm/V$$
$$d_{36}(1.064\ \mu m) = 0.74\ pm/V$$

NCPM 温度的实验值

相互作用的波长/μm	T/℃	参考文献
SFG, o + o ⇒ e		
1.047 + 0.241 14 ⇒ 0.196	4	[19]
1.047 + 0.241 6 ⇒ 0.196 3	34	[20]
1.047 + 0.243 1 ⇒ 0.197 3	150	[19]

相位匹配温度带宽、内角带宽和温度带宽的实验值

相互作用的波长/μm	$\theta_{pm}/(°)$	T/℃	$\Delta\theta^{int}/(°)$	ΔT/℃	参考文献
SHG, o + o ⇒ e					
0.946 ⇒ 0.473	90	−15		5.0	[21]
0.523 5 ⇒ 0.261 75	64.8	≈160			[22]
0.532 1 ⇒ 0.266 05	62	≈140			[6]
	61.4	20	0.023		[11]
				6.2	[12]
1.064 2 ⇒ 0.532 1	29.5	20	0.043		[11]
				52.7	[12]
1.338 2 ⇒ 0.669 1	27.7	20		68.7	[12]
SFG, o + o ⇒ e					
1.064 2 + 0.266 05 ⇒ 0.212 84	67.3	20		3.6	[12]
1.547 + 0.221 ⇒ 0.193 38	61.7	150			[23]
1.907 9 + 0.212 8 ⇒ 0.191 4	55	20		1.2	[12]
1.064 2 + 0.354 73 ⇒ 0.266 05	50.6	20		6.1	[12]
1.064 2 + 0.532 1 ⇒ 0.354 73	39.1	20		18.0	[12]

续表

相互作用的波长/μm	$\theta_{pm}/(°)$	$T/℃$	$\Delta\theta^{int}/(°)$	$\Delta T/℃$	参考文献
SHG, e + o ⇒ e					
1.064 2 ⇒ 0.532 1	42.4	20		49.4	[12]
SFG, e + o ⇒ e					
1.907 9 + 0.212 8 ⇒ 0.191 4	57.4	20		1.1	[12]
1.064 2 + 0.532 1 ⇒ 0.354 73	48.9	20		17.0	[12]

激光诱导的体损伤阈值

$\lambda/\mu m$	τ_p/ns	$I_{thr}/(GW \cdot cm^{-2})$	参考文献	备 注
0.2	0.000 14	>250	[10]	1 kHz
0.266	8	17~19	[4]	溶液搅拌生长
	0.75	6.4	[8]	
	0.75	9~10	[24]	位错密度 ~1.5×10^4 cm^{-3}
	0.75	15~20	[24]	位错密度 $(0.7~1) \times 10^4$ cm^{-3}
	0.75	25	[25]	溶液搅拌 TSSG 生长
0.511	20	>0.5	[26]	12 kHz
0.527	0.001 5	>47	[27]	1/6 Hz
0.532	70	>0.043	[28]	1 kHz
	7	>0.13	[29]	10 Hz
	0.014	130~520	[30]	80 个脉冲列
0.539 5	7	>0.67	[31]	10 Hz
0.576	8	>0.1	[32]	10 Hz
0.800	CW	>0.000 003 8	[33]	
	0.001 4	>600	[10]	1 kHz
1.053	0.001 5	>100	[27]	
1.064	CW	0.000 088	[11]	
	13	>0.35	[34]	10 Hz
	7	>0.37	[35]	10 Hz
	1.1	16~19	[8]	沿[100]方向
	1.1	29	[8]	沿[001]方向

激光诱导的表面损伤阈值

$\lambda/\mu m$	τ_p/ns	$I_{thr}/(GW \cdot cm^{-2})$	参考文献	备 注
0.266	8	1.4~1.6	[4]	常规晶体
		2.0	[4]	溶液搅拌生长
		1.3~1.5	[36]	常规晶体,机械抛光
		2.3	[36]	常规晶体,离子束刻
		1.9	[36]	高质量晶体,机械抛光
		2.9	[36]	高质量晶体,离子束刻

关于这一晶体

CLBO 晶体是 1995 年在日本首次采用顶端籽晶泡生法生长的[3,7,9]。通常它也可以采用顶端籽晶溶液生长法(TSSG)或溶液搅拌 TSSG(SS-TSSG)法生长[24,25]。这一紫外非线性光学材料比 BBO 和 LBO 晶体容易生长得多。$CsLiB_6O_{10}$ 晶体的主要应用在于可见及近红外激光到紫外范围的频率转换。在 1996—1998 年间,日本科学家将 CLBO 晶体用于纳秒 Nd:YAG 激光辐射的二次、四次和五次谐波发生。他们用其产生的能量分别为 1.55 J、0.5 J 和 0.23 J 的 532 nm、266 nm 和 213 nm 的脉冲[29,35]。分别在 7 mm 和 5 mm 长的 CLBO 晶体中,对 3 ps、1 053 nm 的激光辐射产生了二次谐波和三次谐波,效率分别为 70% 和 20%[37]。将 CLBO 用于 SHG 和 SFG,可能构建高功率的纳秒紫外光源,以千赫兹重复频率运转,例如,在 266 nm 处为 20 W[38],255 nm 处为 15 W[26],242 nm 处为 3 W[19],以及在 196 nm 处为 1.5 W[19]。中国的研究人员研究了以 CLBO 为基础的 ps UV 激光器泵浦(266 nm 或 355 nm)的 OPG/OPA 系统[15]。尽管他们实现了相当宽的 OPO 调谐范围(对于 266 nm 泵浦为 347 ~ 1 137 nm,对 355 nm 泵浦为 447 ~ 1 725 nm),他们观察到在 450 nm 处 CLBO 晶体的放大因子是 BBO 晶体的 1/7,可能是由于它们有效非线性系数的差别。最近,CLBO 晶体用于 1.547 nm 激光源的八次谐波发生,通过基频波辐射与其七次谐波的 SFG 来实现[23]。

$CsLiB_6O_{10}$ 晶体的缺点之一是由于它在室温下吸收水分而造成的潮解。为了防止这一点,CLBO 晶体必须保持在升到 140 ~ 160 ℃ 温度的状态下[26,38,39],根据这一方法进行的实验表明[39]在一个月时间内 CLBO 倍频器的 SHG 运转稳定,没有衰减。另外一个显著增加 CLBO 晶体倍频效率(2.3 倍)的方法是:通过在升温情况下,用室温的氮气气流冷却,以补偿由热引起的相位失配来[40]。在文献[21]中,CLBO 晶体的 SHG 在 -15 ℃ 时实现。在这一情况下,晶体及其真空密闭容器在高温(150 ℃)下处理,然后窗口再回充干的氮气到大气压强并予以密封。

另一个防止 CLBO 晶体受大气中水分的办法是采用 Si-Cd 膜[31]。首先喷镀晶体,然后在炉子中,在常压、120 ℃ 条件下,烘干 24 小时。

参考文献

[1] T. Sasaki, Y. Mori, I. Kuroda, S. Nakajima, K. Yamaguchi, S. Watanabe, S. Nakai: Caesium lithium borate: a new nonlinear optical crystal. Acta Crystallogr. C 51(11), 2222 – 2224(1995).

[2] J.-M. Tu, D. A. Keshler: $CsLiB_6O_{10}$: a noncentrosymmetric polyborate. Mat. Res. Bull. 30(2), 209 – 215(1995).

[3] Y. Mori, I. Kuroda, S. Nakajima, T. Sasaki, S. Nakai: Growth of a new nonlinear optical crystal: cesium lithium borate. J. Cryst. Growth 156(3), 307 – 309(1995).

[4] M. Nishioka, S. Fukumoto, F. Kawamura, M. Yoshimura, Y. Mori, T. Sasaki: Improvement of laser-induced damage tolerance in $CsLiB_6O_{10}$ for high-power UV laser source. In: *Conference on Lasers and Electrooptics CLEO/QELS 2003*, Technical Digest (OSA, Washington DC, 2003), paper CTuF2.

[5] N. A. Pylneva, N. G. Kononova, A. M. Yurkin, A. E. Kokh, G. G. Bazarova, V. I. Danilov, I. A. Lisova, N. L. Tsirkina: Top-seeded solution growth of CLBO crystals. Proc. SPIE 3610, 148 – 155(1999).

[6] T. Sasaki, Y. Mori, M. Yoshimura: Progress in the growth of a $CsLiB_6O_{10}$ crystal and its application to ultraviolet light generation. Opt. Mater. 23(1 – 2), 343 – 351(2003).

[7] Y. Mori, I. Kuroda, S. Nakajima, T. Sasaki, S. Nakai: New nonlinear optical crystal: cesium lithium borate. Appl. Phys. Lett. 67(13), 1818 – 1820(1995).

[8] M. Yoshimura, T. Kamimura, K. Murase, Y. Mori, H. Yoshida, M. Nakatsuka, T. Sasaki: Bulk laser damage in $CsLiB_6O_{10}$ crystal and its dependence on crystal structure. Jpn. J. Appl. Phys. 38(2A), L129-L131(1999).

[9] Y. Mori, S. Nakajima, A. Miyamoto, M. Inagaki, T. Sasaki, H. Yoshida, S. Nakai: Generation of ultraviolet light by using new nonlinear optical crystal $CsLiB_6O_{10}$. Proc. SPIE 2633, 299 – 307(1995).

[10] V. Petrov, F. Noack, F. Rotermund, M. Tanaka, Y. Okada: Sum-frequency generation of femtosecond pulses in $CsLiB_6O_{10}$ down to 175 nm. Appl. Opt. 39(27), 5076 – 5079 (2000).

[11] G. Ryu, C. S. Yoon, T. P. J. Han, H. G. Gallagher: Growth and characterisation of $CsLiB_6O_{10}$ (CLBO) crystals. J. Cryst. Growth 191(3), 492 – 500(1998).

[12] N. Umemura, K. Yoshida, T. Kamimura, Y. Mori, T. Sasaki, K. Kato: New data of phasematching properties of $CsLiB_6O_{10}$. In: *Advanced Solid-State Lasers*, OSA Trends in Optics and Photonics Series, Vol. 26, ed. by M. M. Fejer, H. Injeyan, U. Keller(OSA, Washington DC, 1999), pp. 715 – 719.

[13] N. Umemura, K. Kato: Ultraviolet generation tunable to 0.185 μm in $CsLiB_6O_{10}$. Appl. Opt. 36(27), 6794 – 6796(1997).

[14] Y. Mori, I. Kuroda, S. Nakajima, T. Sasaki, S. Nakai: Nonlinear optical properties of cesium lithium borate. Jpn. J. Appl. Phys. 34(3A), L296-L298(1995).

[15] J.-Y. Zhang, Y. Kong, Z. Xu, D. Shen: Optical parametric properties of ultraviolet-pumped cesium lithium borate crystals. Appl. Opt. 41(3), 475 – 482(2002).

[16] R. C. Eckardt, H. Masuda, Y. X. Fan, R. L. Byer: Absolute and relative nonlinear optical coefficients of KDP, KD*P, BaB_2O_4, $LiIO_3$, $MgO:LiNbO_3$, and KTP measured by phase-matched second-harmonic generation. IEEE J. Quant. Electr. 26 (5),

922 – 933(1990).

[17] J. E. Midwinter, J. Warner: The effects of phase matching method and of uniaxial crystal symmetry on the polar distribution of second-order non-linear optical polarization. Brit. J. Appl. Phys. 16(11), 1135 – 1142(1965).

[18] I. Shoji, H. Nakamura, R. Ito, T. Kondo, M. Yoshimura, Y. Mori, T. Sasaki: Absolute measurement of second-harmonic nonlinear-optical coefficients of $CsLiB_6O_{10}$ for visible-to-ultraviolet second-harmonic wavelengths. J. Opt. Soc. Am. B 18 (3), 302-307 (2001).

[19] J. Sakuma, K. Deki, A. Finch, Y. Ohsako, T. Yokota: All-solid-state, high-power, deep-UV laser system based on cascaded sum-frequency mixing in $CsLiB_6O_{10}$ crystals. Appl. Opt. 39(30), 5505 – 5511(2001).

[20] J. Sakuma, A. Finch, Y. Ohsako, K. Deki, M. Yoshino, M. Horiguchi, T. Yokota, Y. Mori, T. Sasaki: All-solid-state, 1-W, 5-kHz laser source below 200 nm. In: Advanced Solid-State Lasers, OSA Trends in Optics and Photonics Series, Vol. 26, ed. by M. M. Fejer, H. Injeyan, U. Keller(OSA, Washington DC, 1999), pp. 89 – 92.

[21] D. C. Gerstenberger, T. M. Trautmann, M. S. Bowers: Noncritically phase-matched secondharmonic generation in cesium lithium borate. Opt. Lett. 28 (14), 1242 – 1244 (2003).

[22] K. F. Wall, J. S. Smucz, B. Pati, Y. Isyanova, P. Moulton, J. G. Manni: A quasi-continuouswave deep ultraviolet laser source. IEEE J. Quant. Electr. 39 (9), 1160 – 1169(2003).

[23] H. Kitano, H. Kawai, K. Miramitsu, S. Owa, M. Yoshimura, Y. Mori, T. Sasaki: 387-nm generation in $Gd_xY_{1-x}Ca_4O(BO_3)_3$ crystal and its utilization for 193-nm light source. Jpn. J. Appl. Phys. 42(2B), L166-L169(2003).

[24] T. Kamimura, R. Ono, Y. K. Yap, M. Yoshimura, Y. Mori, T. Sasaki: Influence of crystallinity on the bulk laser-induced damage threshold and absorption of laser light in $CsLiB_6O_{10}$ crystals. Jpn. J. Appl. Phys. 40(2A), L111-L113(2001).

[25] R. Ono, T. Kamimura, S. Fukumoto, Y. K. Yap, M. Yoshimura, Y. Mori, T. Sasaki, K. Yoshida: Effect of crystallinity on the bulk laser damage and UV absorption of CLBO crystals. J. Cryst. Growth 237 – 239, 645 – 648(2002).

[26] D. J. W. Brown, M. J. Withford: High-average-power (15-W) 255-nm source based on second-harmonic generation of a copper laser master oscillator power amplifier system in cesium lithium borate. Opt. Lett. 26(23), 1885 – 1887(2001).

[27] L. B. Sharma, H. Daido, Y. Kato, S. Nakai, T. Zhang, Y. Mori, T. Sasaki: Fourth-harmonic generation of picosecond glass laser pulses with cesium lithium borate crystals. Appl. Phys. Lett. 69(25), 3812 – 3814(1996).

[28] S. Konno, Y. Inoue, T. Kojima, S. Fujikawa, K. Yasui: Efficient high-pulse-energy

greenbeam generation by intracavity frequency doubling of a quasi-continuous-wave laser-diode-pumped Nd:YAG laser. Appl. Opt. 40(24), 4341 – 4343(2001).

[29] Y. K. Yap, M. Inagaki, S. Nakajima, Y. Mori, T. Sasaki: High-power fourth-and fifth-harmonic generation of a Nd:YAG laser by means of a $CsLiB_6O_{10}$. Opt. Lett. 21(17), 1348 – 1350(1996).

[30] T. Srinivasan-Rao, M. Babzien, F. Sakai, Y. Mori, T. Sasaki: Conversion efficiency and damage threshold measurements of $CsLiB_6O_{10}$ with a train of laser pulses. Appl. Phys. Lett. 71(14), 1927 – 1929(1997).

[31] A. E. Kokh, N. G. Kononova, I. A. Lisova, S. V. Muraviov: $CsLiB_6O_{10}$ crystal: fourth- and fifth-harmonic generation of Nd:YAG laser. Proc. SPIE 4268, 43 – 48(2001).

[32] S. Chandra, T. H. Allik, J. A. Hutchinson, J. Fox, C. Swim: Tunable ultraviolet laser source based on solid-state dye laser technology and $CsLiB_6O_{10}$ harmonic generation. Opt. Lett. 22(4), 209 – 211(1997).

[33] W.-L. Zhou, Y. Mori, T. Sasaki, S. Nakai: High-efficiency intracavity continuous-wave ultraviolet generation using crystals $CsLiB_6O_{10}$, β-BaB_2O_4 and LiB_3O_5. Opt. Commun. 123(4 – 6), 583 – 586(1996).

[34] H. Kiriyama, F. Nakano, K. Yamakawa: High-efficiency frequency doubling of a Nd:YAG laser in a two-pass quadrature frequency-conversion scheme using $CsLiB_6O_{10}$ crystals. J. Opt. Soc. Am. B 19(8), 1857 – 1864(2002).

[35] Y. K. Yap, S. Hamamura, A. Taguchi, Y. Mori, T. Sasaki: $CsLiB_6O_{10}$ crystal for frequency doubling the Nd:YAG laser. Opt. Commun. 145(1 – 6), 101 – 104(1998).

[36] T. Kamimura, S. Fukumoto, R. Ono, Y. K. Yap, M. Yoshimura, Y. Mori, T. Sasaki, K. Yoshida: Enhancement of $CsLiB_6O_{10}$ surface-damage resistance by improved crystallinity and ion-beam etching. Opt. Lett. 27(8), 616 – 618(2002).

[37] T. Zhang, Y. Motoki, L. B. Sharma, H. Daido, Y. Kato, Y. Mori, T. Sasaki: 351 nm wavelength generation of picosecond laser pulses. Electron. Lett. 32(5), 452 – 454(1996).

[38] T. Kojima, S. Konno, S. Fujikawa, K. Yoshizawa, Y. Mori, T. Sasaki, M. Tanaka, Y. Okada: 20-W ultraviolet-beam generation by fourth-harmonic generation of an all-solid-state laser. Opt. Lett. 25(1), 58 – 60(2000).

[39] Y. K. Yap, T. Inoue, H. Sakai, Y. Kagebayashi, Y. Mori, T. Sasaki, K. Deki, M. Horiguchi: Long-term operation of $CsLiB_6O_{10}$ at elevated crystal temperature. Opt. Lett. 23(1), 34 – 36(1998).

[40] Y. K. Yap, K. Deki, N. Kitatochi, Y. Mori, T. Sasaki: Alleviation of thermally induced phase mismatch in $CsLiB_6O_{10}$ crystal by means of temperature-profile compensation. Opt. Lett. 23(13), 1016 – 1018(1998).

4.5 $MgO:LiNbO_3$，氧化镁掺杂铌酸锂(MgLN)

负单轴晶：$n_o > n_e$

点群：$3m$

居里温度和在化学计量比和同成分比 LN 晶体中作为 MgO 浓度（mol%）函数在 $\alpha = 20 \text{ cm}^{-1}$ 处的 UV 吸收截止边[1]

[MgO]	T_C/K	$\lambda_{截止}$/μm	[MgO]	T_C/K	$\lambda_{截止}$/μm
化学计量比 LN			4.6	1 480 ± 2	
0	1 466 ± 2	0.306	同成分 LN		
0.8	1 479 ± 2	0.304	0	1 411	0.316
2.0	1 486 ± 1	0.301	>5	1 486	
3.3	1 485 ± 1	0.303			

同成分 LN 晶体以"0"透过计的透明范围：$0.32 \sim 5 \text{ μm}$[2,3,4]

线性吸收系数 α

λ/μm	α/cm^{-1}	参考文献	λ/μm	α/cm^{-1}	参考文献
0.532 1	0.02	[5]		<0.003	[6]
1.064 2	<0.01	[5]			

掺 5 mol% MgO 和摩尔比 Li/Nb = 0.97 晶体折射率的实验值[3]

λ/μm	n_o	n_e	λ/μm	n_o	n_e
0.435 8	2.386 3	2.280 2	0.632 8	2.281 6	2.192 2
0.491 6	2.340 3	2.241 6	0.694 3	2.267 8	2.180 5
0.546 1	2.311 4	2.217 2	0.840 0	2.246 0	2.162 2
0.577 0	2.298 8	2.206 8	1.064 2	2.227 2	2.146 3
0.579 0	2.298 0	2.206 2			

掺 5 mol% MgO 和摩尔比 Li/Nb = 0.946（同成分熔体）晶体折射率的实验值[4]

λ/μm	n_o	n_e	λ/μm	n_o	n_e
0.404 7	2.424 7	2.311 1	0.579 0	2.298 2	2.205 6
0.407 8	2.420 2	2.307 3	0.589 3	2.294 5	2.202 7
0.435 8	2.386 3	2.279 5	0.623 4	2.284 0	2.193 8
0.486 1	2.344 1	2.244 4	0.656 3	2.275 6	2.186 7
0.491 6	2.340 4	2.241 2	0.690 7	2.268 1	2.180 2
0.496 2	2.337 6	2.238 9	0.694 3	2.266 9	2.179 3
0.546 1	2.311 2	2.216 7	1.064 0	2.223 7	2.145 6
0.577 0	2.298 9	2.206 3			

文献[7]中给出了在不同温度(293 K、348 K、389 K、428 K)下 5 mol% MgO 掺杂同成分 LiNbO₃ 的折射率。

5 mol% MgO 掺杂同成分 LiNbO₃ 折射率的温度微商[7]

$\lambda/\mu m$	$dn_o/dT \times 10^6/K^{-1}$	$dn_e/dT \times 10^6/K^{-1}$	$\lambda/\mu m$	$dn_o/dT \times 10^6/K^{-1}$	$dn_e/dT \times 10^6/K^{-1}$
0.539 75	16.663	72.763	1.079 5	4.356	54.190
0.632 8	12.121	64.866	1.341 4	5.895	52.665

最佳色散关系方程组(5 mol% MgO,摩尔比 Li/Nb = 0.937,同成分熔体,λ 以 μm 为单位,$0.4~\mu m < \lambda < 5.0~\mu m$,$T = 294$ K)[8]:

$$n_o^2 = 1 + \frac{2.2454\lambda^2}{\lambda^2 - 0.01242} + \frac{1.3005\lambda^2}{\lambda^2 - 0.05313} + \frac{6.8972\lambda^2}{\lambda^2 - 331.33}$$

$$n_e^2 = 1 + \frac{2.4272\lambda^2}{\lambda^2 - 0.01478} + \frac{1.4617\lambda^2}{\lambda^2 - 0.05612} + \frac{9.6536\lambda^2}{\lambda^2 - 371.216}$$

文献中给出了室温下不同 MgO 掺杂同成分 LiNbO₃ 的其他色散关系方程组:[9]是 0~9 mol% MgO;[3]、[4]是 5 mol% MgO;[10]、[11]是 7 mol% MgO。

文献[12]中给出了波长范围 0.44~1.05 μm 中 MgO 掺杂(0~4.6 mol%)化学计量比 LiNbO₃ 的 Sellmeier 方程。

Sellmeier 方程的温度关系(5~7 mol% MgO,同成分熔体,λ 以 μm 为单位,$0.4~\mu m < \lambda < 4.0~\mu m$,$T$ 以 K 为单位,$273~K < T < 673~K$)[13]:

$$n_o^2 = 4.9130 + \frac{1.173 \times 10^5 + 1.65 \times 10^{-2} T^2}{\lambda^2 - (2.12 \times 10^2 + 2.7 \times 10^{-5} T^2)^2} - 2.78 \times 10^{-8} \lambda^2$$

$$n_e^2 = 4.5567 + \frac{0.97 \times 10^5 + 2.7 \times 10^{-2} T^2}{\lambda^2 - (2.01 \times 10^2 + 5.4 \times 10^{-5} T^2)^2} - 2.24 \times 10^{-8} \lambda^2 +$$
$$2.605 \times 10^{-7} T^2 - 2.1432 \times 10^{-4} T_{NCPM} - 4.07 \times T_{NCPM}^2$$

这里 T_{NPCM}(以℃为单位)是 1.064 $\mu m \Rightarrow$ 0.532 μm SHG 相互作用的非临界相位匹配的温度。

文献[7]、[14]、[15]中给出了 5 mol% MgO 掺杂同成分 LiNbO₃ Sellmeier 方程的其他温度关系。

非线性折射率 γ[16]

$\lambda/\mu m$	$\gamma \times 10^{15}/(cm^2 \cdot W^{-1})$	备注	$\lambda/\mu m$	$\gamma \times 10^{15}/(cm^2 \cdot W^{-1})$	备注
0.780	2.0 ± 0.3	[100]方向		2.0 ± 0.3	[010]方向

5 mol% MgO 掺杂同成分 LiNbO₃ 的矫顽场:≈4.5 kV/mm[17]。

5 mol% MgO 掺杂同成分 LiNbO₃ 矫顽场值与温度的关系[18]

T/K	P/(kV·mm^{-1})	T/K	P/(kV·mm^{-1})
298	4.5	393	1.8
353	2.4	443	1.3

普遍情况下有效二阶非线性系数的表达式(Kleinman 对称条件成立时,$d_{15} = d_{24} = d_{31} = d_{32}$)[19]:

$$d_{ooe} = d_{31}\sin(\theta+\rho) - d_{22}\cos(\theta+\rho)\sin 3\phi$$

$$d_{eoe} = d_{oee} = d_{22}\cos^2(\theta+\rho)\cos 3\phi$$

有效二阶非线性系数的简化表达式(小双折射角近似,Kleinman 对称条件成立时,$d_{15} = d_{24} = d_{31} = d_{32}$)[20]:

$$d_{ooe} = d_{31}\sin\theta - d_{22}\cos\theta\sin 3\phi$$

$$d_{eoe} = d_{oee} = d_{22}\cos^2\theta\cos 3\phi$$

二阶非线性系数的绝对值 5 mol% MgO:LiNbO$_3$[21]:

$$|d_{31}(0.852\ \mu m)| = 4.9\ pm/V$$

$$|d_{33}(0.852\ \mu m)| = 28.4\ pm/V$$

$$|d_{31}(1.064\ \mu m)| = 4.4\ pm/V$$

$$|d_{33}(1.064\ \mu m)| = 25.0\ pm/V$$

$$|d_{31}(1.313\ \mu m)| = 3.4\ pm/V$$

$$|d_{33}(1.313\ \mu m)| = 20.3\ pm/V$$

相位匹配角的实验值($T = 293$ K)

相互作用的波长/μm	θ_{exp}/(°)	参考文献	备注
SHG, o+o⇒e			
1.064 2⇒0.532 1	74.5	[4]	5 mol% MgO,同成分 LN
	76	[5]	5 mol% MgO
	76.5	[3]	5 mol% MgO,Li/Nb = 0.97
	82.3	[10]	7 mol% MgO
1.079 5⇒0.539 75	75.1	[7]	5 mol% MgO,同成分 LN
1.079 6⇒0.539 8	74	[3]	5mol% MgO,Li/Nb = 0.97
1.341 4⇒0.670 7	54	[7]	5 mol% MgO,同成分 LN

注:相位匹配角的值与熔体化学计量比强烈相关。

NCPM 温度的实验值

相互作用的波长/μm	T/℃	参考文献	备注
SHG, $o+o\Rightarrow e$			
$1.047\Rightarrow 0.5235$	75.3	[22]	
$1.0642\Rightarrow 0.5321$	25.4	[23]	0.6 mol% MgO，同成分 LN
	78.5	[24]	7 mol% MgO，沿 X
	85~109	[13]	>5 mol% MgO
	107	[5],[6]	5 mol% MgO
		[25],[26]	
	110	[27]	5 mol% MgO
	110.6	[28]	5 mol% MgO
	110.8	[29]	7 mol% MgO
	113	[30]	
	116	[31]	
$1.0795\Rightarrow 0.53975$	115	[7]	5 mol% MgO，同成分 LN

注：相位匹配温度与熔体化学计量比强烈相关。

角度带宽和温度带宽的实验值

相互作用的波长/μm	T/℃	θ_{pm}/(°)	$\Delta\theta^{int}$/(°)	ΔT/℃	参考文献	备注
SHG, $o+o\Rightarrow e$						
$1.0642\Rightarrow 0.5321$	20	76	0.063		[5]	5 mol% MgO
	25.4	90		0.068	[23]	0.6 mol% MgO
	107	90	2.160	0.73	[5]	5 mol% MgO
	110.6	90		0.73	[28]	5 mol% MgO

激光诱导的损伤阈值

λ/μm	τ_p/ns	I_{thr}/(GW·cm^{-2})	参考文献	备注
0.5321	CW	>0.002	[12]	1 mol% MgO, Li/Nb = 1.38
		>0.002	[12]	2 mol% MgO, Li/Nb = 1.0
		0.002	[12]	5 mol% MgO，同成分 LN
		>0.006	[32]	1.8 mol% MgO, Li/Nb = 0.96~0.99
	≈20	0.34	[5]	5 mol% MgO
0.778	0.002	>10	[33]	7 mol% MgO
0.780	0.00015	>15	[16]	
0.78~0.84	0.0001	>130	[34]	1 kHz, 7 mol% MgO
1.0642	25	>0.025	[23]	0.6 mol% MgO，同成分 LN
	≈20	0.61	[5]	5 mol% MgO
	20	>0.039	[10]	10 Hz, 5 mol% MgO
	0.04	>0.8	[23]	0.6 mol% MgO，同成分 LN

续表

$\lambda/\mu m$	τ_p/ns	$I_{thr}/(GW \cdot cm^{-2})$	参考文献	备注
	0.03	>0.14	[10]	5 Hz, 5 mol% MgO
1.56	0.000 08	>1.36	[35]	1 kHz, 5 mol% MgO

注：在 CW 0.532 μm 辐射下，研究了体光折变损伤。

关于这一晶体

普通 $LiNbO_3$ 晶体的最重要的缺点之一就是，它易受光折变损伤（通常是由于蓝色或绿色 CW 激光照射引起折射率的改变）的影响[36]。通常消除这一效应的方法是将 LN 晶体保持在升温的状态（400 K 或更高）。另一条防止光折变损伤的途径是 MgO 掺杂（对于同成分 LN 通常约为 5 mol% 的水平）。这样做的好处是这种 MgO 掺杂的同成分 $LiNbO_3$ 晶体比未掺杂的 LN 晶体有低得多的矫顽场数值。然而由于大量的 MgO 掺杂，随后对于生长高光学质量的晶体造成了困难。最近，实验表明[12]化学计量比的 $LiNbO_3$ 晶体，只要掺入 1 mol% 的 MgO，就具有比 5 mol% MgO 掺杂同成分 LN 样品更高的抗光折变损伤阈值。

让我们简要考虑两个最新纪录，是最近由日本研究组完成的，他们发展了格子周期极小值（降到 1.4 μm）的 PPMgLN 晶体。在文献[37]中，在一块 2 mm 厚 10 mm 长的晶体中通过基波为 141 mW 的一阶准相位匹配倍频产生了 1.2 mW CW UV 激光（λ = 341.5 μm）。在文献[38]中，由激光二极管终端泵浦的 $Nd:GdVO_4$ 激光器单次通过倍频器在一块 2 mm 厚 10 mm 长的 PPMgLN 中在 531nm 处产生了 890 mW 激光。文献[38]的作者们认为，这是在室温下从 QPM SHG 获得的最高 CW 功率。

■ **参考文献**

[1] K. Niwa, Y. Furukawa, S. Takekawa, K. Kitamura: Growth and characterization of MgO doped near stoichiometric $LiNbO_3$ crystals as a new nonlinear optical crystal. J. Cryst. Growth 208(1-4), 493-500(2000).

[2] K. Mizuuchi, A. Morikawa, T. Sugita, K. Yamamoto: Generation of 360-nm ultraviolet light in first-order periodically poled bulk MgO: $LiNbO_3$. Opt. Lett. 28(11), 935-937(2003).

[3] A. L. Aleksandrovskii, G. I. Ershova, G. K. Kitaeva, S. P. Kulik, I. I. Naumova, V. V. Tarasenko: Dispersion of the refractive indices of $LiNbO_3$: Mg and $LiNbO_3$: Y crystals. Kvant. Elektron. 18(2), 254-256(1991) [In Russian, English trans.: Sov. J. Quantum Electron. 21(2), 225-227(1991)].

[4] Y. Chang, J. Wen, H. Wang, B. Li: Refractive index measurement and second harmonic generation in a series of $LiNbO_3$: Mg(5 mol%) crystals. Chin. Phys. Lett. 9

(8), 427-430(1992).

[5] J. L. Nightingale, W. J. Silva, G. E. Reade, A. Rybicki, W. J. Kozlovsky, R. L. Byer: Fifty percent conversion efficiency second harmonic generation in magnesium oxide doped lithium niobate. Proc. SPIE 681, 20-24(1986).

[6] W. J. Kozlovsky, C. D. Nabors, R. L. Byer: Efficient second harmonic generation of a diode-laser-pumped cw Nd: YAG laser using monolithic MgO: $LiNbO_3$ external resonant cavities. IEEE J. Quant. Electr. 24(6), 913-919(1988).

[7] H. Y. Shen, H. Xu, Z. D. Zheng, W. X. Lin, R. F. Wu, G. F. Hu: Measurement of refractive indices and thermal refractive-index coefficients of $LiNbO_3$ crystal doped with 5 mol% MgO. Appl. Opt. 31(31), 6695-6697(1992).

[8] D. E. Zelmon, D. L. Small, D. Jundt: Infrared corrected Sellmeier coefficients for congruently grown lithium niobate and 5 mol% Magnesium oxide-doped lithium niobate. J. Opt. Soc. Am. B 14(12), 3319-3322(1997).

[9] U. Schlarb, K. Betzler: Influence of the defect structure on the refractive indices of undoped and MgO-doped lithium niobate. Phys. Rev. B 50(2), 751-757(1994).

[10] J. Q. Yao, W. Q. Shi, J. E. Millerd, G. F. Hu, E. Garmire, M. Birnbaum: Room temperature 1.06~0.53 μm second harmonic generation with MgO: $LiNbO_3$. Opt. Lett. 15(23), 1339-1341(1990).

[11] S. Lin, Y. Tanaka, S. Takeuchi, T. Suzuki: Improved dispersion equation for MgO: $LiNbO_3$ crystal in the infrared spectral range derived from sum and difference frequency mixing. IEEE J. Quant. Electr. 32(1), 124-126(1996).

[12] M. Nakamura, S. Higuchi, S. Takekawa, K. Terabe, Y. Furukawa, K. Kitamura: Optical damage resistance and refractive indices in near-stoichiometric MgO-doped $LiNbO_3$. Jpn. J. Appl. Phys. 41(1A/B), L49-L51(2002).

[13] J.-Q. Yao, Y.-Z. Yu, P. Wang, T. Wang, B.-G. Zhang, X. Ding, J. Chen, H. J. Peng, H. S. Kwok: Nearly-noncritical phase matching in MgO: $LiNbO_3$ optical parametric oscillators. Chin. Phys. Lett. 18(9), 1214-1217(2001).

[14] R. C. Eckardt, C. D. Nabors, W. J. Kozlovsky, R. L. Byer: Optical parametric oscillator frequency tuning and control. J. Opt. Soc. Am. B 8(3), 646-667(1991).

[15] D. Y. Sugak, A. O. Matkovskii, I. M. Solskii, I. V. Stefanskii, V. M. Gaba, A. T. Mikhalevich, V. V. Grabovski, V. I. Prokhorenko, B. N. Kopko, V. Y. Oliinyk: Growth and investigation of $LiNbO_3$: MgO single crystals. Proc. SPIE 2795, 257-264(1996).

[16] H. P. Li, C. H. Kam, Y. L. Lam, W. Ji: Femtosecond Z-scan measurements of nonlinear refraction in nonlinear optical materials. Opt. Mater. 15(4), 237-242(2001).

[17] H. Ishizuki, T. Taira, S. Kurimura, J. H. Ro, M. Cha: Periodic poling in 3-mm-thick MgO: $LiNbO_3$ crystals. Jpn. J. Appl. Phys. 42(2A), L108-L110(2003).

[18] H. Ishizuki, I. Shoji, T. Taira: Periodical poling of 3 mm-thick MgO: $LiNbO_3$ crystals

for high-power nonlinear wavelength conversion. In: *CLEO/Europe 2003*, *Technical Digest* (OSA, Washington DC, 2003), paper CE2-2-MON.

[19] I. Shoji, H. Nakamura, K. Ohdaira, T. Kondo, R. Ito, T. Okamoto, K. Tatsuki, S. Kubota: Absolute measurement of second-order nonlinear-optical coefficients of β-BaB_2O_4 for visible to ultraviolet second-harmonic wavelengths. J. Opt. Soc. Am. B 16(4), 620–624(1999).

[20] J. E. Midwinter, J. Warner: The effects of phase matching method and of uniaxial crystal symmetry on the polar distribution of second-order non-linear optical polarization. Brit. J. Appl. Phys. 16(11), 1135–1142(1965).

[21] I. Shoji, T. Kondo, A. Kitamoto, M. Shirane, R. Ito: Absolute scale of second-order nonlinear-optical coefficients. J. Opt. Soc. Am. B 14(9), 2268–2294(1997).

[22] G. T. Maker, A. I. Ferguson: Ti: sapphire laser pumped by a frequency-doubled dio depumped Nd: YLF laser. Opt. Lett. 15(7), 375–377(1990).

[23] B. K. Rhee, J.-S. Lee, G.-T. Joo: Room-temperature 1.06–0.53 μm second-harmonic generation in $LiNbO_3$ with 0.6 mole% MgO doping. Proc. SPIE 3610, 156–163(1999).

[24] M. Tsunekane, S. Kimura, M. Kimura, N. Taguchi, H. Inaba: Continuous-wave, broadband tuning from 788 to 1640 nm by a doubly resonant, MgO: $LiNbO_3$ optical parametric oscillator. Appl. Phys. Lett. 72(26), 3414–3416(1998).

[25] W. J. Kozlovsky, C. D. Nabors, R. C. Eckardt, R. L. Byer: Monolithic MgO: $LiNbO_3$ doubly resonant optical parametric oscillator pumped by a frequency-doubled diodelaser-pumped Nd: YAG laser. Opt. Lett. 14(1), 66–68(1989).

[26] C. D. Nabors, R. C. Eckardt, W. J. Kozlovsky, R. L. Byer: Efficient, single-axial-mode operation of a monolithic MgO: $LiNbO_3$ optical parametric oscillator. Opt. Lett. 14(20), 1134–1136(1989).

[27] D. C. Gerstenberger, G. E. Tye, R. W. Wallace: Efficient second-harmonic conversion of CW single-frequency Nd: YAG laser light by frequency locking to a monolithic ring frequency doubler. Opt. Lett. 16(13), 992–994(1991).

[28] R. C. Eckardt, H. Masuda, Y. X. Fan, R. L. Byer: Absolute and relative nonlinear optical coefficients of KDP, KD*P, BaB_2O_4, $LiIO_3$, MgO: $LiNbO_3$, and KTP measured by phasematched second-harmonic generation IEEE J. Quant. Electr. 26(5), 922–933(1990).

[29] A. Porzio, C. Altucci, M. Autiero, A. Chiummo, C. De Lisio, S. Solimeno: Tunable twin beams generated by a type-I OPO. Appl. Phys. B 73(7), 763–766(2001).

[30] D. C. Gerstenberger, R. W. Wallace: Continuous-wave operation of a doubly resonant lithium niobate optical parametric oscillator system tunable from 966 to 1185 nm. J. Opt. Soc. Am. B 10(9), 1681–1683(1993).

[31] G. T. Maker, A. I. Ferguson: Efficient frequency doubling of a diode laser-pumped mode-locked Nd: YAG laser using an external resonant cavity. Opt. Commun. 76(5 – 6), 369 – 375(1990).

[32] Y. Furukawa, K. Kitamura, S. Takekawa, K. Niwa, H. Hatano: Stoichiometric MgO: $LiNbO_3$ as an effective material for nonlinear optics. Opt. Lett. 23 (24), 1892 – 1894 (1998).

[33] S. Lin, T. Suzuki: Tunable picosecond mid-infrared pulses generated by optical parametric generation/amplification in MgO: $LiNbO_3$ crystals. Opt. Lett. 21 (8), 579 – 581 (1996).

[34] V. Petrov, F. Rotermund, F. Noack: Femtosecond traveling-wave optical parametric amplification in MgO: $LiNbO_3$. Appl. Opt. 37(36), 8504 – 8511(1998).

[35] S. Ashihara, T. Shimura, K. Kuroda, N. E. Yu, S. Kurimura, K. Kitamura, J. H. Ro, M. Cha, T. Taira: Group-velocity-matched cascaded quadratic nonlinearities of femtosecond pulses in periodically poled MgO: $LiNbO_3$. Opt. Lett. 28 (16), 1442 – 1444 (2003).

[36] A. M. Glass: The photorefractive effect. Opt. Eng. 17(5), 470 – 479(1978).

[37] K. Mizuuchi, A. Morikawa, T. Sugita, K. Yamamoto: Efficient second-harmonic generation of 340-nm light in a 1.4-μm periodically poled bulk MgO: $LiNbO_3$. Jpn. J. Appl. Phys. 42 (2A), L90-L91(2003).

[38] K. Mizuuchi, A. Morikawa, T. Sugita, K. Yamamoto, N. Pavel, I. Shoji, T. Taira: High-power continuous wave green generation by single-pass frequency doubling of a Nd: $GdVO_4$ laser in a periodically poled MgO: $LiNbO_3$ operating at room temperature. Jpn. J. Appl. Phys. 42(11A), L1296-L1298(2003).

4.6 $KTiOAsO_4$，砷酸钛氧钾(KTA)

正双轴晶：$2V_Z = 40.4°$，$\lambda = 0.5321$ μm
分子量：241.897
密度：3.454 g/cm^3[1]
点群：$mm2$
晶格常数：
    $a = 13.103$ Å[2]；13.125 Å[3]；13.127 Å[4]
    $b = 6.558$ Å[2]；6.5716 Å[3]；6.5713 Å[4]
    $c = 10.746$ Å[2]；10.786 Å[3]；10.789 Å[4]
介电轴和晶体学轴间的变换：X, Y, $Z \Rightarrow a$, b, c
居里温度：1 153 K[2]；1 149 K($\parallel c$)，1 151 K($\parallel a$)，1 153 K($\parallel b$)[5]

298 K < T < 473 K 温度范围内沿 X 轴的热膨胀[6]：
$$L(T) = L_0[1 + \alpha(T-298) + \beta(T-298)^2]$$
其中 T 以 K 为单位，$T_0 = 298$ K，$\alpha = (7.6 \pm 0.6) \times 10^{-6}$ K^{-1}，$\beta = (8.4 \pm 1.2) \times 10^{-9}$ K^{-2}。

以 "0" 透过计的透明范围：0.35~5.2 μm[2,7]；0.35~5.3 μm[8,5]。

UV 透过截止边（$\alpha = 2$ cm^{-1}）为 0.377 μm（$E \parallel X$）；0.385 μm（$E \parallel Y$）；0.393 μm（$E \parallel Z$）[9]。

线性吸收系数 α

λ/μm	α/cm^{-1}	参考文献	备注
0.473	0.008	[9]	$E \parallel X$
	0.014	[9]	$E \parallel Y$
	0.016	[9]	$E \parallel Z$
0.532	0.005	[9]	$E \parallel X$
	0.005	[9]	$E \parallel Y$
	0.005	[9]	$E \parallel Z$
4.0	0.2	[10]	
5.0	1.0	[10]	

折射率的实验值[11]

λ/μm	n_X	n_Y	n_Z
0.632 8	1.808 3	1.814 2	1.904 8

传统 Sellmeier 方程（λ 以 μm 为单位，$T = 293$ K）[12]：

$$n_X^2 = 3.141\,3 + \frac{0.046\,83}{\lambda^2 - 0.040\,55} - 0.010\,23\lambda^2$$

$$n_Y^2 = 3.159\,3 + \frac{0.048\,28}{\lambda^2 - 0.047\,10} - 0.010\,49\lambda^2$$

$$n_Z^2 = 3.443\,5 + \frac{0.065\,71}{\lambda^2 - 0.054\,35} - 0.014\,60\lambda^2$$

更准确的色散关系（λ 以 μm 为单位，0.4 μm < λ < 5.3 μm 对 n_X 和 n_Y，0.4 μm < λ < 3.6 μm 对 n_Z，$T = 293$ K）[13,14]：

$$n_X^2 = 2.149\,5 + \frac{1.020\,3\lambda^{1.995\,1}}{\lambda^{1.995\,1} - 0.042\,378} + \frac{0.553\,1\lambda^{1.956\,7}}{\lambda^{1.956\,7} - 72.304\,5}$$

$$n_Y^2 = 2.130\,8 + \frac{1.056\,4\lambda^{2.001\,7}}{\lambda^{2.001\,7} - 0.042\,523} + \frac{0.692\,7\lambda^{1.726\,1}}{\lambda^{1.726\,1} - 54.850\,5}$$

$$n_z^2 = 2.1931 + \frac{1.2382\lambda^{1.8920}}{\lambda^{1.8920} - 0.059171} + \frac{0.5088\lambda^{2.0000}}{\lambda^{2.0000} - 53.2898}$$

文献中[4]、[11]、[15]、[16]、[17]、[18]给出了其他色散关系方程组。

与红外相关折射率 n_z 的 Sellmeier 方程(λ 以 μm 为单位,$T = 293$ K)[19]:

$$n_z^2 = 1.214331 + \frac{2.225328\lambda^2}{\lambda^2 - (0.178542)^2} + \frac{0.310017\lambda^2}{\lambda^2 - (8.989998)^2} - 0.009381\lambda^2$$

非线性折射率 γ [20,22]

$\lambda/\mu m$	$\gamma \times 10^{15}/(cm^2 \cdot W^{-1})$	备注
0.780	1.7 ± 0.3	[100]方向
	1.7 ± 0.3	[010]方向

室温下低频(远低于 KTA 晶体声学共振频率,即对"自由"晶体)下测量的线性电光系数[17]

$\lambda/\mu m$	$r_{13}^T/(pm \cdot V^{-1})$	$r_{23}^T/(pm \cdot V^{-1})$	$r_{33}^T/(pm \cdot V^{-1})$
0.6328	11.5 ± 1.2	15.4 ± 1.5	37.5 ± 3.8

KTA 晶体主平面上有效二阶非线性系数的表达式(Kleinman 对称条件不成立)[22]:

XY 平面
$$d_{eoe} = d_{oee} = d_{15}\sin^2\phi + d_{24}\cos^2\phi$$

YZ 平面
$$d_{oeo} = d_{eoo} = d_{15}\sin\theta$$

XZ 平面,$\theta < V_z$
$$d_{ooe} = d_{32}\sin\theta$$

XZ 平面,$\theta > V_z$ *
$$d_{oeo} = d_{eoo} = d_{24}\sin\theta$$

KTA 晶体主平面上有效二阶非线性系数的表达式(Kleinman 对称条件成立,$d_{15} = d_{31}$ 和 $d_{24} = d_{32}$)[22]:

XY 平面
$$d_{eoe} = d_{oee} = d_{31}\sin^2\phi + d_{32}\cos^2\phi$$

YZ 平面

---

* 译者注:原书为"$\theta < V_z$",应为"$\theta > V_z$"。

$$d_{oeo} = d_{eoo} = d_{31}\sin\theta$$

XZ 平面，$\theta < V_z$

$$d_{ooe} = d_{32}\sin\theta$$

XZ 平面，$\theta > V_z$

$$d_{oeo} = d_{eoo} = d_{32}\sin\theta$$

文献[22]中给出了 KTA 晶体任意方向上三波相互作用的有效二阶非线性系数。KTA 二阶非线性系数的符号可能都是相同的[23]。

二阶非线性系数的绝对和相对值：

$d_{15}(1.064~\mu m) = 1.3 \times d_{15}(KTP) = (2.5 \pm 0.2)~pm/V^{[4,24]}$

$d_{24}(1.064~\mu m) = (1.8 \pm 0.1) \times d_{15}(KTA) = (4.4 \pm 0.2)~pm/V^{[4,24]}$

$d_{31}(1.064~\mu m) = (2.8 \pm 0.3)~pm/V^{[11]}$

$d_{31}(1.064~\mu m) = (1.3 \pm 0.1) \times d_{31}(KTP) = (2.9 \pm 0.2)~pm/V^{[24,25]}$

$d_{32}(1.064~\mu m) = (4.2 \pm 0.4)~pm/V^{[11]}$

$d_{32}(1.064~\mu m) = (1.8 \pm 0.1) \times d_{31}(KTA) = (5.1 \pm 0.3)~pm/V^{[24,25]}$

$d_{33}(1.064~\mu m) = (16.2 \pm 1.0)~pm/V^{[11]}$

$d_{15}(1.32~\mu m) = 1.2 \times d_{15}(KTP) = (1.7 \pm 0.1)~pm/V^{[16,24]}$

$d_{24}(1.32~\mu m) = 1.7 \times d_{15}(KTP) = (2.4 \pm 0.2)~pm/V^{[16,24]}$

相位匹配角的实验值($T = 293$ K)

相互作用的波长/μm	$\phi_{exp}/(°)$	$\theta_{exp}/(°)$	参考文献
XY 平面，$\theta = 90°$			
SFG，e + o ⇒ e			
1.318 8 + 0.659 4 ⇒ 0.439 6	62.5		[12]
1.064 2 + 1.907 9 ⇒ 0.683 1	15.7		[12]
YZ 平面，$\phi = 90°$			
SHG，o + e ⇒ o			
1.074 5 ⇒ 0.537 25		90	[12]
1.152 3 ⇒ 0.576 15		69.2	[12]
1.318 8 ⇒ 0.659 4		56	[16]
		55.9	[12]
		55.9	[11]
		55.7	[15]
3.391 3 ⇒ 1.695 65		63.5	[12]
SFG，o + e ⇒ o			
1.318 8 + 0.659 4 ⇒ 0.439 6		79.8	[12]
1.064 2 + 1.907 9 ⇒ 0.683 1		72.1	[12]

续表

相互作用的波长/μm	$\phi_{exp}/(°)$	$\theta_{exp}/(°)$	参 考 文 献
XZ 平面，$\phi=0°$，$\theta>V_z$			
SHG，$o+e \Rightarrow o$			
$1.1422 \Rightarrow 0.5711$		90	[12]
$1.1523 \Rightarrow 0.57615$		82.9	[4]
$1.3188 \Rightarrow 0.6594$		65	[15]
		64.6	[5]
		64.2	[4]
		63.1	[12]
$3.3913 \Rightarrow 1.69565$		70.6	[12]

内角带宽的实验值

相互作用的波长/μm	$\theta_{pm}/(°)$	$\Delta\theta^{int}/(°)$	参 考 文 献
YZ 平面，$\phi=90°$			
SHG，$o+e \Rightarrow o$			
$1.3188 \Rightarrow 0.6594$	56	0.086	[16]
	55.9	0.093	[11]

激光诱导的体损伤阈值

$\lambda/\mu m$	τ_p/ns	$I_{thr}/(GW \cdot cm^{-2})$	参 考 文 献	备 注
0.74~0.84	0.0002	>200	[26]	1 kHz
0.77~0.9	0.0012	>0.3	[27]	81 MHz
0.78	0.00015	>20	[21]	76 MHz
0.85	2	>1	[28]	
1.0642	18	>0.12	[29]	100 Hz
	8	>1.2	[10]	20 Hz，1 000 个脉冲

关于这一晶体

和 KTP 晶体不一样，KTA 晶体主要用于双折射相位匹配。与 KTP 晶体相比，KTA 晶体的主要优点在于有稍高的二阶非线性系数[4,25,11,24]、更长的 IR 截止波长以及在 3.5 μm 处不存在吸收[9]。直到目前为止，只有少数工作涉及 PPKTA[19,30]。

■ 参考文献

[1] B. H. T. Chai：Optical Crystals. In：*CRC Handbook of Laser Science and Technology*，

Supplement 2: Optical Materials, ed. by M. J. Weber (CRC Press, Boca Raton, 1995), pp. 3–65.

[2] J. D. Bierlein, H. Vanherzeele, A. A. Ballman: Linear and nonlinear optical properties of flux-grown KTiOAsO$_4$. Appl. Phys. Lett. 54(9), 783–785(1989).

[3] L. K. Cheng, E. M. McCarron III, J. Calabrese, J. D. Bierlein, A. A. Ballman: Development of the nonlinear optical crystal CsTiOAsO$_4$. I. Structural stability. J. Cryst. Growth 132(1–2), 280–288(1993).

[4] K. Kato: Second-harmonic and sum-frequency generation in KTiOAsO$_4$. IEEE J. Quant. Electr. 30 (4), 881–883(1994).

[5] J. Wei, J. Wang, Y. Liu, L. Shi, M. Wang, Z. Shao: Growth, second harmonic and sum frequency generation operations of potassium titanyl arsenate crystal. Chin. Phys. Lett. 11(2), 95–98(1994).

[6] S. Emanueli, A. Arie: Temperature-dependent dispersion equations for KTiOPO$_4$ and KTiOAsO$_4$. Appl. Opt. 42(33), 6661–6665(2003).

[7] A. H. Kung: Efficient conversion of high-power narrow-band Ti: sapphire laser radiation to the mid-infrared in KTiOAsO$_4$. Opt. Lett. 20(10), 1107–1109(1995).

[8] L. K. Cheng, J. D. Bierlein: KTP and isomorphs-recent progress in device and material development. Ferroelectrics 142(1–2), 209–228(1993).

[9] G. Hansson, H. Karlsson, S. Wang, F. Laurell: Transmission measurements in KTP and isomorphic compounds. Appl. Opt. 39(27), 5058–5069(2000).

[10] W. R. Bosenberg, L. K. Cheng, J. D. Bierlein: Optical parametric frequency conversion properties of KTiOAsO$_4$. Appl. Phys. Lett. 65(22), 2765–2767(1994).

[11] L. K. Cheng, L.-T. Cheng, J. D. Bierlein, F. C. Zumsteg, A. A. Ballman: Properties of doped and undoped crystals of single domain KTiOAsO$_4$. Appl. Phys. Lett. 62(4), 346-348(1993).

[12] K. Kato, N. Umemura, E. Tanaka: 90° phase-matched mid-infrared parametric oscillation in undoped KTiOAsO$_4$. Jpn. J. Appl. Phys. 36(4A), L403-L405(1997).

[13] J. P. Feve, B. Boulanger, O. Pacaud, I. Rousseau, B. Menaert, G. Marnier: Refined Sellmeier equations from phase-matching measurements over the complete transparency range of KTiOAsO$_4$, RbTiOAsO$_4$, and CsTiOAsO$_4$. In: *Advanced Solid-State Lasers, OSA Trends in Optics and Photonics Series, Vol. 34*, ed. by H. Injeyan, U. Keller, C. Marshall (OSA, Washington DC, 2000), pp. 575–577.

[14] J.-P. Feve, B. Boulanger, O. Pacaud, I. Rousseau, B. Menaert, G. Marnier, P. Villeval, C. Bonnin, G. M. Loiacono, D. N. Loiacono: Phase-matching measurements and Sellmeier equations over the complete transparency range of KTiOAsO$_4$, RbTiOAsO$_4$, and CsTiO-AsO$_4$. J. Am. Opt. Soc. B 17(5), 775–780(2000).

[15] B. Boulanger, G. Marnier, B. Menaert, X. Gabirol, J. P. Feve, C. Bonnin, P. Villeval:

Collinear L. C. type II phase-matching for SHG in KTiOAsO$_4$: demonstration of its impossibility at 1.064 μm and first experiments at 1.32 μm. Comparison with KTiOPO$_4$. Nonl. Opt. 4(2), 133 – 142(1993).

[16] L.-T. Cheng, L. K. Cheng, J. D. Bierlein: Linear and nonlinear optical properties of the arsenate isomorphs of KTP. Proc. SPIE 1863, 43 – 53(1993).

[17] L. K. Cheng, L. T. Cheng, J. Galperin, P. A. Morris Hotsenpiller, J. D. Bierlein: Crystal growth and characterization of KTiOPO$_4$ isomorphs from the self-fluxes. J. Cryst. Growth 137 (1 – 2), 107 – 115(1994).

[18] D. L. Fenimore, K. L. Schepler, U. B. Ramabadran, S. R. McPherson: Infrared corrected Sellmeier coefficients for potassium titanyl arsenate. J. Opt. Soc. Am. B 12 (5), 794 – 796(1995).

[19] K. Fradkin-Kashi, A. Arie, P. Urenski, G. Rosenman: Mid-infrared difference-frequency generation in periodically poled KTiOAsO$_4$ and application to gas sensing. Opt. Lett. 25(10), 743 – 745(2000).

[20] H. P. Li, C. H. Kam, Y. L. Lam, F. Zhou, W. Ji: Nonlinear refraction of undoped and Fe-doped KTiOAsO$_4$ crystals in the femtosecond regime. Appl. Phys. B 70(3), 385 – 388(2000).

[21] H. P. Li, C. H. Kam, Y. L. Lam, W. Ji: Femtosecond Z-scan measurements of nonlinear refraction in nonlinear optical materials. Opt. Mater. 15(4), 237 – 242(2001).

[22] V. G. Dmitriev, D. N. Nikogosyan: Effective nonlinearity coefficients for three-wave interactions in biaxial crystals of *mm*2 point group symmetry. Opt. Commun. 95(1 – 3), 173 – 182(1993).

[23] A. Anema, T. Rasing: Relative signs of the nonlinear coefficients of potassium titanyl phosphate. Appl. Opt. 36(24), 5902 – 5904(1997).

[24] I. Shoji, T. Kondo, A. Kitamoto, M. Shirane, R. Ito: Absolute scale of second-order nonlinear-optical coefficients. J. Opt. Soc. Am. B 14(9), 2268 – 2294(1997).

[25] J. Wang, J. Wei, Y. Liu, X. Yin, X. Hu, Z. Shao, M. Jiang: A survey of research on KTP and its analogue crystals. Progr. Cryst. Growth Character. Mater. 40(1 – 4), 3 – 15(2000).

[26] V. Petrov, F. Noack, R. Stolzenberger: Seeded femtosecond optical parametric amplification in the mid-infrared spectral region above 3 μm. Appl. Opt. 36(6), 1164 – 1172 (1997).

[27] S. French, A. Miller, M. Ebrahimzade: Picosecond near-to mid-infrared optical parametric oscillator using KTiOAsO$_4$. Opt. Quant. Electron. 29(11 – 12), 999 – 1021 (1997).

[28] A. H. Kung: Narrow band mid-infrared generation using KTiOAsO$_4$. Appl. Phys. Lett. 65 (4), 1082 – 1084(1994).

[29] M. S. Webb, P. F. Moulton, J. J. Kasinski, R. L. Burnham, G. Loiacono, R. Stolzenberger: High-average-power KTiOAsO$_4$ optical parametric oscillator. Opt. Lett. 23(15), 1161-1163(1998).

[30] G. Rosenman, A. Skliar, Y. Findling, P. Urenski, A. Englander, P. A. Thomas, Z. W. Hu: Periodically poled KTiAsO$_4$ crystals for optical parametric oscillation. J. Phys. D 32(14), L49-L52(1999).

4.7 KNbO$_3$，铌酸钾(KN)

负双轴晶：$2V_Z = 66.78°$，$\lambda = 0.532\ 1\ \mu m$[1]

分子量：188.150

密度：4.617 g/cm^3 [2]

点群：$mm2$(223 K $< T <$ 496 K)

晶格常数：
a = 5.689 6 Å[3]；5.697 Å[4]；5.706 1 Å[5]
b = 3.969 2 Å[3]；3.971 Å[4]；3.979 4 Å[5]
c = 5.725 6 Å[3]；5.722 Å[4]；5.731 9 Å[5]

介电轴和晶体学轴间的变换：$X, Y, Z \Rightarrow b, a, c$

熔化温度：1 333 K[6]

居里温度：498 K[7]

p = 0.101 325 MPa 时的比热容：c_p = 767 J/(kg · K)[8]

热导率：$\kappa >$ 3.5 W/(m · K)[9]；κ = 4 W/(m · K)[8]

以"0"透过计的透明范围：$\approx$ 0.4 μm 到 > 4 μm[3,10]

IR 截止波长在 5.5 μm($\parallel a$ 或 $\parallel c$)[11]。

线性吸收系数 α

$\lambda/\mu m$	α/cm^{-1}	参考文献	备注
0.42 ~ 1.06	< 0.05	[12]	
0.423	0.13 ± 0.02	[13]	沿 a 轴，$E \parallel c$
0.458 ~ 0.515	0.04 ~ 0.07	[8]	
0.8 ~ 1.1	0.001 ~ 0.003	[8]	
0.82	0.015	[14]	
0.846	0.000 034 ± 0.000 022	[13]	沿 a 轴，$E \parallel b$
1.064 2	0.001 8 ~ 0.002 5	[9]	沿 b 轴
3.0	0.05	[11]	沿 c 轴
	0.03	[11]	沿 a 轴
3.5	0.05	[11]	沿 c 轴

续表

$\lambda/\mu m$	α/cm^{-1}	参 考 文 献	备 注
	0.02	[11]	沿 a 轴
4.0	0.08	[11]	沿 c 轴
	0.08	[11]	沿 a 轴
4.5	0.27	[11]	沿 c 轴
	0.45	[11]	沿 a 轴
5.0	1.21	[11]	沿 c 轴
	1.85	[11]	沿 a 轴
5.5	7.60	[11]	沿 c 轴
	4.90	[11]	沿 a 轴

双光子吸收系数 β(沿 a 轴)[13]

$\lambda/\mu m$	τ_p/ns	$\beta \times 10^{11}/(cm \cdot W^{-1})$
0.846	CW	320 ± 50

$T = 295$ K 时折射率的实验值[3]

$\lambda/\mu m$	n_X	n_Y	n_Z
0.430	2.497 4	2.414 5	2.277 1
0.488	2.418 7	2.352 7	2.227 4
0.514	2.395 1	2.333 7	2.212 1
0.633	2.329 6	2.280 1	2.168 7
0.860	2.278 4	2.237 2	2.133 8
1.064	2.257 6	2.219 5	2.119 4
1.500	2.234 1	2.199 2	2.102 9
2.000	2.215 9	2.183 2	2.089 9
2.500	2.198 1	2.167 4	2.077 1
3.000	2.178 5	2.149 8	2.063 0

最佳色散方程(λ 以 μm 为单位,$T = 295$ K)[11,15]:

$$n_X^2 = 4.985\ 6 + \frac{0.152\ 66}{\lambda^2 - 0.063\ 31} - 0.028\ 31\lambda^2 + 2.075\ 4 \times 10^{-6}\lambda^4 - 1.213\ 1 \times 10^{-6}\lambda^6$$

$$n_Y^2 = 4.835\ 3 + \frac{0.128\ 08}{\lambda^2 - 0.056\ 74} - 0.025\ 28\lambda^2 + 1.859\ 0 \times 10^{-6}\lambda^4 - 1.068\ 9 \times 10^{-6}\lambda^6$$

$$n_Z^2 = 4.422\ 2 + \frac{0.099\ 72}{\lambda^2 - 0.054\ 96} - 0.019\ 76\lambda^2$$

在文献[1],[3],[16],[17]中给出了其他色散关系方程组。
Sellmeier 方程的温度关系(λ 以 μm 为单位,T 以 K 为单位)[18]:

$$n_X^2 = 1 + \frac{(2.538\,940\,9 + 3.863\,630\,3 \times 10^{-6}F)\lambda^2}{\lambda^2 - (0.137\,163\,9 + 1.767 \times 10^{-7}F)^2}$$
$$+ \frac{(1.445\,184\,2 - 3.909\,336 \times 10^{-6}F - 1.225\,613\,6 \times 10^{-4}G)\lambda^2}{\lambda^2 - (0.272\,542\,9 + 2.38 \times 10^{-7}F - 6.78 \times 10^{-5}G)^2}$$
$$- (2.837 \times 10^{-2} - 1.22 \times 10^{-8}F)\lambda^2 - 3.3 \times 10^{-10}F\lambda^4$$

$$n_Y^2 = 1 + \frac{(2.638\,666\,9 + 1.670\,846\,9 \times 10^{-6}F)\lambda^2}{\lambda^2 - (0.136\,124\,8 + 0.796 \times 10^{-7}F)^2}$$
$$+ \frac{(1.194\,847\,7 - 1.387\,263\,5 \times 10^{-6}F - 0.907\,427\,07 \times 10^{-4}G)\lambda^2}{\lambda^2 - (0.262\,191\,7 + 1.231 \times 10^{-7}F - 1.82 \times 10^{-5}G)^2}$$
$$- (2.513 \times 10^{-2} - 0.558 \times 10^{-8}F)\lambda^2 - 4.4 \times 10^{-10}F\lambda^4$$

$$n_Z^2 = 1 + \frac{(2.370\,517 + 2.837\,354\,5 \times 10^{-6}F)\lambda^2}{\lambda^2 - (0.119\,407\,1 + 1.75 \times 10^{-7}F)^2}$$
$$+ \frac{(1.048\,952 - 2.130\,378\,1 \times 10^{-6}F - 1.825\,852\,1 \times 10^{-4}G)\lambda^2}{\lambda^2 - (0.255\,360\,5 + 1.89 \times 10^{-7}F - 2.48 \times 10^{-5}G)^2}$$
$$- (1.939 \times 10^{-2} - 0.27 \times 10^{-8}F)\lambda^2 - 5.7 \times 10^{-10}F\lambda^4$$

其中 $F = T^2 - 295.15^2$, $G = T - 295.15$。

光谱范围 0.42~5.3 μm 以及温度范围 295~473 K 中折射率的温度微商以及计算由温度引起的折射率改变的相关方程(λ 以 μm 为单位,T 以 K 为单位)[15]:

$$\frac{dn_X}{dT} = \left(\frac{0.304\,1}{\lambda} - 3.101\,2\right) \times 10^{-5}\,\text{K}^{-1}$$

$$\frac{dn_Y}{dT} = \left(\frac{2.592\,9}{\lambda^3} - \frac{4.738\,1}{\lambda^2} + \frac{4.125\,4}{\lambda} + 1.378\,8\right) \times 10^{-5}\,\text{K}^{-1}$$

$$\frac{dn_Z}{dT} = \left(\frac{1.408\,7}{\lambda^3} - \frac{5.152\,3}{\lambda^2} + \frac{8.743\,2}{\lambda} + 2.235\,0\right) \times 10^{-5}\,\text{K}^{-1}$$

$$\Delta n_X = \frac{dn_X}{dT}[(T - 295.15) - 0.32 \times 10^{-3}(T - 295.15)^2]$$

$$\Delta n_Y = \frac{dn_Y}{dT}[(T - 295.15) + 2.20 \times 10^{-3}(T - 295.15)^2]$$

$$\Delta n_Z = \frac{dn_Z}{dT}[(T - 295.15) + 2.71 \times 10^{-3}(T - 295.15)^2]$$

非线性折射率 γ [19]

λ/μm	$\gamma \times 10^{15}$/(cm^2 · W^{-1})	备注
0.850	1.87 ± 0.35	沿 Y 轴

室温下低频(远低于$KNbO_3$晶体声学共振频率,即对"自由"晶体)下测量的线性电光系数[20]

$\lambda/\mu m$	$r_{13}^T/(pm \cdot V^{-1})$	$r_{23}^T/(pm \cdot V^{-1})$	$r_{33}^T/(pm \cdot V^{-1})$	$r_{42}^T/(pm \cdot V^{-1})$	$r_{51}^T/(pm \cdot V^{-1})$
0.6328	+28±2	+1.3±0.5	64±5	380±50	105±13

室温下高频(远高于$KNbO_3$晶体声学共振频率,即对"受夹"晶体)下测量的线性电光系数[20]

$\lambda/\mu m$	$r_{13}^S/(pm \cdot V^{-1})$*	$r_{23}^S/(pm \cdot V^{-1})$	$r_{33}^S/(pm \cdot V^{-1})$	$r_{42}^S/(pm \cdot V^{-1})$	$r_{51}^S/(pm \cdot V^{-1})$
0.6328	10±2	2±1	25±8	270±40	23±3

矫顽场值:0.5 kV/mm[21];0.55 kV/mm[6]

铌酸钾晶体主平面上有效二阶非线系数的表达式(Kleinman 对称条件不成立)[22]:

XY 平面
$$d_{eeo} = d_{32}\sin^2\phi + d_{31}\cos^2\phi$$

YZ 平面
$$d_{ooe} = d_{32}\sin\theta$$

XZ 平面,$\theta < V_z$
$$d_{oeo} = d_{eoo} = d_{15}\sin\theta$$

XZ 平面,$\theta > V_z$
$$d_{ooe} = d_{31}\sin\theta$$

铌酸钾晶体主平面上有效二阶非线性系数的表达式(Kleinman 对称条件成立,$d_{15}=d_{31}$ 和 $d_{24}=d_{32}$)[22]:

XY 平面
$$d_{eeo} = d_{32}\sin^2\phi + d_{31}\cos^2\phi$$

YZ 平面
$$d_{ooe} = d_{32}\sin\theta$$

XZ 平面,$\theta < V_z$
$$d_{oeo} = d_{eoo} = d_{31}\sin\theta$$

XZ 平面,$\theta > V_z$
$$d_{ooe} = d_{31}\sin\theta$$

---

* 译者注:原书为"$R_{13}^S(pm \cdot V^{-1})$",应为"$r_{13}^S(pm \cdot V^{-1})$"。

在文献[22]中给出了任意方向上三波相互作用的有效二阶非线性系数。
$KNbO_3$ 二阶非线性系数的符号可能都是相同的[10,23]。

二阶非线性系数的绝对值[24]：

$$|d_{32}(0.852 \ \mu m)| = (11.0 \pm 0.6) pm/V$$

$$|d_{33}(0.852 \ \mu m)| = (22.3 \pm 1.1) pm/V$$

$$|d_{24}(1.064 \ \mu m)| = (12.5 \pm 0.6) pm/V$$

$$|d_{32}(1.064 \ \mu m)| = (10.8 \pm 0.6) pm/V$$

$$|d_{33}(1.064 \ \mu m)| = (19.6 \pm 1.0) pm/V$$

$$|d_{32}(1.313 \ \mu m)| = (9.2 \pm 0.5) pm/V$$

$$|d_{33}(1.313 \ \mu m)| = (16.1 \pm 0.8) pm/V$$

二阶非线性系数的相对值：

$$|d_{15}(1.064 \ \mu m)| = (41.2 \pm 0.8) \times d_{11}(SiO_2) = (12.4 \pm 0.2) pm/V^{[16,25]}$$

$$|d_{15}(1.064 \ \mu m)| = (9.2 \pm 0.2) pm/V^{[26]}$$

$$|d_{24}(1.064 \ \mu m)| = (42.8 \pm 0.8) \times d_{11}(SiO_2) = (12.8 \pm 0.2) pm/V^{[16,25]}$$

$$|d_{24}(1.064 \ \mu m)| = (13.0 \pm 0.4) pm/V^{[26]}$$

$$|d_{31}(1.064 \ \mu m)| = (39.5 \pm 0.6) \times d_{11}(SiO_2) = (11.9 \pm 0.2) pm/V^{[16,25]}$$

$$|d_{31}(1.064 \ \mu m)| = (8.9 \pm 0.4) pm/V^{[26]}$$

$$|d_{32}(1.064 \ \mu m)| = (45.7 \pm 0.6) \times d_{11}(SiO_2) = (13.7 \pm 0.2) pm/V^{[16,25]}$$

$$|d_{32}(1.064 \ \mu m)| = (12.4 \pm 0.3) pm/V^{[26]}$$

$$|d_{33}(1.064 \ \mu m)| = (68.5 \pm 0.6) \times d_{11}(SiO_2) = (20.6 \pm 0.2) pm/V^{[16,25]}$$

$$|d_{33}(1.064 \ \mu m)| = (21.9 \pm 0.5) pm/V^{[26]}$$

相位匹配角的实验值($T = 293$ K)

相互作用的波长/μm	$\phi_{exp}/(°)$	$\theta_{exp}/(°)$	参 考 文 献
XY 平面，$\theta = 90°$			
SHG，e + e⇒o			
0.946⇒0.473	≈30		[27]
4.759 9⇒2.379 95	69.9		[11]
YZ 平面，$\phi = 90°$			
SHG，o + o⇒e			
0.86⇒0.43		83.5	[28]
0.89⇒0.445		70.7	[28]
0.92⇒0.46		64	[28]
0.94⇒0.47		60.5	[28]
1.064 2⇒0.532 1		46.4	[11]

续表

相互作用的波长/μm	$\phi_{exp}/(°)$	$\theta_{exp}/(°)$	参 考 文 献
		≈47	[1]
1.318 8⇒0.659 4		30.6	[11]
1.338 2⇒0.669 1		29.7	[15]
3.530 3⇒1.765 15		37.3	[11]
4.729 1⇒2.364 55		77.3	[11]
SFG，o+o⇒e			
1.318 8+0.659 4⇒0.439 6		62.3	[11]
1.318 8+1.064 2⇒0.588 9		37.7	[11]
4.776 2+3.184 1⇒1.910 5		46.6	[11]
5.295 5+3.530 3⇒2.118 2		59.5	[11]
XZ 平面，$\phi=0°$，$\theta>V_z$			
SHG，o+o⇒e			
1.064 2⇒0.532 1		70.4	[4]
		71	[1]，[12]，[16]
		71.4	[11]
		71.5	[29]
1.318 8⇒0.659 4		56.8	[11]
1.338 2⇒0.669 1		56.2	[15]
3.530 3⇒1.765 15		58.8	[11]
SFG，o+o⇒e			
1.318 8+1.064 2⇒0.588 9		62.6	[11]
5.295 5+3.530 3⇒2.118 2		86.1	[11]

NCPM 温度的实验值

相互作用的波长/μm	$T/°C$	参 考 文 献
沿 X 轴		
SHG，I 类		
0.972⇒0.486	−20	[30]
0.982⇒0.491	18.7	[31]
	20	[23]
0.986⇒0.493	20	[32]
0.988⇒0.494	20	[12]
1.047⇒0.523 5	162	[33]
1.064 2⇒0.532 1	178	[34]
	181±2	[1]
	182	[4]
	184±2	[35]
	188	[36]

续表

相互作用的波长/μm	T/℃	参 考 文 献
沿 Y 轴		
SHG，I 类		
0.838 5⇒0.419 25	-34.2	[37]
0.840 6⇒0.420 3	-28.3	[38]
0.842⇒0.421	-22.8	[39]
0.846⇒0.423	-11.5	[13]
0.856⇒0.428	15	[40]
0.857⇒0.428 5	20	[41]
0.859 3⇒0.429 65	20	[37]
0.86⇒0.43	22	[32]
0.861 5⇒0.430 75	30	[42]
0.862⇒0.431	34	[43]
0.879⇒0.439 5	70	[43]
0.928 9⇒0.464 45	158	[37]
0.95⇒0.475	180	[32]
SFG，I 类		
0.676 4+1.064 2⇒0.413 55	-4	[44]
0.694 3+1.064 2⇒0.420 17	27.2	[44]

内角带宽的实验值

相互作用的波长/μm	T/℃	θ_{pm}/(°)	$\Delta\theta^{int}$/(°)	$\Delta\phi^{int}$/(°)
XZ 平面，$\phi=0°$				
SHG，o+o⇒e				
1.064 2⇒0.532 1	20	71	0.013~0.014	
沿 Y 轴				
SHG，I 类				
0.857⇒0.428 5	20	90	0.659	1.117

温度带宽的实验值

相互作用的波长/μm	T/℃	θ_{pm}/(°)	ΔT/℃	参考文献
沿 X 轴				
SHG，I 类				
0.982⇒0.491	18.7	90	0.95	[31]
1.064 2⇒0.532 1	181	90	0.27~0.32	[1]
	182	90	0.28	[4]

续表

相互作用的波长/μm	T/℃	θ_{pm}/(°)	ΔT/℃	参考文献
	184	90	0.28~0.29	[35]
	188	90	0.34	[36]
沿 Y 轴				
SHG，I 类				
0.838 5⇒0.419 25	-34.2	90	0.27	[37]
0.842⇒0.421	-22.8	90	0.30	[39]
0.855⇒0.427 5	26.4(?)	90	0.265	[12]
0.92⇒0.46	163.5(?)	90	0.285	[12]
SFG，I 类				
0.676 4 + 1.064 2⇒0.413 55	-4	90	0.35	[44]

在 $T = 295$ K 时温度带宽的实验值[15]

相互作用的波长/μm	θ_{exp}/(°)	ΔT/℃
YZ 平面，$\phi = 90°$		
SHG，o + o⇒e		
1.064 2⇒0.532 1	46.4	0.39
1.338 2⇒0.669 1	29.7	0.59
3.530 3⇒1.765 15	37.1	2.3
SFG，o + o⇒e		
5.295 5 + 3.530 3⇒2.118 2	59.5	2.4
XZ 平面，$\phi = 0°$，$\theta > V_Z$		
SHG，o + o⇒e		
1.064 2⇒0.532 1	71.4	0.77
1.338 2⇒0.669 1	56.2	2.2
3.530 3⇒1.765 15	58.9	10.1

非临界 SHG 的温度调谐[23]：

沿 X 轴：$\lambda_1 = 0.976\ 04 + 2.53 \times 10^{-4} T + 1.146 \times 10^{-6} T^2$

沿 Y 轴：$\lambda_1 = 0.850\ 40 + 2.94 \times 10^{-4} T + 1.234 \times 10^{-6} T^2$

其中 λ_1 以 μm 为单位，T 以 ℃ 为单位。

非临界 SHG 过程双折射的温度变化[12]：

沿 X 轴（1.064 2 μm⇒0.532 1 μm）：

$$\frac{d[n_Z(2\omega) - n_Y(\omega)]}{dT} = 1.10 \times 10^{-4} \text{K}^{-1}$$

沿 Y 轴（0.92 μm⇒0.46 μm）：

$$\frac{d[n_Z(2\omega)-n_X(\omega)]}{dT} = 1.43 \times 10^{-4} K^{-1}$$

激光诱导的表面损伤阈值

$\lambda/\mu m$	τ_p/ns	$I_{thr}/(GW \cdot cm^{-2})$	参 考 文 献	备 注
0.527	0.5	8.8 ~ 9.4	[45]	沿 b 轴，$E \parallel c$
		12 ~ 15	[45]	沿 b 轴，$E \perp c$
0.532 1	25	0.15 ~ 0.18	[35]	
	10	0.055	[34]	
0.8	0.000 2	>200	[46]	1 kHz
1.047	11	>0.03	[33]	4 kHz, 2 000 小时
1.054	0.7	11	[45]	沿 a 轴，$E \perp c$
		18	[45]	沿 b 轴，$E \perp c$
		37	[45]	沿 b 轴，$E \parallel c$
1.064 2	25	0.15 ~ 0.18	[35]	
	0.1	>100	[23]	

关于这一晶体

十年前，铌酸钾被广泛应用于 CW 激光二极管的倍频。到现今，周期性极化的非线性材料，诸如 PPLN 和 PPKTP 主要被人们所采用。最近也有人报道了周期性极化的 KN[6,21]。

参考文献

[1] Y. Uematsu: Nonlinear optical properties of KNbO$_3$ single crystals in the orthorhombic phase. Jpn. J. Appl. Phys. 13(9), 1362 – 1368(1974).

[2] B. H. T. Chai: Optical Crystals. In: CRC Handbook of Laser Science and Technology, Supplement 2: Optical Materials, ed. by M. J. Weber(CRC Press, Boca Raton, 1995), pp. 3 – 65.

[3] B. Zysset, I. Biaggio, P. Günter: Refractive indices of orthorhombic KNbO$_3$. I. Dispersion and temperature dependence. J. Opt. Soc. Am. B 9(3), 380 – 386(1992).

[4] Y. Uematsu, T. Fukuda: Nonlinear optical properties of KNbO$_3$ single crystals. Jpn. J. Appl. Phys. 10(4), 507(1971).

[5] *Chemist's Handbook*, *Vol. I*, ed. by B. P. Nikolskii(Goskhimizdat, Leningrad, 1962)[In Russian].

[6] J. H. Kim, C. S. Yoon: Domain switching characteristics and fabrication of periodically poled potassium niobate for second-harmonic generation. Appl. Phys. Lett. 81(18), 3332-3334(2002).

[7] M. D. Ewbank, M. J. Rosker, G. L. Bennett: Frequency tuning a mid-infrared optical parametric oscillator by the electro-optic effect. J. Opt. Soc. Am. B14(3), 666 – 671 (1997).

[8] L. E. Busse, L. Goldberg, M. R. Surette, G. Mizell: Absorption losses in MgO-doped and undoped potassium niobate. J. Appl. Phys. 75(2), 1102 – 1110(1994).

[9] Y. Uematsu, T. Fukuda: Characteristics and performance of $KNbO_3$-Nd: YAG intracavity second harmonic generation. Jpn. J. Appl. Phys. 12(6), 841 – 844(1973).

[10] W. R. Bosenberg, R. H. Jarman: Type-II phase-matched $KNbO_3$ optical parametric oscillator. Opt. Lett. 18(16), 1323 – 1325(1993).

[11] N. Umemura, K. Yoshida, K. Kato: Phase-matching properties of $KNbO_3$ in the mid-infrared. Appl. Opt. 38(6), 991 – 994(1999).

[12] K. Kato: High-efficiency second-harmonic generation at 4250 – 4680Å in $KNbO_3$. IEEE J. Quant. Electr. QE – 15(6), 410 – 411(1979).

[13] A. D. Ludlow, H. M. Nelson, S. D. Bergeson: Two-photon absorption in potassium niobate. J. Opt. Soc. Am. B 18(12), 1813 – 1820(2001).

[14] J. J. E. Reid: Resonantly enhanced, frequency doubling of an 820 nm GaAlAs diode laser in a potassium lithium niobate crystal. Appl. Phys. Lett. 62(1), 19 – 21(1993).

[15] N. Umemura, K. Yoshida, K. Kato: Thermo-optic dispersion formula of $KNbO_3$ for mid-infrared OPO. Proc. SPIE 3889, 472 – 480(2000).

[16] J.-C. Baumert, J. Hoffnagle, P. Günter: Nonlinear optical effects in $KNbO_3$ crystals at $Al_xGa_{1-x}As$, dye, ruby and Nd: YAG laser wavelengths. Proc. SPIE 492, 374 – 385 (1984).

[17] G. Ghosh: Dispersion of thermo-optic coefficients in a potassium niobate nonlinear crystal. Appl. Phys. Lett. 65(26), 3311 – 3313(1994).

[18] D. H. Jundt, P. Günter, B. Zysset: A temperature-dependent dispersion equation for $KNbO_3$. Nonl. Opt. 4(4), 341 – 345(1993).

[19] M. Sheik-Bahae, M. Ebrahimzadeh: Measurements of nonlinear refraction in the secondorder $\chi^{(2)}$ materials $KTiOPO_4$, $KNbO_3$, β-BaB_2O_4, and LiB_3O_5. Opt. Commun. 142 (4 – 6), 294 – 298(1997).

[20] I. P. Kaminow: Tables of Linear Electrooptic Coefficients. In: *CRC Handbook of Laser Science and Technology*, *Vol. III*, *Optical Materials: Part 2*, ed. by M. J. Weber(CRC Press, Boca Raton, 1986), pp. 253 – 278.

[21] J.-P. Meyn, M. E. Klein, D. Woll, R. Wallenstein, D. Rytz: Periodically poled potassium niobate for second-harmonic generation at 463 nm. Opt. Lett. 24 (16), 1154 – 1156 (1999).

[22] V. G. Dmitriev, D. N. Nikogosyan: Effective nonlinearity coefficients for three-wave interactions in biaxial crystals of *mm*2 point group symmetry. Opt. Commun. 95(1 – 3),

173 – 182(1993).

[23] I. Biaggio, P. Kerkoc, L.-S. Wu, P. Günter, B. Zysset: Refractive indices of orthorhombic $KNbO_3$. II. Phase-matching configurations for NLO interactions. J. Opt. Soc. Am. B 9(4), 507 – 517(1992).

[24] I. Shoji, T. Kondo, A. Kitamoto, M. Shirane, R. Ito: Absolute scale of second-order nonlinear-optical coefficients. J. Opt. Soc. Am. B 14(9), 2268 – 2294(1997).

[25] D. A. Roberts: Simplified characterization of uniaxial and biaxial nonlinear optical crystals: a plea for standardization of nomenclature and conventions. IEEE J. Quant. Electr. 28(10), 2057 – 2074(1992).

[26] M. V. Pack, D. J. Armstrong, A. V. Smith: Measurement of the $\chi^{(2)}$ tensor of the potassium niobate crystal. J. Opt. Soc. Am. 20(10), 2109 – 2116(2003).

[27] W. P. Risk, R. Pon, W. Lenth: Diode laser pumped blue-light source at 473 nm using intracavity frequency doubling of a 946 nm Nd: YAG laser. Appl. Phys. Lett. 54(17), 1625 – 1627(1989).

[28] Y. Lu, Q. Zhao, Y. Li, H. He, Q. Zou, Z. Lu, Z. Geng: Second-harmonic generation in $KNbO_3$ crystals. Opt. Eng. 32(4), 713 – 716(1993).

[29] S. Haidar, H. Ito: Periodically poled lithium niobate optical parametric oscillator pumped at 0.532 μm and use of its output to produce tunable 4.6 ~ 8.3 μm in $AgGaS_2$ crystal. Opt. Commun. 202(1 – 3), 227 – 231(2002).

[30] C. Zimmermann, T. W. Hansch, R. Byer, S. O'Brien, D. Welch: Second harmonic generation at 972 nm using a distributed Bragg reflection semiconductor laser. Appl. Phys. Lett. 61(23), 2741 – 2743(1992).

[31] D. Fluck, T. Pliska, P. Günter: Compact 10 mW all-solid-state 491 nm laser based on frequency doubling a master oscillator power amplifier laser diode. Opt. Commun. 123 (4 – 6), 624 – 628(1996).

[32] P. Günter: Near-infrared noncritically phase-matched second-harmonic generation in $KNbO_3$. Appl. Phys. Lett. 34(10), 650 – 652(1979).

[33] W. Seelert, P. Kortz, D. Rytz, B. Zysset, D. Ellgehausen, G. Mizell: Second-harmonic generation and degradation in critically phase-matched $KNbO_3$ with a diode-pumped Q-switched Nd: YLF laser. Opt. Lett. 17(20), 1432 – 1434(1992).

[34] K. Kato: High-efficiency high-power parametric oscillation in $KNbO_3$. IEEE J. Quant. Electr. QE – 18(4), 451 – 452(1982).

[35] V. A. Dyakov, V. I. Pryalkin, A. I. Kholodnykh: Potassium niobate optical parametric oscillator pumped by the second harmonic of a garnet laser. Kvant. Elektron. 8(4), 715-721(1981) [In Russian, English trans.: Sov. J. Quantum Electron. 11(4), 433 – 436(1981)].

[36] I. Biaggio, H. Looser, P. Günter: Intracavity frequency doubling of a diode pumped

Nd: YAG laser using a KNbO$_3$ crystal. Ferroelectrics 94, 157 – 161 (1989).

[37] J.-C. Baumert, P. Günter, H. Melchior: High efficiency second-harmonic generation in KNbO$_3$ crystals. Opt. Commun. 48(3), 215 – 220 (1983).

[38] A. Hemmerich, D. H. McIntyre, C. Zimmermann, T. W. Hansch: Second-harmonic generation and optical stabilization of a diode laser in an external ring resonator. Opt. Lett. 15(7), 372 – 374 (1990).

[39] M. K. Chun, L. Goldberg, J. F. Weller: Second-harmonic generation at 421 nm using injection-locked GaAlAs laser array and KNbO$_3$. Appl. Phys. Lett. 53 (13), 1170 – 1171 (1988).

[40] W. J. Kozlovsky, W. Lenth, E. E. Latta, A. Moser, G. L. Bona: Generation of 41 mW of blue radiation by frequency doubling of a GaAlAs diode laser. Appl. Phys. Lett. 56 (23), 2291 – 2292 (1990).

[41] J.-C. Baumert, J. Hoffnagle, P. Günter: High-efficiency intracavity doubling of a Styril – 9 dye laser radiation with KNbO$_3$ crystals. Appl. Opt. 24 (9), 1299 – 1301 (1985).

[42] P. Günter, P. M. Asbeck, S. K. Kurtz: Second-harmonic generation with Ga$_{1-x}$Al$_x$As lasers and KNbO$_3$ crystals. Appl. Phys. Lett. 35(6), 461 – 463 (1979).

[43] L. Goldberg, L. Busse, D. Mehuys: Blue light generation by frequency doubling of AlGaAs broad area amplifier emission. Appl. Phys. Lett. 60(9), 1037 – 1039 (1992).

[44] J.-C. Baumert, P. Günter: Noncritically phase-matched sum frequency generation and image up-conversion in KNbO$_3$ crystals. Appl. Phys. Lett. 50(10), 554 – 556 (1987).

[45] U. Ellenberger, R. Weber, J. E. Balmer, B. Zysset, D. Ellgehausen, G. J. Mizell: Pulsed optical damage threshold of potassium niobate. Appl. Opt. 31(36), 7563 – 7569 (1992).

第5章
周期性极化晶体及"衬底"材料

本章由用于准相位匹配的非线性光学晶体构成,除了前面讨论的 LN、KTP、MgLN、KTA 和 KN,还包括了钽酸锂(LT)、砷酸钛氧铷(RTA)、钛酸钡、氟化钡镁和砷化镓。

5.1　$LiTaO_3$,钽酸钾(LT)

负单轴晶:$n_o > n_e$
分子量:235.886
密度:7.43 g/cm^3[1];7.454 g/cm^3[2]
点群:$3m$
晶格常数:
　　$a = 5.143$ Å[3];5.154 3 Å[4]
　　$c = 13.756$ Å[3];13.783 5 Å[4]
莫氏硬度:6[1];6.7[2]
维氏硬度:766[2]

在 100 g 水中的溶解度[3]

T/K	s/g	T/K	s/g
273	0.001 2	248	0.009 0
298	0.002 5	373	0.012 0
323	0.005 4		

熔点：1 923 K[2]

居里温度：874~880 K[4]；874 K(同成分 LT,[Li]/[Ta] = 0.942)[5]；877 K (同成分 LT,[Li]/[Ta] = 0.942)[6,7]；958 K(化学计量比 LT)[5]；960 K(化学计量比 LT)[6]；961 K(化学计量比 LT)[7]；963 K(化学计量比 LT)[8]

线性热膨胀吸收系数[9]

T/K	$\alpha_t \times 10^6/\mathrm{K}^{-1}$, $\parallel c$	$\alpha_t \times 10^6/\mathrm{K}^{-1}$, $\perp c$
300	4.2	12.0

线性热膨胀吸收系数的平均值[2]

T/K	$\alpha_t \times 10^6/\mathrm{K}^{-1}$, $\parallel c$	$\alpha_t \times 10^6/\mathrm{K}^{-1}$, $\perp c$
273~773		15.4~16.1

温度范围 298 K < T < 773 K 内 $\parallel c$ 的热膨胀系数[4]：
$$L(T) = L(T_0)[1 + \alpha(T - 298) + \beta(T - 298)^2]$$
其中 T 以 K 为单位，$T_0 = 298$ K，$\alpha = 2.2 \times 10^{-6}$ K^{-1}，$\beta = -5.9 \times 10^{-9}$ K^{-2}。

温度范围 298 K < T < 773 K 之间 $\perp c$ 的热膨胀系数[4]：
$$L(T) = L(T_0)[1 + \alpha(T - 298) + \beta(T - 298)^2]$$
其中 T 以 K 为单位，$T_0 = 298$ K，$\alpha = 16.2 \times 10^{-6}$ K^{-1}，$\beta = 5.9 \times 10^{-9}$ K^{-2}。

$p = 0.101\ 325$ MPa 时的比热容 c_p[2]

T/K	$c_p/(\mathrm{J} \cdot \mathrm{kg}^{-1} \cdot \mathrm{K}^{-1})$
298	426

热导率[10]

T/K	$\kappa/(\mathrm{W} \cdot \mathrm{m}^{-1} \cdot \mathrm{K}^{-1})$
300	5

室温下带隙能量(直接跃迁): $E_g = 4.9$ eV[11]
室温下带隙能量(间接跃迁): $E_g = 4.1$ eV[11]
化学计量比 LT 的 UV 透过截止边为 0.26 μm[6]
以 1 cm^{-1} 计的 UV 透过截止边为 0.29 μm[12]
以 "0" 透过计的透明范围: 0.28 ~ 5.5 μm[8]

线性吸收系数 α

λ/μm	$\alpha/$cm^{-1}	参 考 文 献	备 注
0.325	1.7	[13]	
0.514 5	0.005 ~ 0.03	[12]	
1.064	0.001 ~ 0.003	[12]	
	0.001 5	[10]	化学计量比 LT

折射率的实验值[14]

λ/μm	n_o	n_e	λ/μm	n_o	n_e
0.45	2.242 0	2.246 8	2.0	2.106 6	2.111 5
0.50	2.216 0	2.220 5	2.2	2.100 9	2.105 3
0.60	2.183 4	2.187 8	2.4	2.095 1	2.099 3
0.70	2.165 2	2.169 6	2.6	2.089 1	2.093 6
0.80	2.153 8	2.157 8	2.8	2.082 5	2.087 1
0.90	2.145 4	2.149 3	3.0	2.075 5	2.079 9
1.00	2.139 1	2.143 2	3.2	2.068 0	2.072 7
1.20	2.130 5	2.134 1	3.4	2.060 1	2.064 9
1.40	2.123 6	2.127 3	3.6	2.051 3	2.056 1
1.60	2.117 4	2.121 3	3.8	2.042 4	2.047 3
1.80	2.112 0	2.117 0	4.0	2.033 5	2.037 7

化学计量比 LT 的 Sellmeier 方程(λ 以 μm 为单位, 0.44 μm < λ < 1.05 μm, T = 293 K)[7]:

$$n_o^2 = 4.528\ 1 + \frac{0.079\ 841}{\lambda^2 - 0.047\ 857} - 0.032\ 690\lambda^2$$

$$n_e^2 = 4.509\ 6 + \frac{0.082\ 712}{\lambda^2 - 0.041\ 306} - 0.031\ 587\lambda^2$$

同成分 LT([Li]/[Ta] = 0.942)的 n_o 几乎与化学计量比 LT 的相同,而同成分 LT 的 n_e 比化学计量比 LT 的大[7]。

在文献[15]、[16]中给出了其他 Sellmeier 方程。

化学计量比 LT 异常光折射率色散关系的温度关系(λ 以 μm 为单位,T 以 K 为单位,0.39 μm < λ < 4.1 μm,303 K < T < 473 K)[8]:

$$n_e^2(\lambda, T) = 4.502\,483 + \frac{0.007\,294 + 3.483\,933 \times 10^{-8} T^2}{\lambda^2 - [0.185\,087 + 1.607\,839 \times 10^{-8} T^2]^2}$$

$$+ \frac{0.073\,423}{\lambda^2 - 0.199\,595^2} + \frac{0.001}{\lambda^2 - 7.997\,24^2} - 0.023\,57\lambda^2$$

同一作者在[17]中给出了化学计量比 LT 异常光折射率稍有不同系数的相似色散关系。

化学计量比 LT 异常光折射率色散关系的温度关系(λ 以 μm 为单位,T 以 K 为单位,0.39 μm < λ < 4.1 μm,303 K < T < 473 K)[17]:

$$n_e^2(\lambda, T) = 4.514\,261 + \frac{0.011\,901 + 1.821\,94 \times 10^{-8} T^2}{\lambda^2 - [0.110\,744 + 1.566\,2 \times 10^{-8} T^2]^2}$$

$$+ \frac{0.076\,144}{\lambda^2 - 0.195\,596^2} - 0.023\,23\lambda^2$$

在文献[13]中给出了异常光折射率色散关系的其他温度关系。

非线性折射率 γ[18]

λ/μm	$\gamma \times 10^{15}$/(cm^2·W^{-1})	备注
0.8	3.0 ± 0.6	e 光
	1.7 ± 0.3	o 光

室温下低频(远低于 LT 晶体声学共振频率,即对"自由"晶体)下测量的线性电光系数

λ/μm	r_{13}^T/(pm·V^{-1})	r_{22}^T/(pm·V^{-1})	r_{33}^T/(pm·V^{-1})	r_{51}^T/(pm·V^{-1})	参考文献
0.632 8	+8.4 ± 0.9	≈0	+30.5 ± 0.9		[19]
		+0.1 ± 0.01			[20]
3.391 3	+4.5	+0.3	+27	+15	[21]

室温下高频(远高于 LT 晶体声学共振频率,即对"受夹"晶体)下测量的线性电光系数

$\lambda/\mu m$	$r_{13}^S/(pm \cdot V^{-1})$	$r_{22}^S/(pm \cdot V^{-1})$	$r_{33}^S/(pm \cdot V^{-1})$	$r_{51}^S/(pm \cdot V^{-1})$	参考文献
0.6328	6.2	≈0	28.5	8.4	[22]
	7.0		30.3		[23]
		1.0±0.1			[20]
1.1523	5.2	≈0	26.7	8.9	[22]
3.3913	4.4		25.2	7.0	[22]

矫顽场值:

  1.7 kV/mm(化学计量比 LT)[17]；1.7~4.5 kV/mm(化学计量比 LT)[8]

  21 kV/mm(同成分 LT)[24]；≈22 kV/mm(同成分 LT)[5]

钽酸锂二阶非线系数的表达式[25]：

$$|d_{33}(0.852\ \mu m)| = 15.1\ pm/V$$
$$|d_{31}(1.064\ \mu m)| = 0.85\ pm/V$$
$$|d_{33}(1.064\ \mu m)| = 13.8\ pm/V$$
$$|d_{33}(1.313\ \mu m)| = 10.7\ pm/V$$

激光诱导的表面损伤阈值[26]

$\lambda/\mu m$	τ_p/ns	$I_{thr}/(GW \cdot cm^{-2})$	备注
1.06	30	0.14	25 个脉冲
		0.22	10 个脉冲
		0.47	1 个脉冲

关于这一晶体

  钽酸锂与铌酸锂十分相似，然而，它的双折射率较低。因此，在这一晶体中不可能实现正常的双折射相位匹配。然而，这一晶体随着准相位匹配的发明而成为非常流行。除了它很低的矫顽场值(化学计量比 LT 是 1.7 kV/mm, 对于 1 mol% MgO 掺杂的化学计量比 LT 小于 1.5 kV/mm)，这种晶体具有更高的 UV 透过率，使其可以位于 UV 范围内在周期性极化 LT(PPLT)中以二次谐波或和频实现不同的准相位匹配过程。例如，在文献[27]和[13]中，分别产生了 340 nm 和 325 nm 的 SHG。在文献[28]中，通过在多重周期的 PPLT 中的 Nd:YVO₄ 激光辐射两阶段三次谐波发生，获得在 355 nm 处的和频。最近，利用准周期极化的 LT(APPLT)和一台 Nd:YVO₄ 激光器，产生 1064 nm 和 1342 nm 的激光，三种非线性过程(两种 SHG 和一种 SFG)都获得准相位匹配[29]。结果就同时产生了三种波长的光，称作交通信号光，位于绿色

(532 nm)、黄色(593 nm)和红色(671 nm)光谱区。也应该注意的是,在 170 ℃ 温度之上没有观察到光折变损伤[13]。

参考文献

[1] B. H. T. Chai: Optical Crystals. In: *CRC Handbook of Laser Science and Technology, Supplement 2: Optical Materials*, ed. by M. J. Weber(CRC Press, Boca Raton, 1995), pp. 3 – 65.

[2] A. A. Blistanov, V. S. Bondarenko, N. V. Perelomova, F. N. Strizhevskaya, V. V. Tchkalova, M. P. Shaskolskaya: *Acoustic Crystals*(Nauka, Moscow, 1982)[In Russian].

[3] Y. S. Kuzminov: *Lithium Niobate and Lithium Tantalate. Materials for Nonlinear Optics* (Nauka, Moscow, 1975).

[4] Y. S. Kim, R. T. Smith: Thermal expansion of lithium tantalate and lithium niobate crystals. J. Appl. Phys. 40(11), 4637 – 4641(1969).

[5] T. Hatanaka, K. Nakamura, T. Taniuchi, H. Ito, Y. Furukawa, K. Kitamara: Quasi-phasematched optical parametric oscillation with periodically poled stoichiometric LiTaO$_3$. Opt. Lett. 25(9), 651 – 653(2000).

[6] G. Ravi, R. Jayavel, S. Takekawa, M. Nakamura, K. Kitamura: Effect of niobium substitution in stoichiometric lithium tantalate(SLT) single crystals. J. Cryst. Growth 250(1 – 2), 146 – 151(2003).

[7] M. Nakamura, S. Higuchi, S. Takekawa, K. Terabe, Y. Furukawa, K. Kitamura: Refractive indices in undoped and MgO-doped near-stoichiometric LiTaO$_3$ crystals. Jpn. J. Appl. Phys. 41 (4B), L465-L467(2002).

[8] A. Bruner, D. Eger, M. B. Oron, P. Blau, M. Katz, S. Ruschin: Temperature-dependent Sellmeier equation for the refractive index of stoichiometric lithium tantalate. Opt. Lett. 28 (3), 194 – 196(2003).

[9] *Physical Quantities. Handbook*, ed. by I. S. Grigoriev, E. Z. Meilikhov(Energoatomizdat, Moscow, 1991)[In Russian].

[10] P. Blau, S. Pearl, A. Englander, A. Bruner, D. Eger: Average power effects in periodically poled crystals. Proc. SPIE 4972, 34 – 41(2003).

[11] S. Kase, K. Ohi: Optical absorption and interband Faraday rotation in LiTaO$_3$ and LiNbO$_3$. Ferroelectrics 8(1 – 2), 419 – 420(1974).

[12] A. L. Alexandrovski, G. Foulon, L. E. Myers, R. K. Route, M. M. Fejer: UV and visible absorption in LiTaO$_3$. Proc. SPIE 3610, 44 – 51(1999).

[13] J. -P. Meyn, M. M. Fejer: Tunable ultraviolet radiation by second-harmonic generation in periodically poled lithium tantalate. Opt. Lett. 22(16), 1214 – 1216(1997).

[14] W. L. Bond: Measurement of the refractive indices of several crystals. J. Appl. Phys. 36

(5), 1674 – 1677(1965).

[15] S. Matsumoto, E. J. Lim, H. M. Hertz, M. M. Fejer: Quasi phase-matched second harmonic generation of blue light in electrically periodically-poled lithium tantalate waveguides. Electron. Lett. 27(22), 2040 – 2042(1991).

[16] K. S. Abedin, H. Ito: Temperature-dependent dispersion relation of ferroelectric lithium tantalate. J. Appl. Phys. 80(11), 6561 – 6563(1996).

[17] A. Bruner, D. Eger, M. Oron, P. Blau, M. Katz, S. Ruschin: Refractive index dispersion measurements of congruent and stoichiometric $LiTaO_3$. Proc. SPIE 4628, 66 – 73 (2002).

[18] S. Ashihara, J. Nishina, T. Shimura, K. Kuroda, T. Sugita, K. Mizuuchi, K. Yamamoto: Nonlinear refraction of femtosecond pulses due to quadratic and cubic nonlinearities in periodically poled lithium tantalate. Opt. Commun. 222(1 – 6), 421 – 427(2003).

[19] K. Onuki, N. Uchida, T. Saku: Interferometric method for measuring electro-optic coefficients in crystals. J. Opt. Soc. Am. 62(9), 1030 – 1032(1972).

[20] M. Abarkan, J. P. Salvestrini, M. D. Fontana, M. Aillerie: Frequency and wavelength dependencies of electro-optic coefficients in inorganic crystals. Appl. Phys. B 76(7), 765 – 769(2003).

[21] A. Yariv, P. Yeh: *Optical Waves in Crystals*(JohnWiley & Sons, NewYork, 1984).

[22] I. P. Kaminow: Tables of Linear Electrooptic Coefficients. In: *CRC Handbook of Laser Science and Technology, Vol. III, Optical Materials: Part 2*, ed. by M. J. Weber(CRC Press, Boca Raton, 1986), pp. 253 – 278.

[23] P. V. Lenzo, E. H. Turner, E. G. Spencer, A. A. Ballman: Electrooptic coefficients and elastic-wave propagation in single-domain ferroelectric lithium tantalate. Appl. Phys. Lett. 8(4), 81 – 82(1966).

[24] J. -P. Meyn, C. Laue, R. Knappe, R. Wallenstein, M. M. Fejer: Fabrication of periodically poled lithium tantalate for UV generation with diode lasers. Appl. Phys. B. 73(2), 111 – 114(2001).

[25] I. Shoji, T. Kondo, A. Kitamoto, M. Shirane, R. Ito: Absolute scale of second-order nonlinear-optical coefficients. J. Opt. Soc. Am. B 14(9), 2268 – 2294(1997).

[26] G. M. Zverev, E. A. Levchuk, V. A. Pashkov, Y. D. Poryadin: Laser-radiation-induced damage to the surface of lithium niobate and tantalate single crystals. Kvant. Elektron. No. 2, 94 – 96 (1972) [In Russian, English trans. : Sov. J. Quantum Electron. 2 (2), 167 – 169 (1972)].

[27] K. Mizuuchi, K. Yamamoto: Generation of 340-nm light by frequency doubling of a laser diode in bulk periodically poled $LiTaO_3$. Opt. Lett. 21(2), 107 – 109(1996).

[28] Z. W. Liu, S. N. Zhu, Y. Y. Zhu, Y. Q. Qin, J. L. He, C. Zhang, H. T. Wang, N. B. Ming, X. Y. Liang, Z. Y. Xu: Quasi-CW ultraviolet generation in a dual-periodic

LiTaO₃ superlattice by frequency tripling. Jpn. J. Appl. Phys. 40(12), 6841 – 6844 (2001).

[29] J. -L. He, J. Liao, H. Liu, J. Du, F. Xu, H. -T. Wang, S. N. Zhu, Y. Y. Zhu, N. B. Ming: Simultaneous CW red, yellow, and green light generation, "traffic signal lights", by frequency doubling and sum-frequency mixing in an aperiodically poled LiTaO₃. Appl. Phys. Lett. 83(2), 228 – 230(2003).

5.2 RbTiOAsO₄, 砷酸钛氧铷(RTA)

正双轴晶: $2V_Z = 39.2°$, $\lambda = 0.532$ μm
分子量: 288.266
密度: 4.018 g/cm³[1]; 4.05 g/cm³[2]
点群: $mm2$
晶格常数:
  $a = 13.2428$ Å[2]; 13.257 Å[3]
  $b = 6.6685$ Å[2]; 6.6780 Å[3]
  $c = 10.7642$ Å[2]; 10.765 Å[3]
介电轴和晶体学轴间的变换: $X, Y, Z \Rightarrow a, b, c$
熔点: 1 383 K[3]
热导率[4]:

T/K	$\kappa/(\mathrm{W} \cdot \mathrm{m}^{-1} \cdot \mathrm{K}^{-1})$
300	1.6

以"0"透过计的透明范围: 0.35 ~ 5.3 μm[5]; 0.35 ~ 5.1 μm[2]; 0.38 ~ 5.1 μm[6]
UV 透过截止边 ($\alpha = 2$ cm⁻¹) 是在 0.358 μm ($E \parallel X$); 0.366 μm ($E \parallel Y$); 0.371 μm ($E \parallel Z$)[7]
线性吸收系数 α[7]

λ/μm	α/cm⁻¹	备注	λ/μm	α/cm⁻¹	备注
0.473	0.012	$E \parallel X$	0.532	0.015	$E \parallel X$
	0.002	$E \parallel Y$		0.002	$E \parallel Y$
	0.005	$E \parallel Z$		0.002	$E \parallel Z$

在 $T = 293$ K 时折射率的实验值

$\lambda/\mu m$	n_X	n_Y	n_Z	参 考 文 献
0.486 13		1.872 0	1.964 3	[2]
0.532 0	1.847 6	1.857 8	1.944 4	[2]
0.546 07	1.844 4	1.854 3	1.939 7	[2]
0.587 56	1.836 4	1.845 6	1.927 9	[2]
0.632 8	1.829 4	1.836 3	1.918 5	[8]
0.656 28	1.826 7	1.835 2	1.914 2	[2]
1.064 00	1.804 1	1.811 4	1.884 6	[2]
			1.880 8	[9]

折射率 n_Z 的温度微商[4]

$\lambda/\mu m$	$dn_Z/dT \times 10^6/K^{-1}$	$\lambda/\mu m$	$dn_Z/dT \times 10^6/K^{-1}$
1.064	2	1.5	0.3

高精确度的 Sellmeier 方程(λ 以 μm 为单位,$T = 293$ K)[10]

$$n_X^2 = 3.219\,92 + \frac{0.047\,63}{\lambda^2 - 0.040\,63} - 0.010\,35\lambda^2$$

$$n_Y^2 = 3.241\,85 + \frac{0.050\,56}{\lambda^2 - 0.045\,32} - 0.010\,62\lambda^2$$

$$n_Z^2 = 7.002\,29 + \frac{0.067\,87}{\lambda^2 - 0.052\,41} + \frac{917.990\,6}{\lambda^2 - 261.362\,9}$$

在文献[5]、[9]、[11]、[12]、[13]、[14]中给出了其他色散关系方程组。
RTA 晶体在温度范围 $T = 293 \sim 393$ K 以及光谱范围 $0.45\ \mu m < \lambda < 1.62\ \mu m$ 中折射率的温度微商(λ 以 μm 为单位)[10]:

$$\frac{dn_X}{dT} = \left(\frac{0.428\,7}{\lambda^3} - \frac{0.918\,1}{\lambda^2} + \frac{0.668\,5}{\lambda} + 1.968\,7\right) \times 10^{-5}\ K^{-1}$$

$$\frac{dn_Y}{dT} = \left(\frac{0.513\,8}{\lambda^3} - \frac{1.105\,4}{\lambda^2} + \frac{0.803\,5}{\lambda} + 1.959\,1\right) \times 10^{-5}\ K^{-1}$$

光谱范围 $0.45\ \mu m < \lambda < 3.2\ \mu m$($\lambda$ 以 μm 为单位)[10]:

$$\frac{dn_Z}{dT} = \left(\frac{1.590\,5}{\lambda^3} - \frac{4.242\,3}{\lambda^2} + \frac{4.216\,1}{\lambda} + 1.735\,5\right) \times 10^{-5}\ K^{-1}$$

室温下低频(远低于 RTA 晶体声学共振频率,即对"自由"晶体)下测量的线性电光系数

λ/μm	r_{13}^T/(pm·V^{-1})	r_{23}^T/(pm·V^{-1})	r_{33}^T/(pm·V^{-1})	r_{51}^T/(pm·V^{-1})	r_{42}^T/(pm·V^{-1})	参考文献
0.6328	13.5±1.4	17.5±1.8	40.5±4.1			[12]
	10.8±1.0	17.3±1.0	40.0±1.5	12.3±1.0	14.6±1.0	[8]

矫顽场值: 1.76 kV/mm[15]

RTA 晶体主平面上有效二阶非线系数的表达式(Kleinman 对称条件不成立)[16]:

XY 平面
$$d_{eoe} = d_{oee} = d_{15}\sin^2\phi + d_{24}\cos^2\phi$$

YZ 平面
$$d_{oeo} = d_{eoo} = d_{15}\sin\theta$$

XZ 平面, $\theta < V_z$
$$d_{ooe} = d_{32}\sin\theta$$

XZ 平面, $\theta > V_z$
$$d_{oeo} = d_{eoo} = d_{24}\sin\theta$$

RTA 晶体主平面上有效二阶非线性系数的表达式(Kleinman 对称条件成立, $d_{15} = d_{31}, d_{24} = d_{32}$)[16]:

XY 平面
$$d_{eoe} = d_{oee} = d_{31}\sin^2\phi + d_{32}\cos^2\phi$$

YZ 平面
$$d_{oeo} = d_{eoo} = d_{31}\sin\theta$$

XZ 平面, $\theta < V_z$
$$d_{ooe} = d_{32}\sin\theta$$

XZ 平面, $\theta > V_z$
$$d_{oeo} = d_{eoo} = d_{32}\sin\theta$$

在文献[16]中给出了 RTA 晶体任意方向上三波相合作用的有效二阶非线性系数。

RTA 二阶非线性系数的符号可能全是相同的[17]。

二阶非线性系数的绝对和相对值:
$$d_{31}(1.064\ \mu m) = (2.3\pm0.5)\ pm/V^{[11]}$$
$$d_{31}(1.064\ \mu m) = 3.55\times d_{36}(KDP) = 1.4\ pm/V^{[2,18]}$$
$$d_{32}(1.064\ \mu m) = (3.8\pm0.7)\ pm/V^{[11]}$$

$$d_{32}(1.064 \ \mu m) = 11.71 \times d_{36}(KDP) = 4.6 \ pm/V^{[2,18]}$$

$$d_{33}(1.064 \ \mu m) = (15.8 \pm 1.6) \ pm/V^{[11]}$$

$$d_{33}(1.064 \ \mu m) = 31.05 \times d_{36}(KDP) = 12.1 \ pm/V^{[2,18]}$$

RTA晶体中特殊相位匹配方向上有效二阶非线性系数的实验值(SHG,I类,1.064 2⇒0.532 1 μm)[2]

相互作用的波长/μm	d_{eff}/(pm·V^{-1})
$\theta = 52.7°$, $\phi = 39.8°$	1.33

相位匹配角和温度带宽的实验值

相互作用的波长/μm	ϕ_{exp}/(°)	θ_{exp}/(°)	ΔT/℃	参考文献
XY 平面, $\theta = 90°$				
SFG, e+o⇒e				
1.613 2 + 0.641 2⇒0.458 8	5.7		14.6	[10]
YZ 平面, $\phi = 90°$				
SHG, o+e⇒o				
1.318 8⇒0.659 4		61.2	35.2	[10]
SFG, o+e⇒o				
1.318 8 + 1.064 2⇒0.588 95		65.3	59.3	[10]
1.613 2 + 1.064 2⇒0.641 2		52.0	39.8	[10]
1.613 2 + 0.641 2⇒0.458 8		64.9	16.9	[10]
SFG, e+o⇒o				
3.127 1 + 1.613 2⇒1.064 2		69.1	61.0	[10]
XZ 平面, $\phi = 0°$, $\theta > V_z$				
SHG, o+e⇒o				
1.318 8⇒0.659 4		73.7		[6]
		73.5	29.3	[10]
SFG, o+e⇒o				
1.318 8 + 1.064 2⇒0.588 95		82.0	47.8	[10]
1.613 2 + 1.064 2⇒0.641 2		62.2	26.8	[10]

续表

相互作用的波长/μm	$\phi_{exp}/(°)$	$\theta_{exp}/(°)$	$\Delta T/℃$	参考文献
SFG，e + o ⇒ o				
3.127 1 + 1.613 2 ⇒ 1.064 2		86.2	54.3	[10]
DFG，o - e ⇒ o				
1.064 2 - 1.510 8 ⇒ 3.6		45.1		[5]
1.064 2 - 1.450 0 ⇒ 4.0		44.8		[5]

激光诱导的损伤阈值

$\lambda/\mu m$	τ_p/ns	$I_{thr}/(GW \cdot cm^{-2})$	参考文献	备注
0.74 ~ 0.84	0.000 2	> 200	[19]	1 kHz
1.064	10 ~ 20	> 0.4	[20]	PPRTA
	5.5	> 0.1	[15]	1 kHz，PPRTA

关于这一晶体

  1996 年报道了熔盐法生长 RTA 晶体的极化获得成功[21]。如同 PPKTP 一样，它比周期极化铌酸锂晶体有一些实际优越性，矫顽场的值要小一个量级，这就可能容许以更大的口径来生产 PPRTA[15]，对于热透镜效应和相失配的敏感度低，并且不存在光折变损伤，这就使得它在室温下运转稳定。与 PPKTA 相比，周期性极化的 RTA 晶体的红外截止边更长(KTP 约为 4.4 μm，而 RTA 约为 5.2 μm)，并且在 3.5 μm 处没有吸收[7]。这样的性质使得 PPRTA 更适用于中红外 OPO[4,15,22-26] 和 DFG[14,20] 系统。目前可用的 PPRTA 晶体的唯一缺点是，比较起 50 mm 标准长度的 PPLN 元件来说，PPRTA 晶体的尺寸太短了(20 mm 以下)。

参考文献

[1] B. H. T. Chai: Optical Crystals. In: CRC Handbook of Laser Science and Technology, Supplement 2: Optical Materials, ed. by M. J. Weber(CRC Press, Boca Raton, 1995), pp. 3 - 65.

[2] J. Han, Y. Liu, M. Wang, D. Nie: Flux growth and properties of RbTiOPO$_4$ (RTA) crystals. J. Cryst. Growth 128(1 - 4), 864 - 866(1993).

[3] L. K. Cheng, E. M. McCarron III, J. Calabrese, J. D. Bierlein, A. A. Ballman: Development of the nonlinear optical crystal CsTiOAsO$_4$. I. Structural stability. J. Cryst.

Growth 132(1-2), 280 – 288(1993).

[4] A. Carleton, D. J. M. Stothard, I. D. Lindsay, M. Ebrahimzadeh, M. H. Dunn: Compact, continuous-wave, singly resonant optical parametric oscillator based on periodically poled RbTiOAsO$_4$ in a Nd: YVO$_4$ laser. Opt. Lett. 28(7), 555 – 557(2003).

[5] D. L. Fenimore, K. L. Schepler, D. Zelmon, S. Kück, U. B. Ramabadran, P. von Richter, D. Small: Rubidium titanyl arsenate difference-frequency generation and validation of new Sellmeier coefficients. J. Opt. Soc. Am B 13(9), 1935 – 1940(1996).

[6] J. Wang, J. Wei, Y. Liu, X. Yin, X. Hu, Z. Shao, M. Jiang: A survey of research on KTP and its analogue crystals. Progr. Cryst. Growth Character. Mater. 40(1 – 4), 3 – 15 (2000).

[7] G. Hansson, H. Karlsson, S. Wang, F. Laurell: Transmission measurements in KTP and isomorphic compounds. Appl. Opt. 39(27), 5058 – 5069(2000).

[8] X. Yin, J. Y. Wang: Electro-optic property of RbTiOPO$_4$ (RTA) crystal. Nonl. Opt. 23 (2), 93 – 96(2000).

[9] J.-P. Feve, B. Boulanger, O. Pacaud, I. Rousseau, B. Menaert, G. Marnier, P. Villeval, C. Bonnin, G. M. Loiacono, D. N. Loiacono: Phase-matching measurements and Sellmeier equations over the complete transparency range of KTiOAsO$_4$, RbTiOAsO$_4$, and CsTiOAsO$_4$. J. Am. Opt. Soc. B 17(5), 775 – 780(2000).

[10] K. Kato, E. Takaoka, N. Umemura: Thermo-optic dispersion formula for RbTiOAsO$_4$. Jpn. J. Appl. Phys. 42(10), 6420 – 6423(2003).

[11] L.-T. Cheng, L. K. Cheng, J. D. Bierlein: Linear and nonlinear optical properties of the arsenate isomorphs of KTP. Proc. SPIE 1863, 43 – 53(1993).

[12] L. K. Cheng, L. T. Cheng, J. Galperin, P. A. Morris Hotsenpiller, J. D. Bierlein: Crystal growth and characterization of KTiOPO$_4$ isomorphs from the self-fluxes. J. Cryst. Growth 137(1 – 2), 107 – 115(1994).

[13] J. P. Feve, B. Boulanger, O. Pacaud, I. Rousseau, B. Menaert, G. Marnier: Refined Sellmeier equations from phase-matching measurements over the complete transparency range of KTiOAsO$_4$, RbTiOAsO$_4$, and CsTiOAsO$_4$. In: *Advanced Solid-State Lasers*, *OSA Trends in Optics and Photonics Series*, *Vol. 34*, ed. by H. Injeyan, U. Keller, C. Marshall(OSA, Washington DC, 2000), pp. 575 – 577.

[14] K. Fradkin-Kashi, A. Arie, P. Urenski, G. Rosenman: Characterization of optical and nonlinear properties of periodically-poled RbTiOAsO$_4$ in the mid-infrared range via difference-frequency generation. Appl. Phys. B 71(2), 251 – 255(2000).

[15] H. Karlsson, M. Olson, G. Arvidsson, F. Laurell, U. Bäder, A. Borsutzky, R. Wallenstein, S. Wickström, M. Gustafsson: Nanosecond optical parametric oscillator based on large-aperture periodically poled RbTiOAsO$_4$. Opt. Lett. 24(5), 330 – 332(1999).

[16] V. G. Dmitriev, D. N. Nikogosyan: Effective nonlinearity coefficients for three-wave interactions in biaxial crystals of *mm*2 point group symmetry. Opt. Commun. 95(1 – 3), 173 – 182(1993).

[17] A. Anema, T. Rasing: Relative signs of the nonlinear coefficients of potassium titanyl phosphate. Appl. Opt. 36(24), 5902 – 5904(1997).

[18] I. Shoji, T. Kondo, A. Kitamoto, M. Shirane, R. Ito: Absolute scale of second-order nonlinear-optical coefficients. J. Opt. Soc. Am. B 14(9), 2268 – 2294(1997).

[19] V. Petrov, F. Noack, R. Stolzenberger: Seeded femtosecond optical parametric amplification in the mid-infrared spectral region above 3 μm. Appl. Opt. 36(6), 1164 – 1172 (1997).

[20] W. Chen, G. Mouret, D. Boucher, F. K. Tittel: Mid-infrared trace gas detection using continuous-wave difference frequency generation in periodically poled $RbTiOAsO_4$. Appl. Phys. B 72(7), 873 – 876(2001).

[21] H. Karlsson, F. Laurell, P. Henriksson, G. Arvidsson: Frequency doubling in periodically poled $RbTiOAsO_4$. Electron. Lett. 32(6), 556 – 557(1996).

[22] D. T. Reid, Z. Penman, M. Ebrahimzadeh, W. Sibbett, H. Karlsson, F. Laurell: Broadly tunable infrared femtosecond optical parametric oscillator based on periodically poled $RbTiOAsO_4$. Opt. Lett. 22(18), 1397 – 1399(1997).

[23] D. T. Reid, G. T. Kennedy, A. Miller, W. Sibbett, M. Ebrahimzadeh: Widely tunable, nearto mid-infrared femtosecond and picosecond optical parametric oscillators using periodically poled $LiNbO_3$ and $RbTiOAsO_4$. IEEE J. Sel. Topics Quant. Electr. 4(2), 238 – 248(1998).

[24] G. T. Kennedy, D. T. Reid, A. Miller, M. Ebrahimzadeh, H. Karlsson, G. Arvidsson, F. Laurell: Near- to mid-infrared picosecond optical parametric oscillator based on periodically poled $RbTiOAsO_4$. Opt. Lett. 23(7), 503 – 505(1998).

[25] T. J. Edwards, G. A. Turnbull, M. H. Dunn, M. Ebrahimzadeh, H. Karlsson, G. Arvidsson, F. Laurell: Continuous-wave singly resonant optical parametric oscillator based on periodically poled $RbTiOAsO_4$. Opt. Lett. 23(11), 837 – 839(1998).

[26] P. Loza-Alvarez, D. T. Reid, M. Ebrahimzadeh, W. Sibbett, H. Karlsson, P. Henriksson, G. Arvidsson, F. Laurell: Periodically poled $RbTiOAsO_4$ femtosecond optical parametric oscillator tunable from 1.38 to 1.58 μm. Appl. Phys. B 68(2), 177 – 180(1999).

5.3　$BaTiO_3$，钛酸钡

负单轴晶：$n_o > n_e$（278 K < T < 393 K）

分子量：233.208

密度：5.9 g/cm^3[1]；6.02 g/cm^3，278 K < T < 393 K[2]；6.017 g/cm^3，T > 393 K[2]

点群[2]：

 $3m$，T < 183 K

 $mm2$，183 K < T < 278 K

 $4mm$，278 K < T < 393 K

 $m3m$，T > 393 K

点群 $4mm$ 时晶格常数：

 a = 3.985 Å[2]；3.992 0 Å[3]；3.994 Å[4]

 c = 4.020 Å[2]；4.0361 Å[3]；4.038 Å[4]

莫氏硬度：5[5]

维氏硬度：200 ~ 580[4]

在水中的溶解度：不溶[2]

熔点：1 870 K[6]；1 898 K[4,2]

居里温度：393 K[7]；405 K[8]；(406 ± 2) K[9]

线性热膨胀吸收系数

T/K	$\alpha_t \times 10^6$/K^{-1}	参考文献	T/K	$\alpha_t \times 10^6$/K^{-1}	参考文献
300	11.4	[10]	673	11.3	[4]
393	8.6	[4]	773	12.3	[10]
400	8.6	[10]	873	13.2	[4]
473	9.4	[4]	973	14.2	[4]
623	10.8	[4]	1 073	15.1	[4]

线性热膨胀吸收系数的平均值[2]

T/K	$\alpha_t \times 10^6$/K^{-1}	T/K	$\alpha_t \times 10^6$/K^{-1}	T/K	$\alpha_t \times 10^6$/K^{-1}
113 ~ 174	8.8	277 ~ 293	11.4	397 ~ 583	10.3
174 ~ 277	11.4	303 ~ 395	5.1		

p = 0.101 325 MPa 时的比热容 c_p

T/K	c_p/(J·kg^{-1}·K^{-1})	参考文献	T/K	c_p/(J·kg^{-1}·K^{-1})	参考文献
80	135	[10]	125	239	[4]
100	185	[4]	150	286	[10]

续表

T/K	$c_p/(J\cdot kg^{-1}\cdot K^{-1})$	参考文献	T/K	$c_p/(J\cdot kg^{-1}\cdot K^{-1})$	参考文献
	288	[4]	350	454	[2]
175	322	[6]	400	484	[10]
	327	[4]	600	520	[10]
250	409	[10]	1 000	549	[10]
300	439	[5]			

热导率

T/K	$\kappa/(W\cdot m^{-1}\cdot K^{-1})$	参 考 文 献
293	1.34	[6]
401	0.67	[4]

室温下带隙能量：$E_g = 3.1$ eV[11]

以 $\alpha = 1$ cm^{-1} 计的透明范围：$0.4 \sim 9$ μm[9]

线性吸收系数 α

$\lambda/\mu m$	α/cm^{-1}	参 考 文 献	备 注
0.41	1.0	[9]	$T = 306$ K, $\boldsymbol{E} \parallel c$
0.423	1.0	[9]	$T = 306$ K, $\boldsymbol{E} \perp c$
0.514 5	2.14	[12]	$\boldsymbol{E} \perp c$
	1.16	[12]	$\boldsymbol{E} \parallel c$

双光子吸收系数 β[11]

$\lambda/\mu m$	τ_p/ns	$\beta \times 10^{11}/(cm\cdot W^{-1})$	备 注
0.596	0.001	10	$\perp c, \boldsymbol{E} \perp c$

折射率的实验值

$\lambda/\mu m$	n_o	n_e	参 考 文 献
0.5145	2.494	2.431	[13]
0.55	2.458	2.399	[9]
0.589	2.428	2.371	[14]

折射率的温度微商[12]

$\lambda/\mu m$	$\Delta T/K$	$dn_o/dT \times 10^6/K^{-1}$	$dn_e/dT \times 10^6/K^{-1}$
0.514 5	298~319	≈0	
	297~315		140

修正的 Sellmeier 方程(λ 以 μm 为单位)[15]:

$$n_o^2 = 3.058\,40 + \frac{2.273\,26\lambda^2}{\lambda^2 - 0.074\,09} - 0.024\,28\lambda^2$$

$$n_e^2 = 3.024\,79 + \frac{2.140\,62\lambda^2}{\lambda^2 - 0.067\,007} - 0.021\,69\lambda^2$$

文献[9]、[16]中给出了其他色散关系方程组。

室温下低频(远低于 $BaTiO_3$ 晶体声学共振频率,即对"自由"晶体)下测量的线性电光系数

$\lambda/\mu m$	$r_{13}^T/(pm \cdot V^{-1})$	$r_{33}^T/(pm \cdot V^{-1})$	$r_{51}^T/(pm \cdot V^{-1})$	参考文献	备 注
0.514 5	19.5±1			[13]	T = 296 K
		97±7		[13]	T = 296 K
0.546 1			1 640	[7]	T = 298 K

室温下高频(远高于 $BaTiO_3$ 晶体声学共振频率,即对"受夹"晶体)下测量的线性电光系数

$\lambda/\mu m$	$r_{13}^S/(pm \cdot V^{-1})$	$r_{33}^S/(pm \cdot V^{-1})$	$r_{51}^S/(pm \cdot V^{-1})$	参考文献	备 注
0.541 6			820	[17]	T = 295 K
0.632 8	8			[18]	
		28		[18]	

Verdet 常数,$T = 403$ K[19]

$\lambda/\mu m$	$V/[(°) \cdot T^{-1} \cdot m^{-1}]$
0.620	-2 920

矫顽场值:≈0.1 kV/mm[15]

有效二阶非线性系数的表达式(Kleinman 对称条件成立,$d_{15} = d_{24} = d_{31} = d_{32}$)[20]:

$$d_{ooe} = d_{31}\sin\theta$$

二阶非线性系数值[21]以 d_{36}(KDP)的新绝对值重新计算[22]：

$$d_{15}(1.06~\mu m) = (13.7 \pm 1.2)~pm/V$$
$$d_{32}(1.06~\mu m) = (14.4 \pm 1.2)~pm/V$$
$$d_{33}(1.06~\mu m) = (5.5 \pm 0.4)~pm/V$$

激光诱导的体损伤阈值[11]

$\lambda/\mu m$	τ_p/ns	$I_{thr}/(GW \cdot cm^{-2})$
0.596	0.001	>83

关于这一晶体

  钛酸钡晶体的非线性光学性质在 40 年以前就被人们所熟知[21]。然而，由于其双折射率低，在这种晶体中不可能实现正常的相位匹配。重新开始关于这种晶体的兴趣[15]是与其准相位匹配装置的应用性相关的。低的矫顽场值(≈0.1 KV/mm)使其可以制备大口径的波导。另一个优点是它的高透过率，在 IR 范围最高达 9 μm。

■ 参考文献

[1] B. H. T. Chai: Optical Crystals. In: *CRC Handbook of Laser Science and Technology, Supplement 2: Optical Materials*, ed. by M. J. Weber(CRC Press, Boca Raton, 1995), pp. 3 – 65.

[2] A. A. Blistanov, V. S. Bondarenko, N. V. Perelomova, F. N. Strizhevskaya, V. V. Tchkalova, M. P. Shaskolskaya: *Acoustic Crystals* (Nauka, Moscow, 1982) [In Russian].

[3] *Handbook of Optical Constants of Solids II*, ed. by E. D. Palik(Academic Press, Boston, 1991).

[4] E. M. Voronkova, B. N. Grechushnikov, G. I. Distler, I. P. Petrov: *Optical Materials for Infrared Technique* (Nauka, Moscow, 1965) [In Russian].

[5] *Handbook of Optical Materials*, ed. by M. J. Weber (CRC Press, Boca Raton, 2003), pp. 1 – 512.

[6] S. S. Ballard, J. S. Browder: Thermal Properties. In: *CRC Handbook of Laser Science and Technology, Vol. IV, Optical Materials: Part 2*, ed. by M. J. Weber(CRC Press, Boca Raton, 1987), pp. 49 – 54.

[7] A. R. Johnston, J. M. Weingart: Determination of the low-frequency linear electro-optic effect in tetragonal $BaTiO_3$. J. Opt. Soc. Am. 55(7), 828 – 834(1965).

[8] C. J. Johnson: Some dielectric and electro-optic properties of $BaTiO_3$ single crystals.

Appl. Phys. Lett. 7(8), 221–223(1965).

[9] S. H. Wemple, M. DiDomenico, Jr., I. Camlibel: Dielectric and optical properties of melt-grown $BaTiO_3$. J. Phys. Chem. Solids 29(10), 1797–1803(1968).

[10] *Physical Quantities. Handbook*, ed. by I. S. Grigoriev, E. Z. Meilikhov (Energoatomizdat, Moscow, 1991) [In Russian].

[11] T. F. Boggess, J. O. White, G. C. Valley: Two-photon absorption and anisotropic transient energy transfer in $BaTiO_3$ with 1-psec excitation. J. Opt. Soc. Am. B 7(12), 2255–2258(1990).

[12] D. W. Rush, B. M. Dugan, G. L. Burdge: Temperature-dependent index-of-refraction changes in $BaTiO_3$. Opt. Lett. 16(17), 1295–1297(1991).

[13] S. Ducharme, J. Feinberg, R. R. Neurgaonkar: Electrooptic and piezoelectric measurements in photorefractive barium titanate and strontium barium titanate. IEEE J. Quant. Electr. QE-23(12), 2116–2121(1987).

[14] L. V. Deshpande, M. B. Joshi, R. B. Mishra: Angle of polarization and refractive indices of $BaTiO_3$. J. Opt. Soc. Am. 70(9), 1163–1166(1980).

[15] S. D. Setzler, P. G. Schunemann, T. M. Pollak, L. A. Pomeranz, M. J. Missey, D. E. Zelmon: Periodically poled barium titanate as a new nonlinear optical material. In: *Advanced Solid State Lasers, OSA Trends in Optics and Photonics Series*, Vol. 26, ed. by M. M. Fejer, H. Injeyan, U. Keller(OSA, Washington DC, 1999), pp. 676–680.

[16] S. Singh: Nonlinear optical materials. In: *Handbook of Lasers with Selected Data on Optical Technology*, ed. by R. J. Pressley (Chemical Rubber Co., Cleveland, 1971), pp. 489–525.

[17] A. R. Johnston: The strain-free electro-optic effect in single-crystal barium titanate. Appl. Phys. Lett. 7(7), 195–197(1965).

[18] I. P. Kaminow: Tables of Linear Electrooptic Coefficients. In: *CRC Handbook of Laser Science and Technology*, Vol. III, *Optical Materials: Part 2*, ed. by M. J. Weber(CRC Press, Boca Raton, 1986), pp. 253–278.

[19] M. N. Deeter, G. W. Day, A. H. Rose: Magnetooptic Materials. Crystals and Glasses. In: *CRC Handbook of Laser Science and Technology, Supplement 2, Optical Materials*, ed. by M. J. Weber(CRC Press, Boca Raton, 1995), pp. 367–402.

[20] J. E. Midwinter, J. Warner: The effects of phase matching method and of uniaxial crystal symmetry on the polar distribution of second-order non-linear optical polarization. Brit. J. Appl. Phys. 16(11), 1135–1142(1965)

[21] R. C. Miller, D. A. Kleinman, A. Savage: Quantitative studies of optical harmonic generation in CdS, $BaTiO_3$, and KH_2PO_4 type crystals. Phys. Rev. Lett. 11(4), 146–149(1963).

[22] I. Shoji, T. Kondo, A. Kitamoto, M. Shirane, R. Ito: Absolute scale of second-order

nonlinear-optical coefficients. J. Opt. Soc. Am. B 14(9), 2268 – 2294(1997).

5.4 MgBaF$_4$，氟化钡镁

负双轴晶：$2V_Z = 117.5°$，$\lambda = 0.532\ 1\ \mu m$[1]

分子量：237.629

点群：$mm2$

晶格常数[2]：$a = 4.125$ Å；$b = 14.509$ Å；$c = 5.841$ Å

介电轴和晶体学轴间的变换：$X,\ Y,\ Z \Rightarrow b,\ c,\ a$

透明范围：$< 0.14 \sim 10\ \mu m$[2,3]

线性吸收系数 α[3]

$\lambda/\mu m$	α/cm^{-1}	$\lambda/\mu m$	α/cm^{-1}
0.144	2.0	0.178	0.5
0.164	1.0	0.200	0.2

$T = 295$ K 时折射率的实验值

$\lambda/\mu m$	n_X	n_Y	n_Z	参 考 文 献
0.157 629 9	1.587 1	1.613 8		[3]
0.480 125 4	1.452 9	1.470 467		[3]
0.501 707 7	1.451 9	1.469 492		[3]
0.508 724 0	1.445 17	1.469 179		[3]
0.532 1	1.450 8	1.467 8	1.474 2	[1]
0.546 226 0	1.450 4	1.467 8		[3]
0.587 725 4	1.449 2	1.466 6		[3]
0.644 025 0	1.448 0	1.465 2		[3]
1.064 2	1.443 6	1.460 4	1.467 4	[2]

可见范围的 Sellmeier 方程（λ 以 μm 为单位，$T = 293$ K）[1]：

$$n_X^2 = 2.077\ 0 + \frac{0.007\ 60}{\lambda^2 - 0.007\ 9}$$

$$n_Y^2 = 2.123\ 8 + \frac{0.008\ 60}{\lambda^2}$$

$$n_Z^2 = 2.1462 + \frac{0.00736}{\lambda^2 - 0.0090}$$

0.157~1.06 μm 范围内的 Sellmeier 方程(λ 以 μm 为单位,T = 293 K)[3]:

$$n_X^2 = 2.07971 + \frac{0.006897}{\lambda^2 - 0.00914}$$

$$n_Y^2 = 2.12832 + \frac{0.0075537}{\lambda^2 - 0.008979}$$

相位匹配角的实验值[1]

相互作用的波长/μm	ϕ_{exp}/(°)	θ_{exp}/(°)	相互作用的波长/μm	ϕ_{exp}/(°)	θ_{exp}/(°)
XY 平面,$\theta = 90°$			XZ 平面,$\phi = 0°$,$\theta < V_Z$		
SHG, o + o ⇒ e			SHG, e + o ⇒ e		
1.0642 ⇒ 0.5321	9.2		1.0642 ⇒ 0.5321		18.9

内角带宽的实验值[1]

相互作用的波长/μm	ϕ_{pm}/(°)	$\Delta\phi^{int}$/(°)	$\Delta\theta^{int}$/(°)
XY 平面,$\theta = 90°$			
SHG, o + o ⇒ e			
1.0642 ⇒ 0.5321	9.2	0.82	2.25

MgBaF$_4$ 晶体主平面上有效二阶非线系数的表达式(Kleinman 对称条件不成立)[4]:

XY 平面
$$d_{ooe} = d_{31}\cos\phi$$

YZ 平面
$$d_{oeo} = d_{eoo} = d_{24}\cos\theta$$

XZ 平面,$\theta < V_Z$
$$d_{oee} = d_{eoe} = d_{15}\sin^2\theta + d_{24}\cos^2\theta$$

XZ 平面,$\theta > V_Z$
$$d_{eeo} = d_{31}\sin^2\theta + d_{32}\cos^2\theta$$

MgBaF$_4$ 晶体主平面上有效二阶非线性系数的表达式(Kleinman 对称条件成立,$d_{15} = d_{31}$ 以及 $d_{24} = d_{32}$)[4]:

XY 平面
$$d_{ooe} = d_{31}\cos\phi$$

YZ 平面
$$d_{oeo} = d_{eoo} = d_{32}\cos\theta$$

XZ 平面，$\theta < V_Z$
$$d_{oee} = d_{eoe} = d_{31}\sin^2\theta + d_{32}\cos^2\theta$$

XZ 平面，$\theta > V_Z$
$$d_{eeo} = d_{31}\sin^2\theta + d_{32}\cos^2\theta$$

在文献[4]中给出了 $MgBaF_4$ 晶体任意方向上有效二阶非线性系数的表达式。二阶非线性系数的值：

$d_{15}(1.064\ \mu m) = 0.07 \times d_{11}(SiO_2) \pm 20\% = (0.021 \pm 0.004)\ pm/V$ [2,5]

$d_{24}(1.064\ \mu m) = 0.07 \times d_{11}(SiO_2) \pm 20\% = (0.021 \pm 0.004)\ pm/V$ [2,5]

$d_{24}(1.064\ \mu m) = 0.062 \times d_{36}(KDP) \pm 17\% = (0.024 \pm 0.004)\ pm/V$ [1,5]

$d_{31}(1.064\ \mu m) = 0.07 \times d_{11}(SiO_2) \pm 20\% = (0.021 \pm 0.004)\ pm/V$ [2,5]

$d_{31}(1.064\ \mu m) = 0.057 \times d_{36}(KDP) \pm 23\% = (0.022 \pm 0.005)\ pm/V$ [1,5]

$d_{32}(1.064\ \mu m) = 0.13 \times d_{11}(SiO_2) \pm 20\% = (0.039 \pm 0.008)\ pm/V$ [2,5]

$d_{32}(1.064\ \mu m) = 0.085 \times d_{36}(KDP) \pm 12\% = (0.033 \pm 0.012)\ pm/V$ [1,5]

$d_{33}(1.064\ \mu m) = 0.05 \times d_{11}(SiO_2) \pm 20\% = (0.015 \pm 0.003)\ pm/V$ [2,5]

$d_{33}(1.064\ \mu m) = 0.023 \times d_{36}(KDP) \pm 14\% = (0.009 \pm 0.001)\ pm/V$ [1,5]

激光诱导的表面损伤阈值

$\lambda/\mu m$	τ_p/ns	$I_{thr}/(GW \cdot cm^{-2})$	参 考 文 献	备 注
0.157	10	>0.000 2	[3]	>10^9 个脉冲
1.064 2	≈20	>1	[2]	

关于这一晶体

这一晶体的非线性光学晶体自 20 世纪 70 年代中期就为人们知晓。然而，最近人们发现其在 UV 区独特的透过率可至 140 nm[3]。同时也实现了铁电畴的翻转，证明了在这种材料中可能实现准相位匹配[3]。

■ 参考文献

[1] P. S. Bechtold, S. Haussühl: Nonlinear optical properties of orthorhombic barium formate and magnesium barium fluoride. Appl. Phys. 14(4), 403–410(1977).

[2] J. G. Bergman, G. R. Crane, H. Guggenheim: Linear and nonlinear optical properties of ferroelectric $BaMgF_4$ and $BaZnF_4$. J. Appl. Phys. 46(11), 4645–4646(1975).

[3] S. C. Buchter, T. Y. Fan, V. Liberman, J. J. Zayhowski, M. Rothschild, E. J. Mason,

A. Cassanho, H. P. Jenssen, J. H. Burnett: Periodically poled BaMgF$_4$ for ultraviolet frequency conversion. Opt. Lett. 26(21), 1693 - 1695(2001).

[4] V. G. Dmitriev, D. N. Nikogosyan: Effective nonlinearity coefficients for three-wave interactions in biaxial crystals of *mm*2 point group symmetry. Opt. Commun. 95(1 - 3), 173 - 182(1993).

[5] D. A. Roberts: Simplified characterization of uniaxial and biaxial nonlinear optical crystals: a plea for standardization of nomenclature and conventions. IEEE J. Quant. Electr. 28(10), 2057 - 2074(1992).

5.5 GaAs，砷化镓

光学各向同性晶体

分子量：144.645

密度：$(5.316\ 1 \pm 0.000\ 2)\text{g/cm}^3$，$T = 298\ \text{K}$[1]；$5.317\ 0\ \text{g/cm}^3$[2]；$5.32\ \text{g/cm}^3$，$T = 293\ \text{K}$[3]

点群：$\bar{4}3m$

晶格常数：

$a = (5.653\ 4 \pm 0.000\ 2)\text{Å}$[1]；$(5.653\ 21 \pm 0.000\ 03)\text{Å}$[4]；$5.653\ 5\text{Å}$[5]

莫氏硬度：4.5[1]

诺氏硬度：750，在压痕载荷为 25 g 时[1]；750 ± 40[6]；721[7]

水中的溶解度：不溶解[4]

熔点：1 510 K[6]；1 511 K[1,2,4,8]；1 513 K[9,10]；1 520 K[11]

线性热膨胀系数

T/K	$\alpha_t \times 10^6 /\text{K}^{-1}$	参考文献	T/K	$\alpha_t \times 10^6 /\text{K}^{-1}$	参考文献
40	-0.50	[1]	300	5.82	[8]
50	-0.15	[8]		6.0	[2]
55	0.00	[1]	400	6.23	[8]
100	1.9	[12]	500	6.5	[12]
	2.05	[8]	600	6.98	[8]
200	4.93	[8]	800	7.1	[12]
293	5.7	[12]		7.4	[8]

线性热膨胀系数的平均值[1]

T/K	$\alpha_t \times 10^6/K^{-1}$	T/K	$\alpha_t \times 10^6/K^{-1}$
78~290	3.64	560~680	7.44
291~560	5.74		

$p = 0.101\ 325$ MPa 时的比热容 c_p[9]

T/K	$c_p/(\mathrm{J \cdot kg^{-1} \cdot K^{-1}})$
273	318

热导率 κ

T/K	$\kappa/(\mathrm{W \cdot m^{-1} \cdot K^{-1}})$	参考文献	T/K	$\kappa/(\mathrm{W \cdot m^{-1} \cdot K^{-1}})$	参考文献
300	52	[9]		52.3	[1]

n-GaAs 的热导率($n = 2 \times 10^{16}\ \mathrm{cm^{-3}}, T = 77$ K)[8]

T/K	$\kappa/(\mathrm{W \cdot m^{-1} \cdot K^{-1}})$	T/K	$\kappa/(\mathrm{W \cdot m^{-1} \cdot K^{-1}})$
80	270	300	58
150	105		

室温下带隙能量(直接跃迁):$E_g = 1.42$ eV[13-15];1.425 eV[16,17];1.428 eV[2];1.43 eV[8,18,19];1.435 eV[11,20-22]

以 $\alpha = 1\ \mathrm{cm^{-1}}$ 计的透明范围:$1.1 \sim 17\ \mu\mathrm{m}$[23];$0.95 \sim 17\ \mu\mathrm{m}$[24]

线性吸收系数 α

$\lambda/\mu\mathrm{m}$	$\alpha/\mathrm{cm^{-1}}$	参考文献	备注
0.53	80 000	[25]	Si-掺杂 GaAs,$n = 10^{18}\ \mathrm{cm^{-3}}, T = 300$ K
1.06	0.9	[25]	Si-掺杂 GaAs,$n = 10^{18}\ \mathrm{cm^{-3}}, T = 300$ K
	1.2	[26]	未掺杂 GaAs
	1.54	[27]	n-型 GaAs,[111]方向,$n = 2 \times 10^{17}\ \mathrm{cm^{-3}}$
	1.57	[17]	Si-掺杂 GaAs,$n = 1.5 \times 10^{17}\ \mathrm{cm^{-3}}, T = 295$ K
	2.5	[28]	$n = 4 \times 10^{16}\ \mathrm{cm^{-3}}, T = 300$ K
	~3	[29]	O_2-掺杂,$n = 3 \times 10^{14}\ \mathrm{cm^{-3}}$ 和 $\rho_0 = 2.4\ \Omega \cdot \mathrm{cm}$
	4.0	[30]	$n = 4 \times 10^{16}\ \mathrm{cm^{-3}}$

续表

$\lambda/\mu m$	α/cm^{-1}	参考文献	备注
1.064 2	0.7	[31]	n-型 GaAs, $\boldsymbol{E} \perp c$
	0.7	[32]	
	1.2	[3]	Cr-掺杂 GaAs, $n = 10^{16} cm^{-3}$, $\rho_0 > 10^7 \Omega \cdot cm$
	1.50 ± 0.15	[33]	
1.318	0.05	[34]	O_2-掺杂样品
2.7~2.8	0.003 2	[35]	$\rho_0 \sim 10^8 \Omega \cdot cm$, 体吸收
3.8~3.9	0.003	[35]	$\rho_0 \sim 10^8 \Omega \cdot cm$, 体吸收
5~6	0.016	[36]	体吸收
9.2	0.006	[24]	
9.3	0.005	[37]	体吸收
9.6	0.008	[24]	
10.6	0.005	[38]	
	0.006 ± 0.002	[39]	$\rho_0 = 10^4 \sim 10^9 \Omega \cdot cm$
	0.009	[37]	体吸收
	0.01~0.05	[40]	
	0.01~0.20	[41]	
	0.012 ± 0.002	[42]	Cr-掺杂 GaAs, $\rho_0 = 3 \times 10^8 \Omega \cdot cm$
12.4	0.05	[24]	
13.78	0.15	[24]	
15.9	0.36	[24]	
16.75	0.71	[24]	
17.22	1.09	[24]	

双光子吸收系数 β

$\lambda/\mu m$	τ_p/ns	$\beta \times 10^{11}/(cm \cdot W^{-1})$	参考文献	备注
1.06	0.005	4 500 ± 1 000	[17]	$T = 295 K$, Si-掺杂 GaAs, $n = 1.5 \times 10^{17} cm^{-3}$

续表

$\lambda/\mu m$	τ_p/ns	$\beta \times 10^{11}/(cm \cdot W^{-1})$	参考文献	备 注
	0.008	1 500 ± 500	[28]	$n = 4 \times 10^{16}$ cm^{-3}
	0.08	2 600	[43]	
	10	3 500 ± 300	[44]	O_2-掺杂样品
1.0642	0.003	2 500	[45]	[001]方向，$E \parallel$ [110]方向
	0.028	2 200 ± 300	[27]	n-型 GaAs，[111]方向，$n = 2 \times 10^{17}$ cm^{-3}
	0.03	3 000 ± 500	[31]	n-型 GaAs，$E \perp c$
		1 800 ± 360	[46]	[100]方向
		2 400 ± 480	[46]	[110]方向
		2 500 ± 500	[46]	[111]方向
	0.035	2 700	[26]	未掺杂 GaAs
	0.038	2 300	[47]	[111]方向
	0.04	2 600 ± 500	[48]	[110]方向
	0.05	2 900	[14]	
	11	3 000 ± 900	[33]	
	~20	~2 000	[49]	
1.318	~20	3 300 ± 1 500	[34]	O_2-掺杂样品
1.32	0.003	1 100	[45]	[001]方向，$E \parallel$ [110]方向

折射率的实验值

$\lambda/\mu m$	n	参考文献	$\lambda/\mu m$	n	参考文献
0.895	3.603	[50]	0.960	3.534	[50]
0.900	3.595	[50]	0.980	3.520	[50]
0.910	3.581	[50]	1.000	3.509	[50]
0.920	3.569	[50]	1.020	3.498	[50]
0.940	3.550	[50]	1.040	3.488	[50]

续表

$\lambda/\mu m$	n	参考文献	$\lambda/\mu m$	n	参考文献
1.060	3.479	[50]	10.0	3.27	[4]
1.100	3.463	[50]	11.0	3.045	[4]
1.150	3.446	[50]	13.0	2.97	[4]
1.200	3.433	[50]	13.7	2.895	[4]
1.400	3.394	[50]	14.5	2.82	[4]
1.435	3.40	[4]	15.0	2.73	[4]
1.500	3.381	[50]	17.0	2.59	[4]
1.700	3.362	[50]	19.0	2.41	[4]
2.87	3.33	[4]	21.9	2.12	[4]
5.1	3.30	[4]			

折射率的温度微商

$\lambda/\mu m$	T/K	$dn/dT \times 10^6 / K^{-1}$	参考文献	备注
10.6	293	56 ± 3	[42]	Cr-掺杂 GaAs, $\rho_0 = 3 \times 10^8 \Omega \cdot cm$
	300	100	[51]	

非线性折射率 $\gamma^{[48,52]}$

$\lambda/\mu m$	$\gamma \times 10^{20}/(m^2 \cdot W^{-1})$
1.064 2	-3260 ± 600

室温下低频(远低于 GaAs 晶体声学共振频率,即对"自由"晶体)下测量的线性电光系数

$\lambda/\mu m$	$r_{41}^T/(pm \cdot V^{-1})$	参考文献	$\lambda/\mu m$	$r_{41}^T/(pm \cdot V^{-1})$	参考文献
1.064 2	-1.17	[53]	1.50	-1.36	[53]
1.152 3	-1.43	[54]	3.391 3	-1.24 ± 0.04	[55]
1.208	-1.25	[53]	10.6	-1.51 ± 0.05	[55]
1.306	-1.28	[53]		-1.6	[56]

室温下高频(远高于 GaAs 晶体声学共振频率,即对"受夹"晶体)下测量的线性电光系数

$\lambda/\mu m$	$r_{41}^S/(pm \cdot V^{-1})$	参考文献	$\lambda/\mu m$	$r_{41}^S/(pm \cdot V^{-1})$	参考文献
0.95~1.08	−1.2	[57]	1.306	−1.46	[53]
1.064 2	−1.33	[53]	1.50	−1.53	[53]
1.208	−1.41	[53]	3.391 3	−1.5	[58]

Verdet 常数,$T = 293$ K

$\lambda/\mu m$	$V/[(°) \cdot T^{-1} \cdot m^{-1}]$	参考文献	$\lambda/\mu m$	$V/[(°) \cdot T^{-1} \cdot m^{-1}]$	参考文献
1.06	5 850 ± 290	[59]	1.95	1 800 ± 180	[59]
	5 000	[8]			

二阶非线性系数的绝对值:

$$d_{36}(1.064 \ \mu m) = 170 \ pm/V^{[60]}$$

$$d_{36}(1.533 \ \mu m) = 119 \ pm/V^{[60]}$$

$$d_{36}(10.6 \ \mu m) = 83 \ pm/V^{[61]}$$

激光诱导的损伤阈值

$\lambda/\mu m$	τ_p/ns	$I_{thr}/(GW \cdot cm^{-2})$	参考文献	备 注
0.694 3	500 000	0.000 02	[62]	表面损伤
	30	0.072(?)	[63]	表面损伤
	20	0.008 ± 0.002	[64]	表面损伤,[100]方向
1.06	300 000	0.000 11~0.000 24	[11]	表面损伤,束腰直径 1 000 μm
	1 000	0.01	[65]	表面损伤
	60	0.013 ± 0.005	[64]	表面损伤,[100]和[110]方向
	35	0.03 ± 0.01	[59]	表面损伤
	20	0.043 ± 0.008	[66]	表面损伤
1.064 2	45	0.02	[3]	表面损伤,10 个脉冲,[100]方向
		0.05	[3]	表面损伤,1 个脉冲,[100]方向
	20	0.045 ± 0.04	[67]	表面损伤,1 个脉冲,[100]方向
	18	0.04	[68]	

续表

$\lambda/\mu m$	τ_p/ns	$I_{thr}/(GW \cdot cm^{-2})$	参考文献	备 注
	10	0.024	[69]	表面损伤
	0.035	200	[70]	[100]方向
2.76	90	0.08	[71]	
2.8	70	0.079	[72]	
10.6	CW	0.000 006	[71]	
		0.000 003	[40]	束腰直径 > 3 cm
		0.000 03	[40]	束腰直径 < 0.5 cm
	200	0.06	[68]	
	150~200	0.07~0.09	[73]	表面损伤,束腰直径 6 cm
		0.135~0.18	[73]	表面损伤,束腰直径 500 μm
	150	0.05~0.11	[40]	表面损伤,束腰直径 ~2 cm
	100	0.01±0.005	[59]	体损伤,Zn–掺杂 p–型 GaAs, $n \approx 10^{19}$ cm^{-3}
		0.03±0.01	[59]	表面损伤,n–型 GaAs, $n \approx 10^{18}$ cm^{-3}
		0.11~0.8	[71]	
	60	1.6~3.2	[40]	束腰直径 100 μm

关于这一晶体

尽管由于 GaAs 各向同性的本性使其不可能实现双折射相位匹配,然而准相位匹配(QPM)仍然是可能的。早在 1976 年,两个美国研究组[74,75]报道了适当取向 GaAs 晶片的堆积在 CO_2 激光辐射 SHG 方面的应用,并成功地实施了第一次实验。在 1993 年提出了 GaAs 晶片扩散键合更为精巧的方法[76],其后被用于 QPM SHG[77-79]、QPM SHG[78]和 QPM DFG[80]的实验中。

参考文献

[1] E. M. Voronkova, B. N. Grechushnikov, G. I. Distler, I. P. Petrov: *Optical Materials for Infrared Technique*(Nauka,Moscow,1965)[In Russian].

[2] *Physical-Chemical Properties of Semiconductors. Handbook* (Nauka, Moscow, 1979) [In

Russian].

[3] A. L. Huang, M. F. Becker, R. M. Walser: Laser-induced damage and ion emission of GaAs at 1.06μm. Appl. Opt. 25(21), 3864 – 3870(1986).

[4] A. A. Blistanov, V. S. Bondarenko, N. V. Perelomova, F. N. Strizhevskaya, V. V. Tchkalova, M. P. Shaskolskaya: *Acoustic Crystals* (Nauka, Moscow, 1982) [In Russian].

[5] *Handbook of Optical Constants of Solids II*, ed. by E. D. Palik (Academic Press, Boston, 1991).

[6] M. Bertolotti, D. Sette, L. Stagni, G. Vitali: Electron microscope observation of laser damage on GaAs, GaSb and InSb. Radiation Effects 16(3 – 4), 197 – 202(1972).

[7] B. H. T. Chai: Optical Crystals. In: *CRC Handbook of Laser Science and Technology, Supplement 2: Optical Materials*, ed. by M. J. Weber (CRC Press, Boca Raton, 1995), pp. 3 – 65.

[8] *Physical Quantities. Handbook*, ed. by I. S. Grigoriev, E. Z. Meilikhov (Energoatomizdat, Moscow, 1991) [In Russian].

[9] S. S. Ballard, J. S. Browder: Thermal Properties. In: *CRC Handbook of Laser Science and Technology, Vol. IV, Optical Materials: Part 2*, ed. by M. J. Weber (CRC Press, Boca Raton, 1987), pp. 49 – 54.

[10] S. Musikant: *Optical Materials. An Introduction to Selection and Application* (Marcel Dekker, Inc., NewYork, 1985).

[11] A. V. Kuanr, S. K. Bansal, G. P. Srivastava: Laser-induced damage in GaAs at 1.06μm. wavelength: surface effects. Opt. Laser Technol. 28(1), 25 – 34(1996).

[12] G. W. C. Kaye, T. H. Laby: *Tables of Physical and Chemical Constants* (Longman Group Ltd., London, 1995).

[13] E. W. van Stryland, M. A. Woodall, H. Vanherzeele, M. J. Soileau: Energy band-gap dependence of two-photon absorption. Opt. Lett. 10(10), 490 – 492(1985).

[14] E. W. van Stryland, L. L. Chase: Two-Photon Absorption. Inorganic Materials. In: *CRC Handbook of Laser Science and Technology, Supplement 2: Optical Materials*, ed. by M. J. Weber (CRC Press, Boca Raton, 1995), pp. 299 – 328.

[15] K. J. Bachmann: *The Materials Science of Microelectronics* (VCH Publishers, NewYork, 1995).

[16] H. Burkhard, H. W. Dinges, E. Kuphal: Optical properties of $In_{1-x}Ga_xP_{1-y}As_y$, InP, GaAs, and GaP determined by ellipsometry. J. Appl. Phys. 53(1), 655 – 662(1982).

[17] A. Penzkofer, A. A. Bugayev: Two-photon absorption and emission dynamics of bulk GaAs. Opt. Quant. Electron. 21(4), 283 – 306(1989).

[18] A. Z. Grasyuk, I. G. Zubarev, A. B. Mironov, I. A. Poluektov: Spectrum of two-photon interband absorption of laser radiation in GaAs. Fiz. Tekh. Polupr. 10(2), 262 – 270 (1976) [In Russian, English trans.: Sov. Phys. - Semicond. 10(2), 159 – 163(1976)].

[19] C. Kittel: *Introduction to Solid State Physics*, Seventh Edition (John Wiley & Sons, New York, 1996).

[20] S. S. Mitra, L. M. Narducci, R. A. Shatas, Y. F. Tsay, A. Vaidyanathan: Nonlinear absorption in direct-gap semiconductors. Appl. Opt. 14(12), 3038 – 3042(1975).

[21] M. Bertolotti, V. Bogdanov, A. Ferrari, A. Yaskov, N. Nazorova, A. Pikhtin, L. Schirone: Temperature dependence of the refractive index in semiconductors. J. Opt. Soc. Am. B 7(6), 918 – 922(1990).

[22] A. Vaidyanathan, A. H. Guenther, S. S. Mitra: Two-photon absorption in direct-gap crystals—an addendum. Phys. Rev. B 22(12), 6480 – 6483(1980).

[23] A. S. Sonin, A. S. Vasilevskaya: *Electrooptic Crystals* (Atomizdat, Moscow, 1971) [In Russian].

[24] *Handbook of Optical Constants of Solids*, ed. by E. D. Palik (Academic Press, Orlando, 1985).

[25] A. Saissy, A. Azema, J. Botineau, F. Gires: Absolute measurement of the 1.06μm twophoton absorption coefficient in GaAs. Appl. Phys. 15(1), 99 – 102(1978).

[26] G. C. Valley, T. F. Boggess, J. Dubard, A. L. Smirl: Picosecond pump-probe technique to measure deep-level, free-carrier, and two photon cross sections in GaAs. J. Appl. Phys. 66(6), 2407 – 2413(1989).

[27] A. A. Bugaev, T. Y. Dunaeva, V. A. Lukoshkin: Influence of nonlinear refraction, absorption by free carriers, and multiple reflection on the determination of the two-photon absorption coefficient of gallium arsenide. Fiz. Tverd. Tela 31(12), 9 – 14(1989) [In Russian, English trans.: Sov. Phys. - Solid State 31(12), 2031 – 2034(1989)].

[28] B. Bosacchi, J. S. Bessey, F. C. Jain: Two-photon absorption of neodymium laser radiation in gallium arsenide. J. Appl. Phys. 49(8), 4609 – 4611(1978).

[29] T. F. Deutsch: Absorption coefficient of infrared laser window materials. J. Phys. Chem. Solids 34(12), 2091 – 2104(1973).

[30] C. C. Lee, H. Y. Fan: Two-photon absorption and photoconductivity in GaAs and InP. Appl. Phys. Lett. 20(1), 18 – 20(1972).

[31] J. H. Bechtel, W. L. Smith: Two-photon absorption in semiconductors with picosecond light pulses. Phys. Rev. B 13(8), 3515 – 3522(1976).

[32] I. M. Catalano, A. Cingolani: Non-parabolic band effect on two-photon absorption in ZnSe and CdTe. Solid State Commun. 43(3), 213 – 215(1982).

[33] A. F. Stewart, M. Bass: Intensity dependent absorption in semiconductors. Appl. Phys. Lett. 37(11), 1040 – 1043(1980).

[34] D. A. Kleinman, R. C. Miller, W. A. Nordland: Two-photon absorption of Nd laser radiation in GaAs. Appl. Phys. Lett. 23(5), 243 – 244(1973).

[35] A. Hordvik, L. Skolnik: Photoacoustic measurements of surface and bulk absorption in

HF/DF laser window materials. Appl. Opt. 16(11), 2919 –2924(1977).

[36] V. B. Nosov, G. T. Petrovskii, M. V. Serzhantova, A. V. Shatilov: Calorimetric measurements of the volume and surface absorption of infrared materials in the 5 – 6μm spectral region. Opt. Mekh. Promyshl. No. 4, 42 – 44 (1989) [In Russian, English trans. : Sov. J. Opt. Technol. 56(4), 238 –240(1989)].

[37] C. P. Christensen, R. Joiner, S. T. K. Nieh, W. H. Steier: Investigation of infrared loss mechanism in high-resistivity GaAs. J. Appl. Phys. 45(11), 4957 –4960(1974).

[38] A. J. Glass, A. H. Guenther: Laser induced damage of optical elements—a status report. Appl. Opt. 12(4), 637 –649(1973).

[39] J. Comly, E. Garmire, A. Yariv: Infrared absorption at 10.6 μm in GaAs. J. Appl. Phys. 38 (10), 4091 –4092(1968).

[40] N. V. Karlov, E. V. Sisakyan: Optical materials for CO_2 lasers. Izv. Akad. Nauk SSSR, Ser. Fiz. 44(8), 1631 – 1638 (1980) [In Russian, English trans. : Bull. Acad. Sci. USSR, Phys. Ser. 44(8), 63 –68(1980)].

[41] M. A. Ilin, N. V. Ovsyannikova, E. V. Sisakyan: Method of measuring the small absorption coefficients of gallium arsenide single crystals. Opt. Mekh. Promyshl. No. 10, 57 – 59 (1977) [In Russian, English trans. : Sov. J. Opt. Technol. 44 (10), 626 – 627 (1977)].

[42] R. Weil: Interference of 10.6-μm coherent radiation in a 5-cm long gallium arsenide parallelepiped. J. Appl. Phys. 40(7), 2857 –2859(1969).

[43] J. S. Aitchison, M. K. Oliver, E. Kapon, E. Colas, P. W. E. Smith: Role of two-photon absorption in ultrafast semiconductor optical switching devices. Appl. Phys. Lett. 56 (14), 1305 –1307(1990).

[44] I. G. Zubarev, A. B. Mironov, S. I. Mikhailov: Influence of deep impurity levels on nonlinear absorption of light in GaAs. Fiz. Tekh. Poluprov. 11(2), 415 – 417(1977) [In Russian, English trans. : Sov. Phys. - Semicond. 11(2), 239 –240(1977)].

[45] I. B. Zotova, Y. J. Ding: Spectral two-photon absorption in the range of 1.3 – 1.75 μm for GaAs. In: *Conference on Lasers and Electrooptics CLEO/QELS 2003, Technical Digest*(OSA, Washington DC, 2003), paper CTuM21.

[46] R. DeSalvo, M. Sheik-Bahae, A. A. Said, D. J. Hagan, E. W. Van Stryland: Z-scan measurements of the anisotropy of nonlinear refraction and absorption in crystals. Opt. Lett. 18(3), 194 – 196(1993).

[47] E. W. van Stryland, H. Vanherzeele, M. A. Woodall, M. J. Soileau, A. L. Smirl, S. Guha, T. F. Boggess: Two photon absorption, nonlinear refraction, and optical limiting in semiconductors. Opt. Eng. 24(4), 613 –623(1985).

[48] A. A. Said, M. Sheik-Bahae, D. J. Hagan, T. H. Wei, J. Wang, J. Young, E. W. van Stryland: Determination of bound-electronic and free-carrier nonlinearities in ZnSe, GaAs,

CdTe, and ZnSe. J. Opt. Soc. Am. B 9(3), 405 – 414(1992).

[49] J. M. Ralston, R. K. Chang: Nd: laser induced absorption in semiconductors and aqueous $PrCl_3$ and $NdCl_3$. Opto-electron. 1(4), 182 – 188(1969).

[50] D. T. F. Marple: Refractive index of GaAs. J. Appl. Phys. 35(4), 1241 – 1242(1964).

[51] O. A. Kolosovskii, L. N. Ustimenko: Measurement of the temperature coefficient of the refractive index of infrared materials using a CO_2 laser. Opt. Spektrosk. 33(4), 781 – 782(1972) [In Russian, English trans.: Opt. Spectrosc. USSR 33(4), 430 – 431 (1972)].

[52] M. Sheik-Bahae, D. C. Hutchings, D. J. Hagan, E. W. van Stryland: Dispersion of bound electron nonlinear refraction in solids. IEEE J. Quant. Electr. 27(6), 1296 – 1309 (1991).

[53] N. Suzuki, K. Tada: Elastooptic and electrooptic properties of GaAs. Jpn. J. Appl. Phys. 23 (8), 1011 – 1016(1984).

[54] A. Yariv, P. Yeh: *Optical Waves in Crystals.* (John Wiley & Sons, NewYork, 1984).

[55] M. Sugie, K. Tada: Measurements of the linear electrooptic coefficients and analysis of the nonlinear susceptibilities in cubic GaAs and hexagonal CdS. Jpn. J. Appl. Phys. 15 (3), 421 – 431(1976).

[56] I. P. Kaminow: Measurements of the electrooptic effect in CdS, ZnTe and GaAs at 10.6μm. IEEE J. Quant. Electr. QE – 4(1), 23 – 26(1968).

[57] E. H. Turner, I. P. Kaminow: Electro-optic effect in GaAs. J. Opt. Soc. Am. 53 (4), 523(1963).

[58] I. P. Kaminow: Tables of Linear Electrooptic Coefficients. In: *CRC Handbook of Laser Science and Technology*, Vol. III, *Optical Materials: Part 2*, ed. by M. J. Weber (CRC Press, Boca Raton, 1986), pp. 253 – 278.

[59] J. L. Smith, G. A. Tanton: Intense laser flux effect on GaAs. Appl. Phys. 4(4), 313 – 315(1974).

[60] I. Shoji, T. Kondo, A. Kitamoto, M. Shirane, R. Ito: Absolute scale of second-order nonlinear-optical coefficients. J. Opt. Soc. Am. B 14(9), 2268 – 2294(1997).

[61] D. A. Roberts: Simplified characterization of uniaxial and biaxial nonlinear optical crystals: a plea for standardization of nomenclature and conventions. IEEE J. Quant. Electr. 28(10), 2057 – 2074(1992).

[62] M. Bertolotti, F. de Pasquale, P. Marietti, D. Sette, G. Vitali: Laser damage on semiconductor surfaces. J. Appl. Phys. 38(10), 4088 – 4090(1967).

[63] M. Birnbaum, T. L. Stocker: Reflectivity enhancement of semiconductors by Q-switched ruby lasers. J. Appl. Phys. 39(13), 6032 – 6036(1968).

[64] J. L. Smith: Surface damage of GaAs from 0.694 and 1.06μm laser radiations. J. Appl. Phys. 43(8), 3399 – 3402(1972).

[65] A. M. Bonch-Bruevich, V. P. Kovalev, G. S. Romanov, Y. A. Imas, M. N. Libenson: The change in reflectivity properties of some semiconductors upon laser exposure. Zh. Tekh. Fiz. 38 (4), 677 – 685 (1968) [In Russian, English trans. : Sov. Phys. - Tech. Phys. 13 (4), 507 – 513 (1968)].

[66] C. L. Sam: Laser damage of GaAs and ZnTe at 1.06 μm. Appl. Opt. 12 (4), 878 – 879 (1973).

[67] A. Garg, A. Kapoor, K. N. Tripathi: Laser-induced damage in GaAs. Opt. Laser Technol. 35 (1), 21 – 24 (2003).

[68] D. C. Hanna, B. Luther-Davies, H. N. Rutt, R. C. Smith, C. R. Stanley: Q-switched laser damage of infrared nonlinear materials. IEEE J. Quant. Electr. QE – 8 (3), 317 – 324 (1972).

[69] R. Tsu, J. E. Baglin, G. J. Lasher, J. C. Tsang: Laser-induced recrystallization and damage in GaAs. Appl. Phys. Lett. 34 (2), 153 – 155 (1979).

[70] A. P. Singh, A. Kapoor, K. N. Tripathi, G. R. Kumar: Thermal and mechanical damage of GaAs in picosecond regime. Opt. Laser Technol. 33 (6), 363 – 369 (2001).

[71] R. M. Wood: *Laser Damage in Optical Materials* (Adam Hilger, Bristol, 1986).

[72] R. D. Peterson, K. L. Schepler, J. L. Brown, P. G. Schunemann: Damage properties of ZnGeP$_2$ at 2 μm. J. Opt. Soc. Am. B 12 (11), 2142 – 2146 (1995).

[73] N. P. Datskevich, N. V. Karlov, G. P. Kuzmin, A. A. Nesterenko, E. V. Sisakyan: The resistance of infrared optical materials to pulsed CO_2 laser light in large irradiated spots. Kratkie Soobshch. Fiz. No. 6, 3 – 7 (1983) [In Russian, English trans. : Sov. Phys. -Lebedev Institute Reports No. 6, 1 – 5 (1983)].

[74] A. Szilagyi, A. Hordvik, H. Schlossberg: A quasi-phase-matching technique for efficient optical mixing and frequency doubling. J. Appl. Phys. 47 (5), 2025 – 2032 (1976).

[75] D. E. Thompson, J. D. McMullen, D. B. Anderson: Second-harmonic generation in GaAs "stack of plates" using high-power CO_2 laser radiation. Appl. Phys. Lett. 29 (2), 113 – 115 (1976).

[76] L. Gordon, G. L. Woods, R. C. Eckardt, R. R. Route, R. S. Feigelson, M. M. Fejer, R. L. Byer: Diffusion-bonded stacked GaAs for quasi-phase-matched second-harmonic generation of a carbon dioxide laser. Electron. Lett. 29 (22), 1942 – 1944 (1993).

[77] E. Lallier, M. Brevignon, J. Lehoux: Efficient second-harmonic generation of a CO_2 laser with quasi-phase-matched GaAs crystal. Opt. Lett. 23 (19), 1511 – 1513 (1998).

[78] L. Becouarn, E. Lallier, M. Brevignon, J. Lehoux: Cascaded second-harmonic and sumfrequency generation of a CO_2 laser by use of a single quasi-phase-matched GaAs crystal. Opt. Lett. 23 (19), 1508 – 1510 (1998).

[79] A. Romann, M. W. Sigrist: Photoacoustic gas sensing employing fundamental and fre-

quency-doubled radiation of a continuously tunable high-pressure CO_2 laser. Appl. Phys. B 75(2-3), 377-383(2002).

[80] D. Zheng, L. A. Gordon, Y. S. Wu, R. S. Feigelson, M. M. Fejer, R. L. Byer, K. L. Vodopyanov: 16-μm infrared generation by difference-frequency mixing in diffusion-bonded-stacked GaAs. Opt. Lett. 23(13), 1010-1012(1998).

第6章
新发展及有前景的晶体

这一章描述了19种新发展及有前景的非线性光学晶体,诸如三硼酸铋(BIBO)、硼酸铝钾(KABO)、氟硼铍酸钾(KBBF)、硼酸氧钙钆(GdCOB)、硼酸氧钙钇(YCOB)、四硼酸锂(LB4)、硫铟锂(LIS)和其他晶体。

6.1 BiB_3O_6,三硼酸铋(BIBO)

负双轴晶:$2V_z = 53.5°$,$\lambda = 0.53975$ μm
分子量:337.407
密度:4.896 g/cm^3[1];5.01 g/cm^3[2]
点群:2
晶格常数:
$a = (7.116 \pm 0.002)$ Å[2]; (7.1203 ± 0.0007) Å[1]
$b = (4.993 \pm 0.002)$ Å[2]; (4.9948 ± 0.0007) Å[1]
$c = (6.508 \pm 0.003)$ Å[2]; (6.5077 ± 0.0007) Å[1]
$\beta = 105.62°$[2]; $105.59°$[1]

介电轴和晶体学轴间的变换：$X \parallel b$，a 轴和 c 轴位于 YZ 平面，两轴之间夹角 $\beta = 105.6°$，Z 轴和 a 轴间的角约为 $31°$（略微取决于波长和/或温度），Y 轴和 c 轴之间的夹角为 $15.2°$[7]。

莫氏硬度：5~5.5

熔点：999 K[3]，(981 ± 5) K[1]

线性热膨胀系数的平均值 α_i（以 10^{-6} K^{-1} 为单位）

T/K	$\alpha_{11}(\parallel a)$	$\alpha_{22}(\parallel b)$	$\alpha_{33}(\parallel c)$	$\alpha_{13} = \alpha_{33}$	参考文献
173~573	−28.1 ± 0.5	53.7 ± 0.5	8.5 ± 0.5	−5.5 ± 0.5	[4]
298~573	−25.6	50.4	7.7	−5.33	[5]

$p = 0.101\,325$ MPa 时的比热容 c_p[6]

T/K	$c_p/(\text{J} \cdot \text{kg}^{-1} \cdot \text{K}^{-1})$	T/K	$c_p/(\text{J} \cdot \text{kg}^{-1} \cdot \text{K}^{-1})$
323	500	423	570
373	540	473	590

对 0.1 cm 长的 BIBO 晶体以 0.5 透过计的透明范围：0.286~2.7 μm[7]。UV 透过截止边在 270 nm，对 0.47 cm 长的 BIBO 晶体以 0.5 透过计的 IR 透过截止边在 2.63 μm[1,8]。

折射率的计算值

λ/μm	n_X	n_Y	n_Z
0.539 75	1.786 90	1.818 46	1.961 34
1.079 50	1.756 50	1.783 09	1.916 10

注：在文献[9]中，列出了同样波长真空中的折射率。上述折射率是以下面的色散关系计算的。

Sellmeier 方程（λ 以 μm 为单位，$T = 295$ K）[7]：

$$n_X^2 = 3.072\,2 + \frac{0.032\,4}{\lambda^2 - 0.031\,5} - 0.013\,3\lambda^2$$

$$n_Y^2 = 3.166\,9 + \frac{0.037\,2}{\lambda^2 - 0.034\,8} - 0.017\,5\lambda^2$$

$$n_Z^2 = 3.652\,5 + \frac{0.051\,1}{\lambda^2 - 0.037\,0} - 0.022\,6\lambda^2$$

注：在文献[7]中给出了 BIBO 真空中的色散关系。上述的 Sellmeier 方程（空气中的）可通过私人通信从文献[7]的作者那里得到。

二阶非线性系数的实验测量值[7,9]：

$$d_{14}(1.0795\ \mu m) = (2.4 \pm 0.3)\ pm/V$$
$$d_{16}(1.0795\ \mu m) = (2.8 \pm 0.2)\ pm/V$$
$$d_{21}(1.0795\ \mu m) = (2.3 \pm 0.2)\ pm/V$$
$$d_{22}(1.0795\ \mu m) = (2.53 \pm 0.08)\ pm/V$$
$$d_{23}(1.0795\ \mu m) = (1.3 \pm 0.1)\ pm/V$$
$$d_{25}(1.0795\ \mu m) = (2.3 \pm 0.2)\ pm/V$$
$$d_{34}(1.0795\ \mu m) = (0.9 \pm 0.1)\ pm/V$$
$$d_{36}(1.0795\ \mu m) = (2.4 \pm 0.3)\ pm/V$$

系数 d_{14}，d_{16}，d_{21}，d_{22}，d_{25}，d_{36} 的符号相同，而与系数 d_{23}，d_{34} 的符号相反[7,9]。

相位匹配角的实验值

相互作用的波长/μm	$\theta_{pm}/(°)$	参 考 文 献
YZ 平面，$\phi = 90°$		
SHG，o + o $\Rightarrow$ e		
1.064 2 $\Rightarrow$ 0.532 1	11.1	[6]，[8]
	168.9	[6]，[8]
0.946 $\Rightarrow$ 0.473	161.7	[10]

注：根据 BIBO 晶体的对称性（点群2），d_{eff} 的空间分布可以选择两个独立的相邻象限就能完全予以描述，例如（$0° < \theta < 90°$，$0° < \phi < 90°$）和（$90° < \theta < 180°$，$0° < \phi < 90°$）。

对一些特殊方向（SHG，Ⅰ类，1.079 5 $\Rightarrow$ 0.539 75 μm）BIBO 晶体有效二阶非线性系数的计算值[7,9]

相位匹配方向	$d_{eff}/(pm \cdot V^{-1})$	相位匹配方向	$d_{eff}/(pm \cdot V^{-1})$
$\theta \approx 10°$，$\phi \approx 90°$	2.3	$\theta \approx 170°$，$\phi \approx 90°$	3.2

见前面表的注。

对某些特殊方向（SHG，Ⅰ类，0.946 $\Rightarrow$ 0.473 μm）BIBO 晶体有效二阶非线性系数的计算值[10]

相位匹配方向	$d_{eff}/(pm \cdot V^{-1})$
$\theta = 161.7°$，$\phi = 90°$	3.34

对某些特定相位匹配方向上 SHG 转换效率（SHG，Ⅰ类，1.064 2 $\Rightarrow$ 0.532 1 μm，$I =$

$3.6\ \text{GW/cm}^2$)的实验值[6,8]

相位匹配方向	晶体长度/cm	SHG 转换效率/%
$\theta = 11.1°$, $\phi = 90°$(YZ 平面)	0.47	58
$\theta = 168.9°$, $\phi = 90°$(YZ 平面)	0.24	67.7

见前面表的注。

激光诱导的表面损伤阈值

$\lambda/\mu m$	τ_p/ns	$I_{thr}/(\text{GW}\cdot\text{cm}^{-2})$	参考文献
1.064	0.035	>4.7	[8]
		>5	[1]

关于这一晶体

新发现的单斜晶体 BIBO 具有相当高的二阶有效非线性值，分别超过 BBO 和 LBO 值的 1.7 倍和 4 倍。因此，在连续波辐射的 SHG 方面可以找到广泛的应用。在文献[11]中，通过 Nd:YAG 激光器($\lambda = 946$ nm, $P = 4.6$ W)的倍频在 BIBO 晶体中产生了 2.8 W 的 CW 蓝光($\lambda = 473$ nm)，Nd:YAG 是由 808 nm 的 21 W 激光二极管所泵浦的。在文献[10]中，在 BIBO 晶体中准连续 OPO (777~1 036 nm, 10 kHz, 50 ns)的信号光实现了 SHG；其结果是：在 UV 区 450~494 nm 实现调谐，在 470 nm 处最大的蓝光输出为 1.3 W。

参考文献

[1] B. Teng, J. Wang, Z. Wang, H. Jiang, X. Hu, R. Song, H. Liu, Y. Liu, J. Wei, Z. Shao: Growth and investigation of a new nonlinear optical crystal: bismuth borate BiB_3O_6. J. Cryst. Growth 224(3-4), 280-283(2001).

[2] R. Fröhlich, L. Bohaty, J. Liebertz: Die Kristallstruktur von Wismutborat, BiB_3O_6. Acta Crystallogr. C 40(3), 343-344(1984) [In German].

[3] P. Becker, J. Liebertz, L. Bohaty: Top-seeded growth of bismuth triborate, BiB_3O_6. J. Cryst. Growth 203(1-2), 149-155(1999).

[4] P. Becker, L. Bohaty: Thermal expansion of bismuth triborate. Cryst. Res. Technol. 36(11), 1175-1180(2001).

[5] B. Teng, Z. Wang, H. Jiang, X. Cheng, H. Liu, X. Hu, S. Dong, J. Wang, Z. Shao: Anisotropic thermal expansion of BiB_3O_6. J. Appl. Phys. 91(6), 3618-3620(2002).

[6] B. Teng, J. Wang, Z. Wang, X. Hu, H. Jiang, H. Liu, X. Cheng, S. Dong, Y. Liu, Z. Shao: Crystal growth, thermal and optical performance of BiB_3O_6. J. Cryst. Growth 233(1-2), 282-286(2001).

[7] H. Hellwig, J. Liebertz, L. Bohaty: Linear optical properties of the monoclinic bismuth borate BiB_3O_6. J. Appl. Phys. 88(1), 240 – 244(2000).

[8] Z. Wang, B. Teng, K. Fu, X. Xu, R. Song, C. Du, H. Jiang, J. Wang, Y. Liu, Z. Shao: Efficient second harmonic generation of pulsed laser radiation in BiB_3O_3 (BIBO) crystal with different phase matching directions. Opt. Commun. 202(1 – 3), 217 – 220(2002).

[9] H. Hellwig, J. Liebertz, L. Bohaty: Exceptional large nonlinear optical coefficients in the monoclinic bismuth borate BiB_3O_6 (BIBO). Solid State Commun. 109(4), 249 – 251 (1999).

[10] Y. Bi, H. -B. Zhang, Z. -P. Sun, Z. -R. -G. -T. Bao, H. -Q. Li, Y. -P. Kong, X. -C. Lin, G. -L. Wang, J. Zhang, W. Hou, R. -N. Li, D. -F. Cui, Z. -Y. Xu, L. -W. Song, P. Zhang, J. -F. Cui, Z. -W. Fan: High-power blue light generation by external frequency doubling of an optical parametric oscillator. Chin. Phys. Lett. 20(11), 1957 – 1959(2003).

[11] C. Czeranowsky, E. Heumann, G. Huber: All-solid-state continuous-wave frequencydoubled Nd: YAG – BIBO laser with 2.8-W output power at 473 nm. Opt. Lett. 28(6), 432 – 434(2003).

6.2 $K_2Al_2B_2O_7$,硼酸铝钾(KABO)

负单轴晶: $n_o > n_e$

分子量: 265.775

密度: 2.47 g/cm^3 [1]

点群: 32

晶格常数:

$a = 8.530$ Å[2]; (8.55800 ± 0.00002) Å[3]; 8.5598 Å[4];

(8.5657 ± 0.0008) Å[5]; (8.5669 ± 0.0009) Å[6]

$c = 8.409$ Å[2]; (8.45576 ± 0.00003) Å[3]; (8.463 ± 0.001) Å[5];

(8.467 ± 0.001) Å[6]; 8.5048 Å[4]

莫氏硬度: ≈6[7]; 5.5~6.5[1]

在水中的溶解度: 不溶[1]

熔点: 1 383 K[1]

线性热膨胀系数的平均值[1]

T/K	$\alpha_t \times 10^6/K^{-1}$, $\parallel a$	$\alpha_t \times 10^6/K^{-1}$, $\parallel c$
298~573	8.4	16.5

$p = 0.101\,325$ MPa 时的比热容 c_p[1]

T/K	$c_p/(\mathrm{J\cdot kg^{-1}\cdot K^{-1}})$	T/K	$c_p/(\mathrm{J\cdot kg^{-1}\cdot K^{-1}})$
321	1 008.4	568	1 390

以"O"透过计的透明范围：$0.18 \sim 3.6$ μm[2]

折射率值的实验值[2]

λ/μm	n_o	n_e	λ/μm	n_o	n_e
0.404 7	1.570 22	1.496 43	0.546 1	1.555 72	1.485 36
0.407 8	1.569 73	1.496 00	0.578 0	1.553 85	1.483 98
0.435 8	1.565 71	1.492 94	0.589 3	1.553 20	1.483 54
0.486 1	1.560 29	1.488 87	0.623 4	1.551 59	1.482 34
0.491 6	1.559 82	1.488 48	0.656 3	1.550 29	1.481 36
0.496 2	1.559 38	1.488 16	0.694 3	1.548 81	1.480 33

温度范围 $293 \sim 393$ K 以及光谱范围 0.193 μm $< \lambda < 1.338\,2$ μm 内折射率的温度微商（以 10^{-5} K^{-1} 为单位）[8]：

$$\frac{\mathrm{d}n_o}{\mathrm{d}T} = 1.610\,1 + 0.036\,1\lambda$$

$$\frac{\mathrm{d}n_e}{\mathrm{d}T} = 1.990\,5 + \frac{0.095\,6}{\lambda} + \frac{0.008\,3}{\lambda^2} - \frac{0.001\,5}{\lambda^3}$$

Sellmeier 方程（λ 以 μm 为单位，0.193 μm $< \lambda < 1.338\,2$ μm，$T = 293$ K）[8]：

$$n_o^2 = 2.376\,5 + \frac{0.013\,03}{\lambda^2 - 0.018\,52} - 0.013\,17\lambda^2$$

$$n_e^2 = 2.173\,67 + \frac{0.009\,50}{\lambda^2 - 0.015\,30} - 0.008\,32\lambda^2$$

文献[2]、[7]、[9]中给出了其他色散关系。

有效二阶非线性系数的表达式（Kleinman 对称条件成立）[10]：

$$d_{ooe} = d_{11}\cos\theta\cos 3\phi$$

$$d_{eoe} = d_{oee} = d_{11}\cos^2\theta\sin 3\phi$$

二阶非线性系数的值：

$d_{11}(1.064$ μm$) = 0.45$ pm/V[2]；0.47 pm/V[1]；(0.46 ± 0.04) pm/V[8]

相位匹配角的实验值

相互作用的波长/μm	$\theta_{exp}/(°)$	参考文献
SHG, o + o ⇒ e		
1.064 2 ⇒ 0.532 1	27.3	[8]
0.8 ⇒ 0.4	33.7	[11]
0.532 1 ⇒ 0.266 05	57.2	[8], [12]
	58.1	[13]
	58.3	[3]
SFG, o + o ⇒ e		
1.064 2 + 0.532 1 ⇒ 0.354 73	36.9	[12]
	37.2	[8]
1.064 2 + 0.266 05 ⇒ 0.212 84	60.2	[8]
1.064 2 + 0.235 8 ⇒ 0.193 0	68.9	[8], [9]
SHG, e + o ⇒ e		
1.064 2 ⇒ 0.532 1	39.3	[8]

内角带宽、温度带宽和光谱带的实验值

相互作用的波长/μm	$\theta_{pm}/(°)$	$\Delta\theta^{int}/(°)$	$\Delta T/℃$	$\Delta\nu/cm^{-1}$	参考文献
SHG, o + o ⇒ e					
1.064 2 ⇒ 0.532 1	27.3		41.9		[8]
0.8 ⇒ 0.4	33.7	0.052 ± 0.017		9.4 ± 2.2	[11]
0.532 1 ⇒ 0.266 05	57.2		4.1		[8]
SFG, o + o ⇒ e					
1.064 2 + 0.532 1 ⇒ 0.354 73	36.9	0.025			[12]
			13.2		[8]
1.064 2 + 0.266 05 ⇒ 0.212 84	60.2		2.9		[8]
1.064 2 + 0.235 8 ⇒ 0.193	68.9	0.011	2.1		[9]
			2.2		[8]

激光诱导的损伤阈值

$\lambda/\mu m$	τ_p/ns	$I_{thr}/(GW \cdot cm^{-2})$	参考文献	备注
0.532	0.035	>1.3	[3]	
0.8	0.000 05	>170	[11]	5 kHz
1.064	10	>1.0	[2]	10 Hz
		15	[2]	1个脉冲

关于这一晶体

  这种新的非线性硼酸盐晶体是非潮解的,并能够按正常的工艺切割和抛光。尽管 KABO 晶体的二阶非线性系数比 BBO 和 CLBO 晶体的要小一些，然而这种材料对在 UV 区实现三波相互作用可能是有用的。在文献[8]中，由 Nd:YAG 激光辐射($P = 18$ W, $\Delta f = 10$ kHz)和 Nd:YAG 激光泵浦 OPO 的二次谐波的 SFG 在一块 0.7 cm 长的 KABO 晶体($\theta = 68.9°, \phi = 0°$)中获得 0.22 W 的准连续输出($\lambda = 193$ nm)。

参考文献

[1] C. Zhang, J. Wang, X. Cheng, X. Hu, H. Jiang, Y. Liu, C. Chen: Growth and properties of $K_2Al_2B_2O_7$ crystal. Opt. Mater. 23(1-2), 357-362(2003).

[2] N. Ye, W. Zeng, J. Jiang, B. Wu, C. Chen, B. Feng, X. Zhang: New nonlinear optical crystal $K_2Al_2B_2O_7$. J. Opt. Soc. Am. B 17(5), 764-768(2000).

[3] C. Zhang, J. Wang, X. Hu, H. Jiang, Y. Liu, C. Chen: Growth of large $K_2Al_2B_2O_7$ crystals. J. Cryst. Growth 235(1-4), 1-4(2002).

[4] C. Zhang, J. Wang, X. Hu, H. Liu, J. Wei, Y. Liu, Y. Wu, C. Chen: Top-seeded growth of $K_2Al_2B_2O_7$. J. Cryst. Growth 231(4), 439-441(2001).

[5] Z. Hu, T. Higasniyama, M. Yoshimura, Y. K. Yap, Y. Mori, T. Sasaki: A new nonlinear optical borate crystal $K_2Al_2B_2O_7$ (KAB). Jpn. J. Appl. Phys. 37(10A), L1093-L1094(1998).

[6] Z. Hu, Y. Mori, T. Higashiyama, M. Yoshimura, Y. K. Yap, Y. Kagebayashi, T. Sasaki: $K_2Al_2B_2O_7$—a new nonlinear optical crystal. Proc. SPIE 3556, 156-161(1998).

[7] N. Ye, W. Zeng, J. Jiang, B. Wu, C. Chen, B. Feng, X. Zhang: New nonlinear optical crystal $K_2Al_2B_2O_7$: errata. J. Opt. Soc. Am. B 18(1), 122(2001).

[8] N. Umemura, M. Ando, K. Suzuki, E. Takaoka, K. Kato, Z.-G. Hu, M. Yoshimura, Y. Mori, T. Sasaki: 200-mW-average power ultraviolet generation at 0.193 μm in $K_2Al_2B_2O_7$. Appl. Opt. 42(15), 2716-2719(2003).

[9] Z.-G. Hu, M. Yoshimura, Y. Mori, T. Sasaki, K. Kato: Growth of $K_2Al_2B_2O_7$ crystal for UV light generation. Opt. Mater. 23(1-2), 353-356(2003).

[10] J. E. Midwinter, J. Warner: The effects of phase matching method and of uniaxial crystal symmetry on the polar distribution of second-order non-linear optical polarization. Brit. J. Appl. Phys. 16(11), 1135-1142(1965).

[11] P. Kumbhakar, S. Adachi, Z.-G. Hu, M. Yoshimura, Y. Mori, T. Sasaki, T. Kobayashi: Generation of tunable near-UV laser radiation by type-I second-harmonic generation in a new crystal, $K_2Al_2B_2O_7$ (KABO). Jpn. J. Appl. Phys. 42(10B), L1255-L1258(2003).

[12] Z. Hu, N. Ushiyama, Y. K. Yap, M. Yoshimura, Y. Mori, T. Sasaki: The crystal growth and nonlinear optical properties of $K_2Al_2B_2O_7$. J. Cryst. Growth 237-239, 654-657(2002).

[13] J. Lu, G. Wang, Z. Hu, C. Chen, J. Wang, C. Zhang, Y. Liu: Efficient 266 nm ul-

traviolet beam generation in $K_2Al_2B_2O_7$ crystal. Chin. Phys. Lett. 19(5), 680-681 (2002).

6.3 $KBe_2BO_3F_2$，氟硼铍酸钾(KBBF)

负单轴晶：$n_o > n_e$

分子量：153.927

点群：32

晶格常数[1]：$a = (4.472 \pm 0.004)$ Å $\quad c = (18.744 \pm 0.009)$ Å

莫氏硬度：≈0，易沿[001]面开裂[2]

熔点：1 373 K[3]

以"0"透过计的透明范围：0.155~3.7 μm[2]

折射率的实验值[4]

$\lambda/\mu m$	n_o	n_e	$\lambda/\mu m$	n_o	n_e
0.404 7	1.487	1.410	0.589 3	1.479	1.401
0.435 8	1.485	1.408	0.632 8	1.478	1.400
0.486 1	1.482	1.406	0.656 3	1.477	1.400
0.546 1	1.479	1.403			

Sellmeier 方程(λ 以 μm 为单位)[2,4]:

$$n_o^2 = 1 + \frac{1.169\ 725 \lambda^2}{\lambda^2 - 0.006\ 240\ 0} - 0.009\ 904 \lambda^2$$

$$n_e^2 = 1 + \frac{0.956\ 611 \lambda^2}{\lambda^2 - 0.006\ 192\ 6} - 0.027\ 849 \lambda^2$$

有效二阶非线性系数的表达式(Kleinman 对称条件成立)[5]:

$$d_{ooe} = d_{11} \cos\theta \cos 3\phi$$

$$d_{eoe} = d_{oee} = d_{11} \cos^2\theta \sin 3\phi$$

二阶非线性系数的绝对值：

$$d_{11}(1.064\ 2\ \mu m) = 0.49\ pm/V^{[6]}$$

相位匹配角的实验值

相互作用的波长/μm	$\theta_{exp}/(°)$	参考文献	相互作用的波长/μm	$\theta_{exp}/(°)$	参考文献
SHG, o+o⇒e			0.374 3⇒0.187 15	59.4	[2]
0.345⇒0.172 5	~71	[7]	0.384 7⇒0.192 35	56.8	[2]
0.354 6⇒0.177 3	~66.2	[7]	0.41⇒0.205	51.5	[2]
0.369 5⇒0.184 75	61	[2]	0.44⇒0.22	46	[2]

续表

相互作用的波长/μm	$\theta_{exp}/(°)$	参考文献	相互作用的波长/μm	$\theta_{exp}/(°)$	参考文献
0.46⇒0.23	44	[2]	0.68⇒0.34	27.6	[2]
0.48⇒0.24	41.7	[2]	0.77⇒0.385	25.1	[2]
0.5⇒0.25	39.6	[2]	0.85⇒0.425	23.1	[2]
0.532⇒0.266	36.2	[2]	0.9⇒0.45	22	[2]
0.55⇒0.275	34.9	[2]	0.95⇒0.425	21	[2]
0.589⇒0.294 5	32.5	[2]	1.064⇒0.532	20.2	[2]
0.6⇒0.3	32.1	[2]	1.342⇒0.671	18.6	[2]

内角带宽的实验值

相互作用的波长/μm	$\theta_{pm}/(°)$	$\Delta\theta^{int}/(°)$	参考文献
SHG, o+o⇒e			
0.41⇒0.205	51.5	0.011 9	[2]
0.44⇒0.22	46	0.012 7	[2]
0.46⇒0.23	44	0.012 7	[2]
0.48⇒0.24	41.7	0.013 9	[2]
0.5⇒0.25	39.6	0.014 3	[2]
0.532⇒0.266	36.2	0.016 6	[2]
		0.015 2	[8]
0.589⇒0.294 5	32.5	0.024 4	[2]
0.9⇒0.45	22.0	0.057 2	[2]
1.064⇒0.532	20.2	0.059 2	[2]
1.342⇒0.671	18.6	0.064 4	[2]

激光诱导的损伤阈值

$\lambda/\mu m$	τ_p/ns	$I_{thr}/(GW \cdot cm^{-2})$	参考文献	备注
0.354 6	0.01	>0.04	[7]	80 MHz
0.4	0.000 05	>17	[9]	1 kHz
0.532	0.035	>4.2	[10]	
		>11.6	[8]	10 Hz
	0.03	>7	[11]	
1.064	8	>5.0	[2]	沿 c

关于这一晶体

KBBF 晶体是能产生深紫外激光的优越非线性材料。在文献[7]中,通过由直接 SHG 获得 172.5 nm 的辐射以及通过 SFG 获得 163.3 nm 的光。KBBF 的主要缺点在于其层状本性,这使得生长超过 1 mm 厚的晶体非常困难。为了克服切割这么薄晶体的问题,提出了特别的棱镜耦合技术[8,9,11]。

■ 参考文献

[1] C. Chen, Y. Wang, Y. Xia, B. Wu, D. Tang, K. Wu, Z. Wenrong, L. Yu, L. Mei: New development of nonlinear optical crystals for the ultraviolet region with molecular engineering approach. J. Appl. Phys. 77(6), 2268 – 2272(1995).

[2] B. Wu, D. Tang, N. Ye, C. Chen: Linear and nonlinear optical properties of the KBe_2BO_3F(KBBF) crystal. Opt. Mater. 5(1 – 2), 105 – 109(1996).

[3] D. Tang, Y. Xia, B. Wu, C. Chen: Growth of a new UV nonlinear optical crystal: $KBe_2(BO_3)F_2$. J. Cryst. Growth 222(1 – 2), 125 – 129(2001).

[4] C. Chen, Z. Xu, D. Deng, J. Zhang, G. K. L. Wong, B. Wu, N. Ye, D. Tang: The vacuum ultraviolet phase-matching characteristics of nonlinear optical $KBe_2BO_3F_2$ crystal. Appl. Phys. Lett. 68(21), 2930 – 2932(1996).

[5] J. E. Midwinter, J. Warner: The effects of phase matching method and of uniaxial crystal symmetry on the polar distribution of second-order non-linear optical polarization. Brit. J. Appl. Phys. 16(11), 1135 – 1142(1965).

[6] G. Wang, C. Zhang, C. Chen, Z. Xu, J. Wang: Determination of nonlinear optical coefficients of $KBe_2BO_3F_2$ crystals. Chin. Phys. Lett. 20(2), 243 – 245(2003).

[7] T. Togashi, T. Kanai, T. Sekikawa, S. Watanabe, C. Chen, C. Zhang, Z. Xu, J. Wang: Generation of vacuum-ultraviolet light by an optically contacted, prism-coupled $KBe_2BO_3F_2$ crystal. Opt. Lett. 28(4), 254 – 256(2003).

[8] G. Wang, C. Zhang, C. Chen, A. Yao, J. Zhang, Z. Hu, J. Wang: High-efficiency 266-nm output of a $KBe_2BO_3F_2$ crystal. Appl. Opt. 42(21), 4331 – 4334(2003).

[9] C. Chen, J. Lu, T. Togashi, T. Suganuma, T. Sekikawa, S. Watanabe, Z. Xu, J. Wang: Second-harmonic generation from a $KBe_2BO_3F_2$ crystal in the deep ultraviolet. Opt. Lett. 27(8), 637 – 639(2002).

[10] J. Lu, G. Wang, Z. Xu, C. Chen, J. Wang, C. Zhang, Y. Liu: High-efficiency fourthharmonic generation of KBBF crystal. Opt. Commun. 200(1 – 6), 415 – 418(2001).

[11] C. Chen, J. Lu, G. Wang, Z. Xu, J. Wang, C. Zhang, Y. Liu: Deep ultraviolet harmonic generation with $KBe_2BO_3F_2$ crystal. Chin. Phys. Lett. 18(8), 1081(2001).

6.4 $BaAlBO_3F_2$,氟硼酸铝钡(BABF)

负单轴晶:$n_o > n_e$

分子量：261.117

点群：$\bar{6}$

晶格常数[1]：$a = (4.887\,9 \pm 0.000\,6)$ Å，$c = (9.403 \pm 0.001)$ Å

以"0"透过计的透明范围：0.165 μm 到 >1.6 μm[2]

折射率值的实验值[2]

$\lambda/\mu m$	n_o	n_e	$\lambda/\mu m$	n_o	n_e
0.230	1.717 1	1.660 4	0.683	1.626 6	1.583 4
0.244	1.704 5	1.649 2	0.733	1.625 3	1.582 1
0.266	1.688 6	1.636 4	0.783	1.624 0	1.581 0
0.300	1.671 9	1.621 9	0.833	1.622 7	1.580 5
0.355	1.654 8	1.607 3	0.933	1.621 4	1.579 1
0.400	1.646 4	1.600 4	1.064	1.619 3	1.577 5
0.440	1.641 3	1.595 7	1.150	1.618 0	1.577 0
0.488	1.636 9	1.591 7	1.250	1.617 0	1.576 4
0.514	1.634 6	1.590 1	1.350	1.615 4	1.575 6
0.532	1.633 6	1.589 0	1.450	1.614 1	1.574 3
0.580	1.630 7	1.586 6	1.547	1.613 0	1.574 0
0.633	1.628 4	1.585 0			

Sellmeier 方程（λ 以 μm 为单位）[2]：

$$n_o^2 = 2.621\,3 + \frac{0.013\,53}{\lambda^2 - 0.012\,04} - 0.010\,55\lambda^2$$

$$n_e^2 = 2.483\,3 + \frac{0.011\,78}{\lambda^2 - 0.009\,96} - 0.004\,47\lambda^2$$

有效二阶非线性系数的表达式（Kleinman 对称条件成立）[3]：

$$d_{ooe} = d_{11}\cos\theta\cos 3\phi - d_{22}\cos\theta\sin 3\phi$$

$$d_{eoe} = d_{oee} = d_{11}\cos^2\theta\sin 3\phi + d_{22}\cos^2\theta\cos 3\phi$$

相位匹配角的实验值[2]

相互作用的波长/μm	$\theta_{exp}/(°)$
SHG, o + o ⇒ e	
1.064 ⇒ 0.532	34.1

关于这一晶体

发现在粉末状态时，BABF 所产生的二次谐波功率两倍于 KDP 的粉末倍频效应[2]。

参考文献

[1] Z.-G. Hu, M. Yoshimura, Y. Mori, T. Sasaki: Growth of a new nonlinear optical crystal—BaAlBO$_3$F$_2$. J. Cryst. Growth 260(3-4), 287-290(2004).

[2] Z.-G. Hu, M. Yoshimura, K. Muramatsu, Y. Mori, T. Sasaki: A new nonlinear optical crystal—BaAlBO$_3$F$_2$(BABF). Jpn. J. Appl. Phys. 41(10B), L1131-L1133(2002).

[3] J. E. Midwinter, J. Warner: The effects of phase matching method and of uniaxial crystal symmetry on the polar distribution of second-order non-linear optical polarization. Brit. J. Appl. Phys. 16(11), 1135-1142(1965).

6.5 La$_2$CaB$_{10}$O$_{19}$，硼酸钙镧(LCB)

正双轴晶：$2V_Z = 9.7°$，$\lambda = 0.532\,1\,\mu m$

分子量：729.978

密度(计算值)：3.665 g/cm^3 [1]

点群：2

晶格常数：

$a = (11.043 \pm 0.003)\,Å$ [1]； 11.056 Å [2]

$b = (6.563 \pm 0.002)\,Å$ [1]； 6.577 Å [2]

$c = (9.129 \pm 0.002)\,Å$ [1]； 9.119 Å [2]

介电轴和晶体学轴的变换[3]：$Y \parallel b$，a轴和c轴位于XZ平面，两轴夹角β = 91.47°，Z轴和a轴夹角为46.03°，X轴和c轴夹角为47.5°，在文献[4]中给出了相似的认定。

莫氏硬度：6.5 [1]

熔点：1 338 K [1]

以0.5透过计的透明范围：0.185~3.0 μm [3]； 0.28~2.45 μm [4]

Sellmeier 方程(λ以μm为单位)[3]：

$$n_X^2 = 2.781\,22 + \frac{0.016\,318\,6}{\lambda^2 - 0.014\,600\,2} - 0.016\,229\,9\lambda^2$$

$$n_Y^2 = 2.785\,33 + \frac{0.015\,168\,8}{\lambda^2 - 0.020\,607\,9} - 0.015\,547\,5\lambda^2$$

$$n_Z^2 = 2.961\,67 + \frac{0.020\,423\,8}{\lambda^2 - 0.013\,691\,2} - 0.020\,144\,7\lambda^2$$

LCB晶体主平面上有效二阶非线性系数的表达式(小走离角近似，Kleinman对称条件成立，$d_{14} = d_{25} = d_{36}$，$d_{16} = d_{21}$和$d_{23} = d_{34}$)[5]：

XY平面

$$d_{ooe} = d_{23}\cos\phi$$
$$d_{eoe} = d_{oee} = d_{14}\sin 2\phi$$

YZ 平面
$$d_{eeo} = d_{14}\sin 2\theta$$
$$d_{oeo} = d_{eoo} = d_{16}\cos\theta$$

XZ 平面,$\theta < V_Z$
$$d_{eoe} = d_{oee} = d_{16}\cos^2\theta + d_{23}\sin^2\theta \pm d_{14}\sin 2\theta$$

XZ 平面,$\theta > V_Z$
$$d_{eeo} = d_{16}\cos^2\theta + d_{23}\sin^2\theta \pm d_{14}\sin 2\theta$$

LCB 晶体中一个特殊的相位匹配方向(SHG,I 类,1.064 2⇒0.532 1 μm)的有效二阶非线性系数的实验值[3]

相互作用的波长/μm	d_{eff}/(pm·V^{-1})
$\theta = 34.3°$,$\phi = 7.7°$	1.05

激光诱导的体损伤阈值[3]

λ/μm	τ_p/ns	I_{thr}/(GW·cm^{-2})
1.064	0.035	>8

关于这一晶体

这一新发展的单斜晶体具有高的硬度和中等的非线性,它也是不潮解的[2]。获得的最短 SHG 波长为 288 nm[4]。

参考文献

[1] Y. Wu, J. Liu, P. Fu, F. Guo, G. Zhao, J. Qin, C. Chen: A new class of nonlinear optical crystals R$_2$CaB$_{10}$O$_{19}$ (RCB). Proc. SPIE 3556, 8 – 13 (1998).

[2] X. W. Xu, T. C. Chong, G. Y. Zhang, S. D. Cheng, M. H. Li, C. C. Phua: Growth and optical properties of a new nonlinear optical lanthanum borate crystal. J. Cryst. Growth 237-239, 649 – 653 (2002).

[3] G. Wang, J. Lu, D. Cui, Z. Xu, Y. Wu, P. Fu, X. Guan, C. Chen: Efficient second harmonic generation in a new nonlinear La$_2$CaB$_{10}$O$_{19}$ crystal. Opt. Commun. 209 (4 – 6), 481 – 484 (2002).

[4] Y. Wu, P. Fu, F. Zheng, S. Wan, X. Guan: Growth of a nonlinear optical crystal La$_2$CaB$_{10}$O$_{19}$ (LCB). Opt. Mater. 23 (1 – 2), 373 – 375 (2003).

[5] B. V. Bokut: Optical mixing in biaxial crystals. Zh. Prikl. Spektrosk. 7(4), 621 – 624 (1967) [In Russian, English trans.: J. Appl. Spectrosc. 7(4), 425 – 429(1967)].

6.6 GdCa$_4$O(BO$_3$)$_3$，硼酸氧钙钆(GdCOB)

负双轴晶：$2V_z = 120.7°$，$\lambda = 0.546$ μm[1,2]

分子量：509.986

密度计算值：3.736 g/cm^3[3]

点群：m

晶格常数：

  $a = (8.106 \pm 0.002)$Å[1]；(8.095 ± 0.007)Å[4]；

   8.093 7Å[5]；(8.098 ± 0.002)Å[6]

  $b = (16.028 \pm 0.003)$Å[1]；(16.018 ± 0.006)Å[4]；

   16.013Å[5]；(16.019 ± 0.006)Å[6]

  $c = (3.557 \pm 0.001)$Å[1]；(3.558 ± 0.008)Å[4]；

   3.557 9Å[5]；(3.559 ± 0.007)Å[6]

  $\beta = 101.25°$[1]；$101.26° \pm 0.01°$[4]；$101.27°$[5,6]

介电轴和晶体学轴的变换：$Y \parallel b$，a 轴和 c 轴位于 XZ 平面，两轴夹角 $\beta = 101.27°$，Z 轴和 a 轴夹角为 $27.2°$，X 轴和 c 轴夹角为 $16.2°$[7,8]。早先报道了稍有不同的变换：$(a,Z) = 26°$，$(c,X) = 15°$[1,2]。

莫氏硬度：6.5[9]

诺氏硬度：550~715 kg/mm^2[2]

熔点：1 753 K[4,5]；1 756 K(同成分熔化)[10]

线性热膨胀系数的平均值

T/K	$\alpha_t \times 10^6$/K^{-1}, $\parallel a$	$\alpha_t \times 10^6$/K^{-1}, $\parallel b$	$\alpha_t \times 10^6$/K^{-1}, $\parallel c$	参考文献
293~1 133	10.2	8.3	14.3	[11]
293~1 273	10.35	7.78	13.10	[6]

$T = 293$ K 时的热导率

T/K	κ/(W·m^{-1}·K^{-1}), $\parallel X$	κ/(W·m^{-1}·K^{-1}), $\parallel Y$	κ/(W·m^{-1}·K^{-1}), $\parallel Z$	参考文献
287	2.173			[12]
289			2.401	[12]
291		1.32		[12]
293	2.54	1.32	2.06	[9]

续表

T/K	$\kappa/(W\cdot m^{-1}\cdot K^{-1}),\parallel X$	$\kappa/(W\cdot m^{-1}\cdot K^{-1}),\parallel Y$	$\kappa/(W\cdot m^{-1}\cdot K^{-1}),\parallel Z$	参考文献
297	2.539			[12]
324	2.227			[12]
345			1.880	[12]
353	2.016			[12]
394			1.799	[12]
403		1.22		[12]
424	2.237			[12]
445			1.807	[12]
474	2.277			[12]
496			1.852	[12]
525			1.789	[12]
526	2.009			[12]
545		1.18		[12]

高透明范围:0.32~2.6 μm[13]

在 UV 透明范围 0.2~0.32 μm 内有三个尖锐的吸收线,中心位置约在 0.25 μm、0.277 μm 和 0.31 μm 处[2]。

在 IR 透明范围 2.6~3.7 μm 内,在 2.72 μm、2.9 μm 和 3.25 μm 处有吸收带[2,13]。

折射率的实验值[2]

$\lambda/\mu m$	n_X	n_Y	n_Z	$\lambda/\mu m$	n_X	n_Y	n_Z
0.4047	1.7209	1.7476	1.7563	0.5780	1.6966	1.7225	1.7310
0.4358	1.7142	1.7409	1.7493	0.5876	1.6960	1.7218	1.7303
0.4678	1.7089	1.7350	1.7436	0.6439	1.6923	1.7181	1.7265
0.4800	1.7068	1.7333	1.7418	0.6678	1.6910	1.7168	1.7250
0.5086	1.7033	1.7295	1.7379	0.7290	1.6879	1.7133	1.7216
0.5461	1.6992	1.7253	1.7340	0.7960	1.6860	1.7112	1.7197

最佳 Sellmeier 方程($T=293$ K,λ 以 μm 为单位,0.4129 μm $<\lambda<$ 1.3382 μm)[14]:

$$n_X^2 = 2.8063 + \frac{0.02315}{\lambda^2 - 0.01378} - 0.00537\lambda^2$$

$$n_Y^2 = 2.8959 + \frac{0.02398}{\lambda^2 - 0.01389} - 0.01132\lambda^2$$

$$n_z^2 = 2.9248 + \frac{0.02410}{\lambda^2 - 0.01406} - 0.01139\lambda^2$$

在文献[1]、[2]、[5]、[8]、[15]中给出了其他色散关系方程组。

室温下低频(远低于GdCOB晶体声学共振频率,即对"自由"晶体)下测量的线性电光系数(以 pm/V 为单位)[16]

$\lambda/\mu m$	r_{11}^T	r_{21}^T	r_{31}^T	r_{13}^T	r_{23}^T	r_{33}^T	r_{51}^T	r_{53}^T	r_{42}^T	r_{62}^T
0.6328	0.4	0.5	0.6	0.1	0.4	2.0	0.7	1.5	0.5	0.8

GdCOB晶体主平面上有效二阶非线性系数的表达式(小走离角近似,Kleinman对称条件成立,$d_{12}=d_{26},d_{13}=d_{35},d_{15}=d_{31},d_{24}=d_{32}$)[2,17]:

XY 平面, $\theta=90°$
$$d_{ooe} = d_{13}\sin\phi$$
$$d_{eoe} = d_{oee} = d_{31}\sin^2\phi + d_{32}\cos^2\phi$$

YZ 平面, $\phi=90°$
$$d_{eeo} = d_{13}\sin^2\theta + d_{12}\cos^2\theta$$
$$d_{oeo} = d_{eoo} = d_{31}\sin\theta$$

XZ 平面, $\phi=0°$, $V_Z > \theta > 0°$
$$d_{ooe} = d_{12}\cos\theta - d_{32}\sin\theta$$

XZ 平面, $\phi=0°$, $90° > \theta > V_Z$
$$d_{oeo} = d_{eoo} = d_{12}\cos\theta - d_{32}\sin\theta$$

XZ 平面, $\phi=0°$, $180°-V_Z > \theta > 90°$;或 $\phi=180°$, $90° > \theta > V_Z$
$$d_{oeo} = d_{eoo} = d_{12}\cos\theta + d_{32}\sin\theta$$

XZ 平面, $\phi=0°$, $180° > \theta > 180°-V_Z$;或 $\phi=180°$, $V_Z > \theta > 0°$
$$d_{ooe} = d_{12}\cos\theta + d_{32}\sin\theta$$

二阶非线性系数最可靠的实验值:

$d_{11}(1.0642\ \mu m) = 0$[17]

$d_{12}(1.0642\ \mu m) = 0.24$ pm/V[18]; 0.27 pm/V[17]; 0.31 pm/V[19]

$d_{13}(1.0642\ \mu m) = -0.74$ pm/V[18]; -0.85 pm/V[17]; -0.87 pm/V[19]

$d_{31}(1.0642\ \mu m) = 0.20$ pm/V[17]

$d_{32}(1.0642\ \mu m) = 2.23$ pm/V[17]; 2.26 pm/V[19]; 2.39 pm/V[18]

$d_{33}(1.0642\ \mu m) = -1.87$ pm/V[17]

GdCOB晶体主平面上SHG的相位匹配角和内角带宽的实验值

相互作用的波长/μm	$\phi_{pm}/(°)$	$\theta_{pm}/(°)$	$\Delta\phi^{int}/(°)$	$\Delta\theta^{int}/(°)$	参 考 文 献
XY 平面，$\theta = 90°$					
SHG，o+o⇒e					
1.064 2⇒0.532 1	46		0.10		[2],[15],[18],[19]
0.946⇒0.473	55.9		0.11		[20]
XZ 平面，$\phi = 0°$，$\theta < V_Z$					
SHG，o+o⇒e					
1.064 2⇒0.532 1		19.7		0.15	[2],[15],[18],[19]

注：在双轴晶体中，存在两个接收角：一个以 θ 表示；另一个以 ϕ 表示。作者仅给出了最小的一个。

对某些相位匹配方向（SHG，Ⅰ类，1.064 2 μm⇒0.532 1 μm）GdCOB 晶体有效二阶非线性系数的实验值

相位匹配方向	$d_{eff}/(pm \cdot V^{-1})$	参 考 文 献
$\theta = 90°$，$\phi = 46°$（XY 平面）	0.59	[17]
	0.63	[19]
$\theta = 19.7°$，$\phi = 0°$（XZ 平面）	0.48	[19]
	0.50	[17]
$\theta = 160.3°$，$\phi = 0°$（XZ 平面）	1.01	[17]
	1.05	[19]
$\theta = 66.8°$，$\phi = 47.4°$	0.68	[17]
$\theta = 67°$，$\phi = 46°$	0.78	[19]
$\theta = 66.8°$，$\phi = 132.6°$	1.51	[6]
	1.68	[17]
$\theta = 67°$，$\phi = 134°$	1.8	[19]

注：在 GdCOB 晶体情况下，d_{eff} 的性质包含了镜面和倒反对称性[21]。这意味着可以选择两个独立的象限，例如（$0° < \theta < 90°$, $0° < \phi < 90°$）以及（$0° < \theta < 90°$, $90° < \phi < 180°$）就能完全描述 d_{eff} 的空间分布。随后，在这两个象限中每一个（θ,ϕ）方向的 d_{eff} 值等于（$180° - \theta, 180° - \phi$）方向的值，反之亦然。例如（$\theta = 66.8°, \phi = 132.6°$）和（$\theta = 113.2°, \phi = 47.4°$）具有相等的 d_{eff} 值。

激光诱导的体损伤阈值

$\lambda/\mu m$	τ_p/ns	$I_{thr}/(GW \cdot cm^{-2})$	参考文献	备注
0.337	0.015~0.075	>1.35	[10]	
0.532	7	1	[2]	
1.064	6	>1	[15]	10 Hz
	0.035	>6	[17]	
		>8	[22]	
		130	[6]	1 个脉冲

关于这一晶体

　　GdCOB 是在近期发展起来的，在 1996—1997 年间，由法国和日本科学家同时发现[1,4,5]。从那时起发表了数以百计的文章投入研究 GdCOB 及其最相近的同构晶体 YCOB。到现在，这些材料当然是点群 m 的晶体中研究得最为彻底的。这就使我们在低对称性晶体中关于三波相互作用的物理本质方面取得了显著的进步。结果表明，对于这样的晶体 d_{eff} 的空间分布只要选择两个独立的象限（例如 $0°<\theta<90°,0°<\phi<90°$ 以及 $0°<\theta<90°,90°<\phi<180°$）就能完全地描述。测量了 GdCOB 的二阶非线性系数并发现对于 Nd:YAG 激光辐射的 SHG，有效非线性的极大值不是位于第一象限（$0°<\theta<90°,0°<\phi<90°$）[13,17,19]；推导了 GdCOB 在主平面上 d_{eff} 的表达式[17]。

　　在文献[22]和[23]中研究了 4% 原子浓度的 Sr 掺杂的 GdCOB 晶体在一台 CW 波 Nd:YVO$_4$ 激光器（1 064 nm）的腔内 SHG。结果表明以这样掺杂量生长的 6 mm 长的晶体（$\theta=66.8°,\phi=132.6°$）对于 SHG 转换效率有很大改善。以 13 W 的基频能量产生了 2.3 W 绿光输出[22]。同样的 Sr 掺杂晶体用于锁模 Nd:YAG 激光器的腔外倍频，在基频强度为 4~8 GW/cm^2 时能量转换效率达到 55%，这个值超过未掺杂晶体样品 1.4 倍。在文献[24]中，1.2 cm 长的 4 at.% Li 掺杂 GdCOB 晶体（$\theta=66.8°,\phi=132.3°$）用于 13 W Nd:YVO$_4$ 激光器的腔内 SHG，产生了 2.55 W、531 nm 的 CW 激光输出。

　　与 GdCOB 相关的最新技术进展是通过激光二极管终端泵浦 Nd:YVO$_4$ 激光器（$P=6.8$ W）的腔内倍频 1.2 cm 长的晶体（$\theta=66.8°,\phi=132.3°$）产生了 2.8 W 连续绿光（$\lambda=532$ nm）[25]。然而相同长度的 KTP 晶体（$\theta=90°,\phi=25°$）在同一构型中使用，产生约 4.5 W 的绿光。

参考文献

[1] G. Aka, L. Bloch, J. M. Benitez, P. Crochet, A. Kahn-Harari, D. Vivien, F. salin, P. Coquelin, D. Colin: A new non linear oxoborate crystal, characterized by using femtosecond broadband pulses. In: *Advanced Solid-State Lasers*, *OSA Trends in Optics and Photonics Series*, *Vol.1*, ed. by S. A. Payne, C. Pollock (OSA, Washington DC, 1996), pp. 336–340.

[2] G. Aka, A. Kahn-Harari, F. Mougel, D. Vivien, F. Salin, P. Coquelin, P. Colin, D. Pelenc, J. P. Damelet: Linear and nonlinear-optical properties of a new gadolinium calcium oxoborate crystal, Ca$_4$GdO(BO$_3$)$_3$. J. Opt. Soc. Am. B 14(9), 2238–2247(1997).

[3] A. B. Ilyukhin, B. F. Dzhurinskii: Crystal structures of binary oxoborates LnCa$_4$O(BO$_3$)$_3$ (Ln=Gd,Tb,and Lu) and Eu$_2$CaO(BO$_3$)$_2$. Zh. Neorg. Khim. 38(6), 917–920(1993) [In Russian,English trans.:Russ. J. Inorg. Chem. 38(6),847–850(1993)].

[4] G. Aka, A. Kahn-Harari, D. Vivien, J.-M. Benitez, F. Salin, J. Godard: A new nonlinear and neodymium laser self-frequency doubling crystal with congruent melting: $Ca_4GdO(BO_3)_3$ (GdCOB). Eur. J. Solid State Inorg. Chem. 33(8), 727–736(1996).

[5] M. Iwai, T. Kobayashi, H. Furuya, Y. Mori, T. Sasaki: Crystal growth and optical characterization of rare-earth (Re) calcium oxyborate $ReCa_4O(BO_3)_3$ (Re = Y or Gd) as new nonlinear optical material. Jpn. J. Appl. Phys. 36(3A), L276-L279(1997).

[6] J. Zhou, Z. Zhong, J. Xu, J. Luo, W. Hua, S. Fan: Bridgman growth and characterization of nonlinear optical single crystals $Ca_4GdO(BO_3)_3$. Mater. Sci. Eng. B 97(3), 283–287(2003).

[7] Z. Wang, J. Liu, R. Song, X. Xu, X. Sun, H. Jiang, K. Fu, J. Wang, Y. Liu, J. Wei, Z. Shao: The second-harmonic-generation property of $GdCa_4O(BO_3)_3$ crystal with various phase-matching directions. Opt. Commun. 187(4–6), 401–405(2001).

[8] Z. Shao, J. Lu, Z. Wang, J. Wang, M. Jiang: Anisotropic properties of Nd: ReCOB (Re = Y, Gd): a low symmetry self-frequency doubling crystal. Progr. Cryst. Growth Character. Mater. 40(1–4), 63–73(2000).

[9] D. Vivien, G. Aka, A. Kahn-Harari, A. Aron, F. Mougel, J.-M. Benitez, B. Ferrand, R. Klein, G. Kugel, N. Le Nain, M. Jacquet: Crystal growth and optical properties of rare earth calcium oxoborates. J. Cryst. Growth, 237–239, 621–628(2002).

[10] T. Łukasiewicz, A. Majchrowski, I. V. Kityk, J. Kroog: Influence of the rare-earth doping on the photoinduced EOEs in the GdCOB. Mater. Lett. 57(13–14), 2049–2052 (2003).

[11] C. Wang, H. Zhang, X. Meng, L. Zhu, Y. T. Chow, X. Liu, R. Cheng, Z. Yang, S. Zhang, L. Sun: Thermal, spectroscopic properties and laser performance at 1.06 and 1.33 μm of Nd: $Ca_4YO(BO_3)_3$ and Nd: $Ca_4GdO(BO_3)_3$ crystals. J. Cryst. Growth 220 (1–2), 114–120(2000).

[12] F. Auge, F. Druon, F. Balembois, P. Georges, A. Brun, F. Mougel, G. Aka, D. Vivien: Theoretical and experimental investigations of a diode-pumped quasi-three-level laser: the Yb^{3+}-doped $Ca_4GdO(BO_3)_3$ (Yb:GdCOB) laser. IEEE. J. Quant. Electr. 36(5), 598–606 (2000).

[13] S. Zhang, Z. Cheng, J. Lu, G. Li, J. Lu, Z. Shao, H. Chen: Studies of the effective nonlinear coefficient of $GdCa_4O(BO_3)_3$ crystal. J. Cryst. Growth 205(3), 453–456 (1999).

[14] N. Umemura, H. Nakao, H. Furuya, M. Yoshimura, Y. Mori, T. Sasaki, K. Yoshida, K. Kato: 90° phase-matching properties of $YCa_4O(BO_3)_3$ and $Gd_xY_{1-x}Ca_4O(BO_3)_3$. Jpn. J. Appl. Phys. 40(2A), 596–600(2001).

[15] G. Aka, F. Mougel, D. Pelenc, B. Ferrand, D. Vivien: Comparative evaluation of Gd-COB and YCOB nonlinear-optical properties, in principal and out of principal plane con-

figurations, for the 1064 nm Nd: YAG laser frequency conversion. Proc. SPIE 3928, 108 – 114(2000).

[16] X. Yin, J. Y. Wang, H. D. Jiang: Measurement of electro-optic coefficients of low symmetry crystal GdCa$_4$O(BO$_3$)$_3$. Opt. Laser Technol. 33(8), 563 – 566(2001).

[17] Z. P. Wang, J. H. Liu, R. B. Song, H. D. Jiang, S. J. Zhang, K. Fu, C. Q. Wang, J. Y. Wang, Y. G. Liu, J. Q. Wei, H. C. Chen, Z. S. Shao: Anisotropy of nonlinear-optical property of RCOB (R = Gd, Y) crystal. Chin. Phys. Lett. 18 (3), 385 – 387 (2001).

[18] F. Mougel, G. Aka, F. Salin, D. Pelenc, B. Ferrand, A. Kahn-Harari, D. Vivien: Accurate second harmonic generation phase matching angles prediction and evaluation of nonlinear coefficients of YCa$_4$O(BO$_3$)$_3$ (YCOB) crystal. In: *Advanced Solid State Lasers*, *OSA Trends in Optics and Photonics Series*, *Vol. 26*, ed. by M. M. Fejer, H. Injeyan, U. Keller(OSA, Washington DC, 1999), pp. 709 – 714.

[19] G. Aka, F. Mougel, D. Vivien, R. Klein, G. Kugel, B. Ferrand, D. Pelenc: Conversion efficiency and absolute effective nonlinear optical coefficients of YCOB and GdCOB measured for different type I SHG phase matching configurations. In: *Advanced Solid-State Lasers*, *OSA Trends in Optics and Photonics Series*, *Vol. 50*, ed. by C. Marshall (OSA, Washington DC, 2001), pp. 548 – 553.

[20] E. Reino, E. Verdier, G. Aka, J. M. Benitez, D. Vivien: Frequency conversion for blue laser emission in Gd$_{1-x}$Y$_x$COB. In: *Advanced Solid-State Lasers*, *OSA Trends in Optics and Photonics Series*, *Vol. 68*, ed. by M. E. Fermann, L. R. Marshall(OSA, Washington DC, 2002), pp. 32 – 36.

[21] X. Chen, M. Huang, Z. Luo, Y. Huang: Determination of the optimum phase-matching directions for the self-frequency conversion of Nd: GdCOB and Nd: YCOB crystals. Opt. Commun. 196(1 – 6), 299 – 307(2001).

[22] J. Liu, Z. Wang, S. Zhang, J. Wang, H. Chen, Z. Shao, M. Jiang: Second-harmonic generation of 1.06 μm in Sr doped GdCa$_4$O(BO$_3$)$_3$ crystal. Opt. Commun. 195(1 – 4), 267 – 271(2001).

[23] S.-J. Zhang, Z.-X. Cheng, J.-H. Liu, J.-R. Han, J.-Y. Wang, Z.-S. Shao, H.-C. Chen: Effect of strontium ion on the growth and second-harmonic generation properties of GdCa$_4$O(BO$_3$)$_3$ crystal. Chin. Phys. Lett. 18(1), 63 – 64(2001).

[24] J. Liu, Z. Fei, S. Zhang, C. Du, J. Wang, H. Chen, Z. Shao: Investigation on intracavity second-harmonic generation of a new Li-doped GdCa$_4$O(BO$_3$)$_3$ crystal. Opt. Laser Technol. 33(8), 597 – 600(2001).

[25] J. Liu, X. Xu, C. Q. Wang, S. Zhang, J. Wang, H. Chen, Z. Shao, M. Jiang: Intracavity second-harmonic generation of 1.06 μm in GdCa$_4$O(BO$_3$)$_3$. Appl. Phys. B. 72 (2), 163 – 166(2001).

6.7 YCa₄O(BO₃)₃，硼酸氧钙钇(YCOB)

负双轴晶：$2V_z = 121.1°$，$\lambda = 0.546\ \mu m$

分子量：441.642

密度：3.31 g/cm³ [1]

点群：m

晶格常数：

$a = 8.046 Å$ [2]；$(8.077\ 0 \pm 0.000\ 3) Å$ [3]

$b = 15.959 Å$ [2]；$(16.019\ 4 \pm 0.000\ 5) Å$ [3]

$c = 3.517 Å$ [2]；$(3.530\ 8 \pm 0.000\ 1) Å$ [3]

$\beta = 101.19°$ [2]；$101.167° \pm 0.004°$ [3]

介电轴和晶体学轴的变换：$Y \parallel b$，a 轴和 c 轴位于 XZ 平面，两轴夹角 $\beta = 101.167°$，Z 轴和 a 轴夹角为 $24.7°$，X 轴和 c 轴夹角为 $13.5°$ [3]。在文献[1]和[4]中分别报道了稍有不同的变换：$(a,Z) = 23°$，$(c,X) = 12°$ 和 $(a,Z) = 23.6°$，$(c,X) = 12.6°$。

相对于晶体学轴 (a,c)（绕 Y 轴）XZ 平面的热旋转[5]：

$$\frac{d\alpha_{ext}}{dT} = \pm \left(\frac{0.006\ 4}{\lambda^3} - \frac{0.017\ 3}{\lambda^2} + \frac{0.014\ 9}{\lambda} + 0.004\ 3 \right) \times 0.057\ 3\ (°)/K$$

此处，$0.397\ 3\ \mu m < \lambda < 0.669\ 1\ \mu m$，对于 $\phi = 180°$ 及 $\phi = 0°$ 两个传播方向取正号和负号。

莫氏硬度：6～6.5 [6]

熔点：1 783 K [1,7]

线性热膨胀系数的平均值

T/K	$\alpha_t \times 10^6/K^{-1}$，$\parallel a$	$\alpha_t \times 10^6/K^{-1}$，$\parallel b$	$\alpha_t \times 10^6/K^{-1}$，$\parallel c$	参考文献
293～473	8.39	5.18	9.17	[8]
293～1 173	9.9	8.2	12.8	[9]

$p = 0.101\ 325$ MPa 时的比热容 c_p [9]

T/K	$c_p/(J \cdot kg^{-1} \cdot K^{-1})$
373	729.7

$T = 293$ K 时的热导率 [6]

$\kappa/(W\cdot m^{-1}\cdot K^{-1}), \parallel X$	$\kappa/(W\cdot m^{-1}\cdot K^{-1}), \parallel Y$	$\kappa/(W\cdot m^{-1}\cdot K^{-1}), \parallel Z$
2.60	2.33	3.01

$T=373$ K 时的热导率[9]

$\kappa/(W\cdot m^{-1}\cdot K^{-1}), \parallel a$	$\kappa/(W\cdot m^{-1}\cdot K^{-1}), \parallel b$	$\kappa/(W\cdot m^{-1}\cdot K^{-1}), \parallel c$
1.83	1.72	2.17

高透明范围：0.202~2.5 μm[10]

在 IR 透明范围 2.5~3.7 μm 内，在 2.7 μm，2.9 μm 和 3.25 μm 处有吸收带[11]。

线性吸收系数 α

$\lambda/\mu m$	α/cm^{-1}	参考文献	$\lambda/\mu m$	α/cm^{-1}	参考文献
0.21	1.0	[12]	1.06	0.013	[13]

最佳 Sellmeier 方程($T=293$ K, λ 以 μm 为单位, 0.3547 μm $<\lambda<$ 1.9079 μm)[10,14]：

$$n_X^2 = 2.7697 + \frac{0.02034}{\lambda^2 - 0.01779} - 0.00643\lambda^2$$

$$n_Y^2 = 2.8741 + \frac{0.02213}{\lambda^2 - 0.01871} - 0.01078\lambda^2$$

$$n_Z^2 = 2.9107 + \frac{0.02232}{\lambda^2 - 0.01887} - 0.01256\lambda^2$$

在文献[2]、[3]、[4]、[7]、[12]、[15]中给出了其他色散关系方程组。

光谱范围：0.3973~1.3382 μm 和温度区间 293~393 K 内折射率的温度微商 (λ 以 μm 为单位)[5]：

$$\frac{dn_X}{dT} = (8.2058 - 5.0188\lambda) \times 10^{-6} K^{-1}$$

$$\frac{dn_Y}{dT} = (2.8217 + 1.9154\lambda) \times 10^{-6} K^{-1}$$

$$\frac{dn_Z}{dT} = (3.0310 + 1.8399\lambda) \times 10^{-6} K^{-1}$$

YCOB 晶体主平面上有效二阶非线性系数的表达式(小走离角近似，Kleinman 对称条件成立, $d_{12}=d_{26}$, $d_{13}=d_{35}$, $d_{15}=d_{31}$, $d_{24}=d_{32}$)[16,17]：

XY 平面，$\theta = 90°$

$$d_{ooe} = d_{13}\sin\phi$$
$$d_{eoe} = d_{oee} = d_{31}\sin^2\phi + d_{32}\cos^2\phi$$

YZ 平面，$\phi = 90°$
$$d_{eeo} = d_{13}\sin^2\theta + d_{12}\cos^2\theta$$
$$d_{oeo} = d_{eoo} = d_{31}\sin\theta$$

XZ 平面，$\phi = 0°$，$V_Z > \theta > 0°$
$$d_{ooe} = d_{12}\cos\theta - d_{32}\sin\theta$$

XZ 平面，$\phi = 0°$，$90° > \theta > V_Z$
$$d_{oeo} = d_{eoo} = d_{12}\cos\theta - d_{32}\sin\theta$$

XZ 平面，$\phi = 0°$，$180° - V_Z > \theta > 90°$；或 $\phi = 180°$，$90° > \theta > V_Z$
$$d_{oeo} = d_{eoo} = d_{12}\cos\theta + d_{32}\sin\theta$$

XZ 平面，$\phi = 0°$，$180° > \theta > 180° - V_Z$；或 $\phi = 180°$，$V_Z > \theta > 0°$
$$d_{ooe} = d_{12}\cos\theta + d_{32}\sin\theta$$

二阶非线性系数最可靠的实验值：

$d_{11}(1.064\ 2\ \mu m) = 0$[17]； ≈ 0[18]

$d_{12}(1.064\ 2\ \mu m) = 0.24\ pm/V$[17]； $0.34\ pm/V$[19]； $0.43\ pm/V$[3]

$d_{13}(1.064\ 2\ \mu m) = -0.71\ pm/V$[19]； $-0.73\ pm/V$[17]； $-0.92\ pm/V$[3]

$d_{31}(1.064\ 2\ \mu m) = 0.41\ pm/V$[17]

$d_{32}(1.064\ 2\ \mu m) = 2.00\ pm/V$[3]； $2.03\ pm/V$[19]； $2.35\ pm/V$[17]

$d_{33}(1.064\ 2\ \mu m) = -1.60\ pm/V$[17]

$T = 293\ K$ 时 YCOB 晶体主平面上 SHG 和 SFG 相位匹配角的实验值

相互作用的波长/μm	$\phi_{pm}/(°)$	$\theta_{pm}/(°)$	参 考 文 献
XY 平面，$\theta = 90°$			
SHG, o + o ⇒ e			
1.064 2 ⇒ 0.532 1	35.0		[3], [7], [19]
0.737 9 ⇒ 0.368 95	77.3		[10]
SHG，Ⅰ类，沿 Y			
0.724 ⇒ 0.362	90		[20]
SFG, o + o ⇒ e			
1.064 2 + 0.532 1 ⇒ 0.354 7	73.2		[11]
	73.6		[10]
	73.7		[5]
	73.8		[21]
SHG, e + o ⇒ e			
1.064 2 ⇒ 0.532 1	73.4		[7]

续表

相互作用的波长/μm	$\phi_{pm}/(°)$	$\theta_{pm}/(°)$	参 考 文 献
	74.8		[22]
	75.2		[10]
	75.3		[5]
SHG，Ⅱ类，沿 Y			
1.03⇒0.515	90		[10]
SFG，e + o⇒e			
1.907 9 + 1.064 2⇒0.683 1	81.2		[10]
YZ，$\phi = 90°$			
SHG，e + e⇒o			
0.737 9⇒0.368 95		66.9	[10]
SFG，e + e⇒o			
1.064 2 + 0.532 1⇒0.354 7		58.7	[11]
		59.7	[5]
		59.8	[21]
		59.9	[10]
SHG，e + o⇒o			
1.064 2⇒0.532 1		58.7	[3]，[7]
		61.1	[22]
		62.7	[5]，[10]
SFG，e + o⇒o			
1.907 9 + 1.064 2⇒0.683 1		73.5	[10]
XZ 平面，$\phi = 0°$，$\theta < V_Z$			
SHG，Ⅰ类，沿 Z			
0.83⇒0.415		0	[12]
0.832 5⇒0.416 25		0	[10]，[5]
SHG，o + o⇒e			
0.9⇒0.45		18.7	[5]
0.954⇒0.477		24.1	[10]
1.064 2⇒0.532 1		30.8	[10]，[5]
		31.7	[3]，[7]，[19]
1.338 2⇒0.669 1		38.2	[10]
		38.3	[5]
SFG，o + o⇒e			
1.064 2 + 0.737 9⇒0.435 8		17.1	[5]
1.569 + 0.532 1⇒0.397 3		18.6	[5]
1.318 8 + 0.659 4⇒0.439 6		23.0	[10]
1.907 9 + 0.532 1⇒0.416 1		26.6	[10]

YCOB 晶体主平面上 SHG 和 SFG 内角带宽的实验值

相互作用的波长/μm	ϕ_{pm}/(°)	θ_{pm}/(°)	$\Delta\phi^{int}$/(°)	$\Delta\theta^{int}$/(°)	参 考 文 献
XY,$\theta=90°$					
SHG, o+o⇒e					
1.064 2⇒0.532 1	35.0		0.09		[3],[7],[19]
SHG, e+o⇒e					
1.064 2⇒0.532 1	73.4		0.32		[7]
SFG, o+o⇒e					
1.064 2+0.532 1⇒0.354 7	73.2		0.11		[23]
YZ 平面,$\phi=90°$					
SHG, e+o⇒o					
1.064 2⇒0.532 1		58.7		0.74	[3],[7]
SFG, e+e⇒o					
1.064 2+0.532 1⇒0.354 7		58.7		0.19	[23]
XZ 平面,$\phi=0°$,$\theta<V_z$					
SHG, o+o⇒e					
1.064 2⇒0.532 1		31.7		0.08	[3],[7],[19]

注:在双轴晶中,存在两个接收角:一个以 θ 表示,另一个以 ϕ 表示。作者仅给出了最小的一个。

YCOB 晶体在某些特别的相位匹配方向(SHG,Ⅰ类,0.946 μm⇒0.473 μm)内角带宽的实验值[24]

相位匹配方向	Δ/(°)
$\theta=67.9°$,$\phi=136.8°$	0.06

YCOB 晶体主平面上 SHG 和 SFG 温度带宽的实验值

相互作用的波长/μm	ΔT/℃	参考文献	备 注
XY 平面,$\theta=90°$			
SHG, o+e⇒e			
1.064 2⇒0.532 1	32.7	[10]	
	32.8	[5]	$\phi=75.3°$
SFG, o+o⇒e			
1.064 2+0.532 1⇒0.354 7	8.6	[10]	$\phi=73.7°$
	9.7	[25]	
	10	[12]	
YZ 平面,$\phi=90°$			
SHG, o+e⇒o			
1.064 2⇒0.532 1	31.5	[10]	$\theta=62.7°$

续表

相互作用的波长/μm	ΔT/℃	参考文献	备注
	31.7	[14]	
	29.2	[5]	
SFG, e + e ⇒ o			
1.064 2 + 0.532 1 ⇒ 0.354 7	6.2	[10]	
	8.5	[23]	
XZ 平面,$\phi=0°$,$\theta<V_z$ 和 $\theta>180°-V_z$			
SHG,I 类,沿 Z			
0.832 5 ⇒ 0.416 25	21.6	[5]	
	31.5	[10]	
SHG, o + o ⇒ e			
0.9 ⇒ 0.45	24.6	[5]	$\theta=18.7°$
	45.3	[5]	$\theta=161.3°$
1.064 2 ⇒ 0.532 1	75	[5]	$\theta=30.8°$
1.338 2 ⇒ 0.669 1	61	[5]	$\theta=141.7°$
SFG, o + o ⇒ e			
1.064 2 + 0.737 9 ⇒ 0.435 8	36.5	[5]	$\theta=162.9°$
1.569 + 0.532 1 ⇒ 0.397 3	16.9	[5]	$\theta=18.6°$
	33.8	[5]	$\theta=161.4°$

YCOB 晶体对某些相位匹配方向(SHG,I 类,1.064 2 μm ⇒ 0.532 1 μm)上有效二阶非线性系数的实验值

相位匹配方向	d_{eff}/(pm·V^{-1})	参 考 文 献
$\theta=90°$,$\phi=35.3°$(XY 平面)	0.39	[19]
$\theta=90°$,$\phi=35°$(XY 平面)	0.42	[17]
$\theta=31.7°$,$\phi=0°$(XZ 平面)	0.78	[17]
	1.03	[19]
$\theta=148.3°$,$\phi=0°$(XZ 平面)	1.36	[19]
	1.44	[17]
$\theta=65°$,$\phi=36.5°$	1.14	[17]
$\theta=65.9°$,$\phi=36.5°$	0.91	[9]
$\theta=66.3°$,$\phi=143.5°$	1.45	[9]
$\theta=67°$,$\phi=143.5°$	1.73	[17]
$\theta=66°$,$\phi=145°$	1.8	[19]

注:在 YCOB 晶体情况下,d_{eff} 的性质包含了镜面和倒反对称性[26]。这意味着可以选择两个独立的象限,例如(0°<θ<90°,0°<ϕ<90°)以及(0°<θ<90°,90°<ϕ<180°)就能完全描述 d_{eff} 的空间分布。随后,在这两个象限中每一个(θ,ϕ)方向的 d_{eff} 值等于(180°-θ,180°-ϕ)方向的值,反之亦然。例如($\theta=33°$,$\phi=9°$)和($\theta=147°$,$\phi=171°$)两个方向具有相等的 d_{eff} 值。

YCOB 晶体某些相位匹配方向（Ⅰ类，$1.0642\ \mu m + 0.5321\ \mu m \Rightarrow 0.3547\ \mu m$，$I = 0.8\ GW/cm^2, l = 1.04\ cm$）上 THG 转换效率的实验值[21]

相位匹配方向	THG 转换效率/%
$\theta = 65°, \phi = 82.8°$	2
$\theta = 90°, \phi = 73.8°(XY\ 平面)$	7
$\theta = 111°, \phi = 79.6°$	20
$\theta = 106°, \phi = 77.2°$	26

见前面的注。

在文献[27]中给出了 $1.061\ \mu m$ 辐射在 YCOB 中Ⅰ类 THG($\theta = 106°, \phi = 77.2°$) 方向优越性的其他证明。

激光诱导的体损伤阈值

$\lambda/\mu m$	τ_p/ns	$I_{thr}/(GW \cdot cm^{-2})$	参考文献	备 注
0.532	6	1	[7]	
1.064	10	85	[9]	1 个脉冲
	6	>1	[7]	10 Hz
	1.1	18.4	[12]	沿 Y 轴，$E \parallel Z$

关于这一晶体

日本科学家于1997年介绍了 YCOB 晶体[2]，在很短的时间内发表了数以百计的文章，投入这种晶体以及最为相近的同构晶体 GdCOB 的研究。到现在，这些材料当然是点群为 m 的晶体中被研究得最为彻底的。这就使我们在低对称晶体关于三波相互作用的物理本质研究方面取得了显著的进步。结果表明，这样晶体 d_{eff} 的空间分布只要选择两个独立的象限（例如 $0° < \theta < 90°, 0° < \phi < 90°$ 以及 $0° < \theta < 90°, 90° < \phi < 180°$）就能完全描述。测量了 YCOB 的二阶非线性系数并发现其对于 Nd:YAG 激光辐射的 SHG 和 THG 的有效非线性最大值并不位于第一象限($0° < \theta < 90°, 0° < \phi < 90°$)[17,18,19]。推导了 YCOB 在主平面上 d_{eff} 的表达式[17]。最近，有报道说在这一单斜晶体中，由于 XZ 面环绕 Y 轴的热旋转，产生了 SHG 温度带宽的空间各向异性[5]。

与 YCOB 晶体相关的最新技术成就之一是通过激光二极管阵列终端泵浦 $Nd:YVO_4$ 激光器($P = 5.6\ W$)腔内倍频一块 1.2 cm 长的晶体($\theta = 64.5°, \phi = 35.5°$)可产生 2.35 W CW 的绿光输出($\lambda = 532\ nm$)[13]。另一个相似的应用是 $Nd:YVO_4$ 激光辐射的 THG[27]，利用 KTP 晶体作倍频以及一块 1.1 cm 长的 YCOB 晶体($\theta = 106°, \phi = 77.2°$)，作者进行激光运转获得 355 nm、124 mW 的准连续光(重复频率为 20 kHz)。

参考文献

[1] Q. Ye, B. H. T. Chai: Crystal growth of $YCa_4O(BO_3)_3$ and its orientation. J. Cryst. Growth 197(1-2), 228-235(1999).

[2] M. Iwai, T. Kobayashi, H. Furuya, Y. Mori, T. Sasaki: Crystal growth and optical characterization of rare-earth(Re) calcium oxyborate $ReCa_4O(BO_3)_3$ (Re = Y or Gd) as new nonlinear optical material. Jpn. J. Appl. Phys. 36(3A), L276-L279(1997).

[3] F. Mougel, G. Aka, F. Salin, D. Pelenc, B. Ferrand, A. Kahn-Harari, D. Vivien: Accurate second harmonic generation phase matching angles prediction and evaluation of nonlinear coefficients of $YCa_4O(BO_3)_3$ (YCOB) crystal. In: *Advanced Solid State Lasers, OSA Trends in Optics and Photonics Series, Vol. 26*, ed. by M. M. Fejer, H. Injeyan, U. Keller(OSA, Washington DC, 1999), pp. 709-714.

[4] J. Wang, Z. Shao, J. Wei, X. Hu, Y. Liu, B. Gong, G. Li, J. Lu, M. Guo, M. Jiang: Research on growth and self-frequency doubling of Nd: ReCOB (Re = Y or Gd) crystals. Progr. Cryst. Growth Character. Mater. 40(1-4), 17-31(2000).

[5] N. Umemura, M. Ando, K. Suzuki, E. Takaoka, K. Kato, M. Yoshimura, Y. Mori, T. Sasaki: Temperature-insensitive second-harmonic generation at 0.532 1 μm in $YCa_4O(BO_3)_3$. Jpn. J. Appl. Phys. 42(8), 5040-5042(2003).

[6] Q. Ye, L. Shah, J. Eichenholz, D. Hammons, R. Peale, M. Richardson, A. Chin, B. H. T. Chai: Investigation of diode-pumped, self-frequency doubled RGB lasers from Nd: YCOB crystals. Opt. Commun. 164(1-3), 33-37(1999).

[7] G. Aka, F. Mougel, D. Pelenc, B. Ferrand, D. Vivien: Comparative evalution of Gd-COB and YCOB nonlinear-optical properties, in principal and out of principal plane configurations, for the 1064 nm Nd: YAG laser frequency conversion. Proc. SPIE 3928, 108-114(2000).

[8] D. A. Hammons, M. Richardson, B. H. T. Chai, A. K. Chin, R. Jollay: Scaling of longitudinally diode-pumped self-frequency-doubling Nd: YCOB lasers. IEEE J. Quant. Electr. 36(8), 991-999(2000).

[9] J. Luo, S. J. Fan, H. Q. Xie, K. C. Xiao, S. X. Qian, Z. W. Zhong, G. X. Qiang, R. Y. Sun, J. Y. Xu: Thermal and nonlinear optical properties of $Ca_4YO(BO_3)_3$. Cryst. Res. Technol. 36(11), 1215-1221(2001).

[10] N. Umemura, H. Nakao, H. Furuya, M. Yoshimura, Y. Mori, T. Sasaki, K. Yoshida, K. Kato: 90° phase-matching properties of $YCa_4O(BO_3)_3$ and $Gd_xY_{1-x}Ca_4O(BO_3)_3$. Jpn. J. Appl. Phys. 40(2A), 596-600(2001).

[11] H. Furuya, M. Yoshimura, T. Kobayashi, K. Murase, Y. Mori, T. Sasaki: Crystal growth and characterization of $Gd_xY_{1-x}Ca_4O(BO_3)_3$ crystal. J. Cryst. Growth 198-199, 560-563(1999).

[12] M. Yoshimura, T. Kobayashi, H. Furuya, K. Murase, Y. Mori, T. Sasaki: Crystal growth and optical properties of yttrium calcium oxyborate YCa$_4$O(BO$_3$)$_3$. In: *Advanced Solid-State Lasers, OSA Trends in Optics and Photonics Series, Vol. 19*, ed. by W. R. Bosenberg, M. M. Fejer(OSA, Washington DC, 1998), pp. 561 – 564.

[13] J. Liu, C. Wang, S. Zhang, C. Du, J. Lu, J. Wang, H. Chen, Z. Shao, M. Jiang: Investigation on intracavity second-harmonic generation at 1.06 μm in YCa$_4$O(BO$_3$)$_3$ by using an endpumped Nd: YVO$_4$ laser. Opt. Commun. 182(1 – 3), 187 – 191(2000).

[14] N. Umemura, K. Yoshida, H. Furuya, Y. Mori, T. Sasaki, E. Takaoka, K. Kato: New data on the phase-matching properties of YCa$_4$O(BO$_3$)$_3$. In: *Advanced Solid-State Lasers, OSA Trends in Optics and Photonics Series, Vol. 34*, ed. by H. Injeyan, U. Keller, C. Marshall(OSA, Washington DC, 2000), pp. 501 – 505.

[15] Z. Shao, J. Lu, Z. Wang, J. Wang, M. Jiang: Anisotropic properties of Nd: ReCOB (Re = Y, Gd): a low symmetry self-frequency doubling crystal. Progr. Cryst. Growth Character. Mater. 40(1 – 4), 63 – 73(2000).

[16] G. Ak a, A. Kahn-Harari, F. Mougel, D. Vivien, F. Salin, P. Coquelin, P. Colin, D. Pelenc, J. P. Damelet: Linear and nonlinear-optical properties of a new gadolinium calcium oxoborate crystal, Ca$_4$GdO(BO$_3$)$_3$. J. Opt. Soc. Am. B 14(9), 2238 – 2247 (1997).

[17] Z. P. Wang, J. H. Liu, R. B. Song, H. D. Jiang, S. J. Zhang, K. Fu, C. Q. Wang, J. Y. Wang, Y. G. Liu, J. Q. Wei, H. C. Chen, Z. S. Shao: Anisotropy of nonlinear-optical property of RCOB (R = Gd, Y) crystal. Chin. Phys. Lett. 18(3), 385 – 387 (2001).

[18] C. Chen, Z. Shao, J. Jiang, J. Wei, J. Lin, J. Wang, N. Ye, L. Lu, B. Wu, M. Jiang, M. Yoshimura, Y. Mori, T. Sasaki: Determination of the nonlinear optical coefficients of YCa$_4$O(BO$_3$)$_3$ crystal. J. Opt. Soc. Am. B 17(4), 566 – 571(2000).

[19] G. Aka, F. Mougel, D. Vivien, R. Klein, G. Kugel, B. Ferrand, D. Pelenc: Conversion efficiency and absolute effective nonlinear optical coefficients of YCOB and GdCOB measured for different type I SHG phase matching configurations. In: *Advanced Solid-State Lasers, OSA Trends in Optics and Photonics Series, Vol. 50*, ed. by C. Marshall (OSA, Washington DC, 2001), pp. 548 – 553.

[20] H. Nakao, S. Makio, H. Furuya, K. Kawamura, S. Yasuda, Y. K. Yap, M. Yoshimura, Y. Mori, T. Sasaki: Crystal growth of GdYCOB for non-critical phase-matched secondharmonic generation at 860 nm. J. Cryst. Growth 237 – 239, 632 – 836(2002).

[21] Z. Wang, K. Fu, X. Xu, X. Sun, H. Jiang, R. Song, J. Liu, J. Wang, Y. Liu, J. Wei, Z. Shao: The optimum configuration for the third-harmonic generation of 1.064 μm in a YCOB crystal. Appl. Phys. B 72(7), 839 – 842(2001).

[22] M. Yoshimura, H. Furuya, I. Yamada, K. Murase, H. Nakao, M. Yamazaki, Y. Mori,

T. Sasaki: Noncritically phase-matched second-harmonic generation of a Nd: YAG laser in GdYCOB crystal. In: *Advanced Solid-State Lasers*, *OSA Trends in Optics and Photonics Series*, *Vol. 26*, ed. by M. M. Fejer, H. Injeyan, U. Keller (OSA, Washington DC, 1999), pp. 702 – 706.

[23] M. Yoshimura, H. Furuya, T. Kobayashi, K. Murase, Y. Mori, T. Sasaki: Noncritically phase-matched frequency conversion in $Gd_xY_{1-x}Ca_4O(BO_3)_3$ crystal. Opt. Lett. 24 (4) 193 – 197 (1999).

[24] E. Reino, E. Verdier, G. Aka, J. M. Benitez, D. Vivien: Frequency conversion for blue laser emission in $Gd_{1-x}Y_x$COB. In: *Advanced Solid-State Lasers*, *OSA Trends in Optics and Photonics Series*, *Vol. 68*, ed. by M. E. Fermann, L. R. Marshall (OSA, Washington DC, 2002), pp. 32 – 36.

[25] H. Furuya, H. Nakao, I. Yamada, Y. F. Ruan, Y. K. Yap, M. Yoshimura, Y. Mori, T. Sasaki: Alleviation of photoinduced damage in $Gd_xY_{1-x}Ca_4O(BO_3)_3$ at elevated temperature for noncritically phase-matched 355-nm generation. Opt. Lett. 25 (21), 1588 – 1590 (2000).

[26] X. Chen, M. Huang, Z. Luo, Y. Huang: Determination of the optimum phase-matching directions for the self-frequency conversion of Nd: GdCOB and Nd: YCOB crystals. Opt. Commun. 196 (1 – 6), 299 – 307 (2001).

[27] C. Du, Z. Wang, J. Liu, X. Xu, K. Fu, J. Wang, Z. Shao: Investigation of intracavity third-harmonic generation at 1.06 μm in $YCa_4O(BO_3)_3$ crystals. Appl. Phys. B 74 (2), 125 – 127 (2002).

6.8 $Gd_xY_{1-x}Ca_4O(BO_3)_3$，硼酸氧钙钇钆(GdYCOB)

负双轴晶

点群：m

$x = 0.24$ 时的晶格常数[1]：$a = 8.067$ Å；$b = 15.991$ Å；$c = 3.531$ Å；$\beta = 101.18°$

$x = 0.24$ 时介电轴和晶体学轴的变换：$Y \parallel b$，a 轴和 c 轴位于 XZ 平面，两轴夹角 $\beta = 101.18°$，Z 轴和 a 轴夹为 $23.8°$，X 轴和 c 轴夹角为 $12.6°$[1]

熔点：$\approx 1\ 773$ K[2]

高透明范围：$0.32 \sim 2.5$ μm[2]

Sellmeier 方程（$T = 293$ K，λ 以 μm 为单位，$0.412\ 9$ μm $< \lambda < 1.338\ 2$ μm）[3]：

$$n_X^2(Gd_xY_{1-x}COB) = (1-x)n_X^2(YCOB) + xn_X^2(GdCOB)$$
$$n_Y^2(Gd_xY_{1-x}COB) = (1-x)(1+0.001\ 98x^2)^2 n_Y^2(YCOB) + xn_Y^2(GdCOB)$$
$$n_Z^2(Gd_xY_{1-x}COB) = (1-x)(1+0.007\ 32x^2)^2 n_Z^2(YCOB) + xn_Z^2(GdCOB)$$

其中 YCOB 和 GdCOB 的折射率由以下色散关系给出[3]：

YCOB

$$n_X^2 = 2.7697 + \frac{0.02034}{\lambda^2 - 0.01779} - 0.00643\lambda^2$$

$$n_Y^2 = 2.8741 + \frac{0.02213}{\lambda^2 - 0.01871} - 0.01078\lambda^2$$

$$n_Z^2 = 2.9107 + \frac{0.02232}{\lambda^2 - 0.01887} - 0.01256\lambda^2$$

GdCOB

$$n_X^2 = 2.8063 + \frac{0.02315}{\lambda^2 - 0.01378} - 0.00537\lambda^2$$

$$n_Y^2 = 2.8959 + \frac{0.02398}{\lambda^2 - 0.01389} - 0.01132\lambda^2$$

$$n_Z^2 = 2.9248 + \frac{0.02410}{\lambda^2 - 0.01406} - 0.01139\lambda^2$$

GdYCOB 晶体主平面上有效二阶非线性系数的表达式（小走离角近似，Kleinman 对称条件成立，$d_{12} = d_{26}$，$d_{13} = d_{35}$，$d_{15} = d_{31}$，$d_{24} = d_{32}$）[4,5]：

XY 平面，$\theta = 90°$

$$d_{ooe} = d_{13}\sin\phi$$
$$d_{eoe} = d_{oee} = d_{31}\sin^2\phi + d_{32}\cos^2\phi$$

YZ 平面，$\phi = 90°$

$$d_{eeo} = d_{13}\sin^2\theta + d_{12}\cos^2\theta$$
$$d_{oeo} = d_{eoo} = d_{31}\sin\theta$$

XZ 平面，$\phi = 0°$，$V_Z > \theta > 0°$

$$d_{ooe} = d_{12}\cos\theta - d_{32}\sin\theta$$

XZ 平面，$\phi = 0°$，$90° > \theta > V_Z$

$$d_{oeo} = d_{eoo} = d_{12}\cos\theta - d_{32}\sin\theta$$

XZ 平面，$\phi = 0°$，$180° - V_Z > \theta > 90°$；或 $\phi = 180°$，$90° > \theta > V_Z$

$$d_{oeo} = d_{eoo} = d_{12}\cos\theta + d_{32}\sin\theta$$

XZ 平面，$\phi = 0°$，$180° > \theta > 180° - V_Z$；或 $\phi = 180°$，$V_Z > \theta > 0°$

$$d_{ooe} = d_{12}\cos\theta + d_{32}\sin\theta$$

GdYCOB 晶体主平面上相位匹配角的实验值

相互作用的波长/μm	组分参数 x	参考文献	备注
沿 Y 轴，$\phi=90°$，$\theta=90°$			
SHG，Ⅰ类			
$0.7735 \Rightarrow 0.38675$	0.68	[6]	$T=240\ ℃$
SHG，Ⅱ类			
$1.0642 \Rightarrow 0.5321$	0.275	[7]	
	≈0.28	[8]	$T=52\ ℃$
SFG，Ⅰ类			
$1.0642 + 0.5321 \Rightarrow 0.3547$	0.24	[2]	
	0.28	[1]	
沿 Z 轴，$\phi=0°$，$\theta=0°$			
SHG，Ⅰ类			
$0.8435 \Rightarrow 0.42175$	0.15	[9]	
$0.8612 \Rightarrow 0.4306$	0.32	[9]	
$0.925 \Rightarrow 0.4625$	0.48	[10]	
$0.9293 \Rightarrow 0.46465$	0.84	[11]	
$0.946 \Rightarrow 0.473$	0.87	[12]	

GdYCOB 晶体沿 Y 轴某些特殊相互作用内角带宽和温度带宽的实验值（SHG 情况下，组分参数 $x=0.275$；以及 SFG 情况下，$x=0.28$）

相互作用的波长/μm	$T/℃$	$\Delta T/℃$	$\Delta\theta^{int}/(°)$	$\Delta\phi^{int}/(°)$	参考文献
SHG，Ⅱ类					
$1.0642 \Rightarrow 0.5321$	27	32.4	6.8	4.0	[7]
SFG，Ⅰ类					
$1.0642 + 0.5321 \Rightarrow 0.3547$	21	6.6	3.8	2.2	[1],[13],[14]

GdYCOB 晶体沿 Z 轴 SHG 内角带宽的实验值（组分参数 $x=0.87$）[12]

相互作用的波长/μm	$\Delta\theta^{int}/(°)$
SHG，Ⅰ类	
$0.946 \Rightarrow 0.473$	0.53

GdYCOB 晶体沿 Y 轴某些特殊相互作用有效二阶非线性系数的实验值

相互作用的波长/μm	组分参数 x	$d_{eff}/(pm\cdot V^{-1})$	参考文献
SHG，Ⅱ类			
$1.0642 \Rightarrow 0.5321$	0.275	$0.35(d_{31})$	[7]
SFG，Ⅰ类			
$1.0642 + 0.5321 \Rightarrow 0.3547$	0.28	$0.55(d_{13})$	[1],[13],[14]

激光诱导的损伤阈值

$\lambda/\mu m$	τ_p/ns	$I_{thr}/(GW \cdot cm^{-2})$	参考文献	备 注
0.355	10	0.002 ~ 0.003	[14]	62.5 kHz,灰迹形成
1.064	5	> 0.45	[8]	5 Hz
	0.035	> 0.45	[8]	10 Hz

注:形成灰迹后可以将GdYCOB晶体保持在升高的温度下($T = 240$ ℃[15])予以消除。这种损伤的消除可通过150 ℃、25小时退火来完全实现[14]。

关于这一晶体

众所周知,YCOB晶体的Ⅰ类NCPM对于基波724 nm是沿Y轴,则对832 nm是沿Z轴的[9];同时,GdCOB晶体分别对于826 nm和961 nm沿Y轴和Z轴有类似的相互作用[9]。很清楚这两种晶体的固溶体,即$Gd_x Y_{1-x} Ca_4 O(BO_3)_3$(GdYCOB),随其组分参数$x$变化,沿着$Y$轴和$Z$轴分别在724 ~ 826 nm以及832 ~ 961 nm波长范围内对任意一个波长都能实现NCPM。利用这一方法,在文献[6]中报道了在GdYCOB($x = 0.68$),沿Y轴实现了773.5 nm的Ⅰ类NCPM SHG。在文献[7]和[8]中,在$x = 0.28$的GdYCOB中晶体中沿Y轴实现了Nd:YAG激光辐射($\lambda = 1.0642$ μm)的Ⅱ类NCPM。最后,在$x = 0.24$的GdYCOB晶体中也演示了沿同一轴Nd:YAG激光辐射的Ⅰ类NCPM[2]。

参考文献

[1] M. Yoshimura, H. Furuya, T. Kobayashi, K. Murase, Y. Mori, T. Sasaki: Noncritically phase-matched frequency conversion in $Gd_x Y_{1-x} Ca_4 O(BO_3)_3$ crystal. Opt. Lett. 24(4), 193 – 197(1999).

[2] H. Furuya, M. Yoshimura, T. Kobayashi, K. Murase, Y. Mori, T. Sasaki: Crystal growth and characterization of $Gd_x Y_{1-x} Ca_4 O(BO_3)_3$ crystal. J. Cryst. Growth 198 – 199, 560 – 563(1999).

[3] N. Umemura, H. Nakao, H. Furuya, M. Yoshimura, Y. Mori, T. Sasaki, K. Yoshida, K. Kato: 90° phase-matching properties of $YCa_4 O(BO_3)_3$ and $Gd_x Y_{1-x} Ca_4 O(BO_3)_3$. Jpn. J. Appl. Phys. 40(2A), 596 – 600(2001).

[4] G. Aka, A. Kahn-Harari, F. Mougel, D. Vivien, F. Salin, P. Coquelin, P. Colin, D. Pelenc, J. P. Damelet: Linear and nonlinear-optical properties of a new gadolinium calcium oxoborate crystal, $Ca_4 GdO(BO_3)_3$. J. Opt. Soc. Am. B 14(9), 2238 – 2247(1997).

[5] Z. P. Wang, J. H. Liu, R. B. Song, H. D. Jiang, S. J. Zhang, K. Fu, C. Q. Wang, J.

Y. Wang, Y. G. Liu, J. Q. Wei, H. C. Chen, Z. S. Shao: Anisotropy of nonlinear-optical property of RCOB(R = Gd, Y) crystal. Chin. Phys. Lett. 18(3), 385 – 387(2001).

[6] H. Kitano, H. Kawai, K. Miramitsu, S. Owa, M. Yoshimura, Y. Mori, T. Sasaki: 387-nm generation in $Gd_x Y_{1-x} Ca_4 O(BO_3)_3$ crystal and its utilization for 193-nm light source. Jpn. J. Appl. Phys. 42(2B), L166-L169(2003).

[7] M. Yoshimura, H. Furuya, I. Yamada, K. Murase, H. Nakao, M. Yamazaki, Y. Mori, T. Sasaki: Noncritically phase-matched second-harmonic generation of a Nd: YAG laser in GdYCOB crystal. In: *Advanced Solid-State Lasers*, *OSA Trends in Optics and Photonics Series*, *Vol. 26*, ed. by M. M. Fejer, H. Injeyan, U. Keller (OSA, Washington DC, 1999), pp. 702 – 706.

[8] A. Zoubir, J. Eichenholz, E. Fujiwara, D. Grojo, E. Baleine, A. Rapaport, M. Bass, B. Chai, M. Richardson: Non-critical phase-matched second harmonic generation in $Gd_{1-x}Y_x COB$. Appl. Phys. B 77(4), 437 – 440(2003).

[9] H. Nakao, S. Makio, H. Furuya, K. Kawamura, S. Yasuda, Y. K. Yap, M. Yoshimura, Y. Mori, T. Sasaki: Crystal growth of GdYCOB for non-critical phase-matched second-harmonic generation at 860 nm. J. Cryst. Growth 237 – 239, 632 – 836(2002).

[10] M. Yoshimura, T. Kobayashi, H. Furuya, K. Murase, Y. Mori, T. Sasaki: Crystal growth and optical properties of yttrium calcium oxyborate $YCa_4O(BO_3)_3$. In: *Advanced Solid-State Lasers*, *OSA Trends in Optics and Photonics Series*, *Vol. 19*, ed. by W. R. Bosenberg, M. M. Fejer (OSA, Washington DC, 1998), pp. 561 – 564.

[11] P. B. W. Burmester, T. Kellner, K. Petermann, G. Huber, R. Uecker, P. Reiche: Type-I noncritically phase-matched second-harmonic generation in $Gd_{1-x}Y_x Ca_4 O(BO_3)_3$. Appl. Phys. B 68(6), 1143 – 1146(1999).

[12] E. Reino, E. Verdier, G. Aka, J. M. Benitez, D. Vivien: Frequency conversion for blue laser emission in $Gd_{1-x}Y_x COB$. In: *Advanced Solid-State Lasers*, *OSA Trends in Optics and Photonics Series*, *Vol. 68*, ed. by M. E. Fermann, L. R. Marshall (OSA, Washington DC, 2002), pp. 32 – 36.

[13] M. Yoshimura, H. Furuya, I. Yamada, K. Murase, H. Nakao, M. Yamazaki, Y. Mori, T. Sasaki: Noncritically phase-matched ultraviolet generation in $Gd_x Y_{1-x} Ca_4 O(BO_3)_3$. In: *Advanced Solid-State Lasers*, *OSA Trends in Optics and Photonics Series*, *Vol. 26*, ed. by M. M. Fejer, H. Injeyan, U. Keller (OSA, Washington DC, 1999), pp. 82 – 88.

[14] H. Furuya, H. Nakao, I. Yamada, Y. F. Ruan, Y. K. Yap, M. Yoshimura, Y. Mori, T. Sasaki: Alleviation of photoinduced damage in $Gd_x Y_{1-x} Ca_4 O(BO_3)_3$ crystal at elevated crystal temperature for noncritically phase-matched 355-nm generation. Opt. Lett. 25(21), 1588 – 1590(2000).

[15] H. Furuya, H. Nakao, I. Yamada, Y. Ruan, Y. K. Yap, M. Yoshimura, Y. Mori,

T. Sasaki: Photoinduced damage in GdYCOB and its circumvention. In: *Advanced Solid-State Lasers*, *OSA Trends in Optics and Photonics Series*, *Vol. 34*, ed. by H. Injeyan, U. Keller, C. Marshall (OSA, Washington DC, 2000), pp. 404–408.

6.9 $Li_2B_4O_7$，四硼酸锂（LB4）

负双轴晶：$n_o > n_e$

分子量：169.118

密度：2.45 g/cm³ [1]

点群：$4mm$

晶格常数：

$a = 9.477$ Å [2]；9.47 Å [3]；9.479 Å [4]

$c = 10.286$ Å [2]；10.26 Å [3]；10.297 Å [4]

莫氏硬度：5 [4]；6 [5]

熔点：1 190 K [1]

线性热膨胀系数 α_t [4]

$\alpha_t \times 10^6/K^{-1}$，$\parallel c$	$\alpha_t \times 10^6/K^{-1}$，$\perp c$
3.74	11.1

以"o"透过计的透明范围：0.16 ~ 3.5 μm [6]

在 298 K 及 10 325 Pa 下的折射率实验值 [6]

$\lambda/\mu m$	n_o	n_e	$\lambda/\mu m$	n_o	n_e
0.184 887	1.774 654	1.699 128	0.706 52	1.606 162	1.549 767
0.202 548	1.733 360	1.662 581	0.852 11	1.602 303	1.546 566
0.214 438	1.714 118	1.645 491	1.013 98	1.598 952	1.543 901
0.253 652	1.674 704	1.610 421	1.128 64	1.596 794	1.542 245
0.365 015	1.632 529	1.572 896	1.529 58	1.589 202	1.536 671
0.435 835	1.621 944	1.563 516	1.970 09	1.579 263	1.529 652
0.546 07	1.612 982	1.555 638	2.325 42	1.569 365	1.522 829
0.632 82	1.608 779	1.551 997			

温度范围 233 ~ 373 K 以及光谱范围 0.435 84 ~ 0.643 85 μm 时折射率的温度微商（以 $10^{-6} K^{-1}$ 为单位）[6]：

T/K

$\lambda/\mu m$	233~253	253~273	273~293	293~313	313~333	333~353	353~373
$\dfrac{dn_o}{dT}$							
0.435 84	3.1	2.6	2.1	1.7	1.2	0.7	0.2
0.479 99	2.9	2.4	1.9	1.4	0.9	0.4	-0.1
0.546 07	2.7	2.2	1.7	1.2	0.7	0.2	-0.4
0.589 29	2.7	2.2	1.6	1.1	0.6	0.1	-0.5
0.632 82	2.6	2	1.5	1.0	0.5	0	-0.5
0.643 85	2.6	2	1.5	1.0	0.5	-0.1	-0.6
$\dfrac{dn_e}{dT}$							
0.435 84	4.6	4.2	3.8	3.4	3.0	2.6	2.2
0.479 99	4.4	4	3.6	3.2	2.8	2.4	2.0
0.546 07	4.3	3.8	3.4	3.0	2.6	2.2	1.8
0.589 29	4.2	3.8	3.3	2.9	2.5	2.1	1.6
0.632 82	4.1	3.7	3.3	2.8	2.4	2.0	1.6
0.643 85	4.1	3.7	3.3	2.8	2.4	2.0	1.5

最佳 Sellmeier 方程($T=298$ K,λ 以 μm 为单位)[6]:

$$n_o^2 = 2.564\,31 + \frac{0.012\,337}{\lambda^2 - 0.013\,103} - 0.019\,075\lambda^2$$

$$n_e^2 = 2.386\,51 + \frac{0.010\,664}{\lambda^2 - 0.012\,878} - 0.012\,813\lambda^2$$

有效二阶非线性系数表达式(Kleinman 对称条件成立,$d_{15}=d_{24}=d_{31}=d_{32}$)[7]:

$$d_{ooe} = d_{31}\sin\theta$$

二阶非线性系数的绝对值[2,8]

$$d_{31}(1.064\,2\ \mu m) = (0.12 \pm 0.03)\ pm/V$$

$$d_{33}(1.064\,2\ \mu m) = (0.47 \pm 0.09)\ pm/V$$

相位匹配角和"走离"角的计算值

相互作用的波长/μm	$\theta_{pm}/(°)$	$\rho_3/(°)$
SHG,o+o$\Rightarrow$e		
0.488$\Rightarrow$0.244	87.83	0.16
0.510 6$\Rightarrow$0.255 3	71.57	1.30
0.514 5$\Rightarrow$0.257 25	70.15	1.38
0.532 1$\Rightarrow$0.266 05	64.95	1.66
0.578 2$\Rightarrow$0.289 1	55.90	2.00

相互作用的波长/μm	$\theta_{pm}/(°)$	$\rho_3/(°)$
0.8⇒0.4	37.63	2.06
1.064 2⇒0.532 1	30.97	1.86
1.318 8⇒0.659 4	29.92	1.81
SFG, o+o⇒e		
1.064 2+0.266 05⇒0.212 84	73.84	1.21
1.064 2+0.354 73⇒0.266 05	52.83	2.11
1.064 2+0.532 1⇒0.354 73	40.28	2.12

激光诱导的体损伤阈值

$\lambda/\mu m$	τ_p/ns	$I_{thr}/(GW \cdot cm^{-2})$	参考文献	备注
0.532	10	>0.1	[6]	10 Hz
1.064	10	40	[5]	10 Hz

激光诱导体损伤阈值[3]

$\lambda/\mu m$	τ_p/ns	$I_{thr}/(GW \cdot cm^{-2})$	备注
0.266	10	0.83	10 Hz
0.532	10	1.9	10 Hz
1.064	10	8.4	10 Hz

关于这一晶体

尽管在20年以前人们就已经知道LB4晶体的声光应用[9]，而直至今日它在非线性光学方面仍然应用不多。其中的主要原因在于$Li_2B_4O_7$的非线性系数相当小。然而，这种材料的UV截止边非常短（在160 nm附近）并有很高的体抗光伤阈值（在1.064 nm处比熔融石英高4倍）[5,6]。$Li_2B_4O_7$的另一个优点是其吸湿性相当小。所以，这种材料在紫外和深紫外有潜在的应用前景。利用LB4晶体，日本科学家用以产生纳秒Nd:YAG激光的四次谐波和五次谐波，其脉冲能量分别为0.16 J和0.07 J[6]。最近，借助于级联FoHG方法（三块四倍频晶体），在266 nm处的总能量为0.43 J，相当于从532 nm辐射的转换效率为30.5%。还有，也实现了15个小时4 W UV功率的运转。

LB4晶体通常用提拉法生长[5]。最近，一个日本研究组采用改良的布里奇曼法生长了高光学质量的LB4晶体，直径达10 cm，长度达20 cm[1,10]。

■ 参考文献

[1] N. Tsutsui, Y. Ino, K. Imai, N. Senguttuvan, M. Ishii: Growth of high quality 4in diameter $Li_2B_4O_7$ single crystals. J. Cryst. Growth 229(1-4), 283-288(2001).

[2] S.-I. Furusawa, O. Chikagawa, S. Tange, T. Ishidate, H. Orihara, Y. Ishibashi, K. Miwa: Second harmonic generation in $Li_2B_4O_7$. J. Phys. Soc. Japan 60(8), 2691-2693(1991).

[3] T. Sugawara, R. Komatsu, S. Uda: Surface damage and radiation resistance of lithium tetraborate single crystals. Opt. Mater. 13(2), 225-229(1999).

[4] Data sheet of Molecular Technology GmbH, Available at www.mt-berlin.com.

[5] R. Komatsu, T. Sugawara, K. Sassa, N. Sarukura, Z. Liu, S. Izumida, Y. Segawa, S. Uda, T. Fukuda, K. Yamanouchi: Growth and ultraviolet application of $Li_2B_4O_7$ crystals: generation of the fourth and fifth harmonics of Nd: $Y_3Al_5O_{12}$ lasers. Appl. Phys. Lett. 70(26), 3492-3494(1997).

[6] T. Sugawara, R. Komatsu, S. Uda: Linear and nonlinear optical properties of lithium tetraborate. Solid State Commun. 107(5), 233-237(1998).

[7] J. E. Midwinter, J. Warner: The effects of phase matching method and of uniaxial crystal symmetry on the polar distribution of second-order non-linear optical polarization. Brit. J. Appl. Phys. 16(11), 1135-1142(1965).

[8] K. Hagimoto, A. Mito: Determination of the second-order susceptibility of ammonium dihydrogen phosphate and α-quartz at 633 and 1064 nm. Appl. Opt. 34(36), 8276-8282(1995).

[9] R. W. Whatmore, N. M. Shorrocks, C. O'Hara, F. W. Ainger, I. M. Young: Lithium tetraborate: a new temperature-compensated SAW substrate material. Electr. Lett. 17(1), 11-12(1981).

[10] N. Tsutsui, Y. Ino, K. Imai, N. Senguttuvan, M. Ishii: Growth of large size LBO ($Li_2B_4O_7$) single crystals by modified Bridgman technique. J. Cryst. Growth 211(1-4), 271-275(2000).

6.10 LiRbB$_4$O$_7$, 四硼酸铷锂(LRB4)

负双轴晶：$2V_z = 130°$, $\lambda = 0.532$ μm[1]

分子量：247.644

密度：2.63 g/cm³（计算值）[2]

点群：222

晶格常数：

$a = (8.625\,7 \pm 0.001\,2)\,\text{Å}^{[2]}$

$b = (11.257\,6 \pm 0.001\,3)\,\text{Å}^{[2]}$

$c = (12.853\,1 \pm 0.001\,5)\,\text{Å}^{[2]}$

介电轴和晶体学轴的变换：$X, Y, Z \Rightarrow b, c, a$。

1.5 cm 长晶体以 0.01 透过计的透明范围：0.187~3.468 μm[1]

室温下折射率的实验值[1]

$\lambda/\mu m$	n_X	n_Y	n_Z	$\lambda/\mu m$	n_X	n_Y	n_Z
0.400 5	1.526 60	1.552 76	1.559 24	0.560 5	1.514 25	1.539 67	1.545 84
0.410 5	1.525 70	1.551 65	1.558 14	0.569 5	1.513 75	1.539 26	1.545 33
0.42	1.524 40	1.549 65	1.556 84	0.580 5	1.513 25	1.538 57	1.544 84
0.431	1.523 45	1.549 56	1.555 53	0.587 5	1.512 90	1.538 26	1.544 33
0.441	1.523 00	1.549 26	1.555 33	0.6	1.512 30	1.537 46	1.543 84
0.452	1.521 45	1.547 66	1.553 33	0.609 5	1.512 00	1.537 06	1.543 54
0.458	1.520 80	1.546 87	1.552 63	0.621 5	1.511 65	1.536 86	1.543 24
0.468 5	1.520 00	1.546 05	1.551 93	0.633	1.511 10	1.536 15	1.542 54
0.479	1.519 15	1.545 07	1.550 93	0.640 5	1.510 95	1.535 86	1.542 54
0.500 5	1.517 70	1.543 67	1.549 43	0.652	1.510 45	1.535 56	1.541 84
0.509	1.516 90	1.542 56	1.548 63	0.661	1.510 25	1.535 16	1.541 84
0.518 5	1.516 40	1.542 17	1.548 03	0.67	1.509 90	1.534 65	1.541 44
0.532	1.515 90	1.541 76	1.547 63	0.678	1.509 70	1.534 76	1.541 14
0.541	1.515 20	1.540 75	1.546 73	0.689 5	1.509 45	1.534 35	1.540 74
0.548	1.514 75	1.540 25	1.546 33	0.700 5	1.509 25	1.533 65	1.540 54

Sellmeier 方程（λ 以 μm 为单位, $T = 293$ K）[1]：

$$n_X^2 = 1 + \frac{1.261\,015\,3\lambda^2}{\lambda^2 - 0.008\,735\,4} - 0.013\,554\,5\lambda^2$$

$$n_Y^2 = 1 + \frac{1.345\,872\,7\lambda^2}{\lambda^2 - 0.008\,039\,4} - 0.033\,091\,8\lambda^2$$

$$n_Z^2 = 1 + \frac{1.351\,071\,1\lambda^2}{\lambda^2 - 0.009\,180\,6} - 0.007\,456\,2\lambda^2$$

LRB4 晶体主平面上有效二阶非线性系数的表达式（Kleinman 对称条件成立，$d_{14} = d_{25} = d_{36}$）[3]：

XY 平面

$$d_{eoe} = d_{oee} = d_{14}\sin 2\phi$$

YZ 平面

XZ 平面，$\theta < V_Z$
$$d_{eeo} = d_{14}\sin 2\theta$$
$$d_{eoe} = d_{oee} = -d_{14}\sin 2\theta$$

XZ 平面，$\theta > V_Z$
$$d_{eeo} = -d_{14}\sin 2\theta$$

二阶非线性系数：
$$d_{14}(1.064 \ \mu m) = 1.15 \times d_{36}(KDP) = 0.45 \ pm/V^{[1,4]}$$

关于这一晶体

最近生长的材料，LB4 的同构晶体，其二阶非线性系数的值比 LB4 稍高一些。

参考文献

[1] R. Komatsu, Y. Ono, T. Kajatani, F. Rotermund, V. Petrov: Optical properties of a new nonlinear borate crystal LiRbB$_4$O$_7$. J. Cryst. Growth 257(1-2), 165-168(2003).

[2] Y. Ono, M. Nakaya, T. Sugawara, N. Watanabe, H. Siraishi, R. Komatsu, T. Kajitani: Structural study of LiKB$_4$O$_7$ and LiRbB$_4$O$_7$: new nonlinear optical crystals. J. Cryst. Growth 229(1-4), 472-476(2001).

[3] B. V. Bokut: Optical mixing in biaxial crystals. Zh. Prikl. Spektrosk. 7(4), 621-624 (1967)[In Russian, English trans.: J. Appl. Spectrosc. 7(4), 425-429(1967)].

[4] D. A. Roberts: Simplified characterization of uniaxial and biaxial nonlinear optical crystals: a plea for standardization of nomenclature and conventions. IEEE J. Quant. Electr. 28(10), 2057-2074(1992).

6.11 CdHg(SCN)$_4$，硫氰酸汞镉(CMTC)

负双轴晶：$n_o > n_e$

分子量：545.221

密度（计算值）：3.25 g/cm$^{3[1]}$；3.54 g/cm$^{3[2]}$

密度（观测值）：3.06 g/cm$^{3[3]}$

点群：$\bar{4}$

晶格常数[1]：
$$a = (11.48 \pm 0.02) \text{Å}^{[3]}; \ (11.487 \pm 0.003) \text{Å}^{[1]}$$
$$c = (4.33 \pm 0.02) \text{Å}^{[3]}; \ (4.218 \pm 0.001) \text{Å}^{[1]}$$

莫氏硬度：2.9$^{[1]}$；2.7($\parallel c$)$^{[4]}$；2.9($\perp c$)$^{[4]}$

分解温度：537 K$^{[4]}$

线性热膨胀系数的平均值[4]

T/K	$\alpha_t \times 10^6/K^{-1}$, $\parallel c$	$\alpha_t \times 10^6/K^{-1}$, $\perp c$
298~473	228	-19.3

$p = 0.101\ 325$ MPa 时的比热容 c_p[4]

T/K	$c_p/(J \cdot kg^{-1} \cdot K^{-1})$
293	758.8

0.22 cm 长晶体以 0.5 透过计的透明范围是：0.4 μm 到 >2.35 μm[1]。
第一个红外吸收带在 2.35 μm 处[1]。
UV 透过截止边在 0.38 μm 处[5,6]。
$T = 293$ K 时折射率的实验值[5]

$\lambda/\mu m$	n_o	n_e	$\lambda/\mu m$	n_o	n_e
0.435 8	2.073	1.806 9	0.589 3	1.981 4	1.758 3
0.447 1	2.061 9		0.656 3	1.963 6	1.748 9
0.546 1	1.997	1.766 8	0.667 8	1.962 1	1.747 6
0.587 5	1.981 9	1.758 6	0.706 5	1.954 3	1.743 9

Sellmeier 方程(λ 以 μm 为单位,$T = 293$ K)[1,5,7]:

$$n_o^2 = 3.661\ 861 + \frac{0.077\ 588}{\lambda^2 - 0.069\ 737} - 0.045\ 487\lambda^2$$

$$n_e^2 = 2.950\ 921 + \frac{0.041\ 337}{\lambda^2 - 0.058\ 791} - 0.007\ 592\lambda^2$$

在文献[1]、[7]中 n_o 表达式最后一个数字的符号有一错误处。
文献[5]给出的 Sellmeier 方程形式不准确。
有效二阶非线性系数表达式(Kleinman 对称条件成立, $d_{15} = d_{31}$ 和 $d_{14} = d_{25} = d_{36}$)[8]:

$$d_{ooe} = d_{36}\sin\theta\sin 2\phi + d_{31}\sin\theta\cos 2\phi$$

$$d_{eoe} = d_{oee} = d_{36}\sin 2\theta\cos 2\phi - d_{31}\sin 2\theta\sin 2\phi$$

二阶非线性系数值:

$|d_{31}(1.064\ \mu m)| = (1.3 \pm 0.1) \times d_{33}(LiIO_3) = (6.0 \pm 0.9)\ pm/V$[3,9,10]

$|d_{31}(1.064\ \mu m)| = (16.0 \pm 3.0) \times d_{36}(KDP) = (6.2 \pm 1.2)\ pm/V$[5,11]

$|d_{36}(1.064\ \mu m)| = (0.3 \pm 0.1) \times d_{33}(LiIO_3) = (1.4 \pm 0.6)\ pm/V$[3,9,10]

$|d_{36}(1.064\ \mu m)| = (3.7 \pm 1.0) \times d_{36}(KDP) = (1.4 \pm 0.4)\ pm/V$[5,11]

相位匹配角的实验值($T = 293$ K)

相互作用的波长/μm	θ_{exp}/(°)	参 考 文 献
SHG，o + o⇒e		
0.809⇒0.404 5	47.7	[1]
	48.4	[5]
SHG，e + o⇒e		
0.946⇒0.473	≈54	[12]
0.809⇒0.404 5	72.7	[5]
SFG，o + e⇒e		
0.946 + 0.938 5⇒0.471 1	≈54	[13]
0.946 + 0.808⇒0.435 8	≈60	[6]

关于这一晶体

从1970年起，就了解了有机金属络合物CMTC晶体的非线性光学性质[3]。最近，这一非线性材料的质量得到很大改进[1,2]，这样就可以将CMTC应用于CW激光二极管的SHG。在文献[7]中报道了从基频功率2 W产生404 nm、11.8 mW的CW蓝光输出。

参考文献

[1] D. Yuan, D. Xu, M. Liu, F. Qi, W. Yu, W. Hou, Y. Bing, S. Sun, M. Jiang: Structure and properties of a complex crystal for laser diode frequency doubling: cadmium mercury thiocyanate. Appl. Phys. Lett. 70(5), 544 – 546(1997).

[2] D. Yuan, Z. Zhong, M. Liu, D. Xu, Q. Fang, Y. Bing, S. Sun, M. Jiang: Growth of cadmium mercury thiocyanate single crystal for laser diode frequency doubling. J. Cryst. Growth 186(1 – 2), 240 – 244(1998).

[3] J. G. Bergman, Jr., J. H. McFee, G. R. Crane: Nonlinear optical properties of CdHg(SCN)$_4$ and ZnHg(SCN)$_4$. Mat. Res. Bull. 5(11), 913 – 918(1970).

[4] D. R. Yuan, D. Xu, G.-H. Zhang, M.-G. Liu, S.-Y. Guo, F.-Q. Meng, M.-K. Lu, Q. Fang, M.-H. Jiang: Thermal and mechanical properties of a complex nonlinear optical material: cadmium mercury thiocyanate crystal. Chin. Phys. Lett. 17(9), 669 – 671 (2000).

[5] G. Zhang, M. Liu, D. Xu, D. Yuan, W. Sheng, J. Yao: Blue-violet light second harmonic generation with CMTC crystals. J. Mat. Sci. Lett. 19(14), 1255 – 1257(2000).

[6] C. Q. Wang, Y. T. Chow, W. A. Gambling, D. R. Yuan, D. Xu, G. H. Zhang, M. H. Jiang: A continuous-wave tunable solid-state blue laser based on intracavity sum-frequen-

cy mixing and pump-wavelength tunung. Appl. Phys. Lett. 75(13), 1821 – 1823 (1999).

[7] J. Jin, S. Guo, F. Lu, Q. Jiao, J. Yao, G. Zhang: Blue-violet light by direct frequency doubling of laser diode. Proc. SPIE 3928, 228 – 231(2000).

[8] J. E. Midwinter, J. Warner: The effects of phase matching method and of uniaxial crystal symmetry on the polar distribution of second-order non-linear optical polarization. Brit. J. Appl. Phys. 16(11), 1135 – 1142(1965).

[9] J. Jerphagnon: Optical nonlinear susceptibilities of lithium iodate. Appl. Phys. Lett. 16(8), 298 – 299(1970).

[10] R. J. Gehr, A. V. Smith: Separated-beam nonphase-matched second-harmonic method of characterizing nonlinear optical coefficients. J. Opt. Soc. Am. B 15(8), 2298 – 2307 (1998).

[11] D. A. Roberts: Simplified characterization of uniaxial and biaxial nonlinear optical crystals: a plea for standardization of nomenclature and conventions. IEEE J. Quant. Electr. 28(10), 2057 – 2074(1992).

[12] C. Q. Wang, Y. T. Chow, W. A. Gambling, D. Yuan, D. Xu, G. Zhang, M. Liu, M. Jiang: Intracavity-frequency-doubling of a 946 nm ND: YAG laser with cadmium mercury thiocyanate crystal. Opt. Laser Technol. 30(5), 291 – 293(1998).

[13] C. Q. Wang, Y. T. Chow, D. R. Yuan, D. Xu, G. H. Zhang, M. G. Liu, J. R. Lu, Z. S. Shao, M. H. Jiang: CW dual-wavelength Nd: YAG laser at 946 and 938.5 nm and intracavity nonlinear frequency conversion with a CMTC crystal. Opt. Commun. 165(4 – 6), 231 – 235(1999).

6.12　Nb:KTiOPO$_4$，铌掺杂 KTP(Nb$_x$K$_{1-x}$Ti$_{1-x}$OPO$_4$ 或 NbKTP)

正双轴晶：$2V_z = 37.8°$，$\lambda = 0.6328$ μm(7.5 mol% Nb)

分子量：198.393(7.5 mol% Nb)

点群：$mm2$

晶格常数：

3.4 mol% Nb 掺杂 KTP 晶体[1]：

　　$a = 12.828$ Å

　　$b = 6.409$ Å

　　$c = 10.592$ Å

7.9 mol% Nb 掺杂 KTP 晶体[2]：

　　$a = 12.819$ Å

$b = 6.411$ Å

$c = 10.599$ Å

介电轴和晶体学轴的变换：$X, Y, Z \Rightarrow a, b, c$

3.4 mol% Nb 掺杂 KTP 晶体 UV 截止波长为 0.35 μm($\parallel a$)或为 0.37 μm($\parallel c$)[1]

3.4 mol% Nb 掺杂 KTP 晶体的 IR 透过截止边在 4.38 μm[1]

$T = 293$ K(7.5 mol% Nb)时折射率的实验值[3,4]

$\lambda/\mu m$	n_X	n_Y	n_Z
0.539 75	1.779 1	1.791 8	1.902 4
0.632 8	1.764 0	1.775 1	1.879 0
1.079 5	1.738 9	1.747 9	1.840 9
1.341 4	1.732 6	1.741 2	1.831 8

Sellmeier 方程(λ 以 μm 为单位，$T = 293$ K)(3.4 mol% Nb)[1]：

$$n_X^2 = 3.002\,8 + \frac{0.041\,13}{\lambda^2 - 0.043\,41} - 0.010\,49\lambda^2$$

$$n_Y^2 = 3.035\,9 + \frac{0.043\,99}{\lambda^2 - 0.048\,43} - 0.010\,70\lambda^2$$

$$n_Z^2 = 3.346\,7 + \frac{0.062\,82}{\lambda^2 - 0.061\,53} - 0.013\,28\lambda^2$$

Sellmeier 方程(λ 以 μm 为单位，$T = 293$ K)(7.5 mol% Nb)[3]：

$$n_X^2 = 3.006\,0 + \frac{0.038\,424}{\lambda^2 - 0.056\,149} - 0.014\,512\lambda^2$$

$$n_Y^2 = 3.035\,1 + \frac{0.041\,414}{\lambda^2 - 0.061\,208} - 0.015\,125\lambda^2$$

$$n_Z^2 = 3.357\,5 + \frac{0.061\,421}{\lambda^2 - 0.061\,847} - 0.020\,850\lambda^2$$

在文献[4]、[5]、[6]、[7]、[8]中给出了 7.5 mol% Nb 掺杂 KTP 晶体的相同数据。在文献[8]中给出了不同 Nb 浓度(3.5、7.5 和 10 mol%)的 Sellmeier 方程。

7.5 mol% NbKTP 折射率的温度微商[3,7]

$\lambda/\mu m$	$dn_X/dT \times 10^5/K^{-1}$	$dn_Y/dT \times 10^5/K^{-1}$	$dn_Z/dT \times 10^5/K^{-1}$
0.539 75	1.45	2.57	4.86
0.632 8	1.35	2.22	4.03
1.079 5	1.01	1.75	3.09
1.341 4	1.04	1.75	3.43

注：在文献[5]、[4]中给出了稍有差别的值。

$T = 293 \sim 416$ K 及光谱范围 $0.539\,75$ μm $< \lambda < 1.341\,4$ μm 中 7.5 mol% NbKTP 折射率的温度微商(λ 以 μm 为单位)[4,6]:

$$\frac{dn_X}{dT} = \left(-\frac{0.422\,91}{\lambda^3} + \frac{1.840\,4}{\lambda^2} - \frac{2.131\,5}{\lambda} + 1.741\,4\right) \times 10^{-5}\,\text{K}^{-1}$$

$$\frac{dn_Y}{dT} = \left(\frac{0.359\,71}{\lambda^3} - \frac{0.389\,11}{\lambda^2} - \frac{0.161\,81}{\lambda} + 1.937\,8\right) \times 10^{-5}\,\text{K}^{-1}$$

$$\frac{dn_Z}{dT} = \left(-\frac{3.068\,0}{\lambda^3} + \frac{13.559\,5}{\lambda^2} - \frac{17.429\,3}{\lambda} + 10.098\,7\right) \times 10^{-5}\,\text{K}^{-1}$$

NbKTP 晶体主平面上有效二阶非线性系数的表达式(Kleinman 对称条件不成立)[9]:

XY 平面
$$d_{eoe} = d_{oee} = d_{15}\sin^2\phi + d_{24}\cos^2\phi$$

YZ 平面
$$d_{oeo} = d_{eoo} = d_{15}\sin\theta$$

XZ 平面,$\theta < V_Z$
$$d_{ooe} = d_{32}\sin\theta$$

XZ 平面,$\theta > V_Z$
$$d_{oeo} = d_{eoo} = d_{24}\sin\theta$$

NbKTP 晶体主平面上有效二阶非线性系数的表达式(Kleinman 对称成立,$d_{15} = d_{31}$ 以及 $d_{24} = d_{32}$)[9]:

XY 平面
$$d_{eoe} = d_{oee} = d_{31}\sin^2\phi + d_{32}\cos^2\phi$$

YZ 平面
$$d_{oeo} = d_{eoo} = d_{31}\sin\theta$$

XZ 平面,$\theta < V_Z$
$$d_{ooe} = d_{32}\sin\theta$$

XZ 平面,$\theta > V_Z$
$$d_{oeo} = d_{eoo} = d_{32}\sin\theta$$

在文献[9]中给出了 NbKTP 晶体中任意方向上三波相互作用有效二阶非线性系数。

二阶非线性系数值(3.4 mol% Nb)[1,10]:
$$d_{15}(1.064\ \mu m) = (0.8 \pm 0.1) \times d_{15}(\text{KTP}) = (1.5 \pm 0.2)\,\text{pm/V}$$
$$d_{24}(1.064\ \mu m) = (2.2 \pm 0.1) \times d_{15}(\text{NbKTP}) = (3.3 \pm 0.4)\,\text{pm/V}$$

二阶非线性系数值(7.9 mol% Nb)[2,10]:
$$d_{15}(1.064\ \mu m) = 0.75 \times d_{15}(\text{KTP}) \pm 10\% = (1.4 \pm 0.2)\,\text{pm/V}$$

$$d_{24}(1.064 \ \mu m) = 1.13 \times d_{24}(KTP) \pm 10\% = (4.2 \pm 0.4) \ pm/V$$
$$d_{33}(1.064 \ \mu m) = 0.9 \times d_{33}(KTP) \pm 10\% = (13.1 \pm 1.3) \ pm/V$$

3.4 mol% Nb 掺杂 KTP 相位匹配角的实验值[1]

相互作用的波长/μm	$\theta_{exp}/(°)$	相互作用的波长/μm	$\theta_{exp}/(°)$
YZ 平面，$\phi = 90°$		1.152 3⇒0.576 15	67.6
SHG，o + e⇒o		1.318 8⇒0.659 4	57.0
1.064 2⇒0.532 1	62.9	1.579 1⇒0.789 55	49.7
SFG，o + e⇒o		SFG，o + e⇒o	
1.318 8 + 0.659 4⇒0.439 6	60.1	1.318 8 + 0.659 4⇒0.439 6	78.1
XZ 平面，$\phi = 0°$，$\theta > V_Z$		1.579 1 + 0.635 8⇒0.453 3	63.3
SHG，o + e⇒o		1.579 1 + 1.064 2⇒0.635 8	51.8
1.064 2⇒0.532 1	80.8		

7.5 mol% Nb 掺杂 KTP 相位匹配角的实验值[3]

相互作用的波长/μm	$\theta_{pm}/(°)$
XZ 平面，$\phi = 0°$，$\theta > V_Z$	
SHG，o + e⇒o	
1.064 2⇒0.532 1	81.4
1.079 5⇒0.539 75	77.6

注：在文献[4]、[5]、[6]、[7]、[8]、[11]中也给出了 7.5 mol% Nb 掺杂 KTP 相位匹配角值的相同数据。

3.4 mol% Nb 掺杂 KTP 温度带宽的实验值[1]

相互作用的波长/μm	$\Delta T/°C$
XZ 平面，$\phi = 0°$，$\theta > V_Z$	
SHG，o + e⇒o	
1.064 2⇒0.532 1	16.4

7.5 mol% Nb 掺杂 KTP 晶体内角带宽、温度带宽和光谱带宽的实验值[6]

相互作用的波长/μm	$\phi_{pm}/(°)$	$\theta_{pm}/(°)$	$\Delta\phi^{int}/(°)$	$\Delta\theta^{int}/(°)$	$\Delta T/°C$	$\Delta\nu/cm^{-1}$
XY 平面，$\theta = 90°$						
SHG，e + o⇒e						
0.98⇒0.49	65.1		0.38	1.70	10.5	1.9
0.965 6⇒0.482 8	90		4.02	1.76	6.6	1.8

续表

相互作用的波长/μm	$\phi_{pm}/(°)$	$\theta_{pm}/(°)$	$\Delta\phi^{int}/(°)$	$\Delta\theta^{int}/(°)$	$\Delta T/℃$	$\Delta\nu/cm^{-1}$
XZ 平面,$\phi=0°$ $\theta>V_z$						
SHG,$o+e\Rightarrow o$						
$1.341\ 4\Rightarrow 0.670\ 7$		56.6	2.72	0.08	42.6	4.4
$1.064\ 2\Rightarrow 0.532\ 1$		81.5	1.89	0.19	20.6	2.2
$1.051\ 1\Rightarrow 0.525\ 55$		90	4.45	1.76	18.3	2.1

关于这一晶体

  KTP 中掺入铌后使晶体的双折射率得以增加并使最短的 SH 波长蓝移[2]。例如,7.5 mol% 掺杂的晶体其 SHG 截止波长从 0.994 μm 移到 0.965 6 μm[6]。这对于像半导体激光器的一些应用是重要的。然而,很高的掺杂量是不恰当的,因为这会同时导致二阶非线性系数的降低[2]。

参考文献

[1] K. Kato, N. Umemura, M. Saga: Second-harmonic and sum-frequency generation in Nb doped KTP. In: *Advanced Solid-State Lasers*, *OSA Trends in Optics and Photonics Series*, *Vol. 19*, ed. by W. R. Bosenberg, M. M. Fejer (OSA, Washington DC, 1998), pp. 82–84.

[2] L. T. Cheng, L. K. Cheng, R. L. Harlow, J. D. Bierlein: Blue light generation using bulk single crystals of niobium-doped $KTiOPO_4$. Appl. Phys. Lett. 64(2), 155–157(1994).

[3] H. Shen, D. Zhang, W. Liu, W. Chen, G. Zhang, G. Zhang, W. Lin: Measurement of refractive indices and thermal refractive-index coefficients of 7.5-mol% Nb: $KTiOPO_4$ crystal. Appl. Opt. 38(6), 987–990(1999).

[4] D. Y. Zhang, H. Y. Shen, W. Liu, G. F. Zhang, W. Z. Chen, G. Zhang, R. R. Zeng, C. H. Huang, W. X. Lin, J. K. Liang: The thermal refractive index coefficients of 7.5 mol% Nb: $KTiOPO_4$ crystals. J. Appl. Phys. 86(7), 3516–3518(1999).

[5] D. Y. Zhang, H. Y. Shen, W. Liu, G. F. Zhang, W. Z. Chen, G. Zhang, R. R. Zeng, C. H. Huang, W. X. Lin, J. K. Liang: The expressions of the principal thermal refractive index coefficients of 7.5 mol% Nb: $KTiOPO_4$ crystals. Opt. Commun. 168(1–4), 111–115(1999).

[6] D. Y. Zhang, H. Y. Shen, W. Liu, G. F. Zhang, W. Z. Chen, G. Zhang, R. R. Zeng, C. H. Huang, W. X. Lin, J. K. Liang: Study of the nonlinear optical properties of 7.5 mol% Nb: KTP crystals. IEEE J. Quant. Electr. 35(10), 1447–1450(1999).

[7] W. Liu, H. Y. Shen, G. F. Zhang, D. Y. Zhang, G. Zhang, W. X. Lin, R. R. Zeng, C. H. Huang: Studies on the phase-matching condition and the cut-off wavelength of Nb:

KTiOPO$_4$ crystal. Opt. Commun. 185(1-3), 191-196(2000).

[8] D. Y. Zhang, H. Y. Shen, W. Liu, G. F. Zhang, W. Z. Chen, G. Zhang, R. R. Zeng, C. H. Huang, W. X. Lin, J. K. Liang: The principal refractive indices and nonlinear optical phase matched properties of Nb:KTP crystals. Opt. Mater. 15(2), 99-102(2000).

[9] V. G. Dmitriev, D. N. Nikogosyan: Effective nonlinearity coefficients for three-wave interactions in biaxial crystals of *mm*2 point group symmetry. Opt. Commun. 95(1-3), 173-182(1993).

[10] I. Shoji, T. Kondo, A. Kitamoto, M. Shirane, R. Ito: Absolute scale of second-order nonlinear-optical coefficients. J. Opt. Soc. Am. B 14(9), 2268-2294(1997).

[11] D. Y. Zhang, H. Y. Shen, W. Liu, W. Z. Chen, G. F. Zhang, G. Zhang, R. R. Zeng, C. H. Huang, W. X. Lin, J. K. Liang: Crystal growth, X-ray diffraction and nonlinear optical properties of Nb:KTiOPO$_4$ crystal. J. Cryst. Growth 218(1), 98-102(2000).

6.13 RbTiOPO$_4$，磷酸钛氧铷(RTP)

负双轴晶：$2V_Z = 39°$，$\lambda = 0.8$ μm[1]

分子量：244.318

密度：3.64 g/cm^3[1]

点群：*mm*2

晶格常数：

$a = 12.964$ Å[2]；12.980 Å[3]

$b = 6.4985$ Å[2]；6.509 Å[3]

$c = 10.563$ Å[2]；10.578 Å[3]

介电轴和晶体学轴的变换：X，Y，$Z \Rightarrow a$，b，c

压痕负载为 50 g 时的维氏硬度：640 kgf/mm^2(沿[100]方向)[3]

熔点：1 213 K[1]

分解温度：1 374 K[2]

以"0"透过计的透明范围：0.35~0.45 μm[1]在 3.5 μm 处有正磷酸盐的谐波吸收[4]

UV 透过截止边($\alpha = 2$ cm^{-1})为 0.360 μm($\boldsymbol{E} \parallel X$)；0.370 μm($\boldsymbol{E} \parallel Y$)；0.384 μm($\boldsymbol{E} \parallel Z$)[4]

线性吸收系数 α[4]

λ/μm	α/cm^{-1}	备 注	λ/μm	α/cm^{-1}	备 注
0.473	0.108	$\boldsymbol{E} \parallel X$	0.532	0.069	$\boldsymbol{E} \parallel X$
	0.163	$\boldsymbol{E} \parallel Y$		0.087	$\boldsymbol{E} \parallel Y$
	0.279	$\boldsymbol{E} \parallel Z$		0.151	$\boldsymbol{E} \parallel Z$

室温下折射率的实验值[1]

$\lambda/\mu m$	n_X	n_Y	n_Z
0.404 7	1.855 1	1.876 5	1.997 2
0.425 4	1.842 9	1.862 1	1.976 4
0.435 8	1.837 7	1.856 0	1.967 2
0.491 6	1.816 9	1.832 1	1.932 8
0.532 2	1.806 7	1.820 5	1.916 0
0.546 1	1.803 7	1.817 2	1.911 7
0.577 0	1.798 1	1.811 0	1.902 9
0.610 4	1.793 0	1.805 3	1.895 2
0.670 8	1.786 0	1.797 5	1.884 3
0.692 5	1.783 9	1.795 2	1.881 1
1.064 4	1.765 2	1.774 9	1.853 6

最佳 Sellmeier 方程（λ 以 μm 为单位，对 n_X，$0.50~\mu m < \lambda < 4.22~\mu m$；对 n_Y，$0.56~\mu m < \lambda < 4.24~\mu m$；对 n_Z，$0.94~\mu m < \lambda < 3.40~\mu m$）[1]：

$$n_X^2 = 2.198\ 2 + \frac{0.899\ 48}{1 - (0.215\ 2/\lambda)^{1.972\ 7}} + \frac{1.543\ 3}{1 - (11.585/\lambda)^{1.950\ 5}}$$

$$n_Y^2 = 2.280\ 4 + \frac{0.845\ 85}{1 - (0.229\ 63/\lambda)^{1.969\ 6}} + \frac{1.100\ 9}{1 - (9.660\ 2/\lambda)^{1.936\ 9}}$$

$$n_Z^2 = 2.341\ 2 + \frac{1.060\ 9}{1 - (0.264\ 61/\lambda)^{2.058\ 5}} + \frac{0.971\ 4}{1 - (8.149/\lambda)^{2.003\ 8}}$$

在文献[5]、[6]、[7]中给出了其他色散关系方程组。

室温下低频（远低于 RTP^* 晶体声光学谐振频率，即对"自由"晶体）下测量的线性电光系数[5,6]

$\lambda/\mu m$	$r_{13}^T/(pm \cdot V^{-1})$	$r_{23}^T/(pm \cdot V^{-1})$	$r_{33}^T/(pm \cdot V^{-1})$
0.632 8	10.9 ± 1.1	15.0 ± 1.5	33.0 ± 3.3

矫顽场值：$3 \sim 3.5~kV/mm$[8]

RTP 晶体主平面上有效二阶非线性系数的表达式（Kleinman 对称条件不成立）[9]：

XY 平面

$$d_{eoe} = d_{oee} = d_{15}\sin^2\phi + d_{24}\cos^2\phi$$

YZ 平面

$$d_{oeo} = d_{eoo} = d_{15}\sin\theta$$

XZ 平面，$\theta < V_Z$

---

* 译者注：原书为"RTA"，应为"RTP"。

$$d_{ooe} = d_{32}\sin\theta$$

XZ 平面，$\theta > V_Z$

$$d_{oeo} = d_{eoo} = d_{24}\sin\theta$$

RTP 晶体主平面上有效二阶非线性系数的表达式（Kleinman 对称条件成立，$d_{15} = d_{31}$ 以及 $d_{24} = d_{32}$）[9]：

XY 平面

$$d_{eoe} = d_{oee} = d_{31}\sin^2\phi + d_{32}\cos^2\phi$$

YZ 平面

$$d_{oeo} = d_{eoo} = d_{31}\sin\theta$$

XZ 平面，$\theta < V_Z$

$$d_{ooe} = d_{32}\sin\theta$$

XZ 平面，$\theta > V_Z$

$$d_{oeo} = d_{eoo} = d_{32}\sin\theta$$

文献[9]中给出了 RTP 晶体任意方向上三波相互作用的有效非线性系数。
RTP 二阶非线性系数的符号可能都是相同的[10]。

二阶非线性系数绝对值[5,6]：

$$d_{31}(1.064\ \mu m) = (3.3 \pm 0.6)\ pm/V$$
$$d_{32}(1.064\ \mu m) = (4.1 \pm 0.8)\ pm/V$$
$$d_{33}(1.064\ \mu m) = (17.1 \pm 3.4)\ pm/V$$

相位匹配角的实验值

相互作用的波长/μm	$\phi_{pm}/(°)$	$\theta_{pm}/(°)$	参 考 文 献
XY 平面，$\theta = 90°$			
SHG，e + o ⇒ e			
1.064 ⇒ 0.532	60.0		[5]
	58.0		[7]
	57.4		[11]
1.079 ⇒ 0.539 5	48.5		[7]
YZ 平面，$\phi = 90°$			
SHG，o + e ⇒ o			
1.064 ⇒ 0.532		76.0	[7]
1.079 ⇒ 0.539 5		73.5	[7]

内角带宽和温度带宽的实验值[7]

相互作用的波长/μm	$\phi_{pm}/(°)$	$\Delta\phi^{int}/(°)$	$\Delta T/℃$
XY，$\theta = 90°$			
SHG，e + o ⇒ e			
1.064 2 ⇒ 0.532 1	58	0.42	40

激光诱导的体损伤阈值

$\lambda/\mu m$	τ_p/ns	$I_{thr}/(GW \cdot cm^{-2})$	参考文献	备注
1.064 2	15	0.9	[7]	10 Hz
	10	>0.2	[11]	10 Hz

关于这一晶体

  作为 KTP 的同构晶体,RTP 晶体在过去十年未引起很大重视,尽管它的激光损伤阈值比 KTP 本身高 1.8 倍[7]。最近,报道了在 PPRTP 中的 QPM SHG[12]。

参考文献

[1] Y. Guillien, B. Menaert, J. P. Feve, P. Segonds, J. Douady, B. Boulanger, O. Pacaud: Crystal growth and refined Sellmeier equations over the complete transparency range of RbTiOPO$_4$. Opt. Mater. 22(2), 155–162(2003).

[2] L. K. Cheng, E. M. McCarron Ⅲ, J. Calabrese, J. D. Bierlein, A. A. Ballman: Development of the nonlinear optical crystal CsTiOAsO$_4$. I. Structural stability. J. Cryst. Growth 132(1–2), 280–288(1993).

[3] C. V. Kannan, S. Ganesa Moorthy, V. Kannan, C. Subramanian, P. Ramasamy: TSSG of RbTiOPO$_4$ single crystals from phosphate flux and their characterization. J. Cryst. Growth 245(3–4), 289–296(2002).

[4] G. Hansson, H. Karlsson, S. Wang, F. Laurell: Transmission measurements in KTP and isomorphic compounds. Appl. Opt. 39(27), 5058–5069(2000).

[5] L.-T. Cheng, L. K. Cheng, J. D. Bierlein: Linear and nonlinear optical properties of the arsenate isomorphs of KTP. Proc. SPIE 1863, 43–53(1993).

[6] L. K. Cheng, L. T. Cheng, J. Galperin, P. A. Morris Hotsenpiller, J. D. Bierlein: Crystal growth and characterization of KTiOPO$_4$ isomorphs from the self-fluxes. J. Cryst. Growth 137(1–2), 107–115(1994).

[7] Y. S. Oseledchik, A. I. Pisarevsky, A. L. Prosvirnin, V. V. Starshenko, N. V. Svitanko: Nonlinear optical properties of the flux grown RbTiOPO$_4$ crystal. Opt. Mater. 3(4), 237-242(1994).

[8] H. Karlsson, F. Laurell: Electric field poling of flux grown KTiOPO$_4$. Appl. Phys. Lett. 71(24), 3474–3476(1997).

[9] V. G. Dmitriev, D. N. Nikogosyan: Effective nonlinearity coefficients for three-wave interactions in biaxial crystals of mm2 point group symmetry. Opt. Commun. 95(1–3), 173–182(1993).

[10] A. Anema, T. Rasing: Relative signs of the nonlinear coefficients of potassium titanyl

phosphate. Appl. Opt. 36(24), 5902-5904(1997).
[11] U. Chatterjee, P. Kumbhakar, A. K. Chaudhary, G. C. Bhar: Tunable mid-infrared generation in rubidium titanyl phosphate crystal by difference frequency mixing. Nonl. Opt. 28(1-2), 95-106(2001).
[12] H. Karlsson, F. Laurell, L. K. Cheng: Periodic poling of RbTiOPO$_4$ for quasi-phase matched blue light generation. Appl. Phys. Lett. 74(11), 1519-1521(1999).

6.14 LiInS$_2$,硫铟锂(LIS)

负双轴晶:$2V_z = 137°$,$\lambda = 0.5321$ μm[1]
分子量:185.881
密度:3.54 g/cm^3[2];3.52 g/cm^3 原生带黄色的晶体[3,4];3.44 g/cm^3 退火后的玫瑰色晶体[3,4]
点群:$mm2$
晶格常数:
原生带黄色的晶体
  $a = (6.890 \pm 0.001)$ Å[3,4]
  $b = (8.053 \pm 0.001)$ Å[3,4]
  $c = (6.478 \pm 0.002)$ Å[3,4]
退火后的玫瑰色晶体
  $a = (6.896 \pm 0.001)$ Å[3,4];6.893 Å[1]
  $b = (8.058 \pm 0.002)$ Å[3,4];8.0578 Å[1]
  $c = (6.484 \pm 0.004)$ Å[3,4];6.4816 Å[1]
介电轴和晶体学轴的变换:$X, Y, Z \Rightarrow b, a, c$
莫氏硬度:3~4[5]
熔点:1 273 K[2,6]
线性热膨胀系数 α_t[7]

T/K	$\alpha_t \times 10^6$/K^{-1}, ∥X	$\alpha_t \times 10^6$/K^{-1}, ∥Y	$\alpha_t \times 10^6$/K^{-1}, ∥Z
293	16.4	9.1	6.8

温度范围 253~393 K 线性热膨胀系数 α_t 的温度关系(T 以 K 为单位)[7]:
$$\alpha_t(\parallel X) = 1.61 \times 10^{-5} + 1.44 \times 10^{-8}(T - 273)$$
$$\alpha_t(\parallel Y) = 0.89 \times 10^{-5} + 0.72 \times 10^{-8}(T - 273)$$
$$\alpha_t(\parallel Z) = 0.66 \times 10^{-5} + 0.93 \times 10^{-8}(T - 273)$$
$p = 0.101\ 325$ MPa 时的比热容 c_p[6]

T/K	c_p/(J·kg^{-1}·K^{-1})
300	500 ± 6

热导率[6]

κ/(W·m^{-1}·K^{-1})，‖X	κ/(W·m^{-1}·K^{-1})，‖Y	κ/(W·m^{-1}·K^{-1})，‖Z
6.2	6	7.6

室温下带隙能量：E_g = 3.56 eV[8]；3.57 eV[1,6]；3.59 eV[2,9]；3.6 eV[3]

以 0.5 透过计的透明范围：0.43 ~ 11.5 μm[4]；0.5 ~ 11 μm[5]；0.57 ~ 8.97 μm[7]

以 0.1 透过计的透明范围：0.4 ~ 12.5 μm[5]

以 "0" 透过计的透明范围：0.34 ~ 13.2 μm[3,5]

线性吸收系数 α

λ/μm	α/cm^{-1}	参考文献	λ/μm	α/cm^{-1}	参考文献
0.6	0.23	[4]		<0.2	[4]
0.76 ~ 0.9	0.15	[4]	1.27	0.09	[4]
1.064	<0.04	[6]	2.53	0.05	[4]
1 ~ 8	0.1 ~ 0.25	[5]		<0.05	[6]
	0.1 ~ 0.15	[6]	9.2 ~ 10.8	1.1 ~ 2.3	[5]

双光子吸收系数 β[1]

λ/μm	τ_p/ns	$\beta \times 10^{11}$/(cm·W^{-1})
0.8	0.000 2	<5

原生 LIS 折射率的实验值[10]

λ/μm	n_X	n_Y	n_Z	λ/μm	n_X	n_Y	n_Z
0.425	2.347 2	2.412 6	2.420 8	0.850	2.146 5	2.184 9	2.192 3
0.450	2.309 6	2.368 5	2.376 6	0.900	2.140 9	2.178 9	2.186 3
0.500	2.258 0	2.309 5	2.317 5	0.950	2.136 4	2.173 7	2.181 2
0.550	2.224 4	2.272 0	2.279 3	1.000	2.132 5	2.169 6	2.176 9
0.600	2.201 1	2.245 5	2.253 6	1.100	2.126 8	2.163 0	2.170 6
0.650	2.184 1	2.226 5	2.234 4	1.200	2.122 5	2.157 9	2.165 5
0.700	2.171 2	2.211 9	2.219 9	1.400	2.115 8	2.150 8	2.158 5
0.750	2.161 0	2.201 0	2.208 5	1.600	2.111 5	1.146 3	2.153 8
0.800	2.153 0	2.191 8	2.199 6	1.800	2.108 2	2.143 0	2.150 1

续表

$\lambda/\mu m$	n_X	n_Y	n_Z	$\lambda/\mu m$	n_X	n_Y	n_Z
2.000	2.105 7	2.140 5	2.147 5	5.500	2.082 8	2.116 6	2.122 9
2.200	2.103 9	2.138 4	2.145 4	6.000	2.078 9	2.112 8	2.118 9
2.400	2.102 6	2.136 7	2.144 0	6.500	2.075 0	2.108 6	2.114 3
2.600	2.101 2	2.135 3	2.142 5	7.000	2.070 1	2.104 0	2.109 6
2.800	2.099 9	2.133 9	2.141 1	7.500	2.065 0	2.099 0	2.104 3
3.000	2.098 7	2.132 5	2.139 8	8.000	2.059 5	2.093 7	2.098 7
3.200	2.097 6	2.131 4	2.138 6	8.500	2.053 4	2.087 6	2.092 4
3.400	2.096 5	2.130 5	2.137 2	9.000	2.047 0	2.081 6	2.085 6
3.600	2.095 4	2.129 1	2.136 1	9.500	2.039 8	2.074 9	2.078 3
3.800	2.094 1	2.128 0	2.134 8	10.000	2.031 9	2.066 6	2.070 3
4.000	2.093 0	2.126 6	2.133 5	10.500	2.023 8	2.058 5	2.061 9
4.500	2.090 0	2.123 7	2.130 4	11.000	2.014 6	2.050 1	2.052 2
5.000	2.086 7	2.120 4	2.127 1				

$\lambda = 1.064~\mu m$ 时折射率的温度微商[6]:

$$dn_X/dT = 3.72 \times 10^{-5} K^{-1}$$
$$dn_Y/dT = 4.55 \times 10^{-5} K^{-1}$$
$$dn_Z/dT = 4.47 \times 10^{-5} K^{-1}$$

最佳 Sellmeier 方程组(λ 以 μm 为单位,$T = 293$ K)[11]:

$$n_X^2 = 6.681~9 + \frac{0.129~4}{\lambda^2 - 0.061~1} + \frac{2~037.53}{\lambda^2 - 897.77}$$

$$n_Y^2 = 7.096~9 + \frac{0.143~3}{\lambda^2 - 0.066~0} + \frac{2~511.13}{\lambda^2 - 988.03}$$

$$n_Z^2 = 7.255~5 + \frac{0.144~3}{\lambda^2 - 0.066~1} + \frac{2~625.82}{\lambda^2 - 9.839~7}$$

Ebbers 推导了不同的 Sellmeier 方程组并发表于文献[1]、[5];在[6]、[12]中给出了其他色散关系。

室温下低频(远低于 LIS 晶体声光学谐振频率,即对"自由"晶体)下测量的线性电光系数[7]

$\lambda/\mu m$	$r_{13}^T/(pm \cdot V^{-1})$	$r_{23}^T/(pm \cdot V^{-1})$	$r_{33}^T/(pm \cdot V^{-1})$
1.064	0.97 ± 0.1	0.42 ± 0.04	−1.33 ± 0.13

LIS 晶体主平面上有效二阶非线性系数的表达式(小走离角近似,Kleinman 对称

条件成立 $d_{15} = d_{31}$ 和 $d_{24} = d_{32}$)[13]:

XY 平面
$$d_{eoe} = d_{oee} = d_{32}\sin^2\phi + d_{31}\cos^2\phi$$

YZ 平面
$$d_{oeo} = d_{eoo} = d_{32}\sin\theta$$

XZ 平面，$\theta < V_z$
$$d_{ooe} = d_{31}\sin\theta$$

XZ 平面，$\theta > V_z$
$$d_{oeo} = d_{eoo} = d_{31}\sin\theta$$

文献[13]中给出了 LIS 晶体任意方向上三波相互作用的有效非线性系数。
二阶非线性系数值：

$d_{31}(2.3\ \mu m) = (7.2 \pm 0.4)\ pm/V$[6]

$d_{32}(2.3\ \mu m) = (5.7 \pm 0.6)\ pm/V$[6]

$d_{33}(2.3\ \mu m) = (-16 \pm 4)\ pm/V$[6]

$d_{31}(10.6\ \mu m) = 0.074 \times d_{36}(GaAs) \pm 15\% = (6.1 \pm 0.9)\ pm/V$[10,14]

$d_{32}(10.6\ \mu m) = 0.064 \times d_{36}(GaAs) \pm 15\% = (5.3 \pm 0.8)\ pm/V$[10,14]

$d_{33}(10.6\ \mu m) = 0.118 \times d_{36}(GaAs) \pm 15\% = (9.8 \pm 1.5)\ pm/V$[10,14]

相位匹配角和内角带宽的实验值

相互作用的波长/μm	$\phi_{pm}/(°)$	$\theta_{pm}/(°)$	$\Delta\phi^{int}/(°)$	$\Delta\theta^{int}/(°)$	参考文献
XY 平面，$\theta = 90°$					
SHG，e + o⇒e					
2.366⇒1.183	82.1				[6]
2.469⇒1.234 5	73.1				[6]
2.481⇒1.240 5	72.4				[4]
2.527⇒1.263 5	69.8				[6]
2.583⇒1.291 5	67.4				[6]
2.590⇒1.295	66.2		0.48		[6]
2.611⇒1.305 5	66.3		0.4		[4]
			0.44		[15]
2.90⇒1.45	57.9				[6]
3.4⇒1.7	50.7				[6]
3.7⇒1.85	48.3				[6]
3.9⇒1.95	49.0				[6]
4.45⇒2.225	51.6				[6]
4.95⇒2.475	56.3				[6]
5.0⇒2.5	57.0		0.92		[9]
5.35⇒2.675	62.8				[6]
5.55⇒2.775	66.0				[6]

续表

相互作用的波长/μm	ϕ_{pm}/(°)	θ_{pm}/(°)	$\Delta\phi^{int}$/(°)	$\Delta\theta^{int}$/(°)	参考文献
5.75⇒2.875	69.1				[6]
5.90⇒2.95	71.4				[6]
YZ 平面，$\phi = 90°$					
SHG, o + e⇒o					
2.542 7⇒1.371 35		35.4			[6]
2.552 7⇒1.276 35		34.0			[6]
2.570 4⇒1.285 2		31.0			[6]
2.582⇒1.291		28.0			[7]
2.587⇒1.293 5		28.7			[6]
2.590⇒1.295		27.9		2.9	[6]
2.602 3⇒1.301 15		25.9			[6]
2.606 7⇒1.303 35		25.1			[6]
2.631 4⇒1.315 7		19.6			[6]

激光诱导的损伤阈值

λ/μm	τ_p/ns	I_{thr}/(GW·cm^{-2})	参考文献	备 注
0.8	0.000 2	>140	[1]	表面损伤
1.064	10	0.1	[6]	10 Hz，体损伤
5.0	0.000 5	0.44	[9]	10 Hz，5 000 个脉冲列，体损伤
		>6	[9]	10 Hz，125 个脉冲列，体损伤
9.55	36	>0.18	[5]	

关于这一晶体

  LIS 是最近新发展的少数几种 IR 非线性材料之一。现在，LIS 是能够一步将钛宝石激光器辐射实现在 5 ~ 11 μm 范围直接下转换的唯一晶体。

■ 参考文献

[1] F. Rotermund, V. Petrov, F. Noack, L. Isaenko, A. Yelisseyev, S. Lobanov: Optical parametric generation of femtosecond pulses up to 9 μm with LiInS$_2$ pumped at 800 nm. Appl. Phys. Lett. 78(18), 2623 – 2625(2001).

[2] L. Isaenko, A. Yelisseyev, S. Lobanov, V. Petrov, F. Rotermund, J.-J. Zondy, G. H. M. Knippels: LiInS$_2$: a new nonlinear crystal for the mid-IR. J. Mater. Sci. Semicond. Process. 4(6), 665 – 668(2002).

[3] L. Isaenko, I. Vasilyeva, A. Yelisseyev, V. Malakhov, L. Dovlitova, J.-J. Zondy, I. Kavun: Growth and characterization of LiInS$_2$ single crystals. J. Cryst. Growth 218(2-4), 313-322(2000).

[4] A. Yelisseyev, L. Isaenko, S. Lobanov, J.-J. Zondy, A. Douillet, I. Thenot, P. Kupecek, G. Mennerat, J. Mangin, S. Fossier, S. Salaün: New ternary sulfide for double application in laser schemes. In: *Advanced Solid-State Lasers*, *OSA Trends in Optics and Photonics Series*, *Vol. 34*, ed. by H. Injeyan, U. Keller, C. Marshall (OSA, Washington DC, 2000), pp. 561-568.

[5] Y. M. Andreev, L. G. Geiko, P. P. Geiko, S. G. Grechin: Optical properties of a nonlinear LiInS$_2$ crystal. Kvant. Elektron. 31(7), 647-648(2001)[In Russian, English trans. :Quantum Electron. 31(7), 647-648(2001)].

[6] S. Fossier, S. Salaün, J. Mangin, O. Bidault, I. Thenot, J.-J. Zondy, W. Chen, F. Rotermund, V. Petrov, P. Petrov, J. Henningsen, A. Yelisseyev, L. Isaenko, S. Lobanov, O. Balach-ninaite, G. Slekys, V. Sirutkaitis: Optical, vibrational, thermal, electrical, damage and phase-matching properties of lithium thioindate. J. Opt. Soc. Am. B 21 (11), 1981-2007(2004).

[7] J. Mangin, S. Salaün, S. Fossier, P. Strimer, J.-J. Zondy, L. Isaenko, A. Yelisseyev: Optical properties of lithium thioindate. Proc. SPIE 4268, 49-57(2001).

[8] A. Eifler, V. Riede, J. Brückner, S. Weise, V. Krämer, G. Lippold, W. Schmitz, K. Bente, W. Grill: Band gap energies and lattice vibrations of the lithium ternary compounds LiInSe$_2$, LiInS$_2$, LiGaSe$_2$ and LiGaS$_2$. Jpn. J. Appl. Phys. 39(Suppl. 1), 279-281(2000).

[9] G. M. H. Knippels, A. F. G. van der Meer, A. M. Macleod, A. Yelisseyev, L. Isaenko, S. Lobanov, I. Thenot, J.-J. Zondy: Mid-infrared(2.75-6.0-μm) second-harmonic generation in LiInS$_2$. Opt. Lett. 26(9), 617-619 (2001).

[10] G. D. Boyd, H. M. Kasper, J. H. McFee: Linear and nonlinear optical properties of LiInS$_2$. J. Appl. Phys. 44(6), 2809-2812(1973).

[11] K. Suzuki, E. Takaoka, T. Mikami, T. Hikoso, K. Kato, N. Umemura: Fourth harmonic generation of the CO$_2$ laser frequency in LiInS$_2$. In: *Proceedings of Autumn Meeting of Japan Society of Applied Physics*(JSAP, Nagoya, 2002)[In Japanese].

[12] V. V. Badikov, V. I. Chizhikov, V. V. Efimenko, T. D. Efimenko, V. L. Panyutin, G. S. Shevyrdyaeva, S. I. Scherbakov: Optical properties of lithium indium selenide. Opt. Mater. 23(3-4), 575-581(2003).

[13] V. G. Dmitriev, D. N. Nikogosyan: Effective nonlinearity coefficients for three-wave interactions in biaxial crystals of *mm*2 point group symmetry. Opt. Commun. 95(1-3), 173-182(1993).

[14] D. A. Roberts: Simplified characterization of uniaxial and biaxial nonlinear optical crystals: a plea for standardization of nomenclature and conventions. IEEE J. Quant. Electr.

28(10), 2057 – 2074(1992).

[15] L. Isaenko, A. Yelisseyev, J.-J. Zondy, G. Knippels, I. Thenot, S. Lobanov: Growth and characterization of single crystals of ternary chalcogenides for laser applications. Opto-Electron. Rev. 9(2), 135 – 141(2001).

6.15 LiInSe$_2$，硒铟锂(LISe)

负双轴晶：$2V_Z = 140°$，$\lambda = 0.5321$ μm[1]

分子量：279.561

点群：$mm2$

晶格常数：

原生的黄色晶体：

$a = (7.1917 \pm 0.0008)$ Å[1]

$b = (8.4116 \pm 0.0010)$ Å[1]

$c = (6.7926 \pm 0.0008)$ Å[1]

介电轴和晶体学轴的变换：$X, Y, Z \Rightarrow b, a, c$

室温下带隙能量：$E_g = 2.83$ eV[2]；2.87 eV[$\boldsymbol{E} \parallel a$][1]；2.86 eV($\boldsymbol{E} \parallel b$)[1]

原生黄色晶体的透明范围：

以 $\alpha = 15$ cm^{-1} 计：0.46 ~ 14 μm[1]

以 $\alpha = 1$ cm^{-1} 计：0.72 ~ 10.4 μm[1]

双光子吸收系数 β[3]

λ/μm	τ_p/ns	$\beta \times 10^{11}$/(cm·W^{-1})
0.82	0.00022	60

原生硒铟锂晶体折射率的实验值[1]

λ/μm	n_X	n_Y	n_Z	λ/μm	n_X	n_Y	n_Z
0.500	2.5228	—	2.6035	0.950	2.3037	—	2.3550
0.525	2.4849	—	2.5594	1.000	2.2977	2.3390	2.3486
0.550	2.4549	2.5178	2.5248	2.000	2.2530	2.2913	2.2988
0.575	2.4313	—	2.4977	3.000	2.2434	2.2842	2.2891
0.600	2.4118	—	2.4758	4.100	2.2398	2.2799	2.2842
0.650	2.3818	2.4331	2.4422	5.000	2.2370	2.2772	2.2818
0.700	2.3601	2.4079	2.4174	6.000	2.2323	2.2718	2.2765
0.750	2.3436	2.3893	2.3989	7.000	2.2271	2.2688	2.2715
0.800	2.3306	2.3746	2.3843	8.000	2.2202	2.2617	2.2649
0.850	2.3196	—	2.3725	10.000	2.2015	2.2522	2.2566
0.900	2.3109	2.3533	2.3632	11.000	2.1935	2.2352	2.2380

Sellmeier 方程 (λ 以 μm 为单位, $T = 293$ K)[1]:

$$n_X^2 = 5.037\,059\,9 + \frac{0.216\,583\,3\lambda^2}{\lambda^2 - 0.085\,692\,9} - 0.001\,853\,4\lambda^2$$

$$n_Y^2 = 5.202\,654\,5 + \frac{0.242\,247\,0\lambda^2}{\lambda^2 - 0.089\,915\,1} - 0.001\,506\,9\lambda^2$$

$$n_Z^2 = 5.239\,914\,2 + \frac{0.241\,417\,8\lambda^2}{\lambda^2 - 0.091\,789\,0} - 0.001\,764\,5\lambda^2$$

在文献[4]中给出了其他色散关系。

$LiInSe_2$ 晶体主平面上有效二阶非线性系数的表达式(小走离角近似, Kleinman 对称条件成立 $d_{15} = d_{31}$ 和 $d_{24} = d_{32}$)[5]:

XY 平面
$$d_{eoe} = d_{oee} = d_{32}\sin^2\phi + d_{31}\cos^2\phi$$

YZ 平面
$$d_{oeo} = d_{eoo} = d_{32}\sin\theta$$

XZ 平面, $\theta < V_Z$
$$d_{ooe} = d_{31}\sin\theta$$

XZ 平面, $\theta > V_Z$
$$d_{oeo} = d_{eoo} = d_{31}\sin\theta$$

文献[5]中给出了 $LiInSe_2$ 晶体任意方向上三波相互作用的有效二阶非线性系数。

二阶非线性系数值:
$$d_{31}(2.8\,\mu m) = 0.76 \times d_{36}(AgGaS_2) = (10.4 \pm 1.7)\,pm/V^{[1,6]}$$
$$d_{32}(2.1 \sim 2.45\,\mu m) = 3 \times d_{24}(KTP) = (7.8 \pm 0.3)\,pm/V^{[1,7]}$$

相位匹配角的实验值[1]

相互作用的波长/μm	$\theta_{pm}/(°)$	相互作用的波长/μm	$\theta_{pm}/(°)$
XZ 平面, $\theta = 0°$		$2.191 \Rightarrow 1.095\,5$	17
SHG, o + o $\Rightarrow$ e		$2.292 \Rightarrow 1.146$	21
$2.119 \Rightarrow 1.059\,5$	10	$2.456 \Rightarrow 1.228$	25

关于这一晶体

LIS 晶体的同构体,其非线性系数 d_{31} 和 d_{32} 的值比 LIS 的稍高。

■ 参考文献

[1] L. Isaenko, A. Yelisseyev, S. Lobanov, V. Petrov, F. Rotermund, G. Slekys, J.-J.

Zondy: LiInSe$_2$: a biaxial ternary chalcogenide crystal for nonlinear optical applications in midinfrared. J. Appl. Phys. 91(12), 9475 – 9480(2002).

[2] A. Eifler, V. Riede, J. Brückner, S. Weise, V. Krämer, G. Lippold, W. Schmitz, K. Bente, W. Grill: Band gap energies and lattice vibrations of the lithium ternary compounds LiInSe$_2$, LiInS$_2$, LiGaSe$_2$ and LiGaS$_2$. Jpn. J. Appl. Phys. 39(Suppl. 1), 279 – 281(2000).

[3] V. V. Petrov, F. Noack, L. Isaenko, A. Yelisseyev, S. Lobanov, A. Titov, F. Rotermund, J. -J. Zondy: Mid-infrared optical parametric generation in lithium-containing ternary compounds LiAB$_2$ (A = Ga,In; B = S,Se). In: *Conference on Lasers and Electrooptics CLEO/QELS 2003, Technical Digest*(OSA, Washington DC, 2003), paper CTuN5.

[4] V. V. Badikov, V. I. Chizhikov, V. V. Efimenko, T. D. Efimenko, V. L. Panyutin, G. S. Shevyrdyaeva, S. I. Scherbakov: Optical properties of lithium indium selenide. Opt. Mater. 23(3 – 4), 575 – 581(2003).

[5] V. G. Dmitriev, D. N. Nikogosyan: Effective nonlinearity coefficients for three-wave interactions in biaxial crystals of *mm*2 point group symmetry. Opt. Commun. 95(1 – 3), 173 – 182(1993).

[6] J. -J. Zondy, D. Touahri, O. Acef: Absolute value of the d_{36} nonlinear coefficient of AgGaS$_2$: prospect for a low-threshold doubly resonant oscillator-based 3:1 frequency divider. J. Opt. Soc. Am. B 14(10), 2481 – 2497(1997).

[7] I. Shoji, T. Kondo, A. Kitamoto, M. Shirane, R. Ito: Absolute scale of second-order nonlinear-optical coefficients. J. Opt. Soc. Am. B 14(9), 2268 – 2294(1997).

6.16 LiGaS$_2$，硫镓锂(LGS)

负双轴晶

分子量：140.781

密度：2.94 g/cm^3(计算值)[1]

点群：*mm*2

晶格常数：

$a = (6.519 \pm 0.006)$ Å[2]; $(6.513\,3 \pm 0.000\,6)$ Å[1]

$b = (7.872 \pm 0.007)$ Å[2]; $(7.862\,9 \pm 0.000\,8)$ Å[1]

$c = (6.238 \pm 0.004)$ Å[2]; $(6.217\,5 \pm 0.000\,5)$ Å[1]

介电轴和晶体学轴的变换：

$X, Y, Z \Rightarrow b, a, c (\lambda < 6.5\ \mu m)$

$X, Y, Z \Rightarrow b, c, a (\lambda > 6.5\ \mu m)$

室温下带隙能量：$E_g = 4.15$ eV[1]; 3.62 eV[3]

以 $\alpha = 5\text{ cm}^{-1}$ 计的透明范围：$0.32 \sim 11.6\text{ }\mu\text{m}$[1]

室温下色散关系方程（λ 以 μm 为单位）[1]：

$$n_X^2 = 4.326\,834 + \frac{0.103\,090\,7}{\lambda^2 - 0.030\,987\,6} - 0.003\,701\,5\lambda^2$$

$$n_Y^2 = 4.478\,907 + \frac{0.120\,426}{\lambda^2 - 0.034\,616\,0} - 0.003\,511\,9\lambda^2$$

$$n_Z^2 = 4.493\,881 + \frac{0.117\,745\,2}{\lambda^2 - 0.033\,700\,4} - 0.003\,776\,7\lambda^2$$

LGS 晶体主平面上有效二阶非线性系数的表达式（小走离角近似，Kleinman 对称条件成立 $d_{15} = d_{31}$ 和 $d_{24} = d_{32}$）[4]：

XY 平面

$$d_{eoe} = d_{oee} = d_{32}\sin^2\phi + d_{31}\cos^2\phi$$

YZ 平面

$$d_{oeo} = d_{eoo} = d_{32}\sin\theta$$

XZ 平面，$\theta < V_Z$

$$d_{ooe} = d_{31}\sin\theta$$

XZ 平面，$\theta > V_Z$

$$d_{oeo} = d_{eoo} = d_{31}\sin\theta$$

在文献[4]中给出了 LGS 晶体任意方向上三波相互作用的有效非线性系数。

关于这一晶体

这是最新提出的 IR 非线性材料，具有方铁矿型结构，UV 透过可低至 $0.32\text{ }\mu\text{m}$。

参考文献

[1] L. Isaenko, A. Yelisseyev, S. Lobanov, A. Titov, V. Petrov, J.-J. Zondy, P. Krinitsin, A. Merkulov, V. Vedenyapin, J. Smirnova：Growth and properties of LiGaX₂ (X = S, Se, Te) single crystals for nonlinear optical applications in the mid-IR. Cryst. Res. Technol. 38(3 – 5), 379 – 387(2003).

[2] J. Leal-Gonzalez, S. S. Melibary, A. J. Smith：Structure of lithium gallium sulfide, LiGaS₂. Acta Crystallogr. C 46(11), 2017 – 2019(1990).

[3] A. Eifler, V. Riede, J. Brückner, S. Weise, V. Krämer, G. Lippold, W. Schmitz, K. Bente, W. Grill：Band gap energies and lattice vibrations of the lithium ternary compounds LiInSe₂, LiInS₂, LiGaSe₂ and LiGaS₂. Jpn. J. Appl. Phys. 39 (Suppl. 1), 279 – 281(2000).

[4] V. G. Dmitriev, D. N. Nikogosyan：Effective nonlinearity coefficients for three-wave interactions

in biaxial crystals of *mm*2 point group symmetry. Opt. Commun. 95(1 – 3), 173 – 182 (1993).

6.17 LiGaSe$_2$，硒镓锂(LGSe)

负双轴晶

分子量：234.461

密度：4.24 g/cm^3(计算值)[1]

点群：*mm*2

晶格常数：

$a = 6.833$ Å[2]；(6.832 ± 0.001) Å[1]

$b = 8.227$ Å[2]；(8.237 ± 0.001) Å[1]

$c = 6.541$ Å[2]；(6.535 ± 0.001) Å[1]

介电轴和晶体学轴的变换：

$X, Y, Z \Rightarrow b, a, c (\lambda < 8 \mu m)$

$X, Y, Z \Rightarrow b, c, a (\lambda > 8 \mu m)$

熔点：1 119 K[2]

室温下带隙能量：$E_g = 3.65$ eV[2]；3.13 eV[3]；3.34 eV[1]

以 $\alpha = 5$ cm^{-1} 计的透明范围：0.37 ~ 13.2 μm[1]

室温下色散方程(λ 以 μm 计)[1]：

$$n_X^2 = 4.995\,92 + \frac{0.151\,30}{\lambda^2 - 0.089\,89} - 0.002\,33\lambda^2$$

$$n_Y^2 = 5.208\,96 + \frac{0.186\,32}{\lambda^2 - 0.076\,87} - 0.002\,11\lambda^2$$

$$n_Z^2 = 5.224\,42 + \frac{0.183\,65}{\lambda^2 - 0.074\,93} - 0.002\,32\lambda^2$$

LiGaSe$_2$ 晶体主平面上有效二阶非线性系数的表达式(小走离角近似，Kleinman 对称条件成立，$d_{15} = d_{31}$ 以及 $d_{24} = d_{32}$)[4]：

XY 平面

$$d_{eoe} = d_{oee} = d_{32} \sin^2\phi + d_{31} \cos^2\phi$$

YZ 平面

$$d_{oeo} = d_{eoo} = d_{32} \sin\theta$$

XZ 平面，$\theta < V_Z$

$$d_{ooe} = d_{31} \sin\theta$$

XZ 平面，$\theta > V_Z$

$$d_{oeo} = d_{eoo} = d_{31}\sin\theta$$

文献[4]中给出了 LiGaSe$_2$ 晶体任意方向上三波相互作用的有效非线性系数。

关于这一晶体

这是最新提出的 IR 非线性材料,具有方铁矿型结构,UV 透过可至 0.37 μm。

参考文献

[1] L. Isaenko, A. Yelisseyev, S. Lobanov, A. Titov, V. Petrov, J. -J. Zondy, P. Krinitsin, A. Merkulov, V. Vedenyapin, J. Smirnova: Growth and properties of LiGaX$_2$(X = S, Se, Te) single crystals for nonlinear optical applications in the mid-IR. Cryst. Res. Technol. 38(3 - 5), 379 - 387(2003).

[2] K. Kuriyama, T. Nozaki: Single-crystal growth and characterization of LiGaSe$_2$. J. Appl. Phys. 52(10), 6441 - 6443(1981).

[3] A. Eifler, V. Riede, J. Brückner, S. Weise, V. Krämer, G. Lippold, W. Schmitz, K. Bente, W. Grill: Band gap energies and lattice vibrations of the lithium ternary compounds LiInSe$_2$, LiInS$_2$, LiGaSe$_2$ and LiGaS$_2$. Jpn. J. Appl. Phys. 39(Suppl. 1), 279 - 281(2000).

[4] V. G. Dmitriev, D. N. Nikogosyan: Effective nonlinearity coefficients for three-wave interactions in biaxial crystals of mm2 point group symmetry. Opt. Commun. 95(1 - 3), 173 - 182(1993).

6.18 AgGa$_x$In$_{1-x}$Se$_2$,硒铟镓银(AGISe)

负单轴晶:$n_o > n_e$

点群:$\bar{4}2m$

$x = 0.58$ 时线性热膨胀系数的平均值[1]

T/K	$\alpha_t \times 10^6/K^{-1}$, ∥ c	$\alpha_t \times 10^6/K^{-1}$, ⊥ c
298 ~ 633	-12.1	16.8

UV 透过截止边为 0.85 μm 以及红外透过截止边为 19 μm($x = 0.65$)[2]。

线性吸收系数 α

$\lambda/\mu m$	α/cm^{-1}	参考文献	备注
1.06	<0.01	[3]	典型晶体
	0.002	[3]	最佳晶体
2.09	0.01	[4]	o 光,$x = 0.474$

续表

$\lambda/\mu m$	α/cm^{-1}	参 考 文 献	备 注
	0.02	[4]	e 光, $x = 0.474$
4.655 ~ 4.82	0.08	[5]	$x = 0.6$
5	0.157	[2]	$x = 0.65$
9.31 ~ 9.64	0.06	[5]	$x = 0.6$
10	0.158	[2]	$x = 0.65$

Sellmeier 方程($x = 0.526$, λ 以 μm 为单位, $T = 293$ K)[4]:

$$n_o^2 = 6.9082 + \frac{0.5586}{\lambda^2 - 0.2870} - 0.00108\lambda^2$$

$$n_e^2 = 6.8262 + \frac{0.6044}{\lambda^2 - 0.3736} - 0.00111\lambda^2$$

其他色散关系在文献[5]中(对 $x = 0.1 \sim 1.0$)和文献[1]中(对 $x = 0.58$)给出。

普遍情况下有效二阶非线性系数的表达式(Kleinman 对称条件成立 $d_{14} = d_{25} = d_{36}$)[6]:

$$d_{ooe} = -d_{36}\sin(\theta + \rho)\sin 2\phi$$
$$d_{eoe} = d_{oee} = 2d_{36}\sin(\theta + \rho)\cos(\theta + \rho)\cos 2\phi$$

有效二阶非线性系数的简化表达式(小双折射角近似, Kleinman 对称条件成立 $d_{14} = d_{25} = d_{36}$)[7]:

$$d_{ooe} = -d_{36}\sin\theta\sin 2\phi$$
$$d_{eoe} = d_{oee} = d_{36}\sin 2\theta\cos 2\phi$$

二阶非线性系数值:

$$d_{36}(9 \sim 10 \ \mu m) \approx 40 \ pm/V, \ 对 \ x = 0.6^{[5]}$$

相位匹配角的实验值($T = 293$ K)

相互作用的波长/μm	$\theta_{exp}/(°)$	参考文献	备 注
SHG, o + o ⇒ e			
9.27 ⇒ 4.635	90	[1]	$x = 0.58$
9.31 ⇒ 4.655	76.8 ± 1.7	[5]	$x = 0.6$
9.55 ⇒ 4.775	83.3 ± 1.7	[5]	$x = 0.6$
9.64 ⇒ 4.82	87.3 ± 1.7	[5]	$x = 0.6$
	90	[8]	$x = 0.65$
SFG, o + o ⇒ e			
9.2714 + 4.6357 ⇒ 3.09047	88.5	[4]	$x = 0.526$
9.5525 + 4.77625 ⇒ 3.18417	84	[4]	$x = 0.526$
10.2466 + 5.1233 ⇒ 3.41553	84.7	[4]	$x = 0.526$
10.591 + 5.2955 ⇒ 3.53033	86.6	[4]	$x = 0.526$

内角带宽和温度带宽的实验值[4]

相互作用的波长/μm	$\Delta\theta^{int}/(°)$	$\Delta T/℃$	备注
SFG, o + o ⇒ e			
10.246 6 + 5.123 3 ⇒ 3.415 53	8.3 ± 0.2	105 ± 5	$x = 0.526$

激光诱导的表面损伤阈值[3]

$\lambda/\mu m$	τ_p/ns	$I_{thr}/(GW \cdot cm^{-2})$	备注
10.7	70	> 30	$x = 0.75$

关于这一晶体

制备了这一黄铜矿型混晶以实现 CO_2 激光辐射的非临界相位匹配(NCPM) SHG[1,5,8] 和 THG[4]。

参考文献

[1] P. G. Schunemann, S. D. Setzler, T. M. Pollak: Phase-matched crystal growth of AgGaSe$_2$ and AgGa$_{1-x}$In$_x$Se$_2$. J. Cryst. Growth 211(1 – 4), 257 – 264(2000).

[2] G. C. Bhar, S. Das, U. Chatterjee, P. K. Datta, Y. N. Andreev: Noncritical second harmonic generation of CO$_2$ laser radiation in mixed chalcopyrite crystal. Appl. Phys. Lett. 63(10), 1316 – 1318(1993).

[3] V. V. Badikov, V. I. Chizhikov, V. B. Laptev, V. L. Panyutin, G. S. Shevyrdyaeva, S. I. Scherbakov: AgGa$_{1-x}$In$_x$Se$_2$ nonlinear crystals for noncritical phase matching processes. Proc. SPIE 4972, 139 – 144(2003).

[4] E. Takaoka, K. Kato: 90° phase-matched third-harmonic generation of CO$_2$ laser frequencies in AgGa$_{1-x}$In$_x$Se$_2$. Opt. Lett. 24(13), 902 – 904 (1999).

[5] Y. M. Andreev, I. S. Baturin, P. P. Geiko, A. I. Gusamov: Frequency doubling of CO$_2$-laser radiation in new nonlinear crystal AgGa$_x$In$_{1-x}$Se$_2$. Kvant. Elektron. 29(1), 66 – 70(1999) [In Russian, English trans.: Quantum Electron. 29(10), 904 – 908(1999)].

[6] R. C. Eckardt, H. Masuda, Y. X. Fan, R. L. Byer: Absolute and relative nonlinear optical coefficients of KDP, KD*P, BaB$_2$O$_4$, LiIO$_3$, MgO:LiNbO$_3$, and KTP measured by phase-matched second-harmonic generation. IEEE J. Quant. Electr. 26(5), 922 – 933(1990).

[7] J. E. Midwinter, J. Warner: The effects of phase matching method and of uniaxial crystal symmetry on the polar distribution of second-order non-linear optical polarization. Brit. J. Appl. Phys. 16(11), 1135 – 1142(1965).

6.19 Tl₄HgI₆，碘汞铊（THI）

正单轴晶：$n_e > n_o$

分子量：1 779.127

点群：$4mm$

熔点：669 K[1]

以"0"透过计的透明范围：1.0~60 μm[1]

在透明范围内线性吸收系数 α 约为 0.5 cm^{-1}[1]

折射率的实验值

$\lambda/\mu m$	n_o	n_e	$\lambda/\mu m$	n_o	n_e
1.2	2.436 0	2.508 7	6.0	2.387 2	2.455 4
1.3	2.428 4	2.500 3	8.0	2.385 0	2.453 2
1.4	2.422 9	2.495 3	9.0	2.383 4	2.452 6
1.5	2.418 3	2.489 7	10.0	2.381 5	2.451 2
1.6	2.415 3	2.486 4	15.0	2.369 6	2.438 5
1.7	2.411 6	2.482 3	20.0	2.353 4	2.422 5
1.8	2.407 5	2.478 0	25.0	2.331 2	2.400 6
2.0	2.403 5	2.473 6	30.0	2.301 5	2.371 2
3.0	2.395 0	2.463 0	35.0	2.262 3	2.331 9
4.0	2.391 2	2.459 2	40.0	2.210 1	2.279 2
5.0	2.389 5	2.457 4			

色散方程（λ 以 μm 为单位，1.2 μm < λ < 40 μm）[1]：

$$n_o^2 = 8.500\,975 + \frac{0.298\,967\,5}{\lambda^2 - 0.137\,902\,3} + \frac{19\,684.543}{\lambda^2 - 7\,043}$$

$$n_e^2 = 8.642\,436 + \frac{0.340\,191\,2}{\lambda^2 - 0.126\,601\,1} + \frac{17\,056.32}{\lambda^2 - 6\,547}$$

有效二阶非线性系数的表达式（小走离角近似，Kleinman 对称条件成立 $d_{15} = d_{24} = d_{31} = d_{32}$）[2]：

$$d_{eoo} = d_{oeo} = d_{31}\sin\theta$$

关于这一晶体

THI 是在红外区具有上至 60 μm 独特透过性的新非线性晶体[1]。

参考文献

[1] K. I. Avdienko, D. V. Badikov, V. V. Badikov, V. I. Chizhikov, V. L. Panyutin, G. S. Shevyrdyaeva, S. I. Scherbakov, E. S. Scherbakova: Optical properties of thallium mercury iodide. Opt. Mater. 23(3-4), 569-573(2003).

[2] J. E. Midwinter, J. Warner: The effects of phase matching method and of uniaxial crystal symmetry on the polar distribution of second-order non-linear optical polarization. Brit. J. Appl. Phys. 16(11), 1135-1142(1965).

第 7 章
自倍频晶体

这一章涉及自倍频晶体，这是由非线性光学晶体掺入激活的三价离子(通常是 Nd^{3+} 或 Yb^{3+})同时具有激光和频率转换性能。结果是它们可以产生红外辐射并同时对这种基频光进行倍频。在这些晶体中，有钕离子掺杂的四硼酸铝钆(NYAB)、镱离子掺杂的四硼酸铝钇(Yb:YAB)、钕和氧化镁钕掺杂的硼酸氧钙钇和其他晶体。

7.1 Nd:MgO:LiNbO$_3$，掺钕掺氧化镁铌酸锂(NdMgLN)

负单轴晶：$n_o > n_e$

点群：$3m$

在 MgO:LN 中 Nd^{3+} 的主吸收带位于 0.52~0.54 μm，0.58~0.61 μm，0.74~0.77 μm 和 0.81~0.82 μm 处[1,2]。

线性吸收系数 α

$\lambda/\mu m$	α/cm^{-1}	参 考 文 献	备 注
0.464	0.9	[3]	$E\perp a$，0.5 wt% Nd_2O_3 和 0.8 wt% MgO
0.752 5	1.2	[4]	$E\parallel c$；0.5 wt% Nd_2O_3
	2.0	[4]	$E\perp c$，0.5 wt% Nd_2O_3
0.809	1.27	[5]	$E\perp c$，0.2 wt% Nd_2O_3 和 5 mol% MgO
0.81	1.39	[3]	$E\perp a$，0.5 wt% Nd_2O_3 和 0.8 wt% MgO
0.813	1.76	[5]	$E\parallel c$，0.2 wt% Nd_2O_3 和 5 mol% MgO
	2.23	[6]	$E\parallel c$，0.2 at.% Nd_2O_3 和 3.3 mol% MgO
1.084	0.42	[3]	$E\perp a$，0.5 wt% Nd_2O_3 和 0.8 wt% MgO

0.34 at.% Nd 和 2.56 mol% MgO 的 NdMgLN 折射率的实验值[7]：

$\lambda/\mu m$	n_o	n_e	$\lambda/\mu m$	n_o	n_e
0.441 6	2.385 4	2.279 8	0.632 8	2.284 2	2.195 8
0.488	2.345 7	2.247 9	0.676 4	2.273 5	2.186 8
0.546 07	2.314	2.221 3	1.064	2.23	2.149 5
0.577	2.301 2	2.210 3			

0.34 mol% Nd 和 2.56 mol% MgO 的 NdMgLN 的色散关系（λ 以 μm 为单位）[7]：

$$n_o^2 = 4.900\ 1 + \frac{0.115\ 737}{\lambda^2 - 0.048\ 182} - 0.030\ 052\lambda^2$$

$$n_e^2 = 4.558\ 1 + \frac{0.097\ 078}{\lambda^2 - 0.044\ 267} - 0.023\ 873\lambda^2$$

在文献[6]中给出了 0.2 at.% Nd 和 3.3 mol% MgO 掺杂，$T = 300$ K 时的色散关系，文献[8]中给出了 0.6 at.% Nd 和 5 mol% MgO 掺杂，$T = 294$ K 时的其他色散关系。

一般情况下有效二阶非线性系数的表达式（Kleinman 对称条件成立，$d_{15} = d_{24} = d_{31} = d_{32}$）[9]：

$$d_{ooe} = d_{31}\sin(\theta+\rho) - d_{22}\cos(\theta+\rho)\sin 3\phi$$
$$d_{eoe} = d_{oee} = d_{22}\cos^2(\theta+\rho)\cos 3\phi$$

有效二阶非线性系数简化表达式（小走离角近似，Kleinman 对称条件成立，$d_{15} = d_{24} = d_{31} = d_{32}$）[10]：

$$d_{ooe} = d_{31}\sin\theta - d_{22}\cos\theta\sin 3\phi$$
$$d_{eoe} = d_{oee} = d_{22}\cos^2\theta\cos 3\phi$$

5 mol% MgO:$LiNbO_3$ 的二阶非线性系数绝对值[11]：

$|d_{31}(0.852\ \mu m)| = 4.9$ pm/V

$|d_{33}(0.852\ \mu m)| = 28.4$ pm/V

$|d_{31}(1.064\ \mu m)| = 4.4$ pm/V

$|d_{33}(1.064\ \mu m)| = 25.0$ pm/V

$|d_{31}(1.313\ \mu m)| = 3.4$ pm/V

$|d_{33}(1.313\ \mu m)| = 20.3$ pm/V

室温下相位匹配角的绝对值[1]

相互作用的波长/μm	θ_{exp}/(°)	参考文献	备 注
SHG, o+o⇒e			
1.093⇒0.546 5	70.8	[6]	0.2 at. % Nd 和 3.3 mol% MgO
		[8]	0.6 at. % Nd 和 5 mol% MgO

NCPM 温度的实验值

相互作用的波长/μm	T/℃	备 注
SHG, o+o⇒e		
1.093⇒0.546 5	152	1.0 at. % Nd 和 5 mol% MgO

$^4F_{3/2}$ 能级的荧光寿命

λ/μm	τ/μs	参考文献	备 注
1.09	80~85	[12]	1 wt. % Nd_2O_3
	100±5	[4]	任何偏振, 0.5 wt% Nd_2O_3
	102	[1]	1.0 at. % Nd 和 5 mol% MgO
	120	[1]	0.5 at. % Nd 和 5 mol% MgO

激光跃迁波长和相应的发射截面(以 10^{-20} cm² 为单位)[1]

跃 迁	λ/μm	$\sigma(E \parallel c)$	$\sigma(E \perp c)$	备 注
$^4F_{3/2} \Rightarrow {}^4I_{11/2}$	1.085	18		1.0 at. % Nd 和 5 mol% MgO
	1.093		5.1	1.0 at. % Nd 和 5 mol% MgO

关于这一晶体

　　LN 是第一个实现自倍频的非线性光学晶体[13,14]。其后, 为了防止光折变损伤, 采用了 MgO 掺杂并且首次实现了 CW 自倍频[1]。现今, NdMgLN 在 QPM SHG[2,15] 和 QPM SFG[3,16] 中找到了它的应用。同时, 还独立发展了 YbMgLN(Yb:MgO:LiNbO₃)[17]。最近, 在体块材料中实现了自倍频(58 mW 的

CW 绿光输出)[18,19],也在周期极化晶体 YbPPMgLN 中实现了自倍频[20],还实现了自泵浦 OPO[21]。

参考文献

[1] T. Y. Fan, A. Cordova-Plaza, M. J. F. Digonnet, R. L. Byer, H. J. Shaw: Nd:MgO:LiNbO$_3$ spectroscopy and laser devices. J. Opt. Soc. Am. B 3(1), 140–147(1986).

[2] Y. Q. Lu, J. J. Zheng, Y. L. Lu, N. B. Ming: Spectral properties and quasi-phase-matched second-harmonic generation in a new active medium: optical superlattice Nd:MgO:LiNbO$_3$. Appl. Phys. B 67(1), 29–32(1998).

[3] G. D. Laptev, A. A. Novikov, V. V. Firsov: Quasi-phase-matched self-frequency summing in a periodically poled Nd:Mg:LiNbO$_3$. Proc. SPIE 4972, 42–49(2003).

[4] I. P. Kaminow, L. W. Stulz: Nd:LiNbO$_3$ laser. IEEE J. Quant. Electr. QE–11(6), 306-308(1975).

[5] A. Cordova-Plaza, T. Y. Fan, M. J. F. Digonnet, R. L. Byer, H. J. Shaw: Nd:MgO:LiNbO$_3$ continuous-wave laser pumped by a laser diode. Opt. Lett. 13(3), 209–211(1988).

[6] S. Ishibashi, H. Itoh, T. Kaino, I. Yokohama, K. Kubodera: New cavity configurations of Nd:MgO:LiNbO$_3$ self-frequency-doubled lasers. Opt. Commun. 125(1–3), 177–185(1996).

[7] G. K. Kitaeva, I. I. Naumova, A. A. Mikhailovsky, P. S. Losevsky, A. N. Penin: Visible and infrared dispersion of the refractive indices in periodically poled and single domain Nd:Mg:LiNbO$_3$. Appl. Phys. B 66(2), 201–205(1998).

[8] M. Gong, G. Xu, K. Han, G. Zhai: Nd:MgO:LiNbO$_3$ self-frequency-doubled laser pumped by a flashlamp at room temperature. Electron. Lett. 26(25), 2062–2063(1990).

[9] I. Shoji, H. Nakamura, K. Ohdaira, T. Kondo, R. Ito, T. Okamoto, K. Tatsuki, S. Kubota: Absolute measurement of second-order nonlinear-optical coefficients of β-BaB$_2$O$_4$ for visible to ultraviolet second-harmonic wavelengths. J. Opt. Soc. Am. B 16(4), 620–624(1999).

[10] J. E. Midwinter, J. Warner: The effects of phase matching method and of uniaxial crystal symmetry on the polar distribution of second-order non-linear optical polarization. Brit. J. Appl. Phys. 16(11), 1135–1142(1965).

[11] I. Shoji, T. Kondo, A. Kitamoto, M. Shirane, R. Ito: Absolute scale of second-order nonlinear-optical coefficients. J. Opt. Soc. Am. B 14(9), 2268–2294(1997).

[12] L. I. Ivleva, A. A. Kaminskii, Y. S. Kuzminov, V. N. Shpakov: Absorption, luminescence, and induced emission of LiNbO$_3$:Nd^{3+} crystals. Doklady AN SSSR 183(5), 1068–1071(1968)[In Russian, English trans.: Sov. Phys. - Doklady 13(12), 1185–

1187(1969)].

[13] L. F. Johnson, A. A. Ballman: Coherent emission from rare earth ions in electrooptic crystals. J. Appl. Phys. 40(1), 297 – 302(1969).

[14] V. G. Dmitriev, E. V. Raevskii, L. N. Rashkovich, N. M. Rubinina, O. O. Selichev, A. A. Fomichev: Simultaneous emission at the fundamental frequency and the second harmonic in an active nonlinear medium: neodymium-doped lithium metaniobate. Pisma Zh. Tech. Fiz. 5(21 – 22), 1400 – 1402(1979) [In Russian, English trans.: Sov. Tech. Phys. Lett. 5(11), 590 – 591(1979)].

[15] N. V. Kravtsov, G. D. Laptev, E. Y. Morozov, I. I. Naumova, V. V. Firsov: Quasi-phase-matched self-doubling of the frequency in an Nd:Mg:LiNbO$_3$ laser with a regular domain structure. Kvant. Elektron. 29(2), 95 – 96(1999) [In Russian, English trans.: Quantum Electron. 29(11), 933 – 934(1999)].

[16] N. V. Kravtsov, G. D. Laptev, I. I. Naumova, A. A. Novikov, V. V. Firsov, A. S. Chirkin: Intracavity quasi-phase matched frequency summing in a laser based on a periodically poled active nonlinear Nd:Mg:LiNbO$_3$ crystal. Kvant. Elektron. 32(10), 923 – 924(2002) [In Russian, English trans.: Quantum Electron. 32(10), 923 – 924(2002)].

[17] E. Montoya, A. Lorenzo, L. E. Bausa: Optical characterization of LiNbO$_3$:Yb^{3+} crystals. J. Phys.: Condens. Matter 11(1), 311 – 320(1999).

[18] E. Montoya, J. Capmany, L. E. Bausa, T. Kellner, A. Diening, G. Huber: Infrared and self-frequency doubled laser action in Yb^{3+}-doped LiNbO$_3$:MgO. Appl. Phys. Lett. 74(21), 3113 – 3115(1999).

[19] E. Montoya, J. A. Sanz-Garcia, J. Capmany, L. E. Bausa, A. Diening, T. Kellner, G. Huber: Continuous wave infrared laser action, self-frequency doubling, and tunability of Yb^{3+}:MgO:LiNbO$_3$. J. Appl. Phys. 87(9), 4056 – 4062(2000).

[20] J. Capmany, E. Montoya, V. Bermudez, D. Callejo, E. Dieguez: Self-frequency doubling in Yb^{3+} doped periodically poled LiNbO$_3$:MgO bulk crystal. Appl. Phys. Lett. 76(11), 1374 – 1376(2000).

[21] J. Capmany, D. Callejo, V. Bermudez, E. Dieguez, D. Artigas, L. Torner: Continuous-wave self-pumped optical parametric oscillator based on Yb^{3+}-doped bulk periodically poled LiNbO$_3$(MgO). Appl. Phys. Lett. 79(3), 293 – 295(2001).

7.2 Nd:YAl$_3$(BO$_3$)$_4$, 掺钕四硼酸铝钇(Nd$_x$Y$_{1-x}$Al$_3$(BO$_3$)$_4$ 或 NYAB)

负单轴晶：$n_o > n_e$

Nd 相对于 Y 的浓度以及相应的 Nd^{3+} 的体积浓度

[Nd]/at. %	[Nd^{3+}] ×10^{-20}/cm^{-3}	[Nd]/at. %	[Nd^{3+}] ×10^{-20}/cm^{-3}
4.0	2.21	10	5.53
4.6	2.54	20	11.06
5.5	3.04		

密度：3.70 g/cm^3（无 Nd 掺杂）[1]；3.72 g/cm^3（无 Nd 掺杂）[2]；3.75 g/cm^3（4 at. % Nd）[3]

点群：32

不同 Nd 离子原子浓度的 Nd$_x$Y$_{1-x}$Al$_3$(BO$_3$)$_4$ 的晶格常数

[Nd]/at. %	a/Å	c/Å	参 考 文 献
0	9.287	7.256	[1]
	9.295 ± 0.003	7.243 ± 0.002	[4]
3~4	9.293	7.245	[5]
4~8	9.293	7.245	[6]
5.6	9.293	7.245	[7]
9	9.295 ± 0.003	7.243 ± 0.001	[2]
24	9.303 ± 0.003	7.281 ± 0.002	[2]
39	9.307 ± 0.003	7.257 ± 0.001	[2]
56	9.314 ± 0.005	7.278 ± 0.001	[2]
63	9.316 ± 0.002	7.294 ± 0.001	[2]
	9.320	7.284	[8]
71	9.322 ± 0.006	7.299 ± 0.001	[2]
72	9.323 ± 0.003	7.294 ± 0.002	[2]
100	9.3416 ± 0.0006	7.3066 ± 0.0008	[4]

莫氏硬度：7.5（无 Nd 掺杂）[1]；7.5~8[9]；8（4 at. % Nd）[5]

熔点（非同成分熔化）：1 463~1 553 K[8]

热导率[10]

T/K	κ/(W·m^{-1}·K^{-1})
300	3~4

NYAB 的 UV 透过截止波长为 0.325 μm（以 e^{-1} 计）[11]

Nd^{3+} 在 YAB 中的主吸收带位于 0.36 μm，0.52~0.53 μm，0.59 μm，0.75 μm，0.80~0.81 μm 和 0.88 μm[11,12]

高透明范围：1.0~2.3 μm[5]

线性吸收系数 α

$\lambda/\mu m$	α/cm^{-1}	参 考 文 献	备 注
0.355	7.6	[9]	5 at. % Nd
0.530	0.94	[5]	4 at. % Nd
0.531	1.2	[13]	3.9 at. % Nd
	1.39	[3]	4 at. % Nd
	2.4	[14]	
	3	[15]	4 at. % Nd
	≈3	[16]	20 at. % Nd
0.532	3.5	[9]	5 at. % Nd
0.588	9.36	[17]	4.6 at. % Nd
0.659	<0.05	[9]	5 at. % Nd
0.748	4.5	[18]	5.5 at. % Nd
0.801	5.03	[5]	4 at. % Nd
0.804	6.8	[19]	4 at. % Nd
	7.8	[13]	(12±4) at. % Nd
0.808	8.3	[19]	4 at. % Nd
	8.4	[13]	(12±4) at. % Nd
1.061	0.04	[5]	4 at. % Nd
1.064	<0.05	[9]	5 at. % Nd
1.318	<0.001	[9]	5 at. % Nd

5 at. % Nd 时折射率的实验值[9]

$\lambda/\mu m$	n_o	n_e	$\lambda/\mu m$	n_o	n_e
0.355	1.821	1.738	0.633	1.780	1.705
0.436	1.797	1.718	0.659	1.778	1.704
0.532	1.786	1.710	1.064	1.765	1.694
0.546	1.784	1.708	1.152	1.763	1.693
0.578	1.782	1.706	1.318	1.762	1.693

10 at. % Nd 时折射率的实验值[14]

$\lambda/\mu m$	n_o	n_e	$\lambda/\mu m$	n_o	n_e
0.404 7	1.806 74	1.729 24	0.589 3	1.776 91	1.701 79
0.435 8	1.792 71	1.714 75	0.656 3	1.772 84	1.698 67
0.486 1	1.785 39	1.708 89	0.706 5	1.772 34	1.699 13
0.546 1	1.779 99	1.704 55			

10 at. % Nd 时的色散关系(λ 以 μm 为单位)[20]：

$$n_o^2 = 1 + \frac{2.08192923\lambda^2}{\lambda^2 - (0.1098684)^2}$$

$$n_e^2 = 1 + \frac{1.83465945\lambda^2}{\lambda^2 - (0.1067225)^2}$$

在文献[6]中给出了 4~8 at.% Nd,以及文献[21]中给出了 5.6 at.% Nd 时的其他色散关系。

5.6 at.% Nd 时色散关系的温度相关性(λ 以 μm 为单位,0.4 μm < λ < 0.7 μm,T 以 K 为单位,293 K < T < 473 K)[21]:

$$n_o^2 = 1 + \frac{172.4727}{(0.10985 + 7.7 \times 10^{-7}T - 2.38 \times 10^{-9}T^2)^{-2} - \lambda^{-2}}$$

$$n_e^2 = 1 + \frac{161.08069}{(0.10669 + 1.3 \times 10^{-6}T - 3.2 \times 10^{-9}T^2)^{-2} - \lambda^{-2}}$$

有效二阶非线性系数的表达式(Kleinman 对称条件成立 $d_{11} = -d_{12} = -d_{26}$)[22]:

$$d_{ooe} = d_{11}\cos\theta\cos 3\phi$$

$$d_{eoe} = d_{oee} = d_{11}\cos^2\theta\sin 3\phi$$

二阶非线性系数值:

$$d_{11}(0.71\ \mu m) = (1.68 \pm 0.34)\ pm/V^{[9]}$$

$$d_{11}(1.062\ \mu m) = 1.43\ pm/V^{[23]}$$

$$d_{11}(1.062\ \mu m) = 3.9 \times d_{36}(KDP) = 1.52\ pm/V^{[14,24]}$$

$$d_{11}(1.062\ \mu m) = 1.7\ pm/V^{[3]}$$

$$d_{11}(1.064\ \mu m) = (1.51 \pm 0.25)\ pm/V^{[9]}$$

$$d_{11}(1.318\ \mu m) = (1.42 \pm 0.17)\ pm/V^{[9]}$$

相位匹配角的实验值(T = 293 K)

相互作用的波长/μm	θ_{exp}/(°)	参 考 文 献	备 注
SHG, o + o ⇒ e			
1.06⇒0.53	30	[21]	5.6 at.% Nd
		[16]	20 at.% Nd
	30.7	[5]	4 at.% Nd
1.061⇒0.5305	30.7	[7]	5.6at.% Nd
1.062⇒0.531	30.7	[15]	4 at.% Nd
	32.9	[13]	(12 ± 4)at.% Nd
1.064⇒0.532	30	[9]	5 at.% Nd
	32.9	[3]	4 at.% Nd
		[6]	4~8 at.% Nd

续表

相互作用的波长/μm	$\theta_{exp}/(°)$	参考文献	备注
	34.5	[14]	10 at.% Nd
1.318⇒0.659	27	[9]	5 at.% Nd
1.338⇒0.669	27	[25]	5.5 at.% Nd
	27	[16]	20 at.% Nd
SHG, e+o⇒e			
1.06⇒0.53	45.6	[5]	4 at.% Nd
1.064⇒0.532	43	[9]	5 at.% Nd
	51	[6]	4~8 at.% Nd
	50.6	[14]	10 at.% Nd
1.318⇒0.659	36	[9]	5 at.% Nd
SFG, o+o⇒e			
1.064+0.532⇒0.355	41	[9]	5 at.% Nd
1.062+0.590⇒0.3793	39.5	[17]	4.6 at.% Nd
1.062+0.750⇒0.4396	≈36	[18]	5.5 at.% Nd
1.062+0.807⇒0.4586	35	[20]	5.5 at.% Nd
1.338+0.807⇒0.4799	30.8	[25]	5.5 at.% Nd
SFG, e+o⇒e			
1.064+0.532⇒0.355	62	[9]	5 at.% Nd

内角带宽和温度带宽的实验值

相互作用的波长/μm	$\Delta\theta^{int}/(°)$	ΔT/℃	$\Delta\nu_2$/cm^{-1}	参考文献	备注
SHG, o+o⇒e					
1.064⇒0.532	0.037	26		[26]	4 at.% Nd
	0.038			[9]	5 at.% Nd
1.318⇒0.659	0.057			[9]	5 at.% Nd
1.338⇒0.669	0.069	50		[27]	5.6 at.% Nd
SHG, e+o⇒e					
1.064⇒0.532	0.058			[9]	5 at.% Nd
1.318⇒0.659	0.087			[9]	5 at.% Nd
SFG, o+o⇒e					
1.064+0.532⇒0.355	0.019			[9]	5 at.% Nd
1.062+0.750⇒0.4396			24.9	[18]	5.5 at.% Nd
SFG, e+o⇒e					
1.064+0.532⇒0.355	0.029			[9]	5 at.% Nd

近室温下相位匹配角的温度变化[21]

相互作用的波长/μm	$\theta_{pm}/(°)$	$d\theta_{pm}/dT/[(°)\cdot K^{-1}]$	备 注
SHG, o + o ⇒ e			
1.06 ⇒ 0.53	30	1.52×10^{-3}	5.6 at.% Nd

$^4F_{3/2}$能级的荧光寿命

λ/μm	τ/μs	参 考 文 献	备 注
1.061	50	[6]	4~8 at.% Nd
	53	[28]	5.6 at.% Nd
	60	[3]	4 at.% Nd
		[5]	4 at.% Nd
		[29]	10 at.% Nd
		[13]	(12±4) at.% Nd
	65	[23]	
1.338	56	[19]	4 at.% Nd

激光跃迁波长和相应的发射截面值(以 10^{-20}cm^2 为单位)

跃迁	λ/μm	σ	$\sigma(E\parallel c)$	$\sigma(E\perp c)$	参考文献	备 注
$^4F_{3/2}\Rightarrow ^4I_{11/2}$	1.061		14	10	[28]	5.6 at.% Nd
		20			[13]	(12±4) at.% Nd
		20.1			[29]	10 at.% Nd
		45			[19]	4 at.% Nd
		100			[6]	4~8 at.% Nd
		100±20			[16]	20 at.% Nd
$^4F_{3/2}\Rightarrow ^4I_{13/2}$	1.338		2.56	2.46	[27]	5.6 at.% Nd
		18±3.6			[16]	20 at.% Nd

激光诱导的损伤阈值

λ/μm	τ_p/ns	$I_{thr}/(\text{GW}\cdot\text{cm}^{-2})$	参 考 文 献
0.8	50	>0.4	[23]
1.06	10	>0.4~0.6	[2]

关于这一晶体

这种自倍频晶体是1981年提出的[16],而且仍然十分流行[15,30],这种晶体有一些缺点,如缺乏高质量的晶体、浓度猝灭、低量子效率、不适当的热致效应和在绿光谱区明显的吸收等。最近，在 0.5 cm 长的 NYAB 晶体中以 807 nm 的 1.6 W 激光二极管泵浦产生 225 mW CW 绿光输出（$\lambda = 530.5$ nm）[15]。在同一块晶体中，以 2.2 W、807 nm 衍射受限的钛宝石激光辐射泵浦，产生了 450 mW CW 的绿光输出[15]。

参考文献

[1] A. A. Filimonov, N. I. Leonyuk, L. B. Meissner, T. I. Timchenko, I. S. Rez: Nonlinear optical properties of isomorphic family of crystals with yttrium-aluminium borate (YAB) structure. Kristall und Technik 9(1), 63–66(1974).

[2] N. I. Leonyuk, L. I. Leonyuk: Growth and characterization of $RM_3(BO_3)_4$ crystals. Progr. Cryst. Growth Character. Mater. 31(3–4), 179–278(1995).

[3] S. Amano, S. Yokoyama, H. Koyama, S. Amano, T. Mochizuki: Diode pumped NYAB green laser. Rev. Laser Eng. 17(12), 895–898(1989)[In Japanese].

[4] N. I. Leonyuk, E. V. Koporulina, Y. Y. Wang, X. B. Hu, A. V. Mokhov: Neodymium and chromium segregation at high-temperature crystallization of $(Nd,Y)Al_3(BO_3)_4$ and $(Nd,Y)Ca_4O(BO_3)_3$ doped with Cr^{3+}. J. Cryst. Growth 252(1–3), 174–179(2003).

[5] Y. X. Fan, R. Schlecht, M. W. Qiu, D. Luo, A. D. Jiang, Y. C. Huang: Spectroscopic and nonlinear-optical properties of a self-frequency-doubling NYAB crystal. In: *OSA Proceedings on Advanced Solid-State Lasers, Vol. 13*, ed. by L. L. Chase, A. A. Pinto (OSA, Washington DC, 1992), pp. 371–375.

[6] Z. D. Luo, J. T. Lin, A. D. Jiang, Y. C. Huang, M. W. Qui: Features and applications of a new self-frequency-doubling laser crystal—NYAB. Proc. SPIE 1104, 132–141(1989).

[7] D. Jaque, J. Capmany, J. Garcia Sole, Z. D. Luo, A. D. Jiang: Continuous-wave laser properties of the self-frequency-doubling $YAl_3(BO_3)_4$:Nd crystal. J. Opt. Soc. Am. B 15(6), 1656–1662(1998).

[8] E. V. Koporulina, N. I. Leonyuk, S. N. Barilo, L. A. Kurnevich, G. L. Bychkov, A. V. Mokhov, G. Bocelli, L. Righi: Flux growth, composition, structural and thermal characteristics of $(R_xY_{1-x})Al_3(BO_3)_4$ (R = Nd, Gd; x = 1, 0.6, 0.65, 0.7, and 0.75) crystals. J. Cryst. Growth 198–199, 460–465(1999).

[9] L. M. Dorozhkin, I. I. Kuratev, V. A. Zhitnyuk, A. V. Shestakov, V. D. Shigorin, G. P. Shipulo: Nonlinear optical properties of neodymium-yttrium-aluminum borate. Kvant. Elektron. 10(7), 1497–1498(1983)[In Russian, English trans.: Sov. J. Quantum Electron. 13(7), 978–980(1983)].

[10] T. Omatsu, Y. Kato, M. Shimosegawa, A. Hasegawa, I. Ogura: Thermal effects in laser diode pumped self-frequency-doubled $Nd_xY_{1-x}Al_3(BO_3)_4$ (NYAB) microchip

laser. Opt. Commun. 118(3 −4), 302 −308(1995).

[11] D. Jaque: Optimum conditions for ultraviolet-laser generation based on self-frequency sum mixing in Nd^{3+}-activated borate crystals. J. Opt. Soc. Am. B 19(6), 1326 −1334 (2002).

[12] I. Schütz. I. Freitag, R. Wallenstein: Miniature self-frequency-doubling CW Nd: YAB laser pumped by a diode-laser. Opt. Commun. 77(2 −3), 221 −225(1990).

[13] H. Hemmati: Diode-pumped self-frequency-doubled neodymium yttrium aluminum borate (NYAB)laser. IEEE J. Quant. Electr. 28(4), 1169 −1171(1992).

[14] B. -S. Lu, J. Wang, H. -F. Pan, M. -H. Jiang, E. -Q. Liu, X. -Y. Hou: Laser self-doubling in neodymium yttrium aluminum borate. J. Appl. Phys. 66(12), 6052 −6054(1989).

[15] J. Bartschke, R. Knappe, K. -J. Boller, R. Wallenstein: Investigation of efficient self-frequency- doubling Nd:YAB lasers. IEEE J. Quant. Electr. 33(12), 2295 −2300(1997).

[16] L. M. Dorozhkin, I. I. Kuratev, N. I. Leonyuk, T. I. Timchenko, A. V. Shestakov: Optical second-harmonic generation in a new nonlinear active medium: neodymium-yttrium-aluminum borate crystals. Pisma Zh. Tekh. Fiz. 7(21), 1297 −1300(1981) [In Russian, English trans. :Sov. Tech. Phys. Lett. 7(11), 555 −556(1981)].

[17] A. Brenier, G. Boulon: Self-frequency summing NYAB laser for tunable UV generation. J. Luminesc. 86(2), 125 −128(2000).

[18] A. Brenier, G. Boulon, D. Jaque, J. Garcia Sole: Self-frequency-summing NYAB laser for tunable blue generation. Opt. Mater. 13(3), 311 −317(1999).

[19] S. Amano, T. Mochizuki: Diode-pumped NYAB green laser. Nonl. Opt. 1(4), 297 −306(1991).

[20] D. Jaque, J. Capmany, F. Molero, J. Garcia Sole: Blue-light laser source by sum-frequency mixing in $Nd:YAl_3(BO_3)_4$. Appl. Phys. Lett. 73(25), 3659 −3661(1998).

[21] D. Jaque, J. Capmany, J. Rams, J. Garcia Sole: Effects of pump heating on laser and spectroscopic properties of the $Nd:[YAl_3(BO_3)_4]$ self-frequency-doubling laser. J. Appl. Phys. 87(3), 1042 −1048(2000).

[22] J. E. Midwinter, J. Warner: The effects of phase matching method and of uniaxial crystal symmetry on the polar distribution of second-order non-linear optical polarization. Brit. J. Appl. Phys. 16(11), 1135 −1142(1965).

[23] R. E. Stone, R. C. Spitzer, S. C. Wang: A Q-switched diode-pumped neodymium yttrium aluminum borate laser. IEEE Photon. Technol. Lett. 2(11), 769 −771(1990).

[24] I. Shoji, T. Kondo, A. Kitamoto, M. Shirane, R. Ito: Absolute scale of second-order nonlinear-optical coefficients. J. Opt. Soc. Am. B 14(9), 2268 −2294(1997).

[25] D. Jaque, J. Capmany, J. Garcia Sole: Red, green, and blue laser light from a single $Nd:YAl_3(BO_3)_4$ crystal based on laser oscillation at 1.3 μm. Appl. Phys. Lett. 75(3), 325 −327(1999).

[26] M.-Y. Hwang, J. T. Lin: Temperature dependence of second harmonic generation in NYAB crystals. Opt. Commun. 95(1-3), 103-108(1993).

[27] D. Jaque, J. Capmany, J. Garcia Sole: Continuous wave laser radiation at 669 nm from a self-frequency-doubled laser of YAl$_3$(BO$_3$)$_4$:Nd^{3+}. Appl. Phys. Lett. 74(13), 1788-1790(1999).

[28] D. Jaque, J. Capmany, Z. D. Luo, J. Garcia Sole: Optical bands and energy levels of Nd^{3+} ion in the YAl$_3$(BO$_3$)$_4$ nonlinear laser crystal. J. Phys.: Condens. Matter 9(44), 9715-9729(1997).

[29] H.-F. Pan, M.-G. Liu, J. Xue, B.-S. Lu: The spectra and sensitization of laser self-frequency-doubling Nd$_x$Y$_{1-x}$Al$_3$(BO$_3$)$_4$ crystal. J. Phys.: Condens. Matter 2(19), 4525-4530(1990).

[30] Y. F. Chen, S. C. Wang, C. F. Kao, T. M. Huang: Investigation of fiber-coupled laser-diode-pumped NYAB green laser performance. IEEE Photon. Technol. Lett. 8(10), 1313-1315(1996).

7.3 Nd:GdAl$_3$(BO$_3$)$_4$，掺钕四硼酸铝钆(Nd$_x$Gd$_{1-x}$Al$_3$(BO$_3$)$_4$ 或 NGAB)

负单轴晶：$n_o > n_e$

Nd 相对于 Gd 的浓度及相应的 Nd^{3+} 的体积浓度

[Nd]/at.%	[Nd^{3+}]×10^{-20}/cm^{-3}	参 考 文 献
3	1.63	[1], [2]
10	5.6	[3]

点群：32

晶格常数：

3 at.% Nd：

　　$a = 9.3416$ Å[1]

　　$c = 7.3066$ Å[1]*

10 at.% Nd：

　　$a = (9.305 \pm 0.008)$ Å[3]

　　$c = (7.258 \pm 0.001)$ Å[3]

---

* 译者注：原书为 "$a = 7.3066$ Å"，应为 "$c = 7.3066$ Å"。

△ 译者注：原书为 "NYAB"，应为 "NGAB"。

NGAB$^{\triangle}$ 的 UV 透过截止波长为 0.32 μm(以 e^{-1} 计)[3]

Nd^{3+} 在 GAB 中的主吸收带位于 0.36 μm、0.53 μm、0.588 μm、0.748 μm、0.807 μm 和 0.88 μm[4]

线性吸收系数 α

λ/μm	α/cm^{-1}	参 考 文 献	备 注
0.353	4.34	[3]	10 at.% Nd
0.432	2.23	[3]	10 at.% Nd
0.475	2.22	[3]	10 at.% Nd
0.537	3.19	[3]	10 at.% Nd
0.588	8.69	[3]	10 at.% Nd
0.689	2.05	[3]	10 at.% Nd
0.749	10.22	[3]	10 at.% Nd
0.808	9.30	[3]	10 at.% Nd
0.811	2.55	[5]	3.35 at.% Nd
0.881	2.61	[3]	10 at.% Nd

3 at.% Nd 时吸收截面(以 10^{-20} cm^2 为单位)[1]

λ/μm	σ($E \parallel c$)	λ/μm	σ($E \parallel c$)
0.44	0.31	0.531	0.45

5 at.% Nd 时折射率的实验值[4]

λ/μm	n_o	n_e	λ/μm	n_o	n_e
0.4368	1.7921	1.7144	0.6328	1.7733	1.6991
0.4861	1.7851	1.7089	0.6563	1.7723	1.6980
0.5321	1.7801	1.7050	0.7065	1.7694	1.6955
0.5461	1.7792	1.7036	1.0641	1.7603	1.6884
0.5893	1.7760	1.7012			

10 at.% Nd 时折射率的实验值[3]

λ/μm	n_o	n_e	λ/μm	n_o	n_e
0.4047	1.8142	1.7352	0.5893	1.7890	1.7122
0.4358	1.8026	1.7253	0.6563	1.7856	1.7099
0.4861	1.7959	1.7178	0.7363	1.7841	1.7087
0.5461	1.7920	1.7141			

在 λ = 0.5893 μm 时对 5 at.% Nd 折射率的温度关系[4]

T/K	n_o	n_e	T/K	n_o	n_e
290	1.776 0	1.701 1	338.5	1.776 4	1.701 6
293	1.776 1	1.701 2	355	1.776 6	1.701 8
303	1.776 1	1.701 3	376	1.776 8	1.702 0
321	1.776 2	1.701 4			

在 5 at.% Nd 时折射率的温度微商[4]

$\lambda/\mu m$	$dn_o/dT \times 10^5/K^{-1}$	$dn_e/dT \times 10^5/K^{-1}$
0.589 3	0.93	1.05

5 at.% Nd 时的色散关系(λ 以 μm 为单位,T = 293 K)[4,6]:

$$n_o^2 = 3.072\,89 + \frac{0.030\,79}{\lambda^2 + 0.032\,65}$$

$$n_e^2 = 2.829\,98 + \frac{0.024\,2}{\lambda^2 + 0.031\,27}$$

10 at.% Nd 时的色散关系(λ 以 μm 为单位)[3]:

$$n_o^2 = 3.208\,7 + \frac{0.003\,4}{\lambda^2 - 0.127\,1} - 0.065\,6\lambda^2$$

$$n_e^2 = 2.915\,0 + \frac{0.004\,8}{\lambda^2 - 0.114\,7} - 0.012\,4\lambda^2$$

有效二阶非线性系数的表达式(Kleinman 对称条件成立 $d_{11} = -d_{12} = -d_{26}$)[7]:

$$d_{ooe} = d_{11}\cos\theta\cos 3\phi$$

$$d_{eoe} = d_{oee} = d_{11}\cos^2\theta\sin 3\phi$$

相位匹配角的实验值(T = 293 K)

相互作用的波长/μm	$\theta_{exp}/(°)$	参考文献	备注
SHG, o + o ⇒ e			
1.062 ⇒ 0.531	30.1	[1], [8]	
	30.3	[9]	
1.064 ⇒ 0.532	30.6	[4]	5 at.% Nd
1.338 ⇒ 0.669	24.4	[2]	
SHG, e + o ⇒ e			
1.064 ⇒ 0.532	44.1	[4]	5 at.% Nd
SFG, o + o ⇒ e			
1.062 + 0.588 ⇒ 0.378 5	38.2	[10]	3 at.% Nd

续表

相互作用的波长/μm	$\theta_{exp}/(°)$	参 考 文 献	备 注
SFG, o + o ⇒ e			
1.062 + 0.748 2 ⇒ 0.438 9	35	[1], [8]	
1.062 + 0.811 ⇒ 0.459 8	34	[5]	3.35 at.% Nd
1.062 + 0.807 ⇒ 0.458 6	34.1	[6]	3.4 at.% Nd
1.338 + 0.748 2 ⇒ 0.479 9	31	[2]	
DFG, e − o ⇒ o			
0.588 − 1.062 ⇒ 1.317 4	27.3	[10]	3 at.% Nd

$^4F_{3/2}$ 能级的荧光寿命

λ/μm	τ/μs	参 考 文 献	备 注
1.062	48 ± 3	[4]	
	55.6	[1]	3 at.% Nd

激光跃迁波长和相应的发射截面值(以 10^{-20} cm^2 为单位)

跃 迁	λ/μm	σ	参 考 文 献	备 注
$^4F_{3/2} \Rightarrow {}^4I_{11/2}$	1.062	20.8	[9]	
		30	[1]	3 at.% Nd
$^4F_{3/2} \Rightarrow {}^4I_{13/2}$	1.338	5.04	[9]	
		5.5	[2]	3 at.% Nd

关于这一晶体

NGAB 是 Nd:YAB 的同构晶体,最近被用于差频发生[10]。泵浦光(接近 588 nm)和得到的 IR($^4F_{3/2} \Rightarrow {}^4I_{11/2}$ 跃迁,1.061 9 μm)辐射在一块 0.43 cm 长掺 3 at.% Nd 的 NGAB 晶体中混频,所得到的调谐范围从 1.305 μm 延伸到 1.365 μm。

参考文献

[1] A. Brenier, C. Tu, M. Qiu, A. Jiang, J. Li, B. Wu: Spectroscopic properties, self-frequency doubling, and self-sum frequency mixing in GdAl$_3$(BO$_3$)$_4$:Nd^{3+}. J. Opt. Soc. Am. B 18(8), 1104−1110(2001).

[2] A. Brenier, C. Tu, J. Li, Z. Zhu, B. Wu: Spectroscopy, laser operation and self-fre-

quency doubling in GdAl$_3$(BO$_3$)$_4$:Nd^{3+}. Opt. Commun. 200(1-6), 355-358(2001).

[3] H.-D. Jiang, J.-Y. Wang, X.-B. Hu, S.-T. Li, B. Teng, C.-Q. Zhang: Absorption spectrum and optical parameters of Nd-doped gadolinium aluminium tetraborate crystals. Jpn. J. Appl. Phys. 40(10), 5981-5984(2001).

[4] C. Tu, M. Qiu, Y. Huang, X. Chen, A. Jiang, Z. Luo: The study of a self-frequency-doubling laser crystal Nd^{3+}:GdAl$_3$(BO$_3$)$_4$. J. Cryst. Growth 208(1-4), 487-492 (2000).

[5] Y. Chen, M. Huang, Y. Huang, Z. Luo: Simultaneous green and blue laser radiation based on a nonlinear laser crystal Nd:GdAl$_3$(BO$_3$)$_4$ and a nonlinear optical crystal KTP. Opt. Commun. 218(4-6), 379-384(2003).

[6] M. Huang, Y. Chen, X. Chen, Y. Huang, Z. Luo: A CW blue laser emission by self-sum-frequency mixing in Nd^{3+}:GdAl$_3$(BO$_3$)$_4$. Opt. Commun. 208(1-3), 163-166 (2002).

[7] J. E. Midwinter, J. Warner: The effects of phase matching method and of uniaxial crystal symmetry on the polar distribution of second-order non-linear optical polarization. Brit. J. Appl. Phys. 16(11), 1135-1142(1965).

[8] G. Aka, A. Brenier: Self-frequency conversion in nonlinear optical crystals. Opt. Mater. 22(2), 89-94(2003).

[9] M. Huang, Y. Chen, X. Chen, Y. Huang, Z. Luo: Study on CW fundamental and self-frequency doubling laser of Nd^{3+}:GdAl$_3$(BO$_3$)$_4$ crystal. Opt. Commun. 204(1-6), 333-338(2002).

[10] A. Brenier, C. Tu, J. Li, Z. Zhu, B. Wu: Self-sum-and-difference-frequency mixing in GdAl$_3$(BO$_3$)$_4$:Nd^{3+} for generation of tunable ultraviolet and infrared radiation. Opt. Lett. 27(4), 240-242(2002).

7.4 Nd:GdCa$_4$O(BO$_3$)$_3$,掺钕硼酸氧钙钆(Nd$_x$Gd$_{1-x}$COB 或 Nd:GdCOB)

负双轴晶

Nd 相对于 Gd 的浓度以及相应的 Nd^{3+} 的体积浓度

[Nd]/at.%	[Nd^{3+}]×10^{-20}/cm^{-3}	[Nd]/at.%	[Nd^{3+}]×10^{-20}/cm^{-3}
4	1.8	7	3.1
5	2.2		

点群:m

晶格常数:

5 at.% Nd 掺杂的 Nd:GdCOB[1]：
  $a = (8.0998 \pm 0.0016)$ Å
  $b = (16.0312 \pm 0.0026)$ Å
  $c = (3.5625 \pm 0.0008)$ Å
  $\beta = 101.242° \pm 0.024°$

7 at.% Nd 掺杂的 Nd:GdCOB 介电轴和晶体学轴的变换：$Y \parallel b$，a 轴和 c 轴位于 XZ 平面，两轴之间夹角为 $\beta = 101.27°$，Z 轴和 a 轴间夹角为 $27.1°$，X 轴和 c 轴间夹角为 $16.1°$[2]。

莫氏硬度：6.5[3]

7 at.% Nd 掺杂的 Nd:GdCOB 晶体线性热膨胀系数[2]

$\alpha_t \times 10^6/\mathrm{K}^{-1}$, $\parallel a$	$\alpha_t \times 10^6/\mathrm{K}^{-1}$, $\parallel b$	$\alpha_t \times 10^6/\mathrm{K}^{-1}$, $\parallel c$
7	5	11.3

5.2 at.% Nd 掺杂的 Nd:GdCOB 晶体线性热膨胀系数的平均值[4]

T/K	$\alpha_t \times 10^6/\mathrm{K}^{-1}$, $\parallel X$	$\alpha_t \times 10^6/\mathrm{K}^{-1}$, $\parallel Y$	$\alpha_t \times 10^6/\mathrm{K}^{-1}$, $\parallel Z$
298~572.5	11.6	5.4	5.9

$p = 0.101\ 325$ MPa 时的比热容（5.2 at.% Nd）[4]

T/K	$c_p/(\mathrm{J \cdot kg^{-1} \cdot K^{-1}})$
330	665

高透明范围：0.9~2.6 μm[2]

线性吸收系数 α

$\lambda/\mu\mathrm{m}$	α/cm^{-1}	参考文献	备注
0.461	0.37~0.42	[5]	5 at.% Nd，取决于偏振方向
0.465	0.43~0.46	[5]	7 at.% Nd，取决于偏振方向
0.53	0.41	[3]	4 at.% Nd
	0.73	[6]	7 at.% Nd
0.81	1.78	[3]	4 at.% Nd
1.06	0.02	[3]	4 at.% Nd

吸收截面 σ（以 $10^{-20} \mathrm{cm}^2$ 为单位）

$\lambda/\mu m$	$\sigma(\boldsymbol{E}\parallel X)$	$\sigma(\boldsymbol{E}\parallel Y)$	$\sigma(\boldsymbol{E}\parallel Z)$	参 考 文 献	备 注
0.530	0.17	0.31	0.27	[3]	4 at.% Nd
	0.22	0.35	0.22	[7]	4 at.% Nd
	0.27	0.21	0.43	[2]	7 at.% Nd
0.545	<0.1	<0.1	<0.1	[7]	4 at.% Nd
0.811	1.86	1.57	2.23	[3],[8]	4 at.% Nd
	0.52	0.35	0.64	[2]	7 at.% Nd

注：文献[3]、[8]中 X 和 Z 轴的标记不对，文献[2]中给出的值是低估的。

9 at.% Nd 掺杂 Nd:GdCOB 晶体的 Sellmeier 方程（λ 以 μm 为单位）[9]：

$$n_X^2 = 2.85005 + \frac{0.00651}{\lambda^2 - 0.11688} - 0.00001\lambda^2$$

$$n_Y^2 = 2.93898 + \frac{0.00674}{\lambda^2 - 0.11711} - 0.00001\lambda^2$$

$$n_Z^2 = 2.96538 + \frac{0.00839}{\lambda^2 - 0.10739} - 0.00001\lambda^2$$

在 Nd 浓度低时，Nd:GdCOB 的折射率非常接近于 GdCOB[5]。

Nd:GdCOB 晶体主平面上有效二阶非线性系数的表达式（Kleinman 对称条件成立 $d_{12}=d_{26}$, $d_{13}=d_{35}$, $d_{15}=d_{31}$, $d_{24}=d_{32}$）[10,11]：

XY 平面，$\theta = 90°$

$$d_{ooe} = d_{13}\sin\phi$$
$$d_{eoe} = d_{oee} = d_{31}\sin^2\phi + d_{32}\cos^2\phi$$

YZ 平面，$\phi = 90°$

$$d_{eeo} = d_{13}\sin^2\theta + d_{12}\cos^2\theta$$
$$d_{oeo} = d_{eoo} = d_{31}\sin\theta$$

XZ 平面，$\phi = 0°$，$V_Z > \theta > 0°$

$$d_{ooe} = d_{12}\cos\theta - d_{32}\sin\theta$$

XZ 平面，$\phi = 0°$，$90° > \theta > V_Z$

$$d_{oeo} = d_{eoo} = d_{12}\cos\theta - d_{32}\sin\theta$$

XZ 平面，$\phi = 0°$，$180° - V_Z > \theta > 90°$；或 $\phi = 180°$，$90° > \theta > V_Z$

$$d_{oeo} = d_{eoo} = d_{12}\cos\theta + d_{32}\sin\theta$$

XZ 平面，$\phi = 0°$，$180° > \theta > 180° - V_Z$；或 $\phi = 180°$，$V_Z > \theta > 0°$

$$d_{ooe} = d_{12}\cos\theta + d_{32}\sin\theta$$

相位匹配角的实验值

相互作用的波长/μm	$\phi_{pm}/(°)$	参 考 文 献
XY 平面，$\theta = 90°$		
SHG，o + o ⇒ e		
1.091 ⇒ 0.545 5	44	[12]
1.064 2 ⇒ 0.532 1	46.02	[5]
1.061 ⇒ 0.530 5	46	[6]，[8]，[9]，[12]，[13]，[14]，[15]
0.936 ⇒ 0.468	58.7	[16]
SFG，o + o ⇒ e		
1.061 + 0.811 ⇒ 0.459 65	60	[12]

Nd:GdCOB 晶体中某些特殊相位匹配方向（SHG，Ⅰ类）上有效二阶非线性系数的实验值

相位匹配方向	$d_{eff}/(\text{pm} \cdot \text{V}^{-1})$	参 考 文 献
$\theta = 90°$，$\phi = 46°$（XY 平面）	0.7	[15]
$\theta = 66.3°$，$\phi = 134.4°$	1.5	[15]
	2.6	[1]

注：在 Nd:GdCOB 晶体情况下，d_{eff} 的性质包含了镜面和倒反对称性，这意味着选择两个独立的象限，例如（$0° < \theta < 90°$，$0° < \phi < 90°$）和（$0° < \theta < 90°$，$90° < \phi < 180°$）就能完全描述 d_{eff} 的空间分布。随后，在两个象限中任何（θ,ϕ）方向上的 d_{eff} 值与（$180° - \theta,180° - \phi$）方向的值相等，反之亦然。例如（$\theta = 66.3°$，$\phi = 134.4°$）和（$\theta = 113.7°$，$\phi = 45.6°$）两个方向具有相等的 d_{eff} 值。

Nd:GdCOB 晶体某些特殊相位匹配方向上 SHG（Ⅰ类，1.064 2 μm ⇒ 0.532 1 μm，$I = 0.03 \text{ GW/cm}^2$，$l = 0.8 \text{ cm}$）的转换效率[13]

相位匹配方向	SHG 转换效率/%
$\theta = 90°$，$\phi = 46°$（XY 平面）	3.9
$\theta = 66.3°$，$\phi = 134.4°$	19.5

见前表注。

$^4F_{3/2}$ 能级的荧光寿命

$\lambda/\mu m$	$\tau/\mu s$	备 注	$\lambda/\mu m$	$\tau/\mu s$	备 注
1.06	98	1 ~ 2 at. % Nd		82	10 at. % Nd
	90	7 at. % Nd		60	20 at. % Nd

激光跃迁波长和相应的发射截面值(以 10^{-20}cm^2 为单位)

跃迁	$\lambda/\mu m$	$\sigma(E \parallel X)$	$\sigma(E \parallel Y)$	$\sigma(E \parallel Z)$	参考文献	备注
$^4F_{3/2} \Rightarrow {}^4I_{9/2}$	0.936	0.54	0.44	0.16	[16]	7 at. % Nd
$^4F_{3/2} \Rightarrow {}^4I_{11/2}$	1.061	2.0	2.1	4.2	[3],[6],[8]	4 at. % Nd

注:文献[3]、[6]、[8]中给出的 X 轴和 Z 轴标记不正确。

关于这一晶体

新发展的 Nd:GdCOB 晶体能用于有效自倍频。在文献[17]、[18]中,在一块 0.4 cm 长的 5 at. % Nd 掺杂的晶体($\theta = 90°, \phi = 46°$)中,吸收由激光二极管发出的 1.3 W 的 810 nm 泵浦光,产生了 115 mW 的 CW 绿光输出($\lambda = 530.5/545$ nm)。545 nm 的线束是由于伴随的自和频过程(1 061 nm + 810 nm $\Rightarrow$ 545 nm)所产生的。在文献[19]中,一块 0.7 cm 长的晶体,以 $\theta = 66.3°, \phi = 134.4°$ 切割,含钕量为 8 at. %,用于 I 类自倍频。在钛宝石激光器泵浦($\lambda = 812$ nm,吸收泵浦功率 1.56 W)下所产生的 530.5 nm CW 绿光的最大功率达到 225 mW。在文献[18]中,通过 1 090 nm 的激光辐射和所剩余的 812 nm 的泵浦光(CW 钛宝石激光器)之间的自和频产生了 1.2 mW 的 CW 蓝光($\lambda = 465$ nm)。

参考文献

[1] A. Brenier, A. Majchrowski, E. Michalski, T. Lukasiewicz: Evaluation of GdCOB: Nd^{3+} for self-frequency doubling in the optimum phase matching direction. Opt. Commun. 217(1-6), 395 – 400(2003).

[2] S. Zhang, Z. Cheng, J. Han, G. Zhou, Z. Shao, C. Wang, Y. T. Chow, H. Chen: Growth and investigation of efficient self-frequency-doubling $Nd_xGd_{1-x}Ca_4O(BO_3)_3$ crystal. J. Cryst. Growth 206(3), 197 – 202(1999).

[3] F. Mougel, G. Aka, A. Kahn-Harari, H. Hubert, J. M. Benitez, D. Vivien: Infrared laser performance and self-frequency doubling of $Nd:Ca_4GdO(BO_3)_3$ (Nd:GdCOB). Opt. Mater. 8(3), 161 – 173(1997).

[4] C. Wang, H. Zhang, X. Meng, L. Zhu, Y. T. Chow, X. Liu, R. Cheng, Z. Yang, S. Zhang, L. Sun: Thermal, spectroscopic properties and laser performance at 1.06 and 1.33 μm of $Nd:Ca_4YO(BO_3)_3$ and $Nd:Ca_4GdO(BO_3)_3$ crystals. J. Cryst. Growth 220(1-2), 114 – 120(2000).

[5] F. Mougel, G. Aka, A. Kahn-Harari, D. Vivien: CW blue laser generation by self sum-frequency mixing in $Nd:Ca_4GdO(BO_3)_3$ (Nd:GdCOB) single crystal. Opt. Mater. 13(3), 293 – 297(1999).

[6] F. Mougel, F. Auge, G. Aka, A. Kahn-Harari, D. Vivien, F. Balembois, P. Georges, A. Brun: New green self-frequency-doubling diode-pumped Nd: $Ca_4GdO(BO_3)_3$ laser. Appl. Phys. B 67(5), 533−535(1998).

[7] C. Maunier, J. L. Doualan, G. Aka, J. Landais, E. Antic-Fidancev, R. Moncorge, D. Vivien: Excited state absorption of the self-frequency doubling laser material: Nd: GdCOB. Opt. Commun. 184(1−4), 209−214(2000).

[8] R. Auge, F. Mougel, G. Aka, A. Kahn-Harari, D. Vivien, F. Balembois, P. Georges, A. Brun: Self-frequency doubling of Nd: $Ca_4GdO(BO_3)_3$ (Nd: GdCOB) laser pumped by CW Ti: sapphire or laser diode. In: *Advanced Solid-State Lasers*, *OSA Trends in Optics and Photonics Series*, Vol. 19, ed. by W. R. Bosenberg, M. M. Fejer (OSA, Washington DC, 1998), pp. 53−55.

[9] J. Wang, Z. Shao, J. Wei, X. Hu, Y. Liu, B. Gong, G. Li, J. Lu, M. Guo, M. Jiang: Research on growth and self-frequency doubling of Nd: ReCOB (Re = Y or Gd) crystals. Progr. Cryst. Growth Character. Mater. 40(1−4), 17−31(2000).

[10] G. Aka, A. Kahn-Harari, F. Mougel, D. Vivien, F. Salin, P. Coquelin, P. Colin, D. Pelenc, J. P. Damelet: Linear and nonlinear-optical properties of a new gadolinium calcium oxoborate crystal, $Ca_4GdO(BO_3)_3$. J. Opt. Soc. Am. B 14(9), 2238−2247 (1997).

[11] Z. P. Wang, J. H. Liu, R. B. Song, H. D. Jiang, S. J. Zhang, K. Fu, C. Q. Wang, J. Y. Wang, Y. G. Liu, J. Q. Wei, H. C. Chen, Z. S. Shao: Anisotropy of nonlinear-optical property of RCOB (R = Gd, Y) crystal. Chin. Phys. Lett. 18(3), 385−387 (2001).

[12] G. Aka, A. Brenier: Self-frequency conversion in nonlinear optical crystals. Opt. Mater. 22 (2), 89−94(2003).

[13] J. Lu, G. Li, J. Liu, S. Zhang, H. Chen, M. Jiang, Z. Shao: Second harmonic generation and self-frequency doubling performance in Nd: $GdCa_4O(BO_3)_3$ crystal. Opt. Commun. 168(5−6), 405−408(1999).

[14] F. Auge, S. Auzanneau, G. Lukas-Leclin, F. Balembois, P. Georges, A. Brun, F. Mougel, G. Aka, A. Kahn-Harari, D. Vivien: Efficient self-frequency-doubling Nd: GdCOB crystal pumped by a high brightness laser diode. In: *Advanced Solid-State Lasers*, *OSA Trends in Optics and Photonics Series*, Vol. 26, ed. by M. M. Fejer, H. Injeyan, U. Keller (OSA, Washington DC, 1999), pp. 77−81.

[15] Z. Shao, J. Lu, Z. Wang, J. Wang, M. Jiang: Anisotropic properties of Nd: ReCOB (Re = Y, Gd): a low symmetry self-frequency doubling crystal. Progr. Cryst. Growth Character. Mater. 40(1−4), 63−73(2000).

[16] F. Auge, G. Lukas-Leclin, F. Balembois, P. Georges, A. Brun, F. Mougel, G. Aka, A. Kahn-Harari, D. Vivien: Blue laser emission by self-frequency-doubling of the $^4F_{3/2}$

[16] ⇒ $^4I_{9/2}$ transition (936 nm) in Nd:GdCOB. In: *Advanced Solid-State Lasers, OSA Trends in Optics and Photonics Series, Vol.34*, ed. by H. Injeyan, U. Keller, C. Marshall (OSA, Washington DC, 2000), pp. 335 – 341.

[17] G. Lucas-Leclin, F. Auge, S. C. Auzanneau, F. Balembois, P. Georges, A. Brun, F. Mougel, G. Aka, D. Vivien: Diode-pumped self-frequency-doubling Nd:GdCa$_4$O(BO$_3$)$_3$ lasers: toward green microchip lasers. J. Opt. Soc. Am. 17(9), 1526 – 1530 (2000).

[18] D. Vivien, F. Mougel, F. Auge, G. Aka, A. Kahn-Harari, F. Balembois, G. Lucas-Leclin, P. Georges, A. Brun, P. Aschehoug, J.-M. Benitez, N. Le Nain, M. Jacquet: Nd:GdCOB: overview of its infrared, green and blue laser performances. Opt. Mater. 16 (1 – 2), 213 – 220 (2001).

[19] C. Q. Wang, Y. T. Chow, W. A. Gambling, S. J. Zhang, Z. X. Cheng, Z. S. Shao, H. C. Chen: Efficient self-frequency doubling of Nd:GdCOB crystal by type-I phase matching out of its principal planes. Opt. Commun. 174(5 – 6), 471 – 474(2000).

7.5 Nd:YCa$_4$O(BO$_3$)$_3$, 掺钕硼酸氧钙钇(Nd$_x$Y$_{1-x}$COB 或 Nd:YCOB)

负双轴晶

Nd 相对于 Y 的浓度及相应的 Nd^{3+} 的体积浓度

[Nd]/at.%	[Nd^{3+}]×10^{-20}/cm^{-3}	[Nd]/at.%	[Nd^{3+}]×10^{-20}/cm^{-3}
4	1.8	7	3.1
5	2.2		

点群：m

4.4 at.% Nd 掺杂 Nd:YCOB 的晶格常数[1]：

$a = (8.076 \pm 0.007)$ Å

$b = (16.020 \pm 0.010)$ Å

$c = (3.527 \pm 0.002)$ Å

$\beta = 101.23°$

莫氏硬度：6 ~ 6.5[2]

线性热膨胀系数的平均值(7 at.% Nd)[3]

T/K	$\alpha_t \times 10^6$/K^{-1}, ∥X	$\alpha_t \times 10^6$/K^{-1}, ∥Y	$\alpha_t \times 10^6$/K^{-1}, ∥Z
298 ~ 572.5	10.9	4.2	5.9

$p = 0.101\ 325$ MPa 时的比热容 (7 at. % Nd)[3]

T/K	$c_p/(\text{J} \cdot \text{kg}^{-1} \cdot \text{K}^{-1})$
330	774

线性吸收系数 α

$\lambda/\mu\text{m}$	α/cm^{-1}	参考文献	备注
0.530	<0.43	[4], [5]	$\boldsymbol{E} \perp Z$, 5 at. % Nd
	0.46	[6]	$\boldsymbol{E} \perp Z$, 5 at. % Nd
0.468	0.03	[4]	$\boldsymbol{E} \perp Z$, 5 at. % Nd
	0.12	[4]	$\boldsymbol{E} \perp Y$, 5 at. % Nd
0.666	0.03	[4]	$\boldsymbol{E} \perp Z$, 5 at. % Nd
	0.12	[4]	$\boldsymbol{E} \perp Y$, 5 at. % Nd
0.794	1.00	[7]	$\boldsymbol{E} \parallel Z$, 2 at. % Nd
	2.36	[1]	4.4 at. % Nd
	3.0	[5]	$\boldsymbol{E} \parallel Z$, 5 at. % Nd
0.812	0.84	[7]	$\boldsymbol{E} \parallel Z$, 2 at. % Nd
	2.88	[1]	4.4 at. % Nd
	1.9	[5], [8]	$\boldsymbol{E} \parallel X$, 5 at. % Nd
	1.55	[5], [8]	$\boldsymbol{E} \parallel Y$, 5 at. % Nd
	2.6	[5], [8]	$\boldsymbol{E} \parallel Z$, 5 at. % Nd
1.061	3.8	[6]	$\boldsymbol{E} \parallel Z$, 5 at. % Nd, $\theta = 90°$, $\phi = 35°$
1.332	2.9	[6]	$\boldsymbol{E} \parallel Z$, 5 at. % Nd, $\theta = 90°$, $\phi = 28°$

吸收截面 σ (以 10^{-20}cm^2 为单位)[4]

$\lambda/\mu\text{m}$	$\sigma(\boldsymbol{E} \parallel Z)$	备注
0.794 5	2.65	5 at. % Nd
0.811 5	2.38	5 at. % Nd

折射率的实验值[2]

$\lambda/\mu\text{m}$	n_X	n_Y	n_Z
1.061	1.684 4	1.715 2	1.725 6

对于 Nd 浓度低时,Nd:YCOB 的折射率非常接近于 YCOB 的折射率。
Nd:YCOB 晶体主平面上有效二阶非线性系数的表达式 (Kleinman 对称条件成立 $d_{12} = d_{26}$, $d_{13} = d_{35}$, $d_{15} = d_{31}$, $d_{24} = d_{32}$)[9,10]:

XY 平面，$\theta = 90°$

$$d_{ooe} = d_{13}\sin\phi$$

$$d_{eoe} = d_{oee} = d_{31}\sin^2\phi + d_{32}\cos^2\phi$$

YZ 平面，$\phi = 90°$

$$d_{eeo} = d_{13}\sin^2\theta + d_{12}\cos^2\theta$$

$$d_{oeo} = d_{eoo} = d_{31}\sin\theta$$

XZ 平面，$\phi = 0°$，$V_Z > \theta > 0°$

$$d_{ooe} = d_{12}\cos\theta - d_{32}\sin\theta$$

XZ 平面，$\phi = 0°$，$90° > \theta > V_Z$

$$d_{oeo} = d_{eoo} = d_{12}\cos\theta - d_{32}\sin\theta$$

XZ 平面，$\phi = 0°$，$180° - V_Z > \theta > 90°$；或 $\phi = 180°$，$90° > \theta > V_Z$

$$d_{oeo} = d_{eoo} = d_{12}\cos\theta + d_{32}\sin\theta$$

XZ 平面，$\phi = 0°$，$180° > \theta > 180° - V_Z$；或 $\phi = 180°$，$V_Z > \theta > 0°$

$$d_{ooe} = d_{12}\cos\theta + d_{32}\sin\theta$$

相位匹配角的实验值

相互作用的波长/μm	$\phi_{pm}/(°)$	参 考 文 献
XY 平面，$\theta = 90°$		
SHG，o + o ⇒ e		
1.332⇒0.666	28	[4]
1.061⇒0.530 5	33	[1]
	33.63	[6]
	33.95	[5]
	35	[4]

Nd:YCOB 晶体某些特殊相位匹配方向上（SHG，I 类，1.064 2 μm ⇒ 0.532 1 μm）二次谐波能量的实验值[11]

相位匹配方向	ε/mJ	相位匹配方向	ε/mJ
$\theta = 90°$，$\phi = 33.6°$（XY 平面）	1.65	$\theta = 66.8°$，$\phi = 144.6°$	3.95
$\theta = 32°$，$\phi = 0°$（XZ 平面）	2.3		

注：Nd:YCOB 晶体 d_{eff} 的性质包含了镜面和倒反对称性，这意味着选择两个独立的象限，例如 ($0° < \theta < 90°$, $0° < \phi < 90°$) 和 ($0° < \theta < 90°$, $90° < \phi < 180°$) 就能完全描述 d_{eff} 的空间分布。随后，在两个象限中任何 (θ, ϕ) 方向上的 d_{eff} 值与 ($180° - \theta, 180° - \phi$) 方向的值相等，反之亦然。例如 ($\theta = 66.8°, \phi = 144.6°$) 和 ($\theta = 113.2°, \phi = 35.4°$) 具有相同的 d_{eff} 值。

$^4F_{3/2}$ 能级的荧光寿命

$\lambda/\mu m$	$\tau/\mu s$	参 考 文 献	备 注
1.06	102	[4], [6]	2 at.% Nd
	100	[4]	5 at.% Nd
	96	[12]	5 at.% Nd
		[4]	10 at.% Nd
	95	[6]	10 at.% Nd

激光跃迁波长以及最强发射线的偏振[4]

跃 迁	$\lambda/\mu m$	偏振	跃 迁	$\lambda/\mu m$	偏振
$^4F_{3/2} \Rightarrow {}^4I_{9/2}$	0.936	$E \parallel Y$	$^4F_{3/2} \Rightarrow {}^4I_{13/2}$	1.332	$E \parallel Z$
$^4F_{3/2} \Rightarrow {}^4I_{11/2}$	1.061	$E \parallel Z$			

关于这一晶体

在文献[5]、[6]中研究了在 812 nm CW 二极管泵浦下在 Nd:YCOB 中 $^4F_{3/2} \Rightarrow {}^4I_{11/12}$ 跃迁($\lambda = 1.061$ μm)的自倍频。在后一个实验中,在 I 类相位匹配,以 $\theta = 90°$,$\phi = 33.6°$ 切割的掺 5 at.% Nd 的 0.5 cm 长的晶体中产生了 245 mW CW 绿光输出。吸收的泵浦功率确定为 3.8 W。在文献[4]中报道了采用同一 AlGaAs 二极管泵浦实现了 $^4F_{3/2} \Rightarrow {}^4I_{13/2}$ 跃迁($\lambda = 1.332$ μm)的自倍频。在 I 类相位匹配($\theta = 90°$,$\phi = 28°$)5 at.% Nd 掺杂 0.5 cm 长的 Nd:YCOB 晶体中,吸收泵浦功率为 0.95 W 时,产生功率约为 16 mW 的 666 nm 波长的 CW 红光。

参考文献

[1] H. J. Zhang, X. L. Meng, L. Zhu, C. Q. Wang, R. P. Cheng, W. T. Yu, S. J. Zhang, L. K. Sun, Y. T. Chow, W. L. Zhang, H. Wang, K. S. Wong: Growth and laser properties of Nd:Ca$_4$YO(BO$_3$)$_3$ crystal. Opt. Commun. 160(4-6), 273-276(1999).

[2] Q. Ye, B. H. T. Chai: Crystal growth of YCa$_4$O(BO$_3$)$_3$ and its orientation. J. Cryst. Growth 197(1-2), 228-235(1999).

[3] C. Wang, H. Zhang, X. Meng, L. Zhu, Y. T. Chow, X. Liu, R. Cheng, Z. Yang, S. Zhang, L. Sun: Thermal, spectroscopic properties and laser performance at 1.06 and 1.33 μm of Nd:Ca$_4$YO(BO$_3$)$_3$ and Nd:Ca$_4$GdO(BO$_3$)$_3$ crystals. J. Cryst. Growth 220(1-2), 114-120(2000).

[4] Q. Ye, L. Shah, J. Eichenholz, D. Hammons, R. Peale, M. Richardson, A. Chin,

B. H. T. Chai: Investigation of diode-pumped, self-frequency doubled RGB lasers from Nd:YCOB crystals. Opt. Commun. 164(1-3), 33-37(1999).

[5] J. M. Eichenholz, D. A. Hammons, L. Shah, Q. Ye, R. E. Peale, M. Richardson, B. H. T. Chai: Diode-pumped self-frequency doubling in a Nd^{3+}:$YCa_4O(BO_3)_3$ laser. Appl. Phys. Lett. 74(14), 1954-1956(1999).

[6] D. A. Hammons, M. Richardson, B. H. T. Chai, A. K. Chin, R. Jollay: Scaling of longitudinally diode-pumped self-frequency-doubling Nd:YCOB lasers. IEEE J. Quant. Electr. 36(8), 991-999(2000).

[7] B. H. T. Chai, J. M. Eichenholz, Q. Ye, D. A. Hammons, W. K. Jang, L. Shah, G. M. Luntz, M. Richardson: Self-frequency doubled Nd:YCOB laser. In: *Advanced Solid-State Lasers*, *OSA Trends in Optics and Photonics Series*, *Vol. 19*, ed. by W. R. Bosenberg, M. M. Fejer(OSA, Washington DC, 1998), pp. 56-58.

[8] Q. Ye, L. Shah, J. M. Eichenholz, D. A. Hammons, R. E. Peale, M. Richardson, B. H. T. Chai, A. Chin: Diode-pumped, self-frequency doubled red Nd:YCOB laser. In: *Advanced Solid-State Lasers*, *OSA Trends in Optics and Photonics Series*, *Vol. 26*, ed. by M. M. Fejer, H. Injeyan, U. Keller (OSA, Washington DC, 1999), pp. 100-103.

[9] G. Aka, A. Kahn-Harari, F. Mougel, D. Vivien, F. Salin, P. Coquelin, P. Colin, D. Pelenc, J. P. Damelet: Linear and nonlinear-optical properties of a new gadolinium calcium oxoborate crystal, $Ca_4GdO(BO_3)_3$. J. Opt. Soc. Am. B 14(9), 2238-2247 (1997).

[10] Z. P. Wang, J. H. Liu, R. B. Song, H. D. Jiang, S. J. Zhang, K. Fu, C. Q. Wang, J. Y. Wang, Y. G. Liu, J. Q. Wei, H. C. Chen, Z. S. Shao: Anisotropy of nonlinear-optical property of RCOB(R=Gd,Y) crystal. Chin. Phys. Lett. 18(3), 385-387 (2001).

[11] J. Wang, Z. Shao, J. Wei, X. Hu, Y. Liu, B. Gong, G. Li, J. Lu, M. Guo, M. Jiang: Research on growth and self-frequency doubling of Nd:ReCOB (Re=Y or Gd) crystals. Progr. Cryst. Growth Character. Mater. 40(1-4), 17-31(2000).

[12] H. Zhang, X. Meng, L. Zhu, P. Wang, X. Liu, R. Cheng, J. Dawes, P. Dekker, S. Zhang, L. Sun: Growth and laser properties of Yb:$Ca_4YO(BO_3)_3$ crystal. J. Cryst. Growth 200(1-2), 335-338(1999).

7.6 Nd:$LaBGeO_5$,掺钕锗酸硼镧($Nd_xLa_{1-x}BGeO_5$ 或 NdLBGO)

正单轴晶:$n_e > n_o$

分子量(对 $LaBGeO_5$):302.213

点群：3

晶格常数：

$a = (7.020 \pm 0.005)$ Å [1]

$c = (6.879 \pm 0.004)$ Å [1]

熔点：1 473 K [2]

0.1 cm 长的 $LaBGeO_5$ 晶体以"0"透过计的透明范围：0.19 ~ 4.5 μm [1]

在 296 K 时未掺杂 $LaBGeO_5$ 晶体折射率的实验值

$\lambda / \mu m$	n_o	n_e	$\lambda / \mu m$	n_o	n_e
0.404 7	1.850 4	1.892 5	0.577 0	1.821 6	1.861 3
0.435 8	1.842 2	1.883 6	0.589 3	1.820 1	1.859 6
0.488 0	1.832 2	1.872 9	0.632 8	1.816 6	1.855 8
0.492 0	1.831 8	1.872 2	1.064 2	1.802 3	1.835 9
0.532 1	1.826 3	1.866 3	1.152 4	1.801 2	1.839 1
0.546 1	1.824 7	1.864 6			

在 296 K 时未掺杂 $LaBGeO_5$ 晶体的色散关系（λ 以 μm 为单位）[1]：

$$n_o^2 = 1 + \frac{2.220\ 9\lambda^2}{\lambda^2 - (0.117\ 3)^2}$$

$$n_e^2 = 1 + \frac{2.356\ 7\lambda^2}{\lambda^2 - (0.119\ 7)^2}$$

有效二阶非线性系数的表达式（Kleinman 对称条件成立，$d_{15} = d_{31}$，$d_{14} = d_{25} = 0$）[3]：

$$d_{eeo} = (d_{11} \sin 3\phi + d_{22} \cos 3\phi) \cos^2 \theta$$

$$d_{oeo} = d_{eoo} = (d_{11} \cos 3\phi - d_{22} \sin 3\phi) \cos \theta + d_{15} \sin \theta$$

利用文献[4]中 $d_{11}(SiO_2)$ 和 $d_{36}(KDP)$ 新的绝对值重新计算的二阶非线性系数值[1]：

$$d_{11}(1.064\ \mu m) = (0.46 \pm 0.07)\ pm/V$$

$$d_{22}(1.064\ \mu m) = (0.23 \pm 0.04)\ pm/V$$

$$d_{31}(1.064\ \mu m) = (0.41 \pm 0.06)\ pm/V$$

$$d_{33}(1.064\ \mu m) = (0.35 \pm 0.05)\ pm/V$$

相位匹配角的实验值（$T = 293$ K）

相互作用的波长/μm	$\theta_{exp}/(°)$	参 考 文 献	备 注
SHG, e + e ⇒ o			
0.849 ⇒ 0.424 5	90	[5]	1.4 at.% Nd

续表

相互作用的波长/μm	$\theta_{exp}/(°)$	参 考 文 献	备 注
SHG, e + e ⇒ o			
1.048 ⇒ 0.524	≈ 54	[5], [6]	1.4 at. % Nd
	54 ± 0.5	[1]	无掺杂
1.314 ⇒ 0.657	≈ 35	[7]	1.4 at. % Nd
1.341 ⇒ 0.670 5	≈ 40	[1]	
1.386 ⇒ 0.693	≈ 35	[7]	1.4 at. % Nd
SHG, e + o ⇒ o			
1.386 ⇒ 0.693	≈ 60	[7]	1.4 at. % Nd

内角带宽和温度带宽的实验值[5]

相互作用的波长/μm	$\Delta\theta^{int}/(°)$	$\Delta T/℃$	备 注
SHG, e + e ⇒ o			
1.048 ⇒ 0.524	0.084	10.1	1.4 at. % Nd

在 300 K 时 $^4F_{3/2}$ 能级的荧光寿命

$\lambda/\mu m$	$\tau/\mu s$	参 考 文 献	备 注
1.048	275 ± 15	[2]	0.1 at. % Nd
	280 ± 5	[1]	0.1 at. % Nd
	280	[8]	

激光跃迁波长和相应的发射截面值(以 10^{-20} cm² 为单位)

跃 迁	$\lambda/\mu m$	$\sigma(\boldsymbol{E} \parallel c)$	$\sigma(\boldsymbol{E} \perp c)$	参 考 文 献	备 注
$^4F_{3/2} \Rightarrow {}^4I_{11/2}$	1.048 2	26	10	[1]	1.4 at. % Nd
		26		[2]	2.0 at. % Nd
		24		[6]	1.4 at. % Nd
	1.071 1		21	[1]	1.4 at. % Nd
			21	[2]	2.0 at. % Nd
			18	[6]	1.4 at. % Nd
$^4F_{3/2} \Rightarrow {}^4I_{13/2}$	1.314 1	7		[1]	1.4 at. % Nd
		9		[7]	1.4 at. % Nd
	1.386 8		6.5	[1]	1.4 at. % Nd
			3	[7]	1.4 at. % Nd

激光诱导的损伤阈值

$\lambda/\mu m$	τ_p/ns	$I_{thr}/(GW \cdot cm^{-2})$	参考文献	备注
1.064 2	10	>0.2	[6]	10 Hz
		>0.5	[1]	

关于这一晶体

在一块 0.4 cm 长的自倍频 NdLBGO 晶体中,吸收的由钛宝石激光器辐射($\lambda = 800$ nm)的功率是 0.6 W,实现了约 0.1 mW 的 CW 绿光输出($\lambda = 524$ nm)[9,10]。在另一个实验中,由同一个西班牙研究组,实现了另一激光跃迁 $^4F_{3/2} \Rightarrow {}^4I_{13/2}$ 的自倍频[7]。在一块 0.2 cm 长的晶体,吸收功率为 1.6 W 时产生约 0.8 mW 的 CW 红光($\lambda = 657$ nm)。

参考文献

[1] A. A. Kaminskii, A. V. Butashin, I. A. Maslyanizin, B. V. Mill, V. S. Mironov, S. P. Rozov, S. E. Sarkisov, V. D. Shigorin: Pure and Nd^{3+}-, Pr^{3+}-ion doped trigonal acentric $LaBGeO_5$ single crystals. Phys. Stat. Solidi A 125(2), 671-696(1991).

[2] A. A. Kaminskii, B. V. Mill, A. V. Butashin: Stimulated emission from Nd^{3+} ions in acentric $LaBGeO_5$ crystals. Phys. Stat. Solidi A 118(1), K59-K64(1990).

[3] J. E. Midwinter, J. Warner: The effects of phase matching method and of uniaxial crystal symmetry on the polar distribution of second-order non-linear optical polarization. Brit. J. Appl. Phys. 16(11), 1135-1142(1965).

[4] I. Shoji, T. Kondo, A. Kitamoto, M. Shirane, R. Ito: Absolute scale of second-order nonlinear-optical coefficients. J. Opt. Soc. Am. B 14(9), 2268-2294(1997).

[5] J. Capmany, J. Garcia Sole: Second harmonic generation in $LaBGeO_5$: Nd^{3+}. Appl. Phys. Lett. 70(19), 2517-2519(1997).

[6] J. Capmany, L. E. Bausa, J. Garcia Sole, R. Moncorge, A. V. Butashin, B. V. Mill, A. A. Kaminskii: Fluorescence and 1.06-0.53 μm second harmonic generation in Nd^{3+} doped $LaBGeO_5$. J. Luminesc. 60-61, 78-80(1994).

[7] J. Capmany, D. Jaque, J. Garcia Sole: Continuous wave laser radiation at 1314 and 1386 nm and infrared to red self-frequency doubling in nonlinear $LaBGeO_5$: Nd^{3+} crystal. Appl. Phys. Lett. 75(18), 2722-2724(1999).

[8] D. Jaque, J. Capmany, Z. D. Luo, J. Garcia Sole: Optical bands and energy levels of Nd^{3+} ion in the $YAl_3(BO_3)_4$ nonlinear laser crystal. J. Phys. : Condens. Matter 9(44), 9715-9729(1997).

[9] J. Capmany, L. E. Bausa, D. Jaque, J. Garsia Sole, A. A. Kaminskii: CW end-pumped Nd^{3+} : $LaBGeO_5$ mini laser for self-frequency doubling. J. Luminesc. 72-74, 816-818

(1997).

[10] J. Capmany, D. Jaque, J. Garcia Sole, A. A. Kaminskii: Continuous wave laser radiation at 524 nm from a self-frequency-doubled laser of $LaBGeO_5:Nd^{3+}$. Appl. Phys. Lett. 72(5), 531−533(1998).

7.7 $Nd:Gd_2(MoO_4)_3$,掺钕钼酸钆($Nd_{2x}Gd_{2-2x}(MoO_4)_3$ 或 NdGMO)

正双轴晶:$2V_Z = 9.9°$,$\lambda = 0.532\ 1\ \mu m$

$Gd_2(MoO_4)_3$ 的分子量:794.313

Nd 相对于 Gd 的浓度及相应的 Nd^{3+} 的体积浓度

[Nd]/at. %	$[Nd^{3+}] \times 10^{-20}/cm^{-3}$	[Nd]/at. %	$[Nd^{3+}] \times 10^{-20}/cm^{-3}$
2.5	1.74	5.0	3.49

密度:4.6 g/cm^3(无 Nd 掺杂)[1,2];4.65 g/cm^3(无 Nd 掺杂)[3]

点群:$mm2$

$Gd_2(MoO_4)_3$ 在 $T = 293$ K 时的晶格常数[4]:

  $a = 10.392$ Å

  $b = 10.416$ Å

  $c = 10.696$ Å

介电极轴和晶体学轴的变换:$X, Y, Z \Rightarrow b, a, c$

居里温度:

对 $x = 0$:432 K[1]

对 $x = 0.03$:432 K[1]

熔点:

$x = 0$:1 438 K[4]

NdGMO:1 428 K[4]

$p = 0.101\ 325$ MPa 时的比热容(无 Nd 掺杂)[5]

T/K	$c_p/(J \cdot kg^{-1} \cdot K^{-1})$	T/K	$c_p/(J \cdot kg^{-1} \cdot K^{-1})$
373	429	473	461

以"0"透过计的透明范围(无掺杂):0.31~5.13 μm[4];0.32~5.5 μm[4];0.32~5.2 μm[3];0.3~6 μm[6,7]

线性吸收系数 α

$\lambda/\mu m$	α/cm^{-1}	参考文献	备 注
0.53	4.34	[8]	15 at.% Nd
0.807	12.35	[2]	3 at.% Nd

$Gd_2(MoO_4)_3$ 折射率的实验值[9]

$\lambda/\mu m$	n_X	n_Y	n_Z
0.4579	1.8758	1.8762	1.9342
0.4765	1.8694	1.8699	1.9270
0.4880	1.8659	1.8663	1.9229
0.4965	1.8634	1.8639	1.9201
0.5017	1.8621	1.8625	1.9185
0.5145	1.8588	1.8593	1.9148
0.5321	1.8545	1.8549	1.9102
0.6328	1.8385	1.8390	1.8915
1.0642	1.8142	1.8146	1.8637

$Gd_2(MoO_4)_3$ 的色散关系(λ 以 μm 为单位,$0.46\ \mu m<\lambda<1.06\ \mu m$)[9]：

$$n_X^2 = 1 + \frac{2.2450\lambda^2}{\lambda^2 - 0.022693}$$

$$n_Y^2 = 1 + \frac{2.24654\lambda^2}{\lambda^2 - 0.0226803}$$

$$n_Z^2 = 1 + \frac{2.41957\lambda^2}{\lambda^2 - 0.0245458}$$

在文献[4]中给出了相同的色散关系，但有一点错误。

NdGMO 晶体主平面上有效二阶非线性系数的表达式（Kleinman 对称条件成立，$d_{15} = d_{31}$ 和 $d_{24} = d_{32}$）[10]：

XY 平面

$$d_{eoe} = d_{oee} = d_{32}\sin^2\phi + d_{31}\cos^2\phi$$

YZ 平面

$$d_{oeo} = d_{eoo} = d_{32}\sin\theta$$

XZ 平面，$\theta < V_Z$

$$d_{ooe} = d_{31}\sin\theta$$

XZ 平面，$\theta > V_Z$

$$d_{oeo} = d_{eoo} = d_{31}\sin\theta$$

在文献[10]中给出了 NdGMO 晶体任意方向上三波相互作用的有效二阶非线性系数。

二阶非线性系数值(无掺杂晶体)[14]，利用$d_{11}(SiO_2)$的新绝对值[11]的重新计算：

$$d_{31}(1.06\ \mu m) = (-2.3 \pm 0.6)\ pm/V$$
$$d_{32}(1.06\ \mu m) = (2.3 \pm 0.6)\ pm/V$$
$$d_{33}(1.06\ \mu m) = (-0.035 \pm 0.009)\ pm/V$$

二阶非线性系数值($x=0.15$)[8]，利用$d_{11}(SiO_2)$的新绝对值[11]重新计算：

$$d_{31}(1.06\ \mu m) = -2.5\ pm/V$$
$$d_{32}(1.06\ \mu m) = 2.5\ pm/V$$

相位匹配角的实验值($T=293$ K，单轴近似)

相互作用的波长/μm	$\theta_{exp}/(°)$	参考文献	备注
SHG, e+e⇒o			
0.974⇒0.487	90	[12]	无掺杂
1.064⇒0.532	68.3	[13]	无掺杂
1.06⇒0.53	65	[4]	无掺杂

注：在文献[12]中可发现未掺杂 GMO 晶体和掺杂($x=0.025$)晶体的相位匹配角没有差别。

未掺杂 GMO* 晶体内角带宽、光谱带宽和温度带宽的实验值

相互作用的波长/μm	$\Delta\theta^{int}/(°)$	$\Delta T/°C$	$\Delta\nu_2/cm^{-1}$	参考文献
SHG, e+e⇒o				
0.974⇒0.487			42	[12]
1.053⇒0.526 5	0.08			[4]
1.064⇒0.532	0.07	5.6		[13]

不同取向 NdGMO 晶体($x=0.03$)在 300 K 时激光跃迁波长[1,2]

跃迁	$\lambda/\mu m$	备注
$^4F_{3/2} \Rightarrow {}^4I_{11/2}$	1.060 6	$\boldsymbol{E} \parallel c$
	1.070 1	$\boldsymbol{E} \perp c$

在 300 K 时 $^4F_{3/2}$ 能级的荧光寿命

$\lambda/\mu m$	$\tau/\mu s$	参考文献	备注
1.070 1	150	[2]	1 at. % Nd
	150 ± 10	[1]	≈1 at. % Nd

---

* 译者注：原书为"NdGMO"，应为"GMO"。

激光诱导的损伤阈值

$\lambda/\mu m$	τ_p/ns	$I_{thr}/(GW \cdot cm^{-2})$	参 考 文 献
1.064 2	6	>0.13	[13]
	0.12	>1 900(?)	[6]

关于这一晶体

  钼酸钆(GMO)的非线性光学性质以及 $Nd^{3+}:Gd_2(MoO_4)_3$ 中产生的红外激光在 20 世纪 70 年代就已被研究[1,8,14]。然而，在 1996—1997 年，Kamiskii 和其多位合作者称 NdGMO 是一种"用于自倍频的新的非线性光学材料"[4,6,7]，尽管在这一晶体中没有实现自倍频。

■ 参考文献

[1] K. S. Bagdasarov, G. A. Bogomolova, A. A. Kaminskii, A. M. Prokhorov, T. M. Prokhortseva: Laser and spectral properties of $Gd_2(MoO_4)_3:Nd^{3+}$ crystal. Doklady AN SSSR 197(3), 557-560(1971)[In Russian, English trans.: Sov. Phys. -Doklady 16(3), 216 – 218(1971)].

[2] A. A. Kamiskii: New room-temperature laser-diode pumped efficient quasi-CW and CW single-mode laser based on ferroelectric and ferroelastic $Gd_2(MoO_4)_3:Nd^{3+}$ crystal. Phys. Stat. Solidi 149(1), K39-K42(1995).

[3] *Handbook of Optical Materials*, ed. by M. J. Weber(CRC Press, Boca Raton, 2003), pp. 1 – 512.

[4] A. A. Kaminskii, A. V. Butashin, H. -J. Eichler, D. Grebe, R. Macdonald, K. Ueda, H. Nishioka, W. Odajima, M. Tateno, J. Song, M. Musha, S. N. Bagaev, A. A. Pavlyuk: Orthorhombic ferroelectric and ferroelastic $Gd_2(MoO_4)_3$ crystal—a new many-purposed nonlinear and optical material: efficient multiple stimulated Raman scattering and CW and tunable second harmonic generation. Opt. Mater. 7(3), 59 – 73(1997).

[5] A. Fouskova: The specific heat of $Gd_2(MoO_4)_3$. J. Phys. Soc. Japan 27(6), 1699(1969).

[6] A. A. Kaminskii, K. Ueda, S. N. Bagaev, A. A. Pavlyuk, J. Song, H. Nishioka, N. Uehara, M. Musha: Orthorhombic dadolinium molybdate—a new nonlinear crystal for frequency doubling of one-micron CW laser emission. Kvant. Elektron. 23(5), 389 – 390(1996)[In Russian, English trans.: Quantum Electron. 26(5), 379 – 380(1996)].

[7] A. A. Kaminskii, H. -J. Eichler, S. N. Bagaev, D. Grebe, R. Macdonald, A. V. Butashin, A. A. Pavlyuk, F. A. Kuznetsov: Orthorhombic $Gd_2(MoO_4)_3$ crystal as a new nonlinear laser material for efficient second-harmonic generation. Kvant. Elektron. 23(2), 99 – 100(1996)[In Russian, English trans.: Quantum Electron. 26(2), 95 – 96(1996)].

[8] R. Bonneville, F. Auzel: Linear and nonlinear susceptibilities of rare earth ferroic molyb-

dates. J. Appl. Phys. 67(10), 4597 – 4602(1977).

[9] S. Singh: Nonlinear optical materials. In: *Handbook of Lasers with Selected Data on Optical Technology*, ed. by R. J. Pressley (Chemical Rubber Co., Cleveland, 1971), pp. 489-525.

[10] V. G. Dmitriev, D. N. Nikogosyan: Effective nonlinearity coefficients for three-wave interactions in biaxial crystals of *mm*2 point group symmetry. Opt. Commun. 95(1 – 3), 173 – 182(1993).

[11] I. Shoji, T. Kondo, A. Kitamoto, M. Shirane, R. Ito: Absolute scale of second-order nonlinear-optical coefficients. J. Opt. Soc. Am. B 14(9), 2268 – 2294(1997).

[12] H. Nishioka, W. Odajima, T. Tateno, K. Ueda, A. A. Kaminskii, A. V. Butashin, S. N. Bagaev, A. A. Pavlyuk: Femtosecond continuously tunable second harmonic generation over entire-visible range in orthorhombic acentric $Gd_2(MoO_4)_3$ crystals. Appl. Phys. Lett. 70(11), 1366 – 1368(1997).

[13] S. I. Kim, J. Kim, S. C. Kim, S. I. Yun, T. Y. Kwon: Second harmonic generation in the $Gd_2(MoO_4)_3$ crystal grown by the Czochralski method. Mat. Lett. 25(5 – 6), 195 – 198(1995).

[14] R. C. Miller, W. A. Nordland, K. Nassau: Nonlinear optical properties of $Gd_2(MoO_4)_3$ and $Tb_2(MoO_4)_3$. Ferroelectrics 2(2), 97 – 99(1971).

7.8　Yb:YAl$_3$(BO$_3$)$_4$，掺镱四硼酸铝钇（Yb$_x$Y$_{1-x}$Al$_3$(BO$_3$)$_4$ 或 Yb:YAB）

负单轴晶：$n_o > n_e$

Yb 相对于 Y 的浓度及相应的 Yb^{3+} 的体积浓度

[Yb]/at. %	[Yb^{3+}] $\times 10^{-20}$/cm^{-3}	[Yb]/at. %	[Yb^{3+}] $\times 10^{-20}$/cm^{-3}
4.0	2.21	10	5.53
4.6	2.54	20	11.06
5.5	3.04		

密度：3.70 g/cm^3（无 Yb 掺杂）[1]；3.72 g/cm^3（无 Yb 掺杂）[2]；3.844 g/cm^3（8 at. % Yb 掺杂）[3]；4.574 g/cm^3（100 at. % Yb）[4]

点群：32

不同 Yb 离子原子浓度的 Yb$_x$Y$_{1-x}$Al$_3$(BO$_3$)$_4$ 的晶格常数

[Nd]/at. %	a/Å	c/Å	参考文献
0	9.287	7.256	[1]
	9.295 ± 0.003	7.243 ± 0.002	[5]
5.6	9.277	7.224	[6]
8	9.931(?)	7.240(?)	[3]
100	9.251 2	7.189 3	[4]

莫氏硬度：7.5(无 Nd 掺杂)[1]

熔点(非同成分熔化)：1 563 K[4]

线性热膨胀系数的平均值[7]

T/K	$\alpha_t \times 10^6$/K^{-1}, $\parallel c$	$\alpha_t \times 10^6$/K^{-1}, $\perp c$	备 注
298 ~ 573	8.1	1.4	1 at. % Yb
	8.5	1.2	10 at. % Yb
	9.7	2.0	25 at. % Yb

$p = 0.101\ 325$ MPa 时的比热容 c_p[7]

T/K	c_p/(J·kg^{-1}·K^{-1})	备 注	T/K	c_p/(J·kg^{-1}·K^{-1})	备 注
298	760	1 at. % Yb	473	1 150	1 at. % Yb
	700	10 at. % Yb		1 050	10 at. % Yb
	680	25 at. % Yb		1 280	25 at. % Yb
373	910	1 at. % Yb	560	1 220	1 at. % Yb
	870	10 at. % Yb		1 080	10 at. % Yb
	750	25 at. % Yb		1 390	25 at. % Yb

热导率[8]

T/K	κ/(W·m^{-1}·K^{-1})	备 注
300	4.7	5.6 at. % Yb

Yb∶YAB 的 UV 透过截止波长为 0.252 μm[3]。

Yb^{3+}在 YAB 中的主吸收带位于 0.938 μm、0.975 μm 和 0.981 μm[4]。

线性吸收系数 α

λ/μm	α/cm^{-1}	参考文献	备 注
0.937	2.3	[6]	5.6 at. % Yb, $E \perp c$
0.975	10.4	[6]	5.6 at. % Yb, $E \perp c$
	6	[6]	5.6 at. % Yb, $E \parallel c$

续表

$\lambda/\mu m$	α/cm^{-1}	参 考 文 献	备 注
0.976	17.05	[3]	8 at.% Yb
	15	[9]	10 at.% Yb, $\boldsymbol{E}\perp c$
	12	[9]	10 at.% Yb, $\boldsymbol{E}\parallel c$
0.98	118	[4]	100 at.% Yb
0.981	8	[6]	5.6 at.% Yb, $\boldsymbol{E}\perp c$
0.998	1.18	[6]	5.5 at.% Yb
1.040	0.12	[6]	5.5 at.% Yb
1.040	0.28	[9]	10 at.% Yb
1.061	<0.07	[9]	10 at.% Yb

8 at.% Yb 的折射率实验值[3]:

$\lambda/\mu m$	n_o	n_e	$\lambda/\mu m$	n_o	n_e
0.404 67	1.801 58	1.729 28	0.589 60	1.774 62	1.701 88
0.435 84	1.785 07	1.718 48	0.656 28	1.771 79	1.698 62
0.486 13	1.780 17	1.709 96	0.706 25	1.769 12	1.697 05
0.546 07	1.776 99	1.704 78			

8 at.% Yb 的色散关系(λ 以 μm 为单位)[3]:

$$n_o^2 = 3.176\,2 + \frac{0.001\,3}{\lambda^2 - 0.148\,0} - 0.097\,1\lambda^2$$

$$n_e^2 = 2.863\,2 + \frac{0.009\,0}{\lambda^2 - 0.093\,7} - 0.008\,3\lambda^2$$

文献[10]中给出了其他色散关系。

有效二阶非线性系数的表达式(Kleinman 对称条件成立, $d_{11} = -d_{12} = -d_{26}$)[11]:

$$d_{ooe} = d_{11}\cos\theta\cos 3\phi$$

$$d_{eoe} = d_{oee} = d_{11}\cos^2\theta\sin 3\phi$$

二阶非线性系数值:

$$d_{11}(1.04\,\mu m) = 1.42\,pm/V^{[12]}$$

相位匹配角的绝对值

相互作用的波长/μm	$\theta_{exp}/(°)$	参 考 文 献	备 注
SHG, o + o ⇒ e			
1.0 ⇒ 0.5	≈31	[9]	10 at.% Yb
1.04 ⇒ 0.52	32.8	[6]	

续表

相互作用的波长/μm	$\theta_{exp}/(°)$	参 考 文 献	备 注
	34.6	[10]	
1.064⇒0.532	31	[3]	8 at. % Yb
SHG, e+o⇒e			
1.04⇒0.52	52.4	[6]	

内角带宽和温度带宽的实验值[13]

相互作用的波长/μm	$\Delta\theta^{int}/(°)$	$\Delta T/℃$	备 注
SHG, o+o⇒e			
1.064⇒0.532	0.077	28	10 at. % Yb

$^2F_{5/2}$能级的荧光寿命

λ/μm	τ/μs	参 考 文 献	λ/μm	τ/μs	参 考 文 献
1.03	1 400	[10]		680	[6]
1.04	600	[12]			

激光跃迁波长和相应的发射截面值[6]

跃迁	λ/μm	σ/cm^2	备 注
$^2F_{5/2}\Rightarrow{}^2F_{7/2}$	1.04	0.8×10^{20}	5.6 at. % Nd

关于这一晶体

Yb:YAB是最成功的自倍频晶体。Yb^{3+}离子的直径(0.870 Å)非常接近于Y^{3+}离子的直径(0.893 Å),因此镱离子很容易进入YAB基质中去。这种晶体没有浓度猝灭,没有激发态吸收,没有倍频波长的吸收。此外,YbYAB晶体中提供高的量子效率、低的量子缺陷、弱的热效应以及潜在的宽增益带宽。Yb:YAB晶体宽的吸收带和高功率InGaAs二极管很好匹配。这些优点和高的二阶非线性系数相结合形成自倍频辐射的宽带调频。

最近,澳大利亚 – 中国研究组通过Ⅰ类Yb:YAB晶体(0.3 cm长,8～10 at. % Yb,θ=31°),以11 W的976 nm InGaAs二极管泵浦产生了1.1 W CW绿光[3,14]。这是迄今为止所有激光二极管泵浦SFD激光器报道中最高的绿光输出。同一研究组报道了50 mW量级水平517～540 nm自倍频激光的调谐[13]。

参考文献

[1] A. A. Filimonov, N. I. Leonyuk, L. B. Meissner, T. I. Timchenko, I. S. Rez: Nonlinear optical properties of isomorphic family of crystals with yttrium-aluminium borate (YAB) structure. Kristall und Technik 9(1), 63–66(1974).

[2] N. I. Leonyuk, L. I. Leonyuk: Growth and characterization of $RM_3(BO_3)_4$ crystals. Progr. Cryst. Growth Character. Mater. 31(3–4), 179–278(1995).

[3] H. Jiang, J. Li, J. Wang, X.-B. Hu, H. Liu, B. Teng, C.-Q. Zhang, P. Dekker, P. Wang: Growth of $Yb:YAl_3(BO_3)_4$ crystals and their optical and self-frequency-doubling properties. J. Cryst. Growth 233(1–2), 248–252(2001).

[4] Y. Xu, X. Gong, Y. Chen, M. Huang, Z. Luo, Y. Huang: Crystal growth and optical properties of $YbAl_3(BO_3)_4$: a promising stoichiometric laser crystal. J. Cryst. Growth 252 (1–3), 241–245(2003).

[5] N. I. Leonyuk, E. V. Koporulina, Y. Y. Wang, X. B. Hu, A. V. Mokhov: Neodymium and chromium segregation at high-temperature crystallization of $(Nd,Y)Al_3(BO_3)_4$ and $(Nd,Y)Ca_4O(BO_3)_3$ doped with Cr^{3+}. J. Cryst. Growth 252(1–3), 174–179(2003).

[6] P. Wang, J. M. Dawes, P. Dekker, D. S. Knowles, J. A. Piper, B. Lu: Growth and evaluation of ytterbium-doped yttrium aluminum borate as a potential self-doubling laser crystal. J. Opt. Soc. Am. B 16(1), 63–69(1999).

[7] J. Li, J. Wang, X. Cheng, X. Hu, P. A. Burns, J. M. Dawes: Thermal and laser properties of $Yb:YAl_3(BO_3)_4$ crystal. J. Cryst. Growth 250(3–4), 458–462(2003).

[8] J. L. Blows, P. Dekker, P. Wang, J. M. Dawes, T. Omatsu: Thermal lensing measurements and thermal conductivity of Yb:YAB. Appl. Phys. B 76(3), 289–292(2003).

[9] P. Wang, P. Dekker, J. M. Dawes, J. A. Piper, Y. Liu, J. Wang: Efficient continuous-wave self-frequency-doubling green diode-pumped $Yb:YAl_3(BO_3)_4$ lasers. Opt. Lett. 25(10), 731–733(2000).

[10] L. Tian, J. Wang, J. Wei, H. Pan, Y. Liu: Growth and optical properties of Yb:YAB crystal. J. Synth. Cryst. 27(3), 225–228(1998) [In Chinese].

[11] J. E. Midwinter, J. Warner: The effects of phase matching method and of uniaxial crystal symmetry on the polar distribution of second-order non-linear optical polarization. Brit. J. Appl. Phys. 16(11), 1135–1142(1965).

[12] P. Dekker, J. Blows, P. Wang, J. Dawes, J. Piper, T. Omatsu, Y. Liu, J. Wang: $Yb:YAl_3(BO_3)_4$: an efficient green self-frequency-doubled laser source. In: *Advanced Solid-State Lasers*, *OSA Trends in Optics and Photonics Series*, *Vol. 50*, ed. by C. Marshall(OSA, Washington DC, 2001), pp. 476–483.

[13] P. Dekker, P. A. Burns, J. M. Dawes, J. A. Piper, J. Li, X. Hu, J. Wang: Widely tunable yellow-green lasers based on the self-frequency-doubling material Yb:

YAB. J. Opt. Soc. Am. B 20(4), 706–712(2003).

[14] P. Dekker, J. M. Dawes, J. A. Piper, Y. Liu, J. Wang: 1.1W CW self-frequency-doubled diode-pumped Yb:YAl$_3$(BO$_3$)$_4$ laser. Opt. Commun. 195(56), 431–436(2001).

7.9 Yb:GdCa$_4$O(BO$_3$)$_3$,掺镱硼酸氧钙钆(Yb$_x$Gd$_{1-x}$COB 或 Yb:GdCOB)

负双轴晶

Yb 相对于 Gd 的浓度及相应的 Yb^{3+} 的体积浓度

[Yb]/at.%	[Yb^{3+}]×10^{-20}/cm^{-3}	[Yb]/at.%	[Yb^{3+}]×10^{-20}/cm^{-3}
4	1.8	7	3.1
5	2.2	15	6.6

点群:m

Yb:GdCOB 的介电轴和晶体学轴变换等同于 GdCOB 的变换[1,2]

莫氏硬度:6.5[3]

熔点:≈1 753 K[3]

线性吸收系数 α

$\lambda/\mu m$	α/cm^{-1}	参考文献	备注
0.902	2.7	[4]	15 at.% Yb, $E \parallel Z$
	3	[5]	15 at.% Yb, $E \parallel Z$
0.976	4	[6]	15 at.% Yb, $E \parallel Z$
	5.5	[5]	

吸收截面 σ(以 10^{-20} cm^2 为单位)

$\lambda/\mu m$	$\sigma(E \parallel X)$	$\sigma(E \parallel Y)$	$\sigma(E \parallel Z)$	参考文献	备注
0.901 5	0.31	0.19	0.41	[7]	7 at.% Yb
	0.38	0.16	0.37	[3]	7 at.% Yb
0.976			1.15	[8]	7 at.% Yb
			1.12	[4]	

Yb:GdCOB 的折射率非常接近于 GdCOB 的折射率[3]。

Yb:GdCOB 晶体主平面上有效二阶非线性系数的表达式(小走离角近似,Kleinman 对称条件成立,$d_{12} = d_{26}, d_{13} = d_{35}, d_{15} = d_{31}, d_{24} = d_{32}$)[9,10]:

XY 平面，$\theta = 90°$

$$d_{ooe} = d_{13}\sin\phi$$
$$d_{eoe} = d_{oee} = d_{31}\sin^2\phi + d_{32}\cos^2\phi$$

YZ 平面，$\phi = 90°$

$$d_{eeo} = d_{13}\sin^2\theta + d_{12}\cos^2\theta$$
$$d_{oeo} = d_{eoo} = d_{31}\sin\theta$$

XZ 平面，$\phi = 0°$，$V_Z > \theta > 0°$

$$d_{ooe} = d_{12}\cos\theta - d_{32}\sin\theta$$

XZ 平面，$\phi = 0°$，$90° > \theta > V_Z$

$$d_{oeo} = d_{eoo} = d_{12}\cos\theta - d_{32}\sin\theta$$

XZ 平面，$\phi = 0°$，$180° - V_Z > \theta > 90°$；或 $\phi = 180°$，$90° > \theta > V_Z$

$$d_{oeo} = d_{eoo} = d_{12}\cos\theta + d_{32}\sin\theta$$

XZ 平面，$\phi = 0°$，$180° > \theta > 180° - V_Z$；或 $\phi = 180°$，$V_Z > \theta > 0°$

$$d_{ooe} = d_{12}\cos\theta + d_{32}\sin\theta$$

相位匹配角的绝对值[3]

相互作用的波长/μm	$\phi_{pm}/(°)$
XY 平面，$\theta = 90°$	
SHG，$o + o \Rightarrow e$	
$1.043 \Rightarrow 0.5215$	≈ 43

Yb:GdCOB 晶体中某些特殊相位匹配方向上有效二阶非线性系数的实验值[11]

相位匹配角	$d_{eff}/(\text{pm}\cdot\text{V}^{-1})$
$\theta = 66.8°$，$\phi = 132.6°$	2.3

室温下 $^2F_{5/2}$ 能级的荧光寿命

$\lambda/\mu m$	$\tau/\mu s$	参考文献	备注
1.032	2300	[5]	15 at.% Yb
	2440	[6]	
	2500	[7]	7 at.% Yb
	2600	[3]	7 at.% Yb

激光跃迁波长和相应的发射截面值（以 10^{-20} cm^2 为单位）

跃迁	$\lambda/\mu m$	$\sigma(E \parallel Z)$	参考文献	备注
$^2F_{5/2} \Rightarrow {}^2F_{7/2}$	1.032	0.55	[3], [7]	7 at.% Yb
		0.36	[12]	

关于这一晶体

  Yb:GdCOB 晶体最近用于 1.04 μm 附近的有效 CW IR 发生[6,13]。利用一块 0.3 cm 长,15 at.% Yb 掺杂的晶体,由 976 nm 光纤耦合二极管泵浦,在吸收泵浦功率为 5.2 W 时,产生 1 043 nm 3.2 W 的输出功率。此外,所产生的红光是连续可调谐的,范围为 1 018 ~ 1 086 nm,在 30 nm 带宽上输出功率大于 1 W。在文献[13,14]中,宽的发射谱被用于发展二极管泵浦 Yb:GdCOB 飞秒激光器($\lambda = 1\ 045$ nm, $\tau = 90$ fs, $P_{av} = 40$ mW, $\Delta f = 100$ MHz)。尽管在早期的工作中[3]已报道过 Yb:GdCOB 的自倍频效应,但至今仍无定量的测定。

参考文献

[1] Z. Wang, J. Liu, R. Song, X. Xu, X. Sun, H. Jiang, K. Fu, J. Wang, Y. Liu, J. Wei, Z. Shao: The second-harmonic-generation property of $GdCa_4O(BO_3)_3$ crystal with various phase-matching directions. Opt. Commun. 187(4 - 6), 401 - 405(2001).

[2] Z. Shao, J. Lu, Z. Wang, J. Wang, M. Jiang: Anisotropic properties of Nd:ReCOB(Re = Y, Gd): a low symmetry self-frequency doubling crystal. Progr. Cryst. Growth Character. Mater. 40(1 - 4), 63 - 73(2000).

[3] F. Mougel, K. Dardenne, G. Aka, A. Kahn-Harari, D. Vivien: Ytterbium-doped $Ca_4GdO(BO_3)_3$: an efficient infrared laser and self-frequency doubling crystal. J. Opt. Soc. Am. B 16(1), 164 - 172(1999).

[4] F. Auge, F. Balembois, P. Georges, A. Brun, F. Mougel, G. Aka, A. Kahn-Harari, D. Vivien: High-efficiency CW diode-pumped lasing and tunability of Yb:GdCOB(Yb^{3+}:$Ca_4GdO(BO_3)_3$). In: *Advanced Solid-State Lasers, OSA Trends in Optics and Photonics Series, Vol. 26*, ed. by M. M. Fejer, H. Injeyan, U. Keller (OSA, Washington DC, 1999), pp. 298 - 302.

[5] F. Auge, F. Druon, F. Balembois, P. Georges, A. Brun, F. Mougel, G. Aka, D. Vivien: Theoretical and experimental investigations of a diode-pumped quasi-three-level laser: theYb^{3+}-doped $Ca_4GdO(BO_3)_3$ (Yb:GdCOB) laser. IEEE. J. Quant. Electr. 36(5), 598-606(2000).

[6] S. Chenais, F. Druon, F. Balembois, G. Lucas-Leclin, P. Georges, A. Brun, M. Zavelani-Rossi, F. Auge, J. P. Chambaret, G. Aka, D. Vivien: Multiwatt, tunable, diode-pumped CW Yb:GdCOB laser. Appl. Phys. B 72(4), 389 - 393(2001).

[7] D. Martrou, F. Mougel, K. Dardenne, G. Aka, A. Kahn-Harari, D. Vivien, B. Viana: Laser performance of an ytterbium doped new single crystal: $Yb^{3+}:Ca_4GdO(BO_3)_3$ (Yb: GdCOB) under end pumped titanium sapphire. In: *Advanced Solid-State Lasers, OSA Trends in Optics and Photonics Series, Vol. 19*, ed. by W. R. Bosenberg, M. M. Fejer (OSA, Washington DC, 1998), pp. 454–458.

[8] F. Auge, F. Balembois, P. Georges, A. Brun, F. Mougel, G. Aka, A. Kahn-Harari, D. Vivien: Efficient and tunable continuous-wave diode-pumped $Yb^{3+}:Ca_4GdO(BO_3)_3$ laser. Appl. Opt. 38(6), 976–979(1999).

[9] G. Aka, A. Kahn-Harari, F. Mougel, D. Vivien, F. Salin, P. Coquelin, P. Colin, D. Pelenc, J. P. Damelet: Linear and nonlinear-optical properties of a new gadolinium calcium oxoborate crystal, $Ca_4GdO(BO_3)_3$. J. Opt. Soc. Am. B 14(9), 2238–2247 (1997).

[10] Z. P. Wang, J. H. Liu, R. B. Song, H. D. Jiang, S. J. Zhang, K. Fu, C. Q. Wang, J. Y. Wang, Y. G. Liu, J. Q. Wei, H. C. Chen, Z. S. Shao: Anisotropy of nonlinear-optical property of RCOB (R = Gd, Y) crystal. Chin. Phys. Lett. 18(3), 385–387 (2001).

[11] S. Zhang, Z. Cheng, S. Zhang, J. Liu, J. Han, J. Wang, H. Chen: Growth and second-harmonic generation properties of Tm^{3+}-, Yb^{3+}-, Bi^{3+}-, and Li^+-doped $GdCa_4O(BO_3)_3$ crystals. Chin. Phys. Lett. 18(3), 388–389(2001).

[12] A. Aron, G. Aka, B. Viana, A. Kahn-Harari, D. Vivien, F. Druon, F. Balembois, P. Georges, A. Brun, N. Lenain, M. Jacquet: Spectroscopic properties and laser performances of Yb:YCOB and potential of the Yb:LaCOB material. Opt. Mater. 16(1–2), 181–188(2001).

[13] F. Druon, S. Chenais, F. Balembois, P. Georges, A. Brun, A. Courjaud, C. Hönninger, F. Salin, M. Zavelani-Rossi, F. Auge, J. P. Chambaret, A. Aron, F. Mougel, G. Aka, D. Vivien: High-power diode-pumped Yb:GdCOB laser: from continuous-wave to femtosecond regime. Opt. Mater. 19(1), 73–80(2002).

[14] F. Druon, F. Balembois, P. Georges, A. Brun, A. Courjaud, C. Hönninger, F. Salin, A. Aron, F. Mougel, G. Aka, D. Vivien: Generation of 90-fs pulses from a mode-locked diodepumped $Yb^{3+}:Ca_4GdO(BO_3)_3$ laser. Opt. Lett. 25(6), 423–425(2000).

7.10 $Yb:YCa_4O(BO_3)_3$,掺镱硼酸氧钙钇($Yb_xY_{1-x}COB$ 或 $Yb:YCOB$)

负双轴晶

Yb 相对于 Y 的浓度及相应的 Yb^{3+} 的体积浓度

[Yb]/at. %	[Yb³⁺] × 10⁻²⁰/cm⁻³	[Yb]/at. %	[Yb³⁺] × 10⁻²⁰/cm⁻³
4	1.8	10	4.5
5	2.3	20	9.0
7	3.2		

密度：3.39 g/cm^3, 10 at. % Yb[1]

点群：m

Yb:YCOB 的介电轴和晶体学轴变换等同于 YCOB 的变换[2]。

吸收截面 σ（以 10^{-20} cm^2 为单位）

$\lambda/\mu m$	$\sigma(\boldsymbol{E} \parallel X)$	$\sigma(\boldsymbol{E} \parallel Y)$	$\sigma(\boldsymbol{E} \parallel Z)$	参考文献	备注
0.900			0.4	[1]	10 at. % Yb
	0.42	0.30	0.53	[3]	18.3 at. % Yb
	0.31	0.13	0.43	[4], [5]	20 at. % Yb
0.976			1.2	[1]	10 at. % Yb
	0.77	0.87	0.81	[4], [5]	20 at. % Yb

Yb:YCOB 的折射率非常接近于 YCOB。

Yb:YCAB 晶体主平面上有效二阶非线性系数的表达式（最小走离角近似，Kleinman 对称条件成立，$d_{12} = d_{26}, d_{13} = d_{35}, d_{15} = d_{31}, d_{24} = d_{32}$）[6,7]：

XY 平面，$\theta = 90°$

$$d_{ooe} = d_{13} \sin \phi$$
$$d_{eoe} = d_{oee} = d_{31} \sin^2 \phi + d_{32} \cos^2 \phi$$

YZ 平面，$\phi = 90°$

$$d_{eeo} = d_{13} \sin^2 \theta + d_{12} \cos^2 \theta$$
$$d_{oeo} = d_{eoo} = d_{31} \sin \theta$$

XZ 平面，$\phi = 0°$，$V_Z > \theta > 0°$

$$d_{ooe} = d_{12} \cos \theta - d_{32} \sin \theta$$

XZ 平面，$\phi = 0°$，$90° > \theta > V_Z$

$$d_{oeo} = d_{eoo} = d_{12} \cos \theta - d_{32} \sin \theta$$

XZ 平面，$\phi = 0°$，$180° - V_Z > \theta > 90°$；或 $\phi = 180°$，$90° > \theta > V_Z$

$$d_{oeo} = d_{eoo} = d_{12} \cos \theta + d_{32} \sin \theta$$

XZ 平面，$\phi = 0°$，$180° > \theta > 180° - V_Z$；或 $\phi = 180°$，$V_Z > \theta > 0°$

$$d_{ooe} = d_{12} \cos \theta + d_{32} \sin \theta$$

Yb:YCOB 晶体主平面上相位匹配角的实验值

相互作用的波长/μm	$\phi_{pm}/(°)$	$\theta_{pm}/(°)$	参考文献
XY 平面，$\theta=90°$			
SHG, o+o⇒e			
1.090⇒0.545	≈36.2		[4]
1.064⇒0.532	34		[8]
XZ 平面，$\phi=0°$			
SHG, o+o⇒e			
1.070⇒0.535		≈31.7	[9]

室温下 $^2F_{5/2}$ 能级的荧光寿命

$\lambda/\mu m$	$\tau/\mu s$	参考文献	备注
1.032	2 100	[10]	2 at. % Yb
	2 500	[1]	1 at. % Yb
	2 700	[10]	5 at. % Yb
	2 800	[10]	10 at. % Yb
	2 850	[1]	10 at. % Yb
	3 000	[4], [10]	20 at. % Yb
		[10]	25 at. % Yb
		[10]	45 at. % Yb

激光跃迁波长及相应的发射截面值（以 $10^{-20}\ cm^2$ 为单位）

跃迁	$\lambda/\mu m$	$\sigma(E\parallel Z)$	参考文献	备注
$^2F_{5/2}\Rightarrow{}^2F_{7/2}$	1.018	0.30	[1]	10 at. % Yb
	1.032	0.36	[3]	18.3 at. % Yb
		0.39	[9]	
	1.050	0.18	[9]	
	1.082	0.12	[1]	10 at. % Yb
	1.084	0.10	[9]	
	1.085	0.76(?)	[3]	18.3 at. % Yb

激光诱导的体损伤阈值

$\lambda/\mu m$	τ_p/ns	$I_{thr}/(GW\cdot cm^{-2})$
1.064	10	>0.06

关于这一晶体

在文献[4]中报道了 1 085 nm 附近 Yb:YCOB 的 CW 红外激光运转。一块

X 切、1.3 cm 长及 20 at. % Yb 掺杂的晶体由调至 900 nm 的钛宝石激光器辐射所泵浦，在吸收的泵浦功率为 1.2 W 时，输出的辐射为 300 mW。文献[3]报道了由 $\lambda = 976$ nm 的 CW 二极管泵浦一块 Y 切、0.186 cm 长 20 at. % Yb 掺杂的晶体，吸收功率为 0.76 W 时，IR 激光输出为 446 mW。

在首次采用 Yb:YCOB 进行的自倍频实验中，在一块 20 at. % Yb 掺杂的晶体（$\theta = 90°, \phi = 36.2°$）中，吸收泵浦功率为 0.9 W 时，产生不到 1 mW 的 543 nm 光[4]。在其后的工作中[9]，用了一块 35 at. % Yb 掺杂的晶体（$\theta = 31.7°, \phi = 0°$），产生了相似量级的 CW 绿光输出。

参考文献

[1] V. A. Lebedev, I. V. Voroshilov, A. N. Gavrienko, B. V. Ignatiev: Kinetic and spectroscopic investigations of Yb:YCa$_4$O(BO$_3$)$_3$ (Yb:YCOB) single crystals. Opt. Mater. 14(2), 171-173(2000).

[2] F. Mougel, G. Aka, F. Salin, D. Pelenc, B. Ferrand, A. Kahn-Harari, D. Vivien: Accurate second harmonic generation phase matching angles prediction and evaluation of nonlinear coefficients of YCa$_4$O(BO$_3$)$_3$ (YCOB) crystal. In: *Advanced Solid State Lasers, OSA Trends in Optics and Photonics Series*, Vol. 26, ed. by M. M. Fejer, H. Injeyan, U. Keller(OSA, Washington DC, 1999), pp. 709 – 714.

[3] P. Wang, J. M. Dawes, P. Dekker, H. Zhang, X. Meng: Spectral characterization and diodepumped laser performance of Yb:YCOB. In: *Advanced Solid State Lasers, OSA Trends in Optics and Photonics Series*, Vol. 26, ed. by M. M. Fejer, H. Injeyan, U. Keller(OSA, Washington DC, 1999), pp. 631 – 634.

[4] D. A. Hammons, J. M. Eichenholz, Q. Ye, B. H. T. Chai, L. Shah, R. E. Peale, M. Richardson, H. Qiu: Laser action in Yb^{3+}:YCOB (Yb:YCa$_4$O(BO$_3$)$_3$). Opt. Commun. 156(4–6), 327-330(1998).

[5] D. A. Hammons, L. Shah, J. Eichenholz, Q. Ye, M. Richardson, B. H. T. Chai, A. Chin, J. Cary: 980 nm diode pumped laser operation and wavelength tunability performance in Yb^{3+}:YCOB. In: Advanced Solid State Lasers, OSA Trends in Optics and Photonics Series, Vol. 26, ed. by M. M. Fejer, H. Injeyan, U. Keller (OSA, Washington DC, 1999), pp. 286 – 290.

[6] G. Aka, A. Kahn-Harari, F. Mougel, D. Vivien, F. Salin, P. Coquelin, P. Colin, D. Pelenc, J. P. Damelet: Linear and nonlinear-optical properties of a new gadolinium calcium oxoborate crystal, Ca$_4$GdO(BO$_3$)$_3$. J. Opt. Soc. Am. B 14(9), 2238 – 2247(1997).

[7] Z. P. Wang, J. H. Liu, R. B. Song, H. D. Jiang, S. J. Zhang, K. Fu, C. Q. Wang, J. Y. Wang, Y. G. Liu, J. Q. Wei, H. C. Chen, Z. S. Shao: Anisotropy of nonlinear-op-

tical property of RCOB(R = Gd, Y) crystal. Chin. Phys. Lett. 18(3), 385 – 387(2001).

[8] W. K. Jang, Q. Ye, J. Eichenholz, M. C. Richardson, B. H. T. Chai: Second harmonic generation in Yb doped $YCa_4O(BO_3)_3$. Opt. Commun. 155(4 – 6), 332 – 334(1998).

[9] A. Aron, G. Aka, B. Viana, A. Kahn-Harari, D. Vivien, F. Druon, F. Balembois, P. Georges, A. Brun, N. Lenain, M. Jacquet: Spectroscopic properties and laser performances of Yb:YCOB and potential of theYb:LaCOB material. Opt. Mater. 16(1 – 2), 181 – 188(2001).

[10] B. H. T. Chai, D. A. Hammons, J. M. Eichenholz, Q. Ye, W. K. Yang, L. Shah, G. M. Luntz, M. Richardson, H. Qiu: Lasing, second harmonic conversion and self-frequency doubling of Yb:YCOB (Yb:$YCa_4B_3O_{10}$) . In: *Advanced Solid-State Lasers, OSA Trends in Optics and Photonics Series*, Vol. 19, ed. by W. R. Bosenberg, M. M. Fejer(OSA, Washington DC, 1998), pp. 59 – 61.

第 8 章

很少用的和传统的晶体

这一章收集了 18 种相对很少用的晶体或者说是老晶体。

8.1　$KB_5O_8 \cdot 4H_2O$，五硼酸钾四水合物（KB5）

正双轴晶：$2V_z = 126.3°$，$\lambda = 0.546\ 1\ \mu m$[1]

分子量：293.210

密度：$1.74\ g/cm^3$[2]

点群：$mm2$

晶格常数[2]：

$a = (11.065 \pm 0.002)\ Å$

$b = (11.171 \pm 0.001)\ Å$

$c = (9.054 \pm 0.006)\ Å$

介电轴和晶体学轴的变换：$X,\ Y,\ Z \Rightarrow a,\ b,\ c$

莫氏硬度：2.5[2]

维氏硬度（以 kgf/mm^2 为单位）[3,4]

压痕负载 5 g	压痕负载 10 g	压痕负载 25 g	备注
64.4	59.7	49.7	沿 a
82.5	74.2	68.8	沿 b
78.7	75.7	68.1	沿 c

以"0"透过计的透明范围：$0.162 \sim 1.5$ μm[5]

线性吸收系数 α

λ/μm	α/cm^{-1}	参考文献	备注
0.212 8	0.18	[6]	o 光, XY 平面, FiHG 方向
	0.14	[7]	o 光, XY 平面, FiHG 方向
0.231 4	0.12	[1]	o 光, XY 平面, THG 方向
0.266 1	0.12	[6]	e 光, XY 平面, FiHG 方向
	0.06	[7]	e 光, XY 平面, FiHG 方向
0.347 2	0.04	[1]	e 光, XY 平面, THG 方向
0.354 7	<0.01	[8]	沿 Y 轴
0.532 1	0.02	[7]	XY 平面, FiHG 方向
	<0.01	[8]	沿 Y 轴
0.694 3	0.03	[1]	e 光, XY 平面, THG 方向
1.064 2	0.06	[7]	e 光, XY 平面, FiHG 方向

双光子吸收系数 β(沿 b 轴)[9]

λ/μm	τ_p/ns	$\beta \times 10^{11}$/(cm·W^{-1})	λ/μm	τ_p/ns	$\beta \times 10^{11}$/(cm·W^{-1})
0.216	0.015	65±10	0.270	0.015	35±5

折射率的实验值

λ/μm	n_X	n_Y	n_Z	参考文献
0.217			1.496 9	[10]
0.220			1.493 8	[10]
0.225			1.489 1	[10]
0.230			1.484 8	[10]
0.234 5		1.493 0		[11]
0.235			1.480 9	[10]
0.240			1.477 4	[10]
0.245			1.474 0	[10]
0.250			1.470 8	[10]
0.390	1.502 1	1.445 7	1.432 7	[11]

续表

$\lambda/\mu m$	n_X	n_Y	n_Z	参考文献
0.400	1.500 5	1.445 3	1.432 0	[11]
0.420	1.498 4	1.443 8	1.430 3	[11]
0.450	1.495 6	1.441 4	1.428 0	[11]
0.500	1.491 7	1.438 0	1.425 1	[11]
0.546	1.488 8	1.435 7	1.423 0	[11]
0.600	1.485 9	1.433 4	1.421 1	[11]
0.650	1.483 9	1.431 9	1.419 6	[11]
0.700	1.482 3	1.430 6	1.418 2	[11]
0.730	1.481 5	1.429 7	1.417 6	[11]
0.765	1.481 3	1.429 2	1.417 1	[11]

最佳 Sellmeier 方程(λ 以 μm 为单位,$T = 293$ K)[12]:

$$n_X^2 = 1.991\ 91 + \frac{0.009\ 253}{\lambda^2 - 0.009\ 329}$$

$$n_Y^2 = 2.029\ 98 + \frac{0.009\ 464}{\lambda^2 - 0.009\ 188}$$

$$n_Z^2 = 2.179\ 08 + \frac{0.010\ 354}{\lambda^2 - 0.008\ 781}$$

在文献[11]、[13]中给出了其他色散关系方程。

KB5 晶体主平面上有效二阶非线性系数的表达式(Kleinman 对称条件不成立)[14]:

XY 平面

$$d_{eeo} = d_{31}\sin^2\phi + d_{32}\cos^2\phi$$

YZ 面

$$d_{ooe} = d_{31}\sin\theta$$

XZ 面,$\theta < V_Z$

$$d_{oeo} = d_{eoo} = d_{24}\sin\theta$$

XZ 面,$\theta > V_Z$

$$d_{ooe} = d_{32}\sin\theta$$

KB5 晶体主平面上有效二阶非线性系数的表达式(Kleinman 对称条件成立,$d_{15} = d_{31}$ 和 $d_{24} = d_{32}$)[14]:

XY 面

$$d_{eeo} = d_{31}\sin^2\phi + d_{32}\cos^2\phi$$

YZ 面

$$d_{ooe} = d_{31}\sin\theta$$

XZ 面，$\theta < V_Z$

$$d_{oeo} = d_{eoo} = d_{32}\sin\theta$$

XZ 面，$\theta > V_Z$

$$d_{ooe} = d_{32}\sin\theta$$

在文献[14]中给出了 KB5 晶体内任意方向上有效二阶非线性系数的表达式。二阶非线性系数[15]：

$$d_{31}(0.532\ 1\ \mu m) = 0.04\ pm/V$$
$$d_{32}(0.532\ 1\ \mu m) = 0.003\ pm/V$$
$$d_{33}(0.532\ 1\ \mu m) = 0.05\ pm/V$$

相位匹配角的实验值（$T = 293$ K）

相互作用的波长/μm	$\phi_{exp}/(°)$	$\theta_{exp}/(°)$	参 考 文 献
XY 面，$\theta = 90°$			
SHG, e+e⇒o			
0.434⇒0.217	90		[10]
0.434 2⇒0.217 1	90		[16]
0.438 4⇒0.219 2	80.5		[17]
0.459 7⇒0.229 85	67.2		[18]
0.476 5⇒0.238 25	60.2		[18]
0.488⇒0.244	56.6		[18]
0.5⇒0.25	52.8		[10]
0.514 5⇒0.257 25	50.2		[18]
0.63⇒0.315	31		[16]
0.694 3⇒0.347 15	26.5		[18]
SFG, e+e⇒o			
0.539 8 + 0.359 87⇒0.215 92	50.4		[19]
0.543 5 + 0.351 1⇒0.213 3	90		[20]
0.694 3 + 0.347 2⇒0.231 4	57		[21]
0.573 7 + 0.334 5⇒0.211 3	90		[20]
0.652 2 + 0.326 1⇒0.217 4	68		[8]
0.621 9 + 0.311 0⇒0.207 3	90		[8]
0.694 3 + 0.305 19⇒0.212 0	70		[22]
0.694 3 + 0.284 09⇒0.201 6	90		[22]
0.789 71 + 0.266 04⇒0.199 0	75		[23]
0.753 22 + 0.266 04⇒0.196 6	90		[23]
0.797 37 + 0.257 25⇒0.194 5	84		[24]

续表

相互作用的波长/μm	$\phi_{exp}/(°)$	$\theta_{exp}/(°)$	参 考 文 献
0.792 35 + 0.257 25 ⇒ 0.194 2	90		[24]
0.9 + 0.232 87 ⇒ 0.185	90		[25]
1.064 15 + 0.266 04 ⇒ 0.212 8	53		[7]
1.064 15 + 0.212 83 ⇒ 0.177 36	80		[12]
1.079 6 + 0.269 9 ⇒ 0.215 92	80		[19]
1.314 17 + 0.19 ⇒ 0.166	90		[26], [27]
YZ 平面，$\phi = 90°$			
SHG, o + o ⇒ e			
0.434 6 ⇒ 0.217 3		90	[21]
0.469 0 ⇒ 0.234 5		17	[21]
0.479 6 ⇒ 0.239 8		0	[16]
SFG, o + o ⇒ e			
0.563 4 + 0.351 1 ⇒ 0.216 3		63	[20]
0.594 8 + 0.334 5 ⇒ 0.214 1		63	[20]
0.626 4 + 0.313 2 ⇒ 0.208 8		68	[8]
0.762 1 + 0.266 04 ⇒ 0.197 2		68	[23]
1.064 15 + 0.212 83 ⇒ 0.177 36		68.5	[12]

NCPM 温度的实验值

相互作用的波长/μm	$T/℃$	参 考 文 献
沿 b 轴		
SFG, I 类		
0.694 3 + 0.283 34 ⇒ 0.201 22	−15	[22]
0.694 3 + 0.283 61 ⇒ 0.201 36	0	[22]
0.694 3 + 0.284 05 ⇒ 0.201 58	20	[22]
0.694 3 + 0.284 49 ⇒ 0.201 80	35	[22]
0.792 02 + 0.257 25 ⇒ 0.194 18	25	[24]
0.793 44 + 0.257 25 ⇒ 0.194 27	40	[24]

激光诱导的表面损伤阈值

$\lambda/\mu m$	τ_p/ns	$I_{thr}/(GW \cdot cm^{-2})$	参 考 文 献	备 注
0.177 4	12	0.000 003	[12]	10 Hz, 50 小时
0.266 1	8	>0.043	[7]	10 Hz
	0.03	>0.48	[28]	1 Hz
0.311	10	>0.013	[8]	10 Hz

续表

$\lambda/\mu m$	τ_p/ns	$I_{thr}/(GW \cdot cm^{-2})$	参 考 文 献	备 注
0.347 2	8	>0.09	[1]	
0.45	7	1	[21]	15 Hz
0.622	10	>0.04	[8]	10 Hz
0.694 3	10	>0.08	[1]	
0.74~0.91	30	>0.05	[25]	
1.064 2	12	>0.085	[7]	10 Hz

关于这一晶体

KB5 于 20 世纪 70 年代在 UV 和深 UV 和频发生方面应用非常普遍。

参考文献

[1] K. Kato: Phase-matched generation of 2 314 Å in $KB_5O_8 \cdot 4H_2O$. Appl. Phys. Lett. 29(9), 562-563(1976).

[2] W. R. Cook, Jr., H. Jaffe: The crystallographic, elastic, and piezoelectric properties of ammonium pentaborate and potassium pentaborate. Acta Crystallogr. 10(11), 705-707(1957).

[3] K. Thamizharasan, S. Xavier Jesu Raja, F. P. Xavier, P. Sagayaraj: Growth, thermal and microhardness studies of single crystals of potassium penta borate (KB5). J. Cryst. Growth 218(2-4), 323-326(2000).

[4] S. A. Rajasekar, K. Thamizharasan, A. Joseph Arul Pragasam, J. Pakiam Julius, P. Sagayaraj: Growth and characterization of pure and doped potassium pentaborate(KB5) single crystals. J. Cryst. Growth 247(1-2), 199-206(2003).

[5] J. A. Paisner, M. L. Spaeth, D. C. Gerstenberger, I. W. Ruderman: Generation of tunable radiation below 2 000 Å by phase-matched sum-frequency mixing in $KB_5O_8 \cdot 4H_2O$. Appl. Phys. Lett. 32(8), 476-478(1978).

[6] K. B. Petrosyan, A. L. Pogosyan, K. M. Pokhsraryan: Generation of ultrashort light pulses in the UV region by up-conversion of radiation in potassium pentaborate. Izv. Akad. Nauk SSSR, Ser. Fiz. 47(8), 1619-1621(1983) [In Russian, English trans.: Bull. Acad. Sci. USSR, Phys. Ser. 47(8), 155-157(1983)].

[7] K. Kato: Phase matched generation of 2 128 Å in $KB_5O_8 \cdot 4H_2O$. Opt. Commun. 19(3), 332-333(1976).

[8] K. Kato: Efficient ultraviolet generation of 2 073-2 174 Å in $KB_5O_8 \cdot 4H_2O$. IEEE J. Quant. Electr. QE-13(7), 544-546(1977).

[9] G. G. Gurzadyan, R. K. Ispiryan: Two-photon absorption in potassium dihydrophosphate,

potassium pentaborate and quartz crystals at 270 and 216 nm. Int. J. Nonl. Opt. Phys. 1 (3), 533–540(1992).

[10] H. Zacharias, A. Anders, J. B. Halpern, K. H. Welge: Frequency doubling and tuning with $KB_5O_8 \cdot 4H_2O$ and application to NO($A^2\Sigma^+$) excitation. Opt. Commun. 19(1), 116–119(1976).

[11] W. R. Cook, Jr., L. M. Hubby, Jr.: Indices of refraction of potassium pentaborate. J. Opt. Soc. Am. 66(1), 72–73(1976).

[12] N. Umemura, K. Kato: Phase-matched UV generation at 0.177 4 μm in $KB_5O_8 \cdot 4H_2O$. Appl. Opt. 35(27), 5332–5335(1996).

[13] F. B. Dunning, R. E. Stickel, Jr.: Sum frequency mixing in potassium pentaborate as a source of tunable coherent radiation at wavelengths below 217 nm. Appl. Opt. 15(12), 3131–3134(1976).

[14] V. G. Dmitriev, D. N. Nikogosyan: Effective nonlinearity coefficients for three-wave interactions in biaxial crystals of $mm2$ point group symmetry. Opt. Commun. 95(1–3), 173–182(1993).

[15] D. A. Roberts: Simplified characterization of uniaxial and biaxial nonlinear optical crystals: a plea for standardization of nomenclature and conventions. IEEE J. Quant. Electr. 28(10), 2057–2074(1992).

[16] H. J. Dewey: Second-harmonic generation in $KB_5O_8 \cdot 4H_2O$ from 217.1 to 315.0 nm. IEEE J. Quant. Electr. QE–12(5), 303–306(1976).

[17] E. Fill, J. Wildenauer: Generation of the fifth and sixth harmonics of iodine laser pulses. Opt. Commun. 47(6), 412–413(1983).

[18] T. S. Chen, W. P. White: Second-harmonic generation in $KB_5O_8 \cdot 4H_2O$. IEEE J. Quant. Electr. QE–12(7), 436–437(1976).

[19] A. G. Arutyunyan, G. G. Gurzadyan, R. K. Ispiryan: Generation of the fifth harmonic of picosecond yttrium aluminate laser radiation. Kvant. Elektron. 16(12), 2493–2495 (1989) [In Russian, English trans.: Sov. J. Quantum Electron. 19(12), 1602–1603 (1989)].

[20] R. E. Stickel, Jr., S. Blit, G. F. Hildebrandt, E. D. Dahl, F. B. Dunning, F. K. Tittel: Generation of coherent UV radiation tunable from 211 nm to 216 nm. Appl. Opt. 17(15), 2270(1978).

[21] C. F. Dewey, Jr., W. R. Cook, Jr., R. T. Hodgson, J. J. Wynne: Frequency doubling in $KB_5O_8 \cdot 4H_2O$ and $NH_4B_5O_8 \cdot 4H_2O$ to 217.3 nm. Appl. Phys. Lett. 26(12), 714–716(1975).

[22] R. E. Stickel, Jr., F. B. Dunning: Generation of coherent radiation tunable from 201 nm to 212 nm. Appl. Opt. 16(9), 2356–2358(1977).

[23] K. Kato: Tunable UV generation in $KB_5O_8 \cdot 4H_2O$ to 1 966Å. Appl. Phys. Lett. 30

(11), 583 – 584(1977).

[24] H. Hemmati, J. C. Bergquist, W. M. Itano: Generation of continuous-wave 194 nm radiation by sum-frequency mixing in an external ring cavity. Opt. Lett. 8(2), 73 – 75 (1983).

[25] R. E. Stickel, Jr., F. B. Dunning: Generation of tunable coherent vacuum UV radiation in KB5. Appl. Opt. 17(7), 981 – 982(1978).

[26] V. Petrov, F. Rotermund, F. Noack: Generation of femtosecond pulses down to 166 nm by sum-frequency mixing in $KB_5O_8 \cdot 4H_2O$. Electron. Lett. 34(18), 1748 – 1750 (1998).

[27] V. Petrov, F. Rotermund, F. Noack, J. Ringling, O. Kittelmann, R. Komatsu: Frequency conversion of Ti: sapphire-based femtosecond laser systems to the 200-nm spectral region using nonlinear optical crystals. IEEE J. Sel. Topics Quant. Electr. 5(6) 1532 – 1542(1999).

[28] A. G. Arutyunyan, V. G. Atanesyan, K. B. Petrosyan, K. M. Pokhsraryan: Frequency multiplication of ultrashort light pulses in potassium pentaborate. Pisma Zh. Tekh. Fiz. 6 (5 – 6), 277 – 280(1980) [In Russian, English trans.: Sov. Tech. Phys. Lett. 6(3), 120 – 121(1980)].

8.2 CsB_3O_5, 三硼酸铯(CBO)

正双轴晶: $2V_Z = 79.0°$, $\lambda = 0.5321$ μm[1]

分子量: 245.335

密度(计算值): 3.357 g/cm³[2]

点群: 222

晶格常数:
$a = (6.213 \pm 0.001)$ Å[2]
$b = (8.521 \pm 0.001)$ Å[2]
$c = (9.170 \pm 0.002)$ Å[2]

介电轴和晶体学轴的变换: $X, Y, Z \Rightarrow c, a, b$

以"0"透过计的透明范围: 0.17 ~ 3.0 μm[3]

折射率的实验值[3]

$\lambda/\mu m$	n_X	n_Y	n_Z
0.354 7	1.549 9	1.584 9	1.614 5
0.476 5	1.537 0	1.575 8	1.603 1
0.488 0	1.536 7	1.573 6	1.600 9
0.496 5	1.536 2	1.571 6	1.599 6

续表

$\lambda/\mu m$	n_X	n_Y	n_Z
0.514 5	1.534 9	1.569 0	1.597 4
0.532 1	1.532 8	1.566 2	1.593 6
0.632 8	1.529 4	1.558 8	1.586 4
1.064 2	1.519 4	1.550 5	1.578 1

Sellmeier 方程(λ 以 μm 为单位,$T = 293$ K)[1]:

$$n_X^2 = 2.303\ 5 + \frac{0.013\ 78}{\lambda^2 - 0.014\ 98} - 0.006\ 12\lambda^2$$

$$n_Y^2 = 2.370\ 4 + \frac{0.015\ 28}{\lambda^2 - 0.015\ 81} - 0.009\ 39\lambda^2$$

$$n_Z^2 = 2.475\ 3 + \frac{0.018\ 06}{\lambda^2 - 0.017\ 52} - 0.016\ 54\lambda^2$$

在文献[3]中给出了其他色散关系方程。

CBO 晶体主平面上有效二阶非线性系数的表达式(Kleinman 对称条件成立,$d_{14} = d_{25} = d_{36}$)[4]:

XY 面

$$d_{eoe} = d_{oee} = d_{14}\sin 2\phi$$

YZ 面

$$d_{eeo} = d_{14}\sin 2\theta$$

XZ 面,$\theta < V_Z$

$$d_{eoe} = d_{oee} = -d_{14}\sin 2\theta$$

XZ 面,$\theta > V_Z$

$$d_{eeo} = -d_{14}\sin 2\theta$$

二阶非线性系数:

$$d_{14}(1.064\ 2\ \mu m) = 0.468 \times d_{22}(BBO) = 1.08\ pm/V^{[3,5]}$$

$$d_{14}(1.064\ 2\ \mu m) = 0.468 \times d_{22}(BBO) = 1.03\ pm/V^{[3,5]}$$

$$d_{14}(1.064\ 2\ \mu m) = (0.53 \pm 0.05) \times d_{22}(BBO) = (1.17 \pm 0.11)\ pm/V^{[1,6]}$$

CBO 晶体主平面上相位匹配角和相位匹配温度带宽的实验值($T = 293$ K)[1]

相互作用的波长/μm	$\phi_{exp}/(°)$	$\theta_{exp}/(°)$	$\Delta T/°C$
XY 面,$\theta = 90°$			
SHG, e + o $\Rightarrow$ e			
1.064 2$\Rightarrow$0.532 1	12.9		18.7

续表

相互作用的波长/μm	$\phi_{exp}/(°)$	$\theta_{exp}/(°)$	$\Delta T/℃$
SFG, e + o ⇒ e			
1.064 2 + 0.532 1 ⇒ 0.354 73	40.3		5.7
YZ 平面, $\phi = 90°$			
SFG, e + e ⇒ o			
1.064 2 + 0.532 1 ⇒ 0.354 73		25.5	
1.064 2 + 0.354 73 ⇒ 0.266 05		52.3	4.0
XZ 平面, $\phi = 0°$, $\theta > V_z$			
SHG, e + e ⇒ o			
1.064 2 ⇒ 0.532 1		58.2	10.8
SFG, e + e ⇒ o			
1.064 2 + 0.532 1 ⇒ 0.354 73		77.9	7.8

内角带宽的实验值[3]

相互作用的波长/μm	$\theta_{pm}/(°)$	$\Delta\theta^{int}/(°)$
XZ 平面, $\phi = 0°$, $\theta > V_z$		
SHG, e + e ⇒ o		
1.064 2 ⇒ 0.532 1	60.2	0.064

激光诱导的损伤阈值

$\lambda/\mu m$	τ_p/ns	$I_{thr}/(GW \cdot cm^{-2})$	参 考 文 献
1.053	1	26	[3]
1.064 2	0.035	>10	[7]

关于这一晶体

在1996年Chen及其合作者研究了CBO的非线性光学性质[3]。然而，这种晶体没有引起很大重视并很快被淡忘了。

■ 参考文献

[1] K. Kato: Tunable UV generation to 0.185 μm in CsB_3O_5. IEEE J. Quant. Electr. 31(1), 169–171(1995).

[2] J. Krogh-Moe: Refinement of the crystal structure of caesium triborate, $Cs_2O \cdot 3B_2O_5$. Acta Crystallogr. B 30(5), 1178–1180(1974).

[3] Y. Wu, T. Sasaki, S. Nakai, A. Yokotani, H. Tang, C. Chen: CsB_3O_5: a new nonlinear crystal. Appl. Phys. Lett. 62(21), 2614 – 2615(1993).

[4] B. V. Bokut: Optical mixing in biaxial crystals. Zh. Prikl. Spektrosk. 7(4), 621 – 624 (1967)[In Russian, English trans.: J. Appl. Spectrosc. 7(4), 425 – 429(1967)].

[5] D. A. Roberts: Simplified characterization of uniaxial and biaxial nonlinear optical crystals: a plea for standardization of nomenclature and conventions. IEEE J. Quant. Electr. 28(10), 2057 – 2074(1992).

[6] I. Shoji, H. Nakamura, K. Ohdaira, T. Kondo, R. Ito, T. Okamoto, K. Tatsuki, S. Kubota: Absolute measurements of second-order nonlinear-optical coefficients of β-BaB_2O_4 for visible to ultraviolet second-harmonic wavelengths. J. Opt. Soc. Am. B 16(4), 620 – 624(1999).

[7] Y. Wu, P. Fu, J. Wang, Z. Xu, L. Zhang, Y. Kong, C. Chen: Characterization of CsB_3O_5 crystal for ultraviolet generation. Opt. Lett. 22(24), 1840 – 1842(1997)

8.3 $C_4H_7D_{12}N_4PO_7$,氘化左旋磷酸精氨酸一水合物(DLAP)

化学式:$C_4H_7D_{12}N_4PO_7$[1] *

负双轴晶:$2V_z = 142.6°$,$\lambda = 0.532\ 1\ \mu m$[2]

分子量:302.286

密度:1.591 g/cm^3 [3]

点群:2

左旋磷酸精氨酸一水合物(LAP)的晶格常数[4]:

  $a = (10.85 \pm 0.02)$ Å

  $b = (7.91 \pm 0.01)$ Å

  $c = (7.32 \pm 0.02)$ Å

  $\beta = 98.0° \pm 0.1°$

氘化左旋磷酸精氨酸一水合物(DLAP)的晶格常数:

  $a = 10.75$ Å[5];10.87 Å[6]

  $b = 7.91$ Å[5];7.92 Å[6]

  $c = 7.32$ Å[5];7.38 Å[6]

介电轴和晶体学轴的变换(对LAP):$Y \parallel b$,a轴和c轴位于XZ平面,两轴间夹角$\beta = 98°$,Z轴和c轴间夹角$\alpha = 35°$[2]。

莫氏硬度:3

---

* 译者注:原书漏掉此化学式,译者后补。

化学分解温度：403 K[2]；380~410 K[5]

线性热膨胀系数 α_l 的平均值(以 10^{-6} K^{-1} 为单位)[3]

T/K	$\alpha_{11}(\parallel a)$	$\alpha_{22}(\parallel b)$	$\alpha_{33}(\parallel c)$	$\alpha_{13}=\alpha_{31}$
298~373	57.4±0.8	8.7±0.5	18.3±0.6	5.0±0.8

以"0"透过计的透明范围：0.22~1.30 μm[2]

线性吸收系数 α

$\lambda/\mu m$	α/cm^{-1}	参考文献	备注
0.266	0.074	[2]	沿 X
	0.131	[2]	沿 Y
	0.184	[2]	沿 Z
0.354 7	0.025	[2]	沿 X
	0.053	[2]	沿 Y
	0.039	[2]	沿 Z
0.526 5	0.01	[7]	
0.532 1	0.01	[1]	
	<0.01	[2]	
0.910	0.028	[2]	沿 X
	0.037	[2]	沿 Y
	0.044	[2]	沿 Z
1.040	0.012	[2]	沿 X
	0.014	[2]	沿 Y
	0.009	[2]	沿 Z
1.053	0.02	[7]	
1.064	0.02	[1]	
	0.012	[2]	沿 X
	0.014	[2]	沿 Y
	0.009	[2]	沿 Z
1.180	0.385	[2]	沿 X
	0.394	[2]	沿 Y
	0.557	[2]	沿 Z

折射率的温度微商[8]

$\lambda/\mu m$	$dn_X/dT \times 10^5/K^{-1}$	$dn_Y/dT \times 10^5/K^{-1}$	$dn_Z/dT \times 10^5/K^{-1}$
0.532 1	-3.64±0.17	-5.34±0.17	-6.69±0.17
1.064 2	-3.73±0.17	-5.30±0.17	-6.30±0.17

Sellmeier 方程（λ 以 μm 为单位，$T = 298$ K）[2]：

$$n_X^2 = 2.235\,2 + \frac{0.011\,8}{\lambda^2 - 0.014\,6} - 0.006\,83\lambda^2$$

$$n_Y^2 = 2.431\,3 + \frac{0.015\,1}{\lambda^2 - 0.021\,4} - 0.014\,3\lambda^2$$

$$n_Z^2 = 2.448\,4 + \frac{0.017\,2}{\lambda^2 - 0.022\,9} - 0.011\,5\lambda^2$$

DLAP 晶体主平面上有效二阶非线性系数的表达式（小走离角近似，Kleinman 对称条件成立，$d_{14} = d_{25} = d_{36}$，$d_{16} = d_{21}$ 和 $d_{23} = d_{34}$）[2,9]：

XY 面

$$d_{ooe} = d_{23}\cos\phi$$
$$d_{eoe} = d_{oee} = d_{14}\sin 2\phi$$

YZ 面

$$d_{eeo} = d_{14}\sin 2\theta$$
$$d_{oeo} = d_{eoo} = d_{16}\cos\theta$$

XZ 面，$\phi = 0°$，$V_Z > \theta > 0°$

$$d_{eoe} = d_{oee} = d_{16}\cos^2\theta + d_{23}\sin^2\theta - d_{14}\sin 2\theta$$

XZ 面，$\phi = 0°$，$90° > \theta > V_Z$

$$d_{eeo} = d_{16}\cos^2\theta + d_{23}\sin^2\theta - d_{14}\sin 2\theta$$

XZ 面，$\phi = 0°$，$180° - V_Z > \theta > 90°$，或 $\phi = 180°$，$90° > \theta > V_Z$

$$d_{eeo} = d_{16}\cos^2\theta + d_{23}\sin^2\theta + d_{14}\sin 2\theta$$

XZ 面，$\phi = 0°$，$180° > \theta > 180° - V_Z$，或 $\phi = 180°$，$V_Z > \theta > 0°$

$$d_{eoe} = d_{oee} = d_{16}\cos^2\theta + d_{23}\sin^2\theta + d_{14}\sin 2\theta$$

二阶非线性系数[2,10]：

$$d_{14}(1.064\,2\ \mu m) = -0.59\ pm/V$$
$$d_{16}(1.064\,2\ \mu m) = 0.40\ pm/V$$
$$d_{22}(1.064\,2\ \mu m) = 0.37\ pm/V$$
$$d_{23}(1.064\,2\ \mu m) = 0.83\ pm/V$$

$T = 297$ K 时相位匹配角的实验值[8]

相互作用的波长/μm	$\phi_{exp}/(°)$	$\theta_{exp}/(°)$	相互作用的波长/μm	$\phi_{exp}/(°)$ $\theta_{exp}/(°)$
XY 面，$\theta = 90°$			XZ 平面，$\phi = 0°$，$\theta < V_Z$	
SHG，$o + o \Rightarrow e$			SHG，$e + o \Rightarrow e$	
$1.064\,2 \Rightarrow 0.532\,1$	22.2		$1.064\,2 \Rightarrow 0.532\,1$	42.8
SHG，$e + o \Rightarrow e$				
$1.064\,2 \Rightarrow 0.532\,1$	37.5			

内角带宽、温度带宽和光谱带宽的实验值[11]

相互作用的波长/μm	$\phi_{pm}/(°)$	$\Delta\phi^{int}/(°)$	$\Delta T/℃$	$\Delta\nu/cm^{-1}$
XY 平面，$\theta = 90°$				
SHG, $o + o \Rightarrow e$				
$1.064\ 2 \Rightarrow 0.532\ 1$	22.2	0.036	5.4	20.2
SHG, $e + o \Rightarrow e$				
$1.064\ 2 \Rightarrow 0.532\ 1$	37.5	0.072	14.6	20.1

激光诱导的损伤阈值

$\lambda/\mu m$	τ_p/ns	$I_{thr}/(GW \cdot cm^{-2})$	参考文献
0.308	17	0.03	[12]
0.526 5	20	38	[7]
	0.6	67	[7]
1.053	25	33	[7]
	1	87	[7]
1.064 2	14	>1.4	[1]
	1	9~13	[2]

关于这一晶体

  DLAP 晶体是属于低对称点群 2 中间首批被深入研究的非线性光学晶体之一。

■ 参考文献

[1] M. Yoshimura, Y. Mori, T. Sasaki, H. Yoshida, M. Nakatsuka: Efficient stimulated Brillouin scattering in the organic crystal deuterated L-arginine phosphate monohydrate. J. Opt. Soc. Am. B 15(1), 446-450(1998).

[2] D. Eimerl, S. Velsko, L. Davis, F. Wang, G. Loiacono, G. Kennedy: Deuterated L-arginine phosphate: a new efficient nonlinear crystal. IEEE J. Quant. Electr. 25(2), 179-193(1989).

[3] D. Eimerl, J. Marion, E. K. Graham, H. A. McKinstry, S. Haussühl: Elastic components and thermal fracture of $AgGaSe_2$ and d-LAP. IEEE J. Ouant. Electr. 27(1), 142-145(1991).

[4] K. Aoki, K. Nagano, Y. Iitaka: The crystal structure of L-arginine phosphate monohydrate. Acta Crystallogr. B 27(1), 11-23(1971).

[5] A. S. Haja Hameed, G. Ravi, R. Jayavel, P. Ramasamy: Nucleation kinetics, growth

and characterization of dLAP, dLAP:KF and dLAP:NaN$_3$ crystals. J. Cryst. Growth 250(1-2), 126-133(2003).

[6] A. S. Haja Hameed, G. Ravi, R. Ilangovan, A. Nixon Azariah, P. Ramasamy: Growth and characterization of deuterated analog of L-arginine phosphate single crystals. J. Cryst. Growth 237-239, 890-893(2002).

[7] A. Yokotani, T. Sasaki, K. Yoshida, S. Nakai: Extremely high damage threshold of a new nonlinear crystal L-arginine phosphate and its deuterium compound. Appl. Phys. Lett. 55(26), 2692-2693(1989).

[8] C. E. Barker, D. Eimerl, S. P. Velsko: Temperature-insensitive phase-matching for secondharmonic generation in deuterated L-arginine phosphate. J. Opt. Soc. Am. B 8(12), 2481-2492(1991).

[9] B. V. Bokut: Optical mixing in biaxial crystals. Zh. Prikl. Spektrosk. 7(4), 621-624 (1967)[In Russian, English trans.: J. Appl. Spectrosc. 7(4), 425-429(1967)].

[10] D. A. Roberts: Simplified characterization of uniaxial and biaxial nonlinear optical crystals: a plea for standardization of nomenclature and conventions. IEEE J. Quant. Electr. 28(10), 2057-2074(1992).

[11] R. B. Andreev, K. V. Vetrov, V. N. Voitsechovskii, V. D. Volosov, I. V. Nikiforuk, B. P. Nikolaeva, V. E. Yakobson: Growth of d-LAP crystals and study of their primary nonlinear optical properties. Izv. Akad. Nauk SSSR, Ser. Fiz. 54(12), 2491-2493 (1990)[In Russian, English trans.: Bull. Acad. Sci. USSR, Phys. Ser. 54(12), 187-189 (1990)].

[12] G. Robertson, M. H. Dunn: Excimer pumped deuterated L-arginine phosphate optical parametric oscillator. Appl. Phys. Lett. 62(26), 3405-3407(1993).

8.4 α-碘酸(α-HIO$_3$)

负双轴晶：$2V_z = 47°$[1]

分子量：175.911

密度：4.63 g/cm^3[1]

点群：222

介电轴和晶体学轴的变换：$X, Y, Z \Rightarrow b, c, a$

以"0"透过计的透明范围：0.32~1.7 μm($\parallel c$)，0.32~2.3 μm($\perp c$)[1]

线性吸收系数 α：<0.5 cm^{-1}，0.35~1.3 μm[2]

$T = 293$ K 时折射率的实验值[3]

$\lambda/\mu m$	n_X	n_Y	n_Z	$\lambda/\mu m$	n_X	n_Y	n_Z
0.35	2.1485	2.1265	1.9612	0.62	1.9884	1.9632	1.8388
0.36	2.1330	2.1077	1.9474	0.64	1.9854	1.9589	1.8368
0.37	2.1171	2.0917	1.9360	0.66	1.9821	1.9560	1.8348
0.38	2.1053	2.0782	1.9257	0.68	1.9791	1.9529	1.8328
0.39	2.0929	2.0662	1.9154	0.70	1.9763	1.9506	1.8311
0.40	2.0808	2.0545	1.9086	0.80	1.9668	1.9409	1.8248
0.41	2.0715	2.0465	1.9020	0.85	1.9634	1.9377	1.8222
0.42	2.0637	2.0394	1.8952	0.90	1.9602	1.9346	1.8202
0.44	2.0494	2.0246	1.8847	0.95	1.9569	1.9314	1.8184
0.46	2.0378	2.0119	1.8753	1.00	1.9541	1.9286	1.8150
0.48	2.0292	2.0026	1.8685	1.10	1.9486	1.9260	1.8114
0.50	2.0194	1.9926	1.8624	1.20	1.9436	1.9229	1.8088
0.52	2.0126	1.9883	1.8562	1.30	1.9390	1.9206	1.8063
0.54	2.0065	1.9829	1.8522	1.40	1.9348	1.9180	1.8038
0.56	2.0010	1.9763	1.8475	1.50	1.9310	1.9157	1.8018
0.58	1.9960	1.9712	1.8436	1.60		1.9132	1.7998
0.60	1.9918	1.9665	1.8405				

$T = 300$ K 时的旋光性[1]

$\lambda/\mu m$	$\rho/[(°) \cdot mm^{-1}]$	$\lambda/\mu m$	$\rho/[(°) \cdot mm^{-1}]$
0.4360	74.5	0.5461	58.7

最佳 Sellmeier 方程(λ 以 μm 为单位,$T = 293$ K)[4]:

$$n_X^2 = 3.739 + \frac{0.07128}{\lambda^2 - 0.05132}$$

$$n_Y^2 = 3.654 + \frac{0.06721}{\lambda^2 - 0.04234}$$

$$n_Z^2 = 3.239 + \frac{0.05353}{\lambda^2 - 0.017226}$$

在文献[3]、[5]中给出了其他色散关系方程。

室温下低频(远低于 α-HIO$_3$ 晶体声学谐振频率,即对"自由"晶体)下测量的线性电光系数[6]

$\lambda/\mu m$	$r_{41}^T/(pm \cdot V^{-1})$	$r_{52}^T/(pm \cdot V^{-1})$	$r_{63}^T/(pm \cdot V^{-1})$
0.6328	6.6±0.3	7.0±0.5	6.0±0.3

α-HIO₃ 晶体主平面上有效二阶非线性系数的表达式（Kleinman 对称条件成立，$d_{14} = d_{25} = d_{36}$）[7]：

XY 面
$$d_{eeo} = -d_{14}\sin 2\phi$$

YZ 面
$$d_{eoe} = d_{oee} = -d_{14}\sin 2\theta$$

XZ 面，$\theta < V_z$
$$d_{eeo} = d_{14}\sin 2\theta$$

XZ 面，$\theta > V_z$
$$d_{eoe} = d_{oee} = d_{14}\sin 2\theta$$

二阶非线性系数值：

$$d_{14}(1.064\ \mu m) = 20 \times d_{11}(SiO_2) \pm 25\% = (6.0 \pm 1.5)\ pm/V^{[1,8]}$$

$$d_{14}(1.152\ 3\ \mu m) = 10.9 \times d_{36}(ADP) \pm 14\% = (5.0 \pm 0.7)\ pm/V^{[9,10]}$$

相位匹配角的实验值（$T = 293$ K）

相互作用的波长/μm	θ_{exp}/(°)	参 考 文 献
YZ 平面，$\phi = 90°$		
SHG, e + o ⇒ e		
0.976 ⇒ 0.488	57.9	[11]
1.029 ⇒ 0.514 5	52.7	[11]
1.064 2 ⇒ 0.532 1	50.4	[12]
1.065 ⇒ 0.532 5	52	[1]
XZ 平面，$\phi = 0°$，$\theta > V_z$		
SHG, e + o ⇒ e		
0.976 ⇒ 0.488	72.2	[11]
1.029 ⇒ 0.514 5	66.1	[11]
1.06 ⇒ 0.53	64.9	[13]
1.065 ⇒ 0.532 5	66	[1]

内角带宽和光谱带宽的实验值[14]

相互作用的波长/μm	θ_{pm}/(°)	$\Delta\theta^{int}$/(°)	$\Delta\nu/cm^{-1}$
XZ 平面，$\phi = 0°$，$\theta > V_z$			
SHG, e + o ⇒ e			
1.06 ⇒ 0.53	66	0.035	3.38

临界 SFG 过程的温度调谐[11]

相互作用的波长/μm	$\theta_{pm}/(°)$	$d\lambda_2/dT/(nm \cdot K^{-1})$
XZ 平面，$\phi=0°$，$\theta>V_z$		
SHG，e + o ⇒ e		
1.922 6 + 0.654 ⇒ 0.488	50	0.055

激光诱导的表面损伤阈值

$\lambda/\mu m$	τ_p/ns	$I_{thr}/(GW \cdot cm^{-2})$	参 考 文 献	备 注
0.488	CW	> 0.000 25	[1]	
0.528	0.007	> 7	[15]	2 Hz
0.53	15	0.055	[13]	
	0.006	> 0.8	[16]	
0.532	0.03	> 0.8	[17]	25 Hz
		> 5.5	[18]	
	0.035	8 ~ 10	[19]	1 Hz
		4 ~ 5	[19]	12.5 Hz

关于这一晶体

α - HIO_3 可能是第一个在非线性光学中应用的 222 点群的双轴晶体。

■ 参考文献

[1] S. K. Kurtz, T. T. Perry, J. G. Bergman, Jr.: Alpha-iodic acid: a solution-grown crystal for nonlinear optical studies and applications. Appl. Phys. Lett. 12(5), 186 – 188 (1967).

[2] V. I. Bespalov, I. A. Batyreva, L. A. Dmitrenko, V. V. Korolikhin, S. P. Kuznetsov, M. A. Novikov: Investigation of the absorption of near infrared radiation in partly deuterated KDP and α-HIO_3 crystals. Kvant. Elektron. 4(7), 1563 – 1566(1977)[In Russian, English trans.: Sov. J. Quantum Electron. 7(7), 885 – 887(1977)].

[3] H. Naito, H. Inaba: Measurement of the refractive indices of α-iodic acid, HIO_3, crystal. Opto-electron. 4(3), 335 – 337(1972).

[4] S. K. Kurtz: Nonlinear Optical Materials. In: *Laser Handbook*, Vol. 1, ed. by F. T. Arecchi, E. O. Schulz-Dubois(North-Holland, Amsterdam,1972), pp. 923 – 974.

[5] R. A. Andrews: IR image parametric up-conversion. IEEE J. Quant. Electr. QE – 6(1), 68 – 80(1970).

[6] E. N. Volkova, V. A. Dianova, A. L. Zueva, A. N. Izrailenko, A. C. Lipatov, V. N. Parygin, L. N. Rashkovich, L. E. Chirkov: Electro-optical and piezoelectric properties of α-HIO_3 crystals. Kristallogr. 16(2), 346 – 349(1971)[In Russian, English trans.: Sov. Phys. -Crystallogr.

16(2),284-287(1971)].

[7] B. V. Bokut: Optical mixing in biaxial crystals. Zh. Prikl. Spektrosk. 7(4), 621-624 (1967)[In Russian, English trans.: J. Appl. Spectrosc. 7(4),425-429(1967)].

[8] D. A. Roberts: Simplified characterization of uniaxial and biaxial nonlinear optical crystals: a plea for standardization of nomenclature and conventions. IEEE J. Quant. Electr. 28 (10), 2057-2074(1992).

[9] J. E. Bjorkholm: Relative measurement of the optical nonlinearities of KDP, ADP, LiNbO$_3$, and α-HIO$_3$. IEEE J. Quant. Electr. QE-4(11), 970-972(1968).

[10] K. Hagimoto, A. Mito: Determination of the second-order susceptibility of ammonium dihydrogen phosphate and α-quartz at 633 and 1064 nm. Appl. Opt. 34 (36), 8276-8282(1995).

[11] V. A. Kiselev, V. F. Kitaeva, L. A. Kulevskii, Y. N. Polivanov, S. N. Poluektov: Investigation of spontaneous parametric emission in biaxial crystal α-HIO$_3$. Zh. Eksp. Teor. Fiz. 62(4), 1291-1301(1972)[In Russian, English trans.: Sov. Phys.-JETP 35(4),687-691(1972)].

[12] H. Ito, H. Naito, H. Inaba: Generalized study on angular dependence of induced secondorder nonlinear optical polarizations and phase matching in biaxial crystals. J. Appl. Phys. 46(9), 3992-3998(1975).

[13] A. I. Izrailenko, A. I. Kovrigin, P. V. Nikles: Parametric generation of light in high-efficiency nonlinear LiIO$_3$ and α-HIO$_3$ crystals. Pisma Zh. Eksp. Teor. Fiz. 12(10), 475-478(1970)[In Russian, English trans.: JETP Lett. 12(10),331-333(1970)].

[14] R. B. Andreev, V. D. Volosov, A. G. Kalintsev: Spectral, angular, and temperature characteristics of HIO$_3$, LiIO$_3$, CDA, DKDP, KDP and ADP non-linear crystals in second-and fourth-harmonic generation. Opt. Spektrosk. 37 (2), 294-299 (1974)[In Russian, English trans.: Opt. Spectrosc. USSR 37(2),169-171(1974)].

[15] G. Dikchyus, E. Zhilinskas, A. Piskarskas, V. Sirutkaitis: Statistical properties and stabilization of a picosecond phosphate-glass laser with 2 Hz repetition frequency. Kvant. Elektron. 6(8), 1610-1619(1979)[In Russian, English trans.: Sov. J. Quantum Electron. 9(8),950-955(1979)].

[16] G. A. Dikchyus, V. I. Kabelka, A. S. Piskarskas, A. Y. Stabinis: Single-pass parametric generation of light in an α-HIO$_3$ crystal pumped with ultrashort pulses. Kvant. Elektron. 1(11), 2513-2515(1974)[In Russian, English trans.: Sov. J. Quantum Electron. 4(11),1402-1403(1974)].

[17] G. Dikchyus, R. Danielius, V. Kabelka, A. Piskarskas, T. Tomkiavichyus, A. Stabinis: Kvant. Elektron. 3(4), 779-784(1976)[In Russian, English trans.: Sov. J. Quantum Electron. 6(4),425-428(1976)].

[18] R. Danielius, G. Dikchyus, V. Kabelka, A. Piskarskas: High efficiency, picosecond

parametric light source with narrow output spectrum and high pulse rate. Zh. Tekh. Fiz. 47(5), 1075 – 1077 (1977) [In Russian, English trans.: Sov. Phys. -Tech. Phys. 22(5), 642 – 643(1977)].

[19] R. Danielius, G. Dikchyus, V. Kabelka, A. Piskarskas, A. Stabinis, Y. Yasevichyute: Parametric excitation of light in the picosecond range. Kvant. Elektron. 4(11), 2379 – 2395 (1977) [In Russian, English trans.: Sov. J. Quantum Electron. 7(11), 1360 – 1368(1977)].

8.5　LiCOOH·H_2O, 甲酸锂一水合物(LFM)

负双轴晶: $2V_Z = 123.8°$, $\lambda = 0.532\ 1\ \mu m$[1]

分子量: 69.974

密度: 1.46 g/cm^3[1]

点群: $mm2$

晶格常数[1]: $a = 4.85$ Å; $b = 6.49$ Å; $c = 10.01$ Å

介电轴和晶体学轴的变换: $X, Y, Z \Rightarrow a, b, c$

以"0"透过计的透明范围: 0.23 ~ 1.56 μm[1,2]

线性吸收系数 $\alpha(\theta = 90°, \phi = 10°)$[3]

$\lambda/\mu m$	α/cm^{-1}	$\lambda/\mu m$	α/cm^{-1}
0.354 7	0.025	1.064 2	0.017
0.532 1	0.012		

折射率的实验值[4]

$\lambda/\mu m$	n_X	n_Y	n_Z	$\lambda/\mu m$	n_X	n_Y	n_Z
0.35	1.318 0	1.507 3	1.554 0	0.60	1.364 3	1.479 6	1.517 4
0.36	1.379 1	1.505 1	1.551 0	0.62	1.363 8	1.478 7	1.516 1
0.37	1.377 7	1.503 4	1.548 4	0.64	1.363 3	1.477 8	1.515 2
0.38	1.376 7	1.501 7	1.545 8	0.66	1.362 8	1.476 8	1.514 4
0.39	1.375 8	1.499 9	1.543 2	0.68	1.362 5	1.476 0	1.513 5
0.40	1.374 8	1.498 1	1.540 5	0.70	1.362 3	1.475 1	1.512 6
0.42	1.372 9	1.495 5	1.536 7	0.80	1.361 4	1.472 9	1.509 9
0.44	1.371 4	1.492 8	1.533 2	0.90	1.360 4	1.471 1	1.507 7
0.46	1.370 5	1.490 2	1.530 1	1.00	1.359 5	1.469 4	1.505 5
0.48	1.369 6	1.488 0	1.527 9	1.10	1.359 0	1.467 5	1.503 2
0.50	1.368 6	1.486 2	1.525 7	1.20	1.358 7	1.465 8	1.501 1
0.52	1.367 7	1.484 5	1.523 6	1.30	1.358 5	1.464 4	1.498 7
0.54	1.366 6	1.482 7	1.521 9	1.40	1.358 3	1.463 0	1.497 0
0.56	1.365 7	1.481 3	1.520 0	1.50	1.358 1	1.461 7	
0.58	1.364 7	1.480 4	1.518 7				

Sellmeier 方程(λ 以 μm 为单位,$T = 293$ K)[4]:

$$n_X^2 = 1.437\ 6 + \frac{0.404\ 5\lambda^2}{\lambda^2 - 0.016\ 926\ 01} - 0.000\ 5\lambda^2$$

$$n_Y^2 = 1.658\ 6 + \frac{0.500\ 6\lambda^2}{\lambda^2 - 0.023\ 409} - 0.012\ 7\lambda^2$$

$$n_Z^2 = 1.671\ 4 + \frac{0.592\ 8\lambda^2}{\lambda^2 - 0.025\ 344\ 64} - 0.015\ 3\lambda^2$$

LFM 晶体主平面上有效二阶非线性系数的表达式(Kleinman 对称条件不成立)[5]:

XY 平面
$$d_{eoe} = d_{oee} = d_{15}\sin^2\phi + d_{24}\cos^2\phi$$

YZ 面
$$d_{oeo} = d_{eoo} = d_{15}\sin\theta$$

XZ 面,$\theta < V_Z$
$$d_{ooe} = d_{32}\sin\theta$$

XZ 面,$\theta > V_Z$
$$d_{oeo} = d_{eoo} = d_{24}\sin\theta$$

LFM 晶体主平面上有效二阶非线性系数的表达式(Kleinman 对称条件成立,$d_{15} = d_{31}$ 和 $d_{24} = d_{32}$)[5]:

XY 平面
$$d_{eoe} = d_{oee} = d_{31}\sin^2\phi + d_{32}\cos^2\phi$$

YZ 面
$$d_{oeo} = d_{eoo} = d_{31}\sin\theta$$

XZ 面,$\theta < V_Z$
$$d_{ooe} = d_{32}\sin\theta$$

XZ 面,$\theta > V_Z$
$$d_{oeo} = d_{eoo} = d_{32}\sin\theta$$

在文献[5]中给出了 LFM 晶体内任意方向上有效二阶非线性系数的表达式。文献[2]中给出的式子不正确。

二阶非线性系数的绝对值[6]:

$$d_{31}(1.064\ 2\ \mu m) = 0.13\ pm/V$$
$$d_{32}(1.064\ 2\ \mu m) = -0.60\ pm/V$$
$$d_{33}(1.064\ 2\ \mu m) = 0.94\ pm/V$$

相位匹配角的实验值

相互作用的波长/μm	$\phi_{exp}/(°)$	$\theta_{exp}/(°)$	参 考 文 献
XY 面，$\theta = 90°$			
SFG, e + o ⇒ e			
1.064 2 + 0.532 1 ⇒ 0.354 7	8.2		[3]
XZ 面，$\phi = 0°$			
SHG, o + o ⇒ e			
0.486 ⇒ 0.243		38.5	[7]
1.064 2 ⇒ 0.532 1		55.1	[1]
SHG, o + e ⇒ e			
1.064 2 ⇒ 0.532 1		82.0	[1]

内角带宽的实验值[3]

相互作用的波长/μm	$\phi_{pm}/(°)$	$\Delta\phi^{int}/(°)$
XY 平面，$\theta = 90°$		
SFG, e + o ⇒ e		
1.064 2 + 0.532 1 ⇒ 0.354 7	8.2	0.04

激光诱导的表面损伤阈值

$\lambda/\mu m$	τ_p/ns	$I_{thr}/(GW \cdot cm^{-2})$	参 考 文 献
0.475	330	0.15	[8]
0.488	CW	>0.000 001	[1]
0.490	330	0.15	[8]

关于这一晶体

    LFM 是属于 $mm2$ 点群的首批双轴非线性光学晶体之一。

■ **参考文献**

[1] S. Singh, W. A. Bonner, J. R. Potopowicz, L. G. van Uitert: Non-linear optical susceptibility of lithium formate monohydrate. Appl. Phys. Lett. 17(7), 292 – 294(1970).

[2] H. Ito, H. Naito, H. Inaba: New phase-matchable nonlinear optical crystals of the formate family. IEEE J. Quant. Electr. QE – 10(2), 247 – 252(1974).

[3] K. Kato: Third-harmonic generation of Nd: YAG laser in lithium formate monohydrate. Opt. Quant. Electron. 8(3), 261 – 262(1976).

[4] H. Naito, H. Inaba: Measurement of the refractive indices of crystalline lithium formate HCOOLi · H$_2$O. Opto-electron. 5(3), 256 – 259(1973).

[5] V. G. Dmitriev, D. N. Nikogosyan: Effective nonlinearity coefficients for three-wave inter-

actions in biaxial crystals of *mm*2 point group symmetry. Opt. Commun. 95(1–3), 173–182(1993).

[6] D. A. Roberts: Simplified characterization of uniaxial and biaxial nonlinear optical crystals: a plea for standardization of nomenclature and conventions. IEEE J. Quant. Electr. 28(10), 2057–2074(1992).

[7] S. J. Bastow, M. H. Dunn: The generation of tunable UV radiation from 238–249 nm by intracavity frequency doubling of Coumarin 102 dye laser. Opt. Commun. 35(2), 259–263(1980).

[8] L. Armstrong, S. E. Neister, R. Adhav: Measuring CFP dye laser damage thresholds on UV doubling crystals. Laser Focus 18(12), 49–53(1982).

8.6 CsH_2AsO_4，砷酸二氢铯(CDA)

负单轴晶：$n_o > n_e$

分子量：273.840

密度：3.53 g/cm^3 [1]

点群：$\bar{4}2m$

晶格常数[2]：

$a = (7.9852 \pm 0.0004)$ Å, $T = 298$ K

$c = (7.8928 \pm 0.0003)$ Å, $T = 298$ K

居里温度：143 K[2]

线性热膨胀系数[2]

$\alpha_t \times 10^6 / K^{-1}$, ∥ c	$\alpha_t \times 10^6 / K^{-1}$, ⊥ c
49	12

以 $\theta = 90°$, $\phi = 45°$ 切割的 17.5 mm 长的晶体以 0.5 透过计的透明范围：0.26~1.43 μm[3]

以"0"透过计的透明范围 UV 截止边：0.216 μm[4]

以"0"透过计的透明范围 IR 截止边：对 o 光 1.87 μm，对 e 光 1.67 μm[5]

线性吸收系数 α

$\lambda/\mu m$	α/cm^{-1}	参考文献	$\lambda/\mu m$	α/cm^{-1}	参考文献
0.35~1.4	0.6	[4]	1.064	0.041	[3]
1.062	0.041	[6]			

双光子吸收系数 β[7]

$\lambda/\mu m$	$\beta \times 10^{11}/(cm \cdot W^{-1})$	备注
0.355	2.81	e 光，$\theta = 90°$，$\phi = 45°$

折射率的实验值[3]

$\lambda/\mu m$	n_o	n_e	$\lambda/\mu m$	n_o	n_e
0.347 2	1.602 7	1.572 2	0.694 3	1.563 2	1.542 9
0.532 1	1.573 3	1.551 4	1.064 2	1.551 6	1.533 0

折射率的温度微商[8]

$\lambda/\mu m$	$dn_o/dT \times 10^5/K^{-1}$	$dn_e/dT \times 10^5/K^{-1}$	$\lambda/\mu m$	$dn_o/dT \times 10^5/K^{-1}$	$dn_e/dT \times 10^5/K^{-1}$
0.405	-3.15	-1.89	0.578	-2.76	-2.39
0.436	-3.05	-2.09	0.633	-2.80	-2.56
0.546	-2.59	-2.12			

最佳色散关系方程（λ 以 μm 为单位，$T = 293$ K）[5]：

$$n_o^2 = 1.877\,632\,8 - 0.036\,022\,22\lambda^2 + 0.005\,234\,121\lambda^4 + \frac{0.550\,395\,1\lambda^2}{\lambda^2 - (0.162\,570\,0)^2}$$

$$n_e^2 = 1.686\,288\,9 - 0.013\,722\,44\lambda^2 + 0.003\,948\,463\lambda^4 + \frac{0.669\,457\,1\lambda^2}{\lambda^2 - (0.146\,471\,2)^2}$$

在文献[8]、[9]中给出了其他色散关系。

普遍情况下有效二阶非线性系数的表达式（Kleinman 对称条件成立，$d_{14} = d_{25} = d_{36}$）[10]：

$$d_{ooe} = -d_{36}\sin(\theta + \rho)\sin 2\phi$$

$$d_{eoe} = d_{oee} = 2d_{36}\sin(\theta + \rho)\cos(\theta + \rho)\cos 2\phi$$

简化的有效二阶非线性系数表达式（小双折射角近似，Kleinman 对称条件成立，$d_{14} = d_{25} = d_{36}$）[11]：

$$d_{ooe} = -d_{36}\sin\theta\sin 2\phi$$

$$d_{eoe} = d_{oee} = d_{36}\sin 2\theta\cos 2\phi$$

二阶非线性系数的绝对值：

$$d_{36}(1.064\,2\ \mu m) = (0.40 \pm 0.05)\ pm/V\ [3]$$

相位匹配角的实验值（$T = 293$ K）

相互作用的波长/μm	θ_{exp}/(°)	参 考 文 献
SHG, o+o⇒e		
1.05⇒0.525	90	[12]
1.052⇒0.526	90	[8]
1.06⇒0.53	87	[13], [14]
1.064 2⇒0.532 1	83.5	[15], [16]
	84.2	[3]
	84.4	[17]

NCPM 温度的实验值

相互作用的波长/μm	T/℃	参 考 文 献	备 注
SHG, o+o⇒e			
1.05⇒0.525	20	[12]	
1.052⇒0.526	20	[8]	
1.06⇒0.53	31	[13]	
1.064 2⇒0.532 1	39.6	[3]	20 Hz
	40.3	[18]	10 Hz
	41	[16]	
	42	[19]	
	43	[17]	
	44.5	[20]	
	45	[6]	
	46	[15]	12.5 Hz
	48	[3]	0.1~1 Hz
	49.2	[21]	10 Hz
1.073⇒0.536 5	61	[19]	
1.078⇒0.539	100	[12]	

内角带宽和温度带宽的实验值

相互作用的波长/μm	T/℃	θ_{pm}/(°)	$\Delta\theta^{int}$/(°)	ΔT/℃	参考文献
SHG, o+o⇒e					
1.06⇒0.53	22	87	≈0.4		[13]
	31	90	≈3.8	≈3	[13]
	20	87	0.43		[14]
	63(?)	90	3.03		[14]
1.062⇒0.531	45	90	2.85	6.5	[6]
1.064 2⇒0.532 1	40.3	90		6.8	[18]
	24	83.5	0.86	~8	[15]

续表

相互作用的波长/μm	T/℃	θ_{pm}/(°)	$\Delta\theta^{int}$/(°)	ΔT/℃	参考文献
	46	90	3.2		[15]
	20	84.15	0.70		[3]
	48	90	2.91	6 ± 0.2	[3]
	20	84.4	0.70		[17]
	43	90	≈3		[17]

光谱带宽的实验值[14]

相互作用的波长/μm	T/℃	θ_{pm}/(°)	$\Delta\nu$/cm^{-1}
SHG, o+o⇒e			
1.06⇒0.53	20	87	199
	63(?)	90	158

相位匹配角的温度变量

相互作用的波长/μm	T/℃	θ_{pm}/(°)	$d\theta_{pm}/dT/[(°)\cdot K^{-1}]$	参考文献
SHG, o+o⇒e				
1.06⇒0.53	20	87	0.085	[14]
	63(?)	90	0.481	[14]
1.064 2⇒0.532 1	24	83.5	0.129	[15]
	20	84.4	0.131	[17]
	35	86.5	0.194	[17]
	39	87.6	0.251	[17]
	41	88.3	0.537	[17]

非临界 SHG 的温度调谐[8]

相互作用的波长/μm	$d\lambda_1/dT/(nm\cdot K^{-1})$
SHG, o+o⇒e	
1.052⇒0.526	0.308

非临界 SHG 过程的双折射温度变化（1.064 2 μm⇒0.532 1 μm, o+o⇒e）:

$$d(n_2^e - n_1^o)/dT = 7.2\times10^{-6} K^{-1} [18]$$

$$d(n_2^e - n_1^o)/dT = (8.0\pm0.2)\times10^{-6} K^{-1} [3]$$

激光诱导的损伤阈值

$\lambda/\mu m$	τ_p/ns	$I_{thr}/(GW \cdot cm^{-2})$	参 考 文 献	备　　注
0.532	10	>0.3	[21]	
1.062	0.007	>4	[6]	
1.064	12	>0.26	[3]	10~20 Hz
	10	0.35	[15]	12.5 Hz
	18	0.4	[18]	2~50 Hz

关于这一晶体

　　CDA(和它的氘化同构晶体 DCDA 一起)在 1970 年广泛用于 Nd:YAG 激光辐射的 NCPM。

■ 参考文献

[1] V. G. Dmitriev, G. G. Gurzadyan, D. N. Nikogosyan: *Handbook of Nonlinear Optical Crystals*, Third Revised Edition(Springer, Berlin, 1999).

[2] W. R. Cook, Jr.: Thermal expansion of crystals with KH_2PO_4 structure. J. Appl. Phys. 38(4), 1637-1642(1967).

[3] K. Kato: Second-harmonic generation in CDA and CD*A. IEEE J. Quant. Electr. QE-10(8), 616-618(1974).

[4] A. S. Sonin, A. S. Vasilevskaya: *Elektrooptic Crystals* (Atomizdat, Moscow, 1971) [In Russian].

[5] D. Eimerl: Electro-optic, linear and nonlinear optical properties of KDP and its isomorphs. Ferroelectrics 72(1-4), 95-139(1987).

[6] T. A. Rabson, H. J. Ruiz, P. L. Shah, F. K. Tittel: Efficient second harmonic generation of picosecond laser pulses. Appl. Phys. Lett. 20(8), 282-284(1972).

[7] P. Liu, W. L. Smith, H. Lotem, J. H. Bechtel, N. Bloembergen, R. S. Adhav: Absolute two-photon absorption coefficients at 355 and 266 nm. Phys. Rev. B 17(12), 4620-4632(1978).

[8] N. P. Barnes, D. J. Gettemy, R. S. Adhav: Variations of the refractive index with temperature and the tuning rate for KDP isomorphs. J. Opt. Soc. Am. 72(7), 895-898(1982).

[9] K. W. Kirby, L. G. DeShazer: Refractive indices of 14 nonlinear crystals isomorphic to KH_2PO_4. J. Opt. Soc. Am. B 4(7), 1072-1078(1987).

[10] R. C. Eckardt, H. Masuda, Y. X. Fan, R. L. Byer: Absolute and relative nonlinear optical coefficients of KDP, KD*P, BaB_2O_4, $LiIO_3$, $MgO:LiNbO_3$, and KTP measured by phasematched second-harmonic generation. IEEE J. Quant. Electr. 26(5), 922-933(1990).

[11] J. E. Midwinter, J. Warner: The effects of phase matching method and of uniaxial crystal symmetry on the polar distribution of second-order non-linear optical polarization. Brit. J. Appl. Phys. 16(11), 1135 – 1142(1965).

[12] R. S. Adhav: Materials for optical harmonic generation. Laser Focus 19(6), 73 – 78 (1983).

[13] V. S. Suvorov, I. S. Rez: Second-harmonic generation without birefringence in CDA (CsH_2AsO_4) crystal at room temperature. Opt. Spektrosk. 27(1), 181 – 183(1969) [English trans.: Opt. Spectrosc. USSR 27(1), 94 – 95(1969)].

[14] R. B. Andreev, V. D. Volosov, A. G. Kalintsev: Spectral, angular, and temperature characteristics of HIO_3, $LiIO_3$, CDA, DKDP, KDP and ADP non-linear crystals in second-and fourth-harmonic generation. Opt. Spektrosk. 37(2), 294 – 299(1974) [English trans.: Opt. Spectrosc. USSR 37(2), 169 – 171(1974)].

[15] Y. D. Golyaev, V. G. Dmitriev, I. Y. Itskhoki, V. N. Krasnyanskaya, I. S. Rez, E. A. Shalaev: Efficient frequency doubler utilizing a cesium dihydrogen arsenate crystal. Kvantovaya Elektron. No. 1, 122 – 123(1973) [English trans.: Sov. J. Quantum Electron. 3(1), 72 – 73(1973)].

[16] R. S. Adhav, A. D. Vlassopoulos: Guide to efficient doubling. Laser Focus 10(5), 47- 48(1974).

[17] K. V. Vetrov, V. D. Volosov, A. G. Kalintsev: Nonlinear characteristics of CDA and DCDA in neodymium-laser second-harmonic generation. Izv. Akad. Nauk SSSR, Ser. Fiz. 52(2), 301 – 303(1988) [English trans.: Bull. Acad. Sci. USSR, Ser. Phys. 52 (2), 78 – 79(1988)].

[18] K. Kato: Efficient second harmonic generation in CDA. Opt. Commun. 9(3), 249 – 251(1973).

[19] R. S. Adhav, R. W. Wallace: Second harmonic generation in 90° phase-matched KDP isomorphs. IEEE J. Quant. Electr. QE – 9(8), 855 – 856(1973).

[20] G. A. Massey, M. D. Jones, J. C. Johnson: Generation of pulse bursts at 212.8 nm by intracavity modulation of an Nd: YAG laser. IEEE J. Quant. Electr. QE – 14(7), 527- 532(1978).

[21] G. A. Massey, R. A. Elliot: Tunable infrared parametric generation in cesium dihydrogen arsenate. IEEE J. Quant. Electr. QE – 10(12), 899 – 900(1974).

8.7　CsD_2AsO_4，氘化砷酸二氢铯(DCDA)

负单轴晶：$n_o > n_e$
分子量：275.853
密度：3.53 g/cm^3 [1]

点群:$\bar{4}2m$

以 $\theta=90°$,$\phi=45°$ 切割的 13.5 mm 长晶体以 0.5 透过计的透明范围:0.27~1.66 μm[2]

以"0"透过计的透明范围 IR 截止边:对 o 光 2.03 μm,对 e 光 1.78 μm[3]

线性吸收系数 α

$\lambda/\mu m$	α/cm^{-1}	参考文献	$\lambda/\mu m$	α/cm^{-1}	参考文献
1.062	0.01	[4]	1.064	0.02	[2]

双光子吸收系数 β[5]

$\lambda/\mu m$	$\beta\times10^{11}/(cm\cdot W^{-1})$	备注
0.355	8.0	o 光,$\theta=90°$,$\phi=45°$
	5.1	e 光,$\theta=90°$,$\phi=45°$

折射率的实验值[2]

$\lambda/\mu m$	n_o	n_e	$\lambda/\mu m$	n_o	n_e
0.347 2	1.589 5	1.568 5	0.694 3	1.559 6	1.541 8
0.532 1	1.568 1	1.549 5	1.064 2	1.550 3	1.532 6

折射率的温度微商[6]

$\lambda/\mu m$	$dn_o/dT\times10^5/K^{-1}$	$dn_e/dT\times10^5/K^{-1}$	$\lambda/\mu m$	$dn_o/dT\times10^5/K^{-1}$	$dn_e/dT\times10^5/K^{-1}$
0.405	-2.26	-1.77	0.578	-2.31	-1.71
0.436	-2.26	-1.51	0.633		-1.70
0.546	-2.47	-1.64			

最佳色散关系方程(λ 以 μm 为单位,$T=293$ K)[3]:

$$n_o^2=1.627\,849\,6-0.018\,220\,310\lambda^2+0.000\,281\,333\,1\lambda^4+\frac{0.780\,817\,0\lambda^2}{\lambda^2-(0.140\,769\,9)^2}$$

$$n_e^2=1.623\,606\,3-0.009\,338\,692\lambda^2+0.001\,965\,413\,0\lambda^4+\frac{0.724\,958\,9\lambda^2}{\lambda^2-(0.141\,485\,0)^2}$$

在文献[6]、[7]中给出了其他色散关系。

普遍情况下有效二阶非线性系数的表达式(Kleinman 对称条件成立,$d_{14}=d_{25}=d_{36}$)[8]:

$$d_{ooe}=-d_{36}\sin(\theta+\rho)\sin2\phi$$

$$d_{eoe} = d_{oee} = 2d_{36}\sin(\theta+\rho)\cos(\theta+\rho)\cos 2\phi$$

有效二阶非线性系数的表达式(小双折射角近似,Kleinman 对称条件成立,$d_{14} = d_{25} = d_{36}$)[9]:

$$d_{ooe} = -d_{36}\sin\theta\sin 2\phi$$

$$d_{eoe} = d_{oee} = d_{36}\sin 2\theta\cos 2\phi$$

二阶非线性系数的绝对值:

$$d_{36}(1.064\ 2\ \mu m) = (0.40 \pm 0.05)\ pm/V^{[2]}$$

相位匹配角的实验值($T = 293$ K)

相互作用的波长/μm	$\theta_{exp}/(°)$	参 考 文 献
SHG, o + o⇒e		
1.034⇒0.517	90	[10]
1.037⇒0.518 5	90	[6]
1.064 2⇒0.532 1	79.35	[2]
	80.8	[11]

NCPM 温度的实验值

相互作用的波长/μm	T/℃	参考文献	备 注
SHG, o + o⇒e			
1.034⇒0.517	20	[10]	
1.037⇒0.518 5	20	[6]	
1.064 2⇒0.532 1	96.4	[11]	70%氘化度
	102	[12],[13]	
	108	[10]	
	109.8	[2]	90%氘化度,20 Hz
	112.3	[2]	90%氘化度,<1 Hz

内角带宽和温度带宽的实验值

相互作用的波长/μm	T/℃	$\theta_{pm}/(°)$	$\Delta\theta^{int}/(°)$	ΔT/℃	参考文献
SHG, o + o⇒e					
1.064 2⇒0.532 1	20	79.35	0.41		[2]
	20	80.8	0.50		[11]
	96.4	90	≈3.5		[11]
	112.3	90	2.90	6.1 ± 0.1	[2]

相位匹配角的温度变化[11]

相互作用的波长/μm	$T/℃$	$\theta_{pm}/(°)$	$d\theta_{pm}/dT/(°)\cdot K^{-1}$
SHG, o+o⇒e			
1.064 2⇒0.532 1	20	80.8	0.042
	66.3	84.3	0.081
	80	86.4	0.270
	87.7	88.1	0.533

非临界 SHG 的温度调谐[6]

相互作用的波长/μm	$d\lambda_1/dT/(nm\cdot K^{-1})$
SHG, o+o⇒e	
1.037⇒0.518 5	0.317

非临界 SHG 过程的双折射温度变化(1.064 2 μm⇒0.532 1 μm, o+o⇒e):

$$\frac{d(n_2^e - n_1^o)}{dT} = (7.8 \pm 0.2) \times 10^{-6} K^{-1}\,^{[2]}$$

激光诱导的损伤阈值

$\lambda/\mu m$	τ_p/ns	$I_{thr}/(GW\cdot cm^{-2})$	参考文献	备注
1.064	12	>0.26	[2]	10~20 Hz
		>0.25	[14]	0.1~20 Hz

关于这一晶体

DCDA(和其同构晶体 CDA 一起)在 1970 年广泛用于 Nd:YAG 激光辐射的 NCPM。

参考文献

[1] V. G. Dmitriev, G. G. Gurzadyan, D. N. Nikogosyan: *Handbook of Nonlinear Optical Crystals*, *Third Revised Edition*(Springer, Berlin, 1999).

[2] K. Kato: Second-harmonic generation in CDA and CD*A. IEEE J. Quant. Electr. QE-10(8), 616-618(1974).

[3] D. Eimerl: Electro-optic, linear and nonlinear optical properties of KDP and its isomorphs. Ferroelectrics 72(1-4), 95-139(1987).

[4] T. A. Rabson, H. J. Ruiz, P. L. Shah, F. K. Tittel: Efficient second harmonic generation of picosecond laser pulses. Appl. Phys. Lett. 20(8), 282-284(1972).

[5] P. Liu, W. L. Smith, H. Lotem, J. H. Bechtel, N. Bloembergen, R. S. Adhav: Abso-

lute two-photon absorption coefficients at 355 and 266 nm. Phys. Rev. B 17(12), 4620-4632(1978).

[6] N. P. Barnes, D. J. Gettemy, R. S. Adhav: Variations of the refractive index with temperature and the tuning rate for KDP isomorphs. J. Opt. Soc. Am. 72(7), 895 – 898 (1982).

[7] K. W. Kirby, L. G. DeShazer: Refractive indices of 14 nonlinear crystals isomorphic to KH_2PO_4. J. Opt. Soc. Am. B 4(7), 1072 – 1078(1987).

[8] R. C. Eckardt, H. Masuda, Y. X. Fan, R. L. Byer: Absolute and relative nonlinear optical coefficients of KDP, KD*P, BaB_2O_4, $LiIO_3$, $MgO:LiNbO_3$, and KTP measured by phasematched second-harmonic generation. IEEE J. Quant. Electr. 26(5), 922 – 933 (1990).

[9] J. E. Midwinter, J. Warner: The effects of phase matching method and of uniaxial crystal symmetry on the polar distribution of second-order non-linear optical polarization. Brit. J. Appl. Phys. 16(11), 1135 – 1142(1965).

[10] R. S. Adhav: Materials for optical harmonic generation. Laser Focus 19(6), 73 – 78 (1983).

[11] K. V. Vetrov, V. D. Volosov, A. G. Kalintsev: Nonlinear characteristics of CDA and DCDA in neodymium-laser second-harmonic generation. Izv. Akad. Nauk SSSR, Ser. Fiz. 52(2), 301 – 303(1988) [In Russian, English trans.: Bull. Acad. Sci. USSR, Ser. Phys. 52(2), 78 – 79(1988)].

[12] R. S. Adhav, R. W. Wallace: Second harmonic generation in 90° phase-matched KDP isomorphs. IEEE J. Quant. Electr. QE – 9(8), 855 – 856(1973).

[13] R. S. Adhav, A. D. Vlassopoulos: Guide to efficient doubling. Laser Focus 10(5), 47-48(1974).

[14] K. Kato: Conversion of high power Nd:YAG laser radiation to the UV at 2 661 Å. Opt. Commun. 13(4), 361 – 362(1975).

8.8　RbH_2PO_4，磷酸二氢铷(RDP)

负单轴晶：$n_o > n_e$

分子量：182.454

密度：2.805 g/cm^3 [1]

点群：$\bar{4}2m$

晶格常数[2]：$a = (7.608 \pm 0.008)$ Å；$c = (7.296 \pm 0.007)$ Å

居里温度：147 K [3,4]

线性热膨胀系数[4]

$\alpha_t \times 10^6/\text{K}^{-1}$, $\parallel c$	$\alpha_t \times 10^6/\text{K}^{-1}$, $\perp c$
42.5	19

以 $\theta = 50°$，$\phi = 45°$ 切割的 15.3 mm 长晶体以 0.5 透过计的透明范围：0.19 ~ 1.38 μm[5]

以 "0" 透过计的透明范围 IR 截止边：对 o 光 1.65 μm，对 e 光 1.87 μm[6]

线性吸收系数 α

$\lambda/\mu m$	α/cm^{-1}	参考文献	备注
0.25 ~ 1.25	< 0.03	[7]	
0.354 7	0.015	[5]	$\theta = 50°$，$\phi = 45°$
0.532 1	0.01	[5]	$\theta = 50°$，$\phi = 45°$
1.064 2	0.041	[5]	$\theta = 50°$，$\phi = 45°$

双光子吸收系数 β[8]

$\lambda/\mu m$	$\beta \times 10^{11}/(\text{cm} \cdot \text{W}^{-1})$	备注
0.355	0.59	e 光，$\theta = 90°$，$\phi = 45°$

折射率的实验值

$\lambda/\mu m$	n_o	n_e	参考文献	$\lambda/\mu m$	n_o	n_e	参考文献
0.347 2	1.528 4	1.496 9	[9]	0.532 1	1.510 6	1.481 1	[10]
0.435 8	1.516 5	1.485 7	[9]	0.546 8	1.508 2	1.479 0	[9]
0.476 5	1.514 0	1.486 1	[10]	0.550 0	1.509 3	1.480 4	[3]
0.488 0	1.513 2	1.483 2	[10]	0.589 3	1.505 3	1.476 5	[9]
0.496 5	1.512 6	1.482 7	[10]	0.600 0	1.506 7	1.478 4	[3]
0.500 0	1.512 5	1.481 3	[3]	0.650 0	1.504 6	1.476 7	[3]
0.501 7	1.512 1	1.482 5	[10]	0.694 3	1.502 0	1.473 5	[9]
0.514 5	1.511 6	1.482 0	[10]	1.064 2	1.492 6	1.470 0	[10]

$\lambda/\mu m$	n_o	参考文献	$\lambda/\mu m$	n_e	参考文献
0.469 9	1.514 8	[11]	0.465 8	1.485 1	[11]
0.495 0	1.512 8	[11]	0.478 0	1.484 5	[11]
0.512 0	1.511 7	[11]	0.495 0	1.483 3	[11]
0.532 9	1.510 4	[11]	0.532 4	1.481 0	[11]

续表

$\lambda/\mu m$	n_o	参考文献	$\lambda/\mu m$	n_e	参考文献
0.585 1	1.507 4	[11]	0.557 7	1.479 8	[11]
0.598 0	1.506 9	[11]	0.587 8	1.478 7	[11]
0.624 5	1.505 6	[11]	0.616 5	1.477 6	[11]
0.647 4	1.504 7	[11]	0.652 1	1.476 6	[11]
0.666 2	1.504 2	[11]	0.664 0	1.476 3	[11]

折射率的温度微商[12]

$\lambda/\mu m$	$dn_o/dT \times 10^5/K^{-1}$	$dn_e/dT \times 10^5/K^{-1}$	$\lambda/\mu m$	$dn_o/dT \times 10^5/K^{-1}$	$dn_e/dT \times 10^5/K^{-1}$
0.405	−3.69	−2.67	0.578	−3.72	−2.80
0.436	−3.86	−2.76	0.633	−3.72	−2.89
0.546	−3.72	−2.54			

最佳色散关系方程(λ 以 μm 为单位,$T = 293$ K)[13]:

$$n_o^2 = 2.249\,885 + \frac{3.688\,005\lambda^2}{\lambda^2 - (11.278\,29)^2} + \frac{0.010\,560}{\lambda^2 - (0.088\,207)^2}$$

$$n_e^2 = 2.159\,913 + \frac{0.988\,431\lambda^2}{\lambda^2 - (11.300\,13)^2} + \frac{0.009\,515}{\lambda^2 - (0.092\,076)^2}$$

在文献[6]、[10]、[12]中给出了其他色散关系。

$T = 295$ K 时低频(远低于 RDP 晶体的声学谐振频率,即对"自由"晶体)下测量的线性电光系数[3]

$\lambda/\mu m$	$r_{41}^T/(pm \cdot V^{-1})$	$r_{63}^T/(pm \cdot V^{-1})$
0.632 8	12.5 ± 0.2	7.7 ± 0.3

普遍情况下有效二阶非线性系数的表达式(Kleinman 对称条件成立,$d_{14} = d_{25} = d_{36}$)[14]:

$$d_{ooe} = -d_{36}\sin(\theta + \rho)\sin 2\phi$$

$$d_{eoe} = d_{oee} = 2d_{36}\sin(\theta + \rho)\cos(\theta + \rho)\cos 2\phi$$

有效二阶非线性系数的表达式(小双折射角近似,Kleinman 对称条件成立,$d_{14} = d_{25} = d_{36}$)[15]:

$$d_{ooe} = -d_{36}\sin\theta\sin 2\phi$$

$$d_{eoe} = d_{oee} = d_{36}\sin 2\theta\cos 2\phi$$

二阶非线性系数的绝对值:

$$d_{36}(0.694\ 3\ \mu m) = 1.04 \times d_{36}(KDP) \pm 15\% = (0.41 \pm 0.06)\ pm/V^{[16,17]}$$
$$d_{36}(0.694\ 3\ \mu m) = 0.92 \times d_{36}(KDP) \pm 10\% = (0.36 \pm 0.04)\ pm/V^{[17,18]}$$

相位匹配角的实验值($T = 293\ K$)

相互作用的波长/μm	$\theta_{exp}/(°)$	参 考 文 献
SHG, o + o ⇒ e		
0.626 ⇒ 0.313	90	[12]
0.627 ⇒ 0.313 5	90	[19]
0.627 5 ⇒ 0.313 75	90	[20]
0.629 4 ⇒ 0.314 7	86.6	[20]
0.632 8 ⇒ 0.316 4	83.2	[21]
0.638 6 ⇒ 0.319 3	78.9	[20]
0.655 0 ⇒ 0.327 5	73.9	[20]
0.670 0 ⇒ 0.335 0	70.8	[20]
0.694 3 ⇒ 0.347 15	66	[9]
1.064 2 ⇒ 0.532 1	50.8	[5], [22]
	50.9	[23]
1.152 3 ⇒ 0.576 15	51	[21]
SHG, e + o ⇒ e		
1.064 2 ⇒ 0.532 1	83.1	[22]
1.152 3 ⇒ 0.576 15	77.1	[21]
THG, o + o ⇒ e		
1.064 2 + 0.532 1 ⇒ 0.354 7	61.2	[5]

NCPM 温度的实验值

相互作用的波长/μm	$T/°C$	参 考 文 献
SHG, o + o ⇒ e		
0.627 ⇒ 0.313 5	20	[19], [23]
0.627 5 ⇒ 0.313 75	20	[20]
0.635 ⇒ 0.317 5	100	[19], [23]
0.637 ⇒ 0.318 5	98	[20]

$T = 293\ K$ 时内角带宽的实验值

相互作用的波长/μm	$\theta_{pm}/(°)$	$\Delta\theta^{int}/(°)$	参 考 文 献
SHG, o + o ⇒ e			
0.627 5 ⇒ 0.313 75	90	1.73	[20]
0.694 3 ⇒ 0.347 15	66	0.14	[24]
1.064 2 ⇒ 0.532 1	50.8	0.10	[22]
		0.11	[5]

续表

相互作用的波长/μm	$\theta_{pm}/(°)$	$\Delta\theta^{int}/(°)$	参 考 文 献
SHG, e+o⇒e			
1.064 2⇒0.532 1		0.40	[25]
	83.1	0.54	[22]
THG, o+o⇒e			
1.064 2+0.532 1⇒0.354 7	61.2	0.08	[5]

非临界 SHG 的温度调谐

相互作用的波长/μm	$d\lambda_1/dT/(nm \cdot K^{-1})$	参 考 文 献
SHG, o+o⇒e		
0.626⇒0.313	0.12	[12]
0.627 5⇒0.313 75	0.123	[20]

非临界 SHG 过程温度带宽的实验值(0.627 5 μm⇒0.313 75 μm, o+o⇒e)

$$\Delta T = (2.5 \pm 0.3)℃^{[20]}$$

非临界 SHG 过程双折射的温度变化(0.627 5 μm⇒0.313 75 μm, o+o⇒e):

$$\frac{d(n_2^e - n_1^o)}{dT} = (1.1 \pm 0.1) \times 10^{-5} K^{-1\,[20]}$$

激光诱导的体损伤阈值

$\lambda/\mu m$	τ_p/ns	$I_{thr}/(GW \cdot cm^{-2})$	参 考 文 献	备 注
0.628 1	330	0.55	[26]	
0.694 3	10	>0.18	[24]	
1.064 2	12	>0.26	[22]	10~20 Hz

关于这一晶体

RDP 在 20 世纪 60 年代末期到 70 年代中期相当普遍地用于红宝石和染料激光辐射的 SHG。因为这些激光器风光不再，RDP 的应用也就停止了。

■ 参考文献

[1] V. G. Dmitriev, G. G. Gurzadyan, D. N. Nikogosyan: *Handbook of Nonlinear Optical Crystals*, Third Revised Edition(Springer, Berlin, 1999).

[2] S. Haussühl: Elastische und thermoelastische Eigenschaften von KH_2PO_4, KH_2AsO_4, $NH_4H_2PO_4$, $NH_4H_2AsO_4$ und RbH_2PO_4. Z. Kristallogr. 120(6), 401–414(1964)[In German].

[3] E. N. Volkova, B. M. Berezhnoi, A. N. Izrailenko, A. V. Mishchenko, L. N. Rashkovich: Electro-optic and optical properties of partly deuterated rubidium dihydrogen phosphate crystals. Izv. Akad. Nauk SSSR, Ser. Fiz. 35(9), 1858 – 1861(1971) [In Russian, English trans.: Bull. Acad. Sci. USSR, Ser. Phys. 35(9), 1690 – 1693(1971)].

[4] W. R. Cook, Jr.: Thermal expansion of crystals with KH_2PO_4 structure. J. Appl. Phys. 38(4), 1637 – 1642(1967).

[5] K. Kato: Efficient UV generation at 3 547 Å in RDP. Appl. Phys. Lett. 25(6), 342 – 343(1974).

[6] D. Eimerl: Electro-optic, linear and nonlinear optical properties of KDP and its isomorphs. Ferroelectrics 72(1 – 4), 95 – 139(1987).

[7] A. S. Sonin, A. S. Vasilevskaya: *Elektrooptic Crystals* (Atomizdat, Moscow, 1971) [In Russian].

[8] P. Liu, W. L. Smith, H. Lotem, J. H. Bechtel, N. Bloembergen, R. S. Adhav: Absolute two-photon absorption coefficients at 355 and 266 nm. Phys. Rev. B 17(12), 4620-4632(1978).

[9] A. S. Vasilevskaya, M. F. Koldobskaya, L. G. Lomova, V. P. Popova, T. A. Regulskaya, I. S. Rez, Y. P. Sobesskii, A. S. Sonin, V. S. Suvorov: Some physical properties of rubidium dihydrogen phosphate single crystals. Kristallogr. 12(3), 447 – 450(1967) [In Russian, English trans.: Sov. Phys. -Crystallogr. 12(3), 383 – 385(1967)].

[10] S. Singh: Nonlinear Optical Materials. In: *Handbook of Lasers*, ed. by R. G. Pressley (The Chemical Rubber Co., Cleveland, 1971), pp. 489 – 525.

[11] E. N. Volkova, S. L. Faerman: Refractive indices of $KD_{2x}H_{2(1-x)}PO_4$ and $RbD_{2x}H_{2(1-x)}PO_4$ crystals. Kvant. Elektron. 3(11), 2508 – 2511(1976) [In Russian, English trans.: Sov. J. Quantum Electron. 6(11), 1380 – 1382(1976)].

[12] N. P. Barnes, D. J. Gettemy, R. S. Adhav: Variations of the refractive index with temperature and the tuning rate for KDP isomorphs. J. Opt. Soc. Am. 72(7), 895 – 898 (1982).

[13] K. W. Kirby, L. G. DeShazer: Refractive indices of 14 nonlinear crystals isomorphic to KH_2PO_4. J. Opt. Soc. Am. B 4(7), 1072 – 1078(1987).

[14] R. C. Eckardt, H. Masuda, Y. X. Fan, R. L. Byer: Absolute and relative nonlinear optical coefficients of KDP, KD*P, BaB_2O_4, $LiIO_3$, $MgO:LiNbO_3$, and KTP measured by phasematched second-harmonic generation. IEEE J. Quant. Electr. 26(5), 922 – 933 (1990).

[15] J. E. Midwinter, J. Warner: The effects of phase matching method and of uniaxial crystal symmetry on the polar distribution of second-order non-linear optical polarization. Brit. J. Appl. Phys. 16(11), 1135 – 1142(1965).

[16] V. S. Suvorov, A. S. Sonin, I. S. Rez: Some nonlinear optical properties of crystals of

the KDP group. Zh. Eksp. Teor. Fiz. 53(1), 49 – 55(1967)[In Russian, English trans.: Sov. Phys. -JETP 26(1),33 – 37(1968)].

[17] D. A. Roberts: Simplified characterization of uniaxial and biaxial nonlinear optical crystals: a plea for standardization of nomenclature and conventions. IEEE J. Quant. Electr. 28(10), 2057 – 2074(1992).

[18] J. E. Pearson, G. A. Evans, A. Yariv: Measurement of the relative nonlinear coefficients of KDP, RDP, RDA, and LiIO$_3$. Opt. Commun. 4(5), 366 – 367(1972).

[19] R. S. Adhav: Materials for optical harmonic generation. Laser Focus 19(6), 73 – 78 (1983).

[20] K. Kato: Highly efficient frequency doubling of visible dye laser radiation in RDP. J. Appl. Phys. 46(6), 2721 – 2722(1975).

[21] M. P. Golovey, I. N. Kalinkina, G. I. Kosourov: On the nonlinear properties of RDP crystal. Opt. Spektrosk. 28(5), 991 – 992(1970)[In Russian, English trans.: Opt. Spectrosc. USSR 28(5),535 – 536(1970)].

[22] K. Kato, S. Nakao: Frequency doubling of Nd: YAG laser radiation in RDP. Jpn. J. Appl. Phys. 13(10), 1681 – 1682(1974).

[23] R. S. Adhav, A. D. Vlassopoulos: Guide to efficient doubling. Laser Focus 10(5), 47-48(1974).

[24] K. Kato, A. J. Alcock, M. C. Richardson: Conversion of high power ruby laser radiation to the UV in RDP. Opt. Commun. 11(1), 5 – 7(1974).

[25] E. V. Nilov, I. L. Yachnev: Some results on investigating RDP(RbH$_2$PO$_4$) crystal as laser frequency doubler. Zh. Prikl. Spektrosk. 7(6), 943 – 945(1967)[In Russian, English trans.: J. Appl. Spectrosc. 7(6),628 – 630(1967)].

[26] L. Armstrong, S. E. Neister, R. Adhav: Measuring CFP dye laser damage thresholds on UV doubling crystals. Laser Focus 18(12), 49 – 53(1982).

8.9 CsTiOAsO$_4$，砷酸钛氧铯(CTA)

正双轴晶：$2V_Z = 52.9°$，$\lambda = 0.532\,1\ \mu m$[1]

分子量：335.704

密度：4.511 g/cm^3 [2]

点群：$mm2$

晶格常数：

    $a = 13.486$ Å[3]; 13.494 Å[4]

    $b = 6.861\,6$ Å[3]; 6.862\,7 Å[4]

    $c = 10.688$ Å[3]; 10.699 Å[4]

介电轴和晶体学轴的变换：$X,Y,Z \Rightarrow a,b,c$

居里温度：917 K[5]

熔点：1 322 K[6]

以"0"透过计的透明范围：$0.35 \sim 5.3~\mu m$[5]；$0.37 \sim 5.3~\mu m$[4]；$0.38 \sim 5.3~\mu m$[7]

室温下折射率的实验值[8]

$\lambda/\mu m$	n_X	n_Y	n_Z
0.66	1.877 1	1.893 9	1.951 9
1.32	1.844 1	1.859 0	1.915 0

最佳色散关系方程（λ 以 μm 为单位，$T=293$ K）[1]：

$$n_X^2 = 2.344\,98 + \frac{1.048\,63\lambda^2}{\lambda^2 - (0.220\,44)^2} - 0.014\,83\lambda^2$$

$$n_Y^2 = 2.744\,40 + \frac{0.707\,33\lambda^2}{\lambda^2 - (0.260\,33)^2} - 0.015\,26\lambda^2$$

$$n_Z^2 = 2.536\,66 + \frac{1.106\,00\lambda^2}{\lambda^2 - (0.249\,88)^2} - 0.017\,11\lambda^2$$

更精确的色散关系（λ 以 μm 为单位，对 n_X 和 n_Y，$0.4~\mu m < \lambda < 5.3~\mu m$；对 n_Z，$0.4~\mu m < \lambda < 2.1~\mu m$；$T=293$ K）[9,10]：

$$n_X^2 = 2.040\,8 + \frac{1.292\,4\lambda^{2.000\,8}}{\lambda^{2.000\,8} - 0.047\,575} + \frac{1.930\,4\lambda^{1.987\,4}}{\lambda^{1.987\,4} - 156.504\,9}$$

$$n_Y^2 = 2.433\,0 + \frac{0.959\,1\lambda^{1.985\,3}}{\lambda^{1.985\,3} - 0.068\,339} + \frac{4.229\,2\lambda^{1.933\,8}}{\lambda^{1.933\,8} - 305.922\,4}$$

$$n_Z^2 = 2.572\,3 + \frac{1.053\,2\lambda^{2.029\,7}}{\lambda^{2.029\,7} - 0.080\,077} + \frac{0.617\,8\lambda^{1.993\,4}}{\lambda^{1.993\,4} - 40.780\,6}$$

在文献[11]、[12]、[13]中给出了其他色散关系方程组。

室温下低频（远低于 CTA 晶体的声学谐振频率，即对"自由"晶体）下测量的线性电光系数[1]

$\lambda/\mu m$	$r_{13}^T/(\text{pm}\cdot\text{V}^{-1})$	$r_{23}^T/(\text{pm}\cdot\text{V}^{-1})$	$r_{33}^T/(\text{pm}\cdot\text{V}^{-1})$
0.632 8	14.2±1.4	18.5±1.9	38±3.8

CTA 晶体主平面上有效二阶非线性系数的表达式（小走离角近似，Kleinman 对称条件成立，$d_{15}=d_{31}$ 和 $d_{24}=d_{32}$）[14]：

XY 平面

$$d_{eoe} = d_{oee} = d_{31}\sin^2\phi + d_{32}\cos^2\phi$$

YZ 平面
$$d_{oeo} = d_{eoo} = d_{31}\sin\theta$$

XZ 平面，$\theta < V_z$
$$d_{ooe} = d_{32}\sin\theta$$

XZ 平面，$\theta > V_z$
$$d_{oeo} = d_{eoo} = d_{32}\sin\theta$$

在文献[14]中给出了 CTA 晶体任意方向上三波相互作用的有效二阶非线性系数。

CTA 二阶非线性系数的符号可能都是相同的[15]。

二阶非线性系数的绝对值：

$$d_{31}(1.064\ \mu m) = (2.1 \pm 0.4)\ pm/V^{[1]}$$
$$d_{32}(1.064\ \mu m) = (3.4 \pm 0.7)\ pm/V^{[1]}$$
$$d_{33}(1.064\ \mu m) = (18.1 \pm 1.8)\ pm/V^{[1]}$$
$$d_{31}(1.32\ \mu m) = (1.1 \pm 0.1)\ pm/V^{[8]}$$
$$d_{32}(1.32\ \mu m) = (1.7 \pm 0.6)\ pm/V^{[8]}$$

内角带宽和温度带宽的实验值

相互作用的波长/μm	$\phi_{pm}/(°)$	$\theta_{pm}/(°)$	$\Delta\phi^{int}/(°)$	$\Delta\theta^{int}/(°)$	参考文献
XY 平面，$\theta = 90°$					
SHG，e + o ⇒ e					
1.318 8 ⇒ 0.659 4	64.5		0.52		[11]
	64				[12]
	59		0.60		[8]
DFG，e − o ⇒ e					
0.530 9 − 0.782 2 ⇒ 1.652 5	41				[16]
YZ 平面，$\phi = 90°$					
SHG，o + e ⇒ o					
1.318 8 ⇒ 0.659 4		76			[12]
		73.1		0.29	[8]

关于这一晶体

作为 KTA 和 RTA 的同构晶体，CTA 尚未得到实际应用。

■ 参考文献

[1] L. T. Cheng, L. K. Cheng, J. D. Bierlein, F. C. Zumsteg: Nonlinear optical and elec-

trooptical properties of single crystal CsTiOAsO$_4$. Appl. Phys. Lett. 63(19), 2618 – 2620 (1993).

[2] B. H. T. Chai: Optical Crystals. In: *CRC Handbook of Laser Science and Technology, Supplement 2: Optical Materials*, ed. by M. J. Weber(CRC Press, Boca Raton, 1995), pp. 3 – 65.

[3] J. Protas, G. Marnier, B. Boulanger, B. Menaert: Structure crystalline de CsTiOAsO$_4$. Acta Crystallogr. C 45(8), 1123 – 1125(1989).

[4] D. T. Reid, M. Ebrahimzade, W. Sibbett: Design criteria and comparison of femtosecond optical parametric oscillators based on KTiOPO$_4$ and RbTiOAsO$_4$. J. Opt. Soc. Am. B 12 (11), 2168 – 2179(1995).

[5] L. K. Cheng, J. D. Bierlein: KTP and isomorphs—recent progress in device and material development. Ferroelectrics 142(1 – 2), 209 – 228(1993).

[6] L. K. Cheng, E. M. McCarron III, J. Calabrese, J. D. Bierlein, A. A. Ballman: Development of the nonlinear optical crystal CsTiOAsO$_4$. I. Structural stability. J. Cryst. Growth 132(1 – 2), 280 – 288(1993).

[7] J. Nordborg, G. Svensson, R. J. Bolt, J. Albertson: Top seeded solution growth of [Rb,Cs]TiOAsO$_4$. J. Cryst. Growth 224(3 – 4), 256 – 268(2001).

[8] B. Boulanger, J. P. Feve, G. Marnier, G. M. Loiacono, D. N. Loiacono C. Bonnin: SHG and internal conical refraction experiments in CsTiOAsO$_4$: comparison with KTiOPO$_4$ and KTiOAsO$_4$ for 1.32-μm type II SHG. IEEE J. Quant. Electr. 33 (6), 945 – 949 (1997).

[9] J. P. Feve, B. Boulanger, O. Pacaud, I. Rousseau, B. Menaert, G. Marnier: Refined Sellmeier equations from phase-matching measurements over the complete transparency range of KTiOAsO$_4$, RbTiOAsO$_4$, and CsTiOAsO$_4$. In: *Advanced Solid-State Lasers, OSA Trends in Optics and Photonics Series, Vol. 34*, ed. by H. Injeyan, U. Keller, C. Marshall(OSA, Washington DC, 2000), pp. 575 – 577.

[10] J. -P. Feve, B. Boulanger, O. Pacaud, I. Rousseau, B. Menaert, G. Marnier, P. Villeval, C. Bonnin, G. M. Loiacono, D. N. Loiacono: Phase-matching measurements and Sellmeier equations over the complete transparency range of KTiOAsO$_4$, RbTiOAsO$_4$, and CsTiOAsO$_4$. J. Am. Opt. Soc. B 17(5), 775 – 780(2000).

[11] L. K. Cheng, L. T. Cheng, F. C. Zumsteg, J. D. Bierlein, J. Galperin: Development of the nonlinear optical crystal CsTiOAsO$_4$. II. Crystal growth and characterization. J. Cryst. Growth 132(1 – 2), 289 – 296(1993).

[12] L. -T. Cheng, L. K. Cheng, J. D. Bierlein: Linear and nonlinear optical properties of the arsenate isomorphs of KTP. Proc. SPIE 1863, 43 – 53(1993).

[13] L. K. Cheng, L. T. Cheng, J. Galperin, P. A. Morris Hotsenpiller, J. D. Bierlein: Crystal growth and characterization of KTiOPO$_4$ isomorphs from the self-fluxes.

J. Cryst. Growth 137(1 − 2), 107 − 115(1994).

[14] V. G. Dmitriev, D. N. Nikogosyan: Effective nonlinearity coefficients for three-wave interactions in biaxial crystals of *mm*2 point group symmetry. Opt. Commun. 95(1 − 3), 173 − 182(1993).

[15] A. Anema, T. Rasing: Relative signs of the nonlinear coefficients of potassium titanyl phosphate. Appl. Opt. 36(24), 5902 − 5904(1997).

[16] B. Lai, N. C. Wong, L. K. Cheng: Continuous-wave tunable light source at 1.6 μm by difference-frequency mixing in $CsTiOAsO_4$. Opt. Lett. 20(17), 1779 − 1781(1995).

8.10 $Ba_2NaNb_5O_{15}$,铌酸钡钠(BNN)

负双轴晶:$2V_Z = 13°$[1]

分子量:100 2.173

密度:5.4 g/cm^3[2];5.407 6 g/cm^3[1],5.42 g/cm^3[3]

点群:*mm*2

298 K 时的晶格常数[4]:

$a = (17.625\ 60 \pm 0.000\ 05)$ Å

$b = (17.591\ 82 \pm 0.000\ 01)$ Å

$c = (3.994\ 915 \pm 0.000\ 004)$ Å

介电轴和晶体学轴的变换:$X, Y, Z \Rightarrow a, b, c$

熔点:1 703 K[2]

居里温度:833 K[2]

热导率[3]:$\kappa = 3.5$ W/mK

以"0"透过计的透明范围:0.37 ~ 5 μm[1,5]

线性吸收系数 α

$\lambda/\mu m$	α/cm^{-1}	参考文献	备 注
0.532 1	0.04	[3]	NCSHG 方向
	0.051 ~ 0.067	[6]	沿 *a* 轴
1.064 2	<0.002	[3]	NCSHG 方向
	0.003	[6]	沿 *a* 轴
	0.002	[7]	沿 *b* 轴

折射率的实验值[1]

$\lambda/\mu m$	n_X	n_Y	n_Z
0.457 9	2.428 4	2.426 6	2.293 1

续表

$\lambda/\mu m$	n_X	n_Y	n_Z
0.476 5	2.409 4	2.407 6	2.279 9
0.488 0	2.399 1	2.397 4	2.272 7
0.496 5	2.392 0	2.390 3	2.267 8
0.501 7	2.387 9	2.386 2	2.264 9
0.514 5	2.378 6	2.376 7	2.258 3
0.532 1	2.367 2	2.365 5	2.250 2
0.632 8	2.322 2	2.320 5	2.217 7
1.064 2	2.258 0	2.256 7	2.170 0

$\lambda = 1.064~\mu m$ 时 n_X 和 n_Z 的温度微商(n_Y 仅与 T 稍有关系)[1]：

$$\frac{dn_X}{dT} = -2.5 \times 10^{-5}~K^{-1}$$

$$\frac{dn_Z}{dT} = +8.0 \times 10^{-5}~K^{-1}$$

最佳色散关系方程(λ 以 μm 为单位，$T = 293~K$)[1]：

$$n_X^2 = 1 + \frac{3.949~5\lambda^2}{\lambda^2 - 0.040~388~94}$$

$$n_Y^2 = 1 + \frac{3.949~5\lambda^2}{\lambda^2 - 0.040~140~12}$$

$$n_Z^2 = 1 + \frac{3.600~8\lambda^2}{\lambda^2 - 0.032~198~71}$$

在文献[8]中给出了其他色散关系方程组。

BNN 晶体主平面上有效二阶非线性系数的表达式(Kleinman 对称条件不成立)[9]：

XY 平面

$$d_{eeo} = d_{31}\sin^2\phi + d_{32}\cos^2\phi$$

YZ 平面

$$d_{ooe} = d_{31}\sin\theta$$

XZ 平面，$\theta < V_Z$

$$d_{oeo} = d_{eoo} = d_{24}\sin\theta$$

XZ 平面，$\theta > V_Z$

$$d_{ooe} = d_{32}\sin\theta$$

BNN 晶体主平面上有效二阶非线性系数的表达式(Kleinman 对称条件成立，$d_{15} = d_{31}$ 和 $d_{24} = d_{32}$)[9]：

XY 平面
$$d_{eeo} = d_{31}\sin^2\phi + d_{32}\cos^2\phi$$

YZ 平面
$$d_{ooe} = d_{31}\sin\theta$$

XZ 平面，$\theta < V_Z$
$$d_{oeo} = d_{eoo} = d_{32}\sin\theta$$

XZ 平面，$\theta > V_Z$
$$d_{ooe} = d_{32}\sin\theta$$

在文献[9]中给出了 BNN 晶体内任意方向上的有效二阶非线性系数的表达式。二阶非线性系数的值：

$$d_{31}(1.064\ \mu m) = 40 \times d_{11}(SiO_2) \pm 5\% = (12 \pm 0.6)\ pm/V^{[1,10]}$$

$$d_{32}(1.064\ \mu m) = 40 \times d_{11}(SiO_2) \pm 10\% = (12 \pm 1.2)\ pm/V^{[1,10]}$$

$$d_{33}(1.064\ \mu m) = 55 \times d_{11}(SiO_2) \pm 7\% = (16.5 \pm 1.2)\ pm/V^{[1,10]}$$

相位匹配角的实验值（$T = 293$ K）[1]

相互作用的波长/μm	$\theta_{exp}/(°)$	相互作用的波长/μm	$\theta_{exp}/(°)$
YZ 平面，$\phi = 90°$		XZ 平面，$\phi = 0°$，$\theta > V_Z$	
SHG，o + o⇒e		SHG，o + o⇒e	
1.064 2⇒0.532 1	73.8	1.064 2⇒0.532 1	75.4

注：PM 角的值与熔体化学计量比强烈相关。

NCPM 温度及温度带宽的实验值

相互作用的波长/μm	$T/℃$	$\Delta T/℃$	参 考 文 献
沿 a 轴			
SHG，o + o⇒e			
1.064 2⇒0.532 1	85	0.45 ~ 0.47	[6]
	85		[11]
	86 ~ 87	0.45	[12]
	89	0.5	[1]
1.08⇒0.54		0.42	[13]
沿 b 轴			
SHG，o + o⇒e			
1.064 2⇒0.532 1	97		[14]
	101	0.5	[1]

注：NCPM 的温度值与熔体化学计量比强烈相关。

非临界SHG过程双折射的温度变化[1]：

沿 b 轴（$1.064\ 2\ \mu m \Rightarrow 0.532\ 1\ \mu m$）：

$$\frac{d[n_Z(2\omega)-n_X(\omega)]}{dT}=1.05\times10^{-4}\mathrm{K}^{-1}$$

激光诱导的损伤阈值

$\lambda/\mu m$	τ_p/ns	$I_{thr}/(GW \cdot cm^{-2})$	参考文献	备注
0.532 1	CW	>0.000 05	[14]	
	450	0.000 2	[15]	2 kHz
	0.05	0.072	[16]	1 kHz
1.064 2	450	0.004	[15]	2 kHz
	0.08	>0.002 5	[6]	500 MHz

关于这一晶体

由于BNN晶体具有很高的二阶非线性，这种晶体在20世纪60年代末和70年代很为人们所重视。然而，由于很难生长出完美的铌酸钡钠晶体来，晶体中存在的一些人们不希望的缺陷，如开裂和孪晶，限制了它的实际应用。最近，发现钕掺杂的BNN晶体（$Nd_xBa_{2-2x}Na_{1-x}Nb_5O_{15}$, $x=0.025$）可以获得高质量的晶体[17]。Nd:BNN的其他性质如下：点群 $4mm$；晶格常数，$a=(12.446\pm0.001)$Å，$c=(3.991\pm0.001)$Å；莫氏硬度，5；密度，5.43 g/cm³，比热容，300 J/(kg·K)；熔点，1 773 K；居里温度，810 K[17]。在文献[18]中，Nd:BNN被成功地用于自倍频。

■ **参考文献**

[1] S. Singh, D. A. Draegert, J. E. Geusic: Optical and ferroelectric properties of barium sodium niobate. Phys. Rev. B 2(7), 2709 – 2724(1970).

[2] L. G. van Uitert, J. J. Rubin, W. A. Bonner: Growth of BaNaNb₅O₁₅ single crystals for optical applications. IEEE J. Quant. Electr. QE – 4(10), 622 – 627(1968).

[3] J. D. Barry, C. J. Kennedy: Thermo-optical effects of intracavity Ba₂Na(NbO₃)₅ on a frequency-doubled Nd:YAG laser. IEEE J. Quant. Electr. QE – 11(8), 575 – 579 (1975).

[4] M. Ferriol: Crystal growth and structure of pure and rare-earth doped barium sodium niobate(BNN). Progr. Cryst. Growth Character. Mater. 43(2 – 3), 221 – 244(2001).

[5] J. E. Geusic, H. J. Levinstein, J. J. Rubin, S. Singh, L. G. van Uitert: The nonlinear optical properties of Ba₂NaNb₅O₁₅. Appl. Phys. Lett. 11(9), 269 – 271(1967).

[6] J. E. Murray, R. J. Pressley, J. H. Boyden, R. B. Webb: CW mode-locked source at 0.532 μm. IEEE J. Quant. Electr. QE-10(2), 263-267(1974).

[7] Y. Uematsu, T. Fukuda: Characteristics and performance of $KNbO_3$-Nd:YAG intracavity second harmonic generation. Jpn. J. Appl. Phys. 12(6), 841-844(1973).

[8] R. A. Andrews: IR image parametric up-conversion. IEEE J. Quant. Electr. QE-6(1), 68-80(1970).

[9] V. G. Dmitriev, D. N. Nikogosyan: Effective nonlinearity coefficients for three-wave interactions in biaxial crystals of *mm*2 point group symmetry. Opt. Commun. 95 (1-3), 173-182(1993).

[10] D. A. Roberts: Simplified characterization of uniaxial and biaxial nonlinear optical crystals: a plea for standardization of nomenclature and conventions. IEEE J. Quant. Electr. 28(10), 2057-2074(1992).

[11] J. E. Geusic, H. J. Levinstein, S. Singh, R. G. Smith, L. G. van Uitert: Continuous 0.532 μm solid-state source using $Ba_2NaNb_5O_{15}$. Appl. Phys. Lett. 12(9), 306-308 (1968).

[12] V. A. Dyakov, V. I. Pryalkin, A. I. Kholodnykh: Potassium niobate optical parametric oscillator pumped by the second harmonic of a garnet laser. Kvant. Elektron. 8(4), 715-721(1981) [In Russian, English trans.: Sov. J. Quantum Electron. 11(4), 433-436(1981)].

[13] F. R. Nash, E. H. Turner, P. M. Bridenbaugh, J. M. Dziedzic: Measurements of second-harmonic generation and the variations in the free and clamped values of the dielectric constants and electro-optic coefficients in barium sodium niobate. J. Appl. Phys. 43(1), 1-9(1972).

[14] R. G. Smith, J. E. Geusic, H. J. Levinstein, J. J. Rubin, S. Singh, L. G. van Uitert: Continuous optical parametric oscillation in $Ba_2NaNb_5O_{15}$. Appl. Phys. Lett. 12(9), 308-310(1968).

[15] R. B. Chesler, M. A. Karr, J. E. Geusic: An experimental and theoretical study of high repetition rate Q-switched Nd:YAG lasers. Proc. IEEE 58(12), 1899-1914(1970).

[16] A. Piskarskas, V. Smilgevichius, A. Umbrasas: The parametric generation of bandwidth-limited picosecond light pulses. Opt. Commun. 73(4), 322-324(1989).

[17] H. R. Xia, L. J. Hu, C. J. Wang, L. X. Li, S. B. Yue, X. L. Meng, L. Zhu, Z. H. Yang, J. Y. Wang: Energy state of Nd^{3+} doped in barium sodium niobate. J. Appl. Phys. 83(5), 2560-2562(1998).

[18] A. A. Kaminskii, D. Jaque, S. N. Bagayev, K.-I. Ueda, S. J. Garsia, J. Capmany: New nonlinear-laser properties of ferroelectric Nd^{3+}: $Ba_2NaNb_5O_{15}$-CW stimulated emission ($^4F_{3/2} \Rightarrow {}^4I_{11/2}$ and $^4F_{3/2} \Rightarrow {}^4I_{13/2}$), collinear and diffuse self-frequency doubling and summation. Kvant. Elektron. 26(2), 95-97(1999) [In Russian, English trans.: Quantum Electron. 29(2), 95-97(1999)].

8.11 $K_3Li_2Nb_5O_{15}$，铌酸锂钾(KLN)

负单轴晶：$n_o > n_e$

点群：$4mm$

分子量：4.3 g/cm^3[1]；$(4.42 \pm 0.07) \text{ g/cm}^3$[2]

晶格常数

$a/\text{Å}$	$c/\text{Å}$	参考文献	备注
12.583	4.041	[3]	$[K_2O]:[Li_2O]:[Nb_2O_5] = 31\%:26\%:43\%$
12.542	4.033	[4]	$[K_2O]:[Li_2O]:[Nb_2O_5] = 32\%:24\%:44\%$
12.58	4.01	[5]	$[K_2O]:[Li_2O]:[Nb_2O_5] = 33.4\%:17.8\%:48.8\%$
12.60	3.99	[6]	$[K_2O]:[Li_2O]:[Nb_2O_5] = 28.9\%:18.1\%:53.0\%$

居里温度：

678 K（摩尔比$[K_2O]:[Li_2O]:[Nb_2O_5] = 33.4\%:17.8\%:48.8\%$）[5]

725 K（摩尔比$[K_2O]:[Li_2O]:[Nb_2O_5] = 30\%:25\%:45\%$）[7]

765 K（摩尔比$[K_2O]:[Li_2O]:[Nb_2O_5] = 32\%:23\%:45\%$）[7]

771 K（摩尔比$[K_2O]:[Li_2O]:[Nb_2O_5] = 33\%:23\%:44\%$）[8]

786 K（摩尔比$[K_2O]:[Li_2O]:[Nb_2O_5] = 32\%:24\%:44\%$）[4]

794 K（摩尔比$[K_2O]:[Li_2O]:[Nb_2O_5] = 32\%:24\%:44\%$）[7]

813 K（摩尔比$[K_2O]:[Li_2O]:[Nb_2O_5] = 31\%:26\%:43\%$）[3]

室温下带隙能量：$E_g = 3.2 \text{ eV}$[9]

透明范围：$0.35 \sim 5 \text{ μm}$[10,11]；$0.4 \sim 5 \text{ μm}$[12]

线性吸收系数 α

$\lambda/\text{μm}$	α/cm^{-1}	参考文献	$\lambda/\text{μm}$	α/cm^{-1}	参考文献
0.38	3.0	[6]	1.064	0.004	[11]

$T = 303$ K 时折射率的实验值[10,13]

$\lambda/\text{μm}$	n_o	n_e	$\lambda/\text{μm}$	n_o	n_e
0.4500	2.4049	2.2512	0.5321	2.3260	2.1975
0.4750	2.3751	2.2315	0.5500	2.3156	2.1900
0.5000	2.3546	2.2144	0.5750	2.3016	2.1801
0.5250	2.3349	2.2010	0.6000	2.2899	2.1720

续表

$\lambda/\mu m$	n_o	n_e	$\lambda/\mu m$	n_o	n_e
0.625 0	2.279 9	2.164 5	0.675 0	2.263 1	2.152 9
0.632 8	2.277 0	2.163 0	1.064 2	2.208 0	2.112 0
0.650 0	2.271 1	2.158 6			

Sellmeier 方程($T=303$ K)[13]:

$$n_o^2 = 1 + \frac{3.708\lambda^2}{\lambda^2 - 0.046\ 01}$$

$$n_e^2 = 1 + \frac{3.349\lambda^2}{\lambda^2 - 0.035\ 46}$$

有效二阶非线性系数的表达式(Kleinman 对称条件成立,$d_{15} = d_{24} = d_{31} = d_{32}$)[14]:

$$d_{ooe} = d_{31}\sin\theta$$

二阶非线性系数:

$d_{31}(0.8\ \mu m) = 11.8\ pm/V$[11]

$d_{31}(1.06\ \mu m) = (1.7 \pm 0.3) \times d_{31}(LN) = (7.8 \pm 1.4)\ pm/V$[15,16]

$d_{31}(1.064\ 2\ \mu m) = 19.3 \times d_{11}(SiO_2) \pm 20\% = (5.8 \pm 1.2)\ pm/V$[10,17]

$d_{33}(1.064\ 2\ \mu m) = 35 \times d_{11}(SiO_2) \pm 15\% = (10.5 \pm 1.5)\ pm/V$[10,17]

非临界相位匹配时相互作用波长的实验值($T = 293$ K)

相互作用的波长/μm	参考文献	备注
SHG, $o + o \Rightarrow e$		
$0.82 \Rightarrow 0.41$	[11]	
$0.827\ 4 \Rightarrow 0.413\ 7$	[7]	$[K_2O]:[Li_2O]:[Nb_2O_5] = 32\%:25\%:43\%$
$0.833 \Rightarrow 0.416\ 5$	[7]	$[K_2O]:[Li_2O]:[Nb_2O_5] = 31\%:26\%:43\%$
$0.833\ 4 \Rightarrow 0.416\ 7$	[3]	$[K_2O]:[Li_2O]:[Nb_2O_5] = 31\%:26\%:43\%$
$0.859\ 5 \Rightarrow 0.429\ 75$	[7]	$[K_2O]:[Li_2O]:[Nb_2O_5] = 32\%:24\%:44\%$
$0.870 \Rightarrow 0.435$	[7]	$[K_2O]:[Li_2O]:[Nb_2O_5] = 31\%:25\%:44\%$
$0.920\ 3 \Rightarrow 0.460\ 15$	[7]	$[K_2O]:[Li_2O]:[Nb_2O_5] = 32\%:23\%:45\%$
$0.929 \Rightarrow 0.464\ 5$	[7]	$[K_2O]:[Li_2O]:[Nb_2O_5] = 31.5\%:23.5\%:45\%$
$0.953 \Rightarrow 0.476\ 5$	[7]	$[K_2O]:[Li_2O]:[Nb_2O_5] = 31\%:24\%:45\%$
$0.959 \Rightarrow 0.479\ 5$	[7]	$[K_2O]:[Li_2O]:[Nb_2O_5] = 30.5\%:24.5\%:45\%$
$0.974 \Rightarrow 0.487$	[7]	$[K_2O]:[Li_2O]:[Nb_2O_5] = 30\%:25\%:45\%$

非临界相位匹配时温度带宽和光谱带宽的实验值

相互作用的波长/μm	T/℃	ΔT/℃	$\Delta \nu$/cm^{-1}	参考文献
SHG, $o+o \Rightarrow e$				
0.833 4 ⇒ 0.416 7	20		1.9(?)	[3]
0.838 2 ⇒ 0.419 1	50	0.4(?)		[3]
0.859 5 ⇒ 0.429 75	20		3.9	[7]
0.869 5 ⇒ 0.434 75	60	0.8	3.2	[4]
0.889 8 ⇒ 0.444 9	20		≈3.0	[18]
0.920 3 ⇒ 0.460 15	20		4.2	[7]

关于这一晶体

KLN 是"老的"非线性材料之一:它在 20 世纪 60 年代中期和 LN 及 BNN 同时被发现[10,1]。然而,直到现在生长质量和尺寸过得去的晶体仍然很困难。最近,由新加坡和日本的科学家终于生长了高质量、无开裂的 KLN 晶体[3,4,7,18]。

■ **参考文献**

[1] L. G. van Uitert, J. J. Rubin, W. A. Bonner: Growth of BaNaNb$_5$O$_{15}$ single crystals for optical applications. IEEE J. Quant. Electr. QE-4(10), 622-627(1968).

[2] S. C. Abrahams, P. B. Jamieson, J. L. Bernstein: Ferroelectric tungsten bronze-type crystal structures. III. Potassium lithium niobate K$_{(6-x-y)}$Li$_{(4+x)}$Nb$_{(10+y)}$O$_{30}$. J. Chem. Phys. 54(6), 2355-2364(1971).

[3] T. C. Chong, X. W. Xu, G. Y. Zhang, H. Kumagai: Blue SHG characteristics and homogeneity of the TSSG grown potassium lithium niobate (KLN) crystal with high Li$_2$O content. J. Cryst. Growth 225(2-4), 489-494(2001).

[4] X.-W. Xu, T.-C. Chong, G.-Y. Zhang, H. Kumagai: Second-harmonic generation of ferroelectric potassium lithium niobate crystals. Jpn. J. Appl. Phys. 40(7), 4540-4543 (2001).

[5] M. Adachi, A. Kawabata: Elastic and piezoelectric properties of potassium lithium niobate (KLN) crystals. Jpn. J. Appl. Phys. 17(11), 1969-1973(1978).

[6] J. Xu, S. Fan, Y. Lin, X. Xu: Bridgeman growth and properties of potassium lithium niobate single crystals. Progr. Cryst. Growth Character. Mater. 40(1-4), 137-144(2000).

[7] X. W. Xu, T. C. Chong, G. Y. Zhang, H. Kumagai: Influence of [K]/[Li] and [Li]/[Nb] ratios in melts on the TSSG growth and SHG characteristics of potassium lithium niobate crystals. J. Cryst. Growth 225(2-4), 458-464(2001).

[8] G. Y. Kang, J. K. Yoon: The growth of potassium lithium niobate (KLN) with low Nb$_2$O$_5$ content. J. Cryst. Growth 193(4), 615-622(1998).

[9] J. Xu, S. Fan, Y. Lin, Y. Fei: Growth and characterization of potassium lithium niobate

crystals. Proc. SPIE 3556, 24 – 30(1998).

[10] L. G. van Uitert, S. Singh, H. J. Levinstein, J. E. Geusic, W. A. Bonner: A new and stable nonlinear optical material. Appl. Phys. Lett. 11(5), 161 – 163(1967); Erratum. Appl. Phys. Lett. 12(6), 224(1968).

[11] J. J. E. Reid: Resonantly enhanced, frequency doubling of an 820 nm GaAlAs diode laser in a potassium lithium niobate crystal. Appl. Phys. Lett. 62(1), 19 – 21(1993).

[12] T. Fukuda: Growth and crystallographic characteristics of $K_3Li_2Nb_5O_{15}$ single crystals. Jpn. J. Appl. Phys. 8(1), 122(1969).

[13] S. Singh: "Nonlinear Optical Materials" in *Handbook of Lasers*, ed. by R. G. Pressley (The Chemical Rubber Co., Cleveland, 1971), pp. 489 – 525.

[14] J. E. Midwinter, J. Warner: The effects of phase matching method and of uniaxial crystal symmetry on the polar distribution of second-order non-linear optical polarization. Brit. J. Appl. Phys. 16(11), 1135 – 1142(1965).

[15] A. W. Smith, G. Burns, B. A. Scott, H. D. Edmonds: Nonlinear optical properties of potassium-lithium niobates. J. Appl. Phys. 42(2), 684 – 686(1971).

[16] I. Shoji, T. Kondo, A. Kitamoto, M. Shirane, R. Ito: Absolute scale of second-order nonlinear-optical coefficients. J. Opt. Soc. Am. B 14(9), 2268 – 2294(1997).

[17] D. A. Roberts: Simplified characterization of uniaxial and biaxial nonlinear optical crystals: a plea for standardization of nomenclature and conventions. IEEE J. Quant. Electr. 28(10), 2057 – 2074(1992).

[18] L. Li, T. C. Chong, X. W. Wu, H. Kumagai, M. Hirano: Growth of potassium lithium niobate (KLN) single crystals for second harmonic generation (SHG) application. J. Cryst. Growth 211(1 – 4), 281 – 285(2000).

8.12 $CO(NH_2)_2$,尿素

正单轴晶: $n_e > n_o$

分子量: 60.055

密度: 1.318 g/cm^3 [1]

点群: $\bar{4}2m$

莫氏硬度: <2.5

以 $\theta = 74°$ 切割 0.5 cm 长晶体以 0.5 透过计的透明范围: 0.2 ~ 1.43 μm [2]

线性吸收系数 α [2]

$\lambda/\mu m$	α/cm^{-1}	备注	$\lambda/\mu m$	α/cm^{-1}	备注
0.213	0.10	o 光, FiHG 方向	1.064	0.02	e 光, FiHG 方向
0.266	0.04	e 光, FiHG 方向			

在文献[3]、[4]中给出了 n_o 和 n_e 与波长关系的图。

最佳色散关系方程组(λ 以 μm 为单位，$T = 293$ K)[5,6]：

$$n_o^2 = 2.1548 + \frac{0.01310}{\lambda^2 - 0.0318}$$

$$n_e^2 = 2.5527 + \frac{0.01784}{\lambda^2 - 0.0294} + \frac{0.0288(\lambda - 1.5)}{(\lambda - 1.5)^2 + 0.03371}$$

在文献[7]、[8]中给出了其他色散关系方程组。

普遍情况下有效二阶非线性系数的表达式(Kleinman 对称条件成立，$d_{14} = d_{25} = d_{36}$)[9,10]：

$$d_{eeo} = 2d_{36}\sin(\theta + \rho)\cos(\theta + \rho)\cos 2\phi$$

$$d_{oeo} = d_{eoo} = -d_{36}\sin(\theta + \rho)\sin 2\phi$$

有效二阶非线性系数的简化表达式(小双折射角近似，Kleinman 对称条件成立，$d_{14} = d_{25} = d_{36}$)[10]：

$$d_{eeo} = d_{36}\sin 2\theta \sin 2\phi$$

$$d_{oeo} = d_{eoo} = -d_{36}\sin\theta \sin 2\phi$$

二阶非线性系数的绝对值：

$$d_{36}(1.0642~\mu m) \approx 3 \times d_{36}(KDP) = 1.2~\text{pm/V}^{[3,11]}$$

$$d_{36}(0.6328~\mu m) = 2.4 \times d_{36}(ADP) \pm 8\% = (1.3 \pm 0.1)\text{pm/V}^{[6,12]}$$

相位匹配角的实验值($T = 293$ K)

相互作用的波长/μm	$\theta_{exp}/(°)$	参 考 文 献
SHG, e + e ⇒ o		
0.476 ⇒ 0.238	90	[7]
0.500 ⇒ 0.250	67.6	[7]
0.550 ⇒ 0.275	54	[7]
0.600 ⇒ 0.300	46.6	[7]
SFG, e + e ⇒ o		
0.6943 + 0.34715 ⇒ 0.23143	77	[2]
1.0642 + 0.26605 ⇒ 0.21284	72	[2]
SHG, o + e ⇒ o		
0.597 ⇒ 0.2985	90	[7]
0.650 ⇒ 0.325	63.6	[7]
0.700 ⇒ 0.350	55.6	[7]
SFG, o + e ⇒ o		
1.0642 + 0.29146 ⇒ 0.2288	90	[7]
1.0642 + 0.29668 ⇒ 0.2320	80	[7]
1.0642 + 0.30656 ⇒ 0.2380	70.4	[7]

续表

相互作用的波长/μm	$\theta_{exp}/(°)$	参 考 文 献
1.064 2 + 0.427 92⇒0.305 2	47.5	[7]
1.064 2 + 0.635 01⇒0.397 7	37.7	[7]
0.720 + 0.537 64⇒0.307 8	63	[13]
0.646 + 0.587 93⇒0.307 8	69	[14]
0.628 75 + 0.532 1⇒0.288 2	90	[7]
0.639 80 + 0.532 1⇒0.290 5	80.5	[7]
0.664 06 + 0.532 1⇒0.295 4	73.4	[7]
SFG, e + o⇒o		
1.064 2 + 0.507 87⇒0.343 8	90	[7]
1.064 2 + 0.53⇒0.353 8	72.2	[7]
1.064 2 + 0.575⇒0.373 3	62.5	[7]
1.064 2 + 0.631 95⇒0.396 5	53.5	[7]

内角带宽的实验值[2]

相互作用的波长/μm	$\Delta\theta^{int}/(°)$
FiHG, e + e⇒o	
1.064 + 0.266⇒0.213	0.017

非临界 SHG 的温度调谐[7]

相互作用的波长/μm	$d\lambda_1/dT/(nm·K^{-1})$
SHG, e + o⇒e	
0.597⇒0.298 5	−0.013

激光诱导的体损伤阈值

$\lambda/\mu m$	τ_p/ns	$I_{thr}/(GW·cm^{-2})$	参 考 文 献	备 注
0.266	10	0.5	[15]	单脉冲
0.355	10	1.4	[15]	单脉冲
		0.15	[16]	3 000 个脉冲
0.532	10	3	[15]	单脉冲
1.064	10	5	[15]	单脉冲

关于这一晶体

尿素是 1970 年发展的几种有机晶体之一,在过去 15 年中这种材料没有实际应用。

参考文献

[1] V. G. Dmitriev, G. G. Gurzadyan, D. N. Nikogosyan: Handbook of Nonlinear Optical Crystals, Third Revised Edition(Springer, Berlin, 1999).

[2] K. Kato: High-efficiency high-power UV generation at 2128 Å in urea. IEEE J. Quant. Electr. QE-16(8), 810-811(1980).

[3] D. Bauerle, K. Betzler, H. Hesse, S. Kapphan, P. Loose: Phase-matched second harmonic generation in urea. Phys. Status Solidi A 42(2), K119-K121(1977).

[4] K. Betzler, H. Hesse, P. Loose: Optical second harmonic generation in organic crystals: urea and ammonium-malate. J. Mol. Struct. 47, 393-396(1978).

[5] W. R. Donaldson, C. L. Tang: Urea optical parametric oscillator. Appl. Phys. Lett. 44(1), 25-27(1984).

[6] M. J. Rosker, C. L. Tang: Widely tunable optical parametric oscillator using urea. J. Opt. Soc. Am. B 2(5), 691-696(1985).

[7] J.-M. Halbout, S. Blit, W. Donaldson, C. L. Tang: Efficient phase-matched second-harmonic generation and sum-frequency mixing in urea. IEEE J. Quant. Electr. QE-15(10), 1176-1180(1979).

[8] M. J. Rosker, K. Cheng, C. L. Tang: Practical urea optical parametric oscillator for tunable generation throughout the visible and near-infrared. IEEE J. Quant. Electr. QE-21(10), 1600-1606(1985).

[9] R. C. Eckardt, H. Masuda, Y. X. Fan, R. L. Byer: Absolute and relative nonlinear optical coefficients of KDP, KD*P, BaB_2O_4, $LiIO_3$, $MgO:LiNbO_3$, and KTP measured by phase-matched second-harmonic generation. IEEE J. Quant. Electr. 26(5), 922-933(1990).

[10] J. E. Midwinter, J. Warner: The effects of phase matching method and of uniaxial crystal symmetry on the polar distribution of second-order non-linear optical polarization. Brit. J. Appl. Phys. 16(11), 1135-1142(1965).

[11] D. A. Roberts: Simplified characterization of uniaxial and biaxial nonlinear optical crystals: a plea for standardization of nomenclature and conventions. IEEE J. Quant. Electr. 28(10), 2057-2074(1992).

[12] K. Hagimoto, A. Mito: Determination of the second-order susceptibility of ammonium dihydrogen phosphate and α-quartz at 633 and 1064 nm. Appl. Opt. 34(36), 8276-8282(1995).

[13] M. Ebrahimzadeh, M. H. Dunn, F. Akerboom: Highly efficient visible urea optical parametric oscillator pumped by a XeCl excimer laser. Opt. Lett. 14(11), 560-562(1989).

[14] M. Ebrahimzadeh, M. H. Dunn: Optical parametric fluorescence and oscillation in urea using an excimer laser. Opt. Commun. 69(2), 161-165(1988).

[15] C. Cassidy, J. M. Halbout, W. Donaldson, C. L. Tang: Nonlinear optical properties of urea. Opt. Commun. 29(2), 243-246(1979).

[16] M. J. Rosker: Recent developments in urea. Proc. SPIE 681, 10-11(1986).

8.13 $LiIO_3$，碘酸锂

负单轴晶：$n_o > n_e$

分子量：181.844

密度：4.48 g/cm³, T = 293 K[1]；4.487 g/cm³[2]；4.49 g/cm³[3]

点群：6

晶格常数：

a = (5.481 5 ± 0.000 3) Å[1]；5.481 3 Å[4]

c = (5.170 9 ± 0.000 4) Å[1]；5.171 7 Å[4]

莫氏硬度：3.5[2]；3.5~4.0[1]；4.0[4]

100 g 水中的溶解度[1]

T/K	s/g	T/K	s/g
283.1	89.4	313.2	79.0
293.4	84.7	348.7	74.9
298.1	82.9		

熔点：692 K[1]

线性热膨胀系数

T/K	$\alpha_t \times 10^6/K^{-1}$, $\parallel c$	$\alpha_t \times 10^6/K^{-1}$, $\perp c$	参 考 文 献
100	25	14	[5]
150	32	17	[5]
200	40	21	[5]
250	47	25	[5]
273	45	25	[1]
298	48	28	[6]
300	50	25	[5]
323	49	26	[7]
350	51	25	[5]
373	51	26	[7]

续表

T/K	$\alpha_t \times 10^6/\text{K}^{-1}$, $\parallel c$	$\alpha_t \times 10^6/\text{K}^{-1}$, $\perp c$	参 考 文 献
400	51	28	[5]
423	54	27	[7]
450	53	29	[5]
473	56	31	[7]

线性热膨胀系数的温度关系(T 以 K 为单位)[5]：

温度范围：80~253 K

$$\alpha_t(\parallel c) = 2.5 \times 10^{-5} + 1.5 \times 10^{-7} T$$

$$\alpha_t(\perp c) = 1.4 \times 10^{-5} + 7.5 \times 10^{-8} T$$

温度范围：273~470 K

$$\alpha_t(\parallel c) = 4.9 \times 10^{-5} + 3 \times 10^{-8} T$$

温度范围：353~470 K

$$\alpha_t(\perp c) = 2.7 \times 10^{-5} + 1.4 \times 10^{-8} T$$

$p = 0.101\ 325$ MPa 时的比热容 c_p：365 J/(kg·K)[3]；569 J/(kg·K)[1]

热导率：

$$\kappa = 1.47 \text{W}/(\text{m}\cdot\text{K})^{[3]}$$

T/K	$\kappa/(\text{W}\cdot\text{m}^{-1}\cdot\text{K}^{-1})$, $\parallel c$	$\kappa/(\text{W}\cdot\text{m}^{-1}\cdot\text{K}^{-1})$, $\perp c$	参 考 文 献
300	0.65	1.27	[5]
400	0.70	1.20	[5]

室温下带隙能量：$E_g = 4.0$ eV[2]；(4.37 ± 0.03) eV[8]

以 "0" 透过计的透明范围：0.28~6 μm[9,10]

线性吸收系数 α

$\lambda/\mu\text{m}$	α/cm^{-1}	参 考 文 献	备 注
0.325	≈0.4	[11]	
0.347 15	0.1	[12]	$\parallel c$
	0.3	[12]	e 光，$\perp c$
0.514 5	0.002 4	[13]	$\parallel c$
	0.002 5	[13]	e 光，$\perp c$
0.532 1	0.3	[14]	e 光
0.542 2	0.37	[10]	
0.650	≈0.001	[11]	
0.659 4	0.000 7~0.002 3	[13]	$\parallel c$

续表

$\lambda/\mu m$	α/cm^{-1}	参考文献	备注
1.064 2	0.000 6 ~ 0.001 7	[13]	e 光, $\perp c$
	0.1	[14]	o 光
	0.25	[14]	e 光
	<0.000 2	[13]	$\parallel c$
	0.000 8	[13]	e 光, $\perp c$
1.084 5	0.06	[10]	
1.315	0.000 5	[3]	
1.318 8	0.000 8 ~ 0.003 6	[13]	$\parallel c$
	0.000 7 ~ 0.001 0	[13]	e 光, $\perp c$

双光子吸收系数 β

$\lambda/\mu m$	τ_p/ns	$\beta \times 10^{11}/(cm \cdot W^{-1})$	参考文献
0.53	0.03 ~ 0.1	<3	[15]
0.532 1	10	<40	[16]

折射率的实验值

$\lambda/\mu m$	n_o	n_e	参考文献	$\lambda/\mu m$	n_o	n_e	参考文献
0.354 7	1.982 2	1.811 3	[17]	0.546 1	1.895 0	1.745 5	[19]
0.366 9	1.970 6	1.802 6	[17]	0.560 0	1.892 1	1.743 3	[17]
0.371 2	1.967 1	1.800 0	[17]	0.579 1	1.889 4	1.741 3	[18]
0.379 5	1.960 0	1.794 7	[17]	0.580 0	1.888 9	1.740 3	[17]
0.387 7	1.954 4	1.790 5	[17]	0.589 6	1.887 5	1.740 0	[18]
0.399 6	1.946 4	1.784 2	[17]	0.600 0	1.885 9	1.738 3	[17]
0.404 7	1.944 3	1.782 6	[18]	0.620 0	1.882 8	1.736 1	[17]
0.435 8	1.927 5	1.770 2	[17]	0.632 8	1.881 5	1.735 1	[20]
0.454 5	1.918 4	1.763 8	[8]	0.643 8	1.880 7	1.734 6	[18]
0.457 9	1.917 0	1.763 0	[8]	0.656 0	1.878 9	1.733 2	[19]
0.465 8	1.914 1	1.761 1	[8]	0.700 0	1.874 6	1.730 0	[19]
0.472 7	1.912 2	1.760 0	[8]	0.766 0	1.869 4	1.726 1	[19]
0.476 5	1.910 0	1.758 3	[8]	0.800 0	1.867 3	1.724 5	[19]
0.480 0	1.910 9	1.757 9	[17]	0.863 0	1.864 0	1.722 0	[19]
0.488 0	1.908 3	1.755 6	[8]	0.900 0	1.862 3	1.720 7	[19]
0.501 7	1.905 3	1.753 7	[8]	1.000 0	1.858 7	1.718 0	[19]
0.508 6	1.903 1	1.751 4	[17]	1.100 0	1.855 9	1.716 0	[19]
0.514 5	1.901 2	1.748 7	[8]	1.200 0	1.853 6	1.714 3	[19]
0.532 0	1.897 5	1.747 5	[17]	1.300 0	1.851 7	1.713 0	[19]

续表

$\lambda/\mu m$	n_o	n_e	参考文献	$\lambda/\mu m$	n_o	n_e	参考文献
1.367 4	1.850 8	1.712 2	[18]	2.500 0	1.837 8	1.703 7	[20]
1.529 6	1.848 2	1.710 1	[18]	3.000 0	1.831 9	1.700 1	[20]
1.692 0	1.846 4	1.708 9	[18]	3.500 0	1.826 6	1.697 1	[20]
1.970 1	1.843 1	1.707 2	[18]	4.000 0	1.814 0	1.689 7	[20]
2.249 3	1.838 5	1.705 0	[18]	5.000 0	1.794 0	1.678 3	[20]

$T=300$ K 时的旋光性

$\lambda/\mu m$	$\rho/(°)$	参考文献	$\lambda/\mu m$	$\rho/(°)$	参考文献
0.286	1 052.9	[21]	0.429	222.46	[21]
0.290	964.99	[21]	0.448	198.72	[21]
0.295	886.65	[21]	0.470	175.75	[21]
0.299	814.39	[21]	0.492	153.61	[21]
0.304	748.76	[21]	0.520	133.02	[21]
0.310	687.46	[21]	0.546	117.42	[21]
0.317	630.44	[21]	0.551	113.36	[21]
0.324	579.01	[21]	0.600	95.27	[21]
0.331	532.44	[21]	0.628	86.80	[21]
0.339	489.47	[21]	1.084	25.0	[10]
0.347	448.42	[21]	1.1	23.83	[22]
0.355	410.37	[21]	1.6	11.00	[22]
0.363	374.34	[21]	2.1	6.33	[22]
0.374	340.18	[21]	2.6	4.12	[22]
0.386	308.07	[21]	3.1	2.89	[22]
0.399	277.45	[21]	3.6	2.32	[22]
0.412	249.32	[21]			

$T=300$ K 时折射率的温度微商

$\lambda/\mu m$	$dn_o/dT \times 10^6/K^{-1}$	$dn_e/dT \times 10^6/K^{-1}$	参考文献
0.532 1	-96	-86	[13]
0.657	-79	-71	[7]
0.659 4	-95	-84	[13]
1.064 2	-89	-75	[13]
1.318 8	-94	-85	[13]

最佳色散关系方程组(λ 以 μm 为单位,$T=293$ K)[23,24]:

$$n_o^2 = 3.4132 + \frac{0.0476}{\lambda^2 - 0.0338} - 0.0077\lambda^2$$

$$n_e^2 = 2.9211 + \frac{0.0346}{\lambda^2 - 0.0320} - 0.0042\lambda^2$$

在文献[17]、[20]、[25]、[26]、[27]、[28]中给出了其他色散关系方程组。

室温下高频(远高于 $LiIO_3$ 晶体声学共振频率,即对"受夹"晶体)下测量的线性电光系数[10]

$\lambda/\mu m$	$r_{13}^S/(pm \cdot V^{-1})$	$r_{33}^S/(pm \cdot V^{-1})$	$r_{41}^S/(pm \cdot V^{-1})$	$r_{51}^S/(pm \cdot V^{-1})$
0.6328	+4.1±0.6	+6.4±1.0	1.4±0.2	+3.3±0.7

Verdet 常数($\parallel c$)[29]

$\lambda/\mu m$	T/K	$V/[(°) \cdot T^{-1} \cdot m^{-1}]$
0.6328	295	757

普通情况下有效二阶非线性系数的表达式(Kleinman 对称条件成立,$d_{15} = d_{24} = d_{31} = d_{32}$)[30]:

$$d_{ooe} = d_{31}\sin(\theta + \rho)$$

有效二阶非线性系数的表达式(小双折射角近似,Kleinman 对称条件成立,$d_{15} = d_{24} = d_{31} = d_{32}$)[31]:

$$d_{ooe} = d_{31}\sin\theta$$

二阶非线性系数的绝对值和相对值:

$d_{31}(1.319\ \mu m) = (3.9 \pm 0.2)\ pm/V$ [32]

$d_{31}(1.0642\ \mu m) = (4.1 \pm 0.4)\ pm/V$ [30];$(4.4 \pm 0.3)\ pm/V$ [33]

$d_{31}(0.806\ \mu m) = (5.2 \pm 0.5)\ pm/V$ [32]

$d_{33}(1.318\ \mu m) = 0.99 \times d_{31}(1.318\ \mu m) = (3.9 \pm 0.2)\ pm/V$ [20,32]

$d_{33}(1.0642\ \mu m) = 1.04 \times d_{31}(1.0642\ \mu m) = (4.6 \pm 0.3)\ pm/V$ [14,33]

相位匹配角的实验值($T = 293\ K$)

相互作用的波长/μm	$\theta_{exp}/(°)$	参 考 文 献
SHG,$o + o \Rightarrow e$		
0.586⇒0.293	90	[23]
0.5863⇒0.29315	90	[28]
0.6⇒0.3	75.6	[28]

续表

相互作用的波长/μm	$\theta_{exp}/(°)$	参 考 文 献
0.62⇒0.31	68.2	[28]
0.694 3⇒0.347 15	52	[9], [34]
0.946⇒0.473	34.3	[35]
1.06⇒0.53	30	[36]
1.064 2⇒0.532 1	30.2	[30], [37]
	30	[27], [38]
1.084 5⇒0.542 25	28.9	[10]
1.152 3⇒0.576 15	27.2	[10]
1.388 6⇒0.694 3	23.1	[39]
1.746⇒0.873	20	[40]
SFG, o+o⇒e		
5.33+1.329 69⇒1.064 2	21	[27]
4.44+1.399 68⇒1.064 2	20.2	[27]
5.2+0.801 29⇒0.694 3	19.5	[41]
2.5+0.961 26⇒0.694 3	21	[42]
5.0+0.662 51⇒0.585	20.3	[43]
2.0+0.826 86⇒0.585	25.1	[43]
4.16+0.610 15⇒0.532 1	21.6	[44]
2.66+0.665 14⇒0.532 1	24.5	[44]
0.946+0.548 4⇒0.347 15	50	[45]
2.67+0.694 3⇒0.551 02	24.4	[46]
1.98+0.694 3⇒0.514 05	27.4	[46]
1.201 3+0.694 3⇒0.44	35.1	[39]
3.391 3+0.514 5⇒0.446 73	24	[47]
2.38+0.488 0⇒0.404 97	30.5	[48]
1.064 2+0.532 1⇒0.354 73	47.5	[37]

内角带宽、温度带宽和光谱带宽的实验值($T=293$ K)

相互作用的波长/μm	$\theta_{pm}/(°)$	$\Delta\theta^{int}/(°)$	$\Delta T/℃$	$\Delta\nu/cm^{-1}$	参考文献
SHG, o+o⇒e					
0.586⇒0.293	90	0.5~0.58		2.04	[23]
0.694 3⇒0.347 15	52	0.018			[9]
1.06⇒0.53	30	0.019		6.27	[19]
1.064 2⇒0.532 1	30	0.022			[38]
	30	0.022	40		[49]
	30	0.024	52.4		[50]
	30	0.026			[30]
1.084 5⇒0.542 25	29	0.020			[10]

相位匹配角的温度变化

$\lambda/\mu m$	$\theta_{pm}/(°)$	$d\theta_{pm}/dT/[(°)\cdot K^{-1}]$	参 考 文 献
SHG, o + o ⇒ e			
1.084 5 ⇒ 0.542 25	29	$< -1.3 \times 10^{-3}$	[10]
1.064 2 ⇒ 0.532 1	30	-8.4×10^{-4}	[50]

激光诱导的体损伤阈值

$\lambda/\mu m$	τ_p/ns	$I_{thr}/(GW\cdot cm^{-2})$	参 考 文 献	备 注
0.44 ~ 0.62	200 ~ 300	0.01	[51]	
0.53	20	0.07 ~ 0.08	[52]	
	15	0.04 ~ 0.05	[36]	
0.532 1	12	0.03	[27]	
	0.1	1	[15]	
	0.035	4 ~ 5	[53]	12.5 Hz
		8 ~ 10	[53]	1 Hz
	0.032	10 ~ 12	[54]	25 Hz
	0.031	5	[15]	
0.64	330	0.004	[55]	
0.694 3	20	0.025	[39]	500 个脉冲
		0.13	[12]	10 个脉冲
	10	0.12	[42]	
1.064 2	180 000	> 0.05	[42]	50 Hz
	300	0.002	[38]	1 kHz
	12	0.12	[27]	
	10	0.12	[56]	100 Hz
	0.13	8	[15]	
	0.045	19	[15]	

激光诱导的表面损伤阈值[57]

$\lambda/\mu m$	τ_p/ns	$I_{thr}/(GW\cdot cm^{-2})$	备 注
1.064 2	12	3.2	$\theta = 30°$,束腰直径 30 μm

关于这一晶体

这一晶体,由于晶体生长工艺简单,二阶非线性系数较高,在 20 世纪 60 年代和 70 年代非常流行。然而,碘酸锂是水溶性的,并且它的激光诱导的体损伤阈值相当低。这些情况限止了这一材料在现代激光技术中的应用。

参考文献

[1] K. I. Avdienko, S. V. Bogdanov, S. M. Arkhipov, B. I. Kidyarov, V. V. Lebedev, Y. E. Nevskii, V. I. Trunov, D. V. Sheloput, R. M. Shklovskaya: *Lithium Iodate. Growth, Properties and Applications* (Nauka, Novosibirsk, 1980) [In Russian].

[2] B. H. T. Chai: Optical Crystals. In: *CRC Handbook of Laser Science and Technology, Supplement 2: Optical Materials*, ed. by M. J. Weber (CRC Press, Boca Raton, 1995), pp. 3 – 65.

[3] G. D. Hager, S. A. Hanes, M. A. Dreger: Continuous wave frequency doubling of a high energy 1315 nm laser. IEEE J. Quant. Electr. 28(11), 2573 – 2576 (1992).

[4] Data sheet of Molecular Technology GmbH. Available at www.mt-berlin.com.

[5] Y. V. Burak, K. Y. Borman, I. S. Girnyk: Characteristics of the temperature dependences of thermal properties of α-LiIO$_3$. Fiz. Tverd. Tela 26(12), 3692 – 3694 (1984) [In Russian, English trans.: Sov. Phys. -Solid State 26(12), 2223 – 2224 (1984)].

[6] Data sheet of Cleveland Crystals Inc. Available at www.clevelandcrystals.com.

[7] J. M. Thierry, E. Coquet, J. M. Crettez: Interferometric measurement of thermal expansion coefficients and birefringence of α-LiIO$_3$ with temperature. Opt. Commun. 16(3), 417 – 419 (1976).

[8] J. M. Crettez, J. Comte, E. Coquet: Optical properties of α- and β-lithium iodate in the visible range. Opt. Commun. 6(1), 26 – 29 (1972).

[9] G. Nath, S. Haussühl: Strong second harmonic generation of a ruby laser in lithium iodate. Phys. Lett. A 29(2), 91 – 92 (1969).

[10] F. R. Nash, J. G. Bergman, G. D. Boyd, E. H. Turner: Optical nonlinearities in LiIO$_3$. J. Appl. Phys. 40(13), 5201 – 5206 (1969).

[11] T. Laurila, R. Hernberg: Frequency-doubled diode laser for ultraviolet absorption spectroscopy at 325 nm. Appl. Phys. Lett. 83(5), 845 – 847 (2003).

[12] G. Nath, H. Mehmanesch, M. Gsänger: Efficient conversion of a ruby laser radiation to 0.347 μm in low-loss lithium iodate. Appl. Phys. Lett. 17(7), 286 – 288 (1970).

[13] D. J. Gettemy, W. C. Harker, G. Lindholm, N. P. Barnes: Some optical properties of KTP, LiIO$_3$, and LiNbO$_3$. IEEE J. Quant. Electr. 24(11), 2231 – 2237 (1988).

[14] J. Jerphagnon: Optical nonlinear susceptibilities of lithium iodate. Appl. Phys. Lett. 16(8), 298 – 299 (1970).

[15] E. W. van Stryland, W. E. Williams, M. J. Soileau, A. L. Smirl: Laser-induced damage, nonlinear absorption and doubling efficiency of LiIO$_3$. IEEE J. Quant. Electr. QE – 20(4), 434 – 439 (1984).

[16] N. M. Bityurin, V. I. Bredikhin, V. N. Genkin: Nonlinear optical absorption and energy structure of LiNbO$_3$ and α-LiIO$_3$ crystals. Kvant. Elektron. 5(11), 2453 – 2457 (1978)

[In Russian, English trans. : Sov. J. Quantum Electron. 8(11), 1377 – 1379(1978)].

[17] K. Takizawa, M. Okada, S. Ieiri: Refractive indices of paratellurite and lithium iodate in the visible and ultraviolet regions. Opt. Commun. 23(2), 279 – 281(1977).

[18] S. Umegaki, S. I. Tanaka, T. Uchiyama, S. Yabumoto: Refractive indices of lithium iodate between 0.4 and 2.2 μm. Opt. Commun. 3(4), 44 – 245(1971).

[19] R. B. Andreev, V. D. Volosov, A. G. Kalintsev: Spectral, angular, and temperature characteristics of HIO_3, $LiIO_3$, CDA, DKDP, KDP and ADP non-linear crystals in second- and fourth-harmonic generation. Opt. Spektrosk. 37(2) 294 – 299 (1974) [In Russian, English trans. : Opt. Spectrosc. USSR 37(2), 169 – 171(1974)].

[20] M. M. Choy, R. L. Byer: Accurate second-order susceptibility measurements of visible and infrared nonlinear crystals. Phys. Rev. B 14(4), 1693 – 1706(1976).

[21] Z. B. Perekalina, G. F. Dobrzhansky, I. A. Spilko: The rotation of the plane of polarization in $LiIO_3$ crystals. Kristallogr. 15(6), 1252 – 1253 (1970) [In Russian, English trans. : Sov. Phys. -Crystallogr. 15(6), 1095(1970)].

[22] V. A. Kizel, V. I. Burkov: Gyrotropy of Crystals (Nauka, Moscow, 1980) [In Russian].

[23] I. M. Beterov, V. I. Stroganov, V. I. Trunov, B. Y. Yurshin: Excitation of optical harmonics in lithium iodate and formate crystals by CW dye laser radiation. Kvant. Elektron. 2 (11), 2440 – 2443 (1975) [In Russian, English trans. : Sov. J. Quantum Electron. 5 (11), 1329 – 1331(1975)].

[24] S. G. Karpenko, N. E. Kornienko, V. L. Strizhevskii: Use of diverging and nonmonochromatic pump waves in nonlinear spectroscopy of infrared radiation. Kvant. Elektron. 1 (8), 1768 – 1779(1974) [In Russian, English trans. : Sov. J. Quantum Electron. 4(8), 979 – 985(1974)].

[25] V. I. Kabelka, V. G. Kolomiets, A. S. Piskarskas, A. Y. Stabinis: Features of parametric interaction of ultra-short light packets in a $LiIO_3$ crystal. Zh. Prikl. Spektrosk. 21(5), 947 – 950 (1974) [In Russian, English trans. : J. Appl. Spectrosc. 21(5), 582 – 585 (1974)].

[26] V. I. Kabelka, A. S. Piskarskas, A. Y. Stabinis, R. L. Sher: Group phase matching of interacting light pulses in nonlinear crystals. Kvant. Elektron. 2(2), 434 – 436(1975) [In Russian, English trans. : Sov. J. Quantum Electron. 5(2), 255 – 256(1975)].

[27] K. Kato: High-power difference-frequency generation at 4.4 – 5.7 μm. IEEE J. Quant. Electr. QE – 21(2), 119 – 120(1985).

[28] H. Buesener, A. Renn, M. Brieger, F. von Moers, A. Hese: Frequency doubling of CW ring-dye-laser radiation in lithium iodate crystals. Appl. Phys. B 39(2), 77 – 81 (1986).

[29] M. Koralewski: Magnetooptical phenomena in $LiIO_3$ crystals. Phys. Stat. SolidiA 61(2), K151 – K154(1980).

[30] R. C. Eckardt, H. Masuda, Y. X. Fan, R. L. Byer: Absolute and relative nonlinear optical coefficients of KDP, KD*P, BaB_2O_4, $LiIO_3$, $MgO:LiNbO_3$, and KTP measured by phasematched second-harmonic generation. IEEE J. Quant. Electr. 26(5), 922–933(1990).

[31] J. E. Midwinter, J. Warner: The effects of phase matching method and of uniaxial crystal symmetry on the polar distribution of second-order non-linear optical polarization. Brit. J. Appl. Phys. 16(11), 1135–1142(1965).

[32] W. J. Alford, A. V. Smith: Wavelength variation of the second-order nonlinear coefficients of $KNbO_3$, $KTiOPO_4$, $KTiOAsO_4$, $LiNbO_3$, $LiIO_3$, β-BaB_2O_4, KH_2PO_4, and LiB_3O_5 crystals: a test of Miller wavelength scaling. J. Opt. Soc. Am. B 18(4), 524–533(2001).

[33] R. J. Gehr, A. V. Smith: Separated-beam nonphase-matched second-harmonic method of characterizing nonlinear optical coefficients. J. Opt. Soc. Am. B 15(8), 2298–2307(1998).

[34] J. E. Pearson, G. A. Evans, A. Yariv: Measurement of the relative nonlinear coefficient of KDP, RDP, RDA and $LiIO_3$. Opt. Commun. 4(5), 366–367(1972).

[35] T. Kellner, F. Heine, G. Huber: Efficient laser performance of Nd:YAG at 946 nm and intracavity frequency doubling with $LiJO_3$, β-BaB_2O_4, and LiB_3O_5. Appl. Phys. B 65(6), 789–792(1997).

[36] A. I. Izrailenko, A. I. Kovrigin, P. V. Nikles: Parametric generation of light in high-efficiency nonlinear $LiIO_3$ and α-HIO_3 crystals. Pisma Zh. Eksp. Teor. Fiz. 12(10), 475–478(1970) [In Russian, English trans.: JETP Lett. 12(10), 331–333(1970)].

[37] M. Okada, S. Ieiri: Kleinman's symmetry relation in nonlinear optical coefficient of $LiIO_3$. Phys. Lett. A 34(1), 63–64(1971).

[38] R. B. Chesler, M. A. Karr, J. E. Geusic: Repetitively Q-switched Nd:YAG-$LiIO_3$ 0.53 μm harmonic source. J. Appl. Phys. 41(10), 4125–4127(1970).

[39] A. J. Campillo, C. L. Tang: Extending the tuning range of tunable oscillators by upconversion. Appl. Phys. Lett. 19(2), 36–38(1971).

[40] A. J. Campillo: Properties of a pulsed $LiIO_3$ doubly resonant parametric oscillator. IEEE J. Quant. Electr. QE-8(10), 809–811(1972).

[41] D. W. Meltzer, L. S. Goldberg: Tunable IR difference frequency generation in $LiIO_3$. Opt. Commun. 5(3), 209–211(1972).

[42] L. S. Goldberg: Optical parametric oscillation in lithium iodate. Appl. Phys. Lett. 17(11), 489–491(1970).

[43] T. M. Jedju, L. Rothberg: Tunable femtosecond radiation in the mid-infrared for timeresolved absorption in semiconductors. Appl. Opt. 27(3), 615–618(1988).

[44] F. Huisken, A. Kulcke, D. Voelkel, C. Laush, J. M. Lisy: New infrared injection-see-

ded optical parametric oscillator with high energy and narrow bandwidth output. Appl. Phys. Lett. 62(8), 805–807(1993).

[45] G. Nath, G. Pauli: Efficient pulsed optical parametric oscillator with a tuning range from 0.415 to 2.1 μm. Appl. Phys. Lett. 22(2), 75–76(1973).

[46] D. Malz, J. Bergmann, J. Heise: Up-conversion of thermal IR-radiation in $LiIO_3$ with a pulsed ruby-laser. Exp. Techn. Phys. 23(4), 379–388(1975).

[47] Y. C. See, J. Falk: Lithium iodate, intracavity upconversion. Appl. Phys. Lett. 36(7), 503–505(1980).

[48] D. Malz, J. Bergmann, J. Heise: Up-conversion of thermal IR-radiation in $LiIO_3$ with a CW argon laser. Exp. Techn. Phys. 23(5), 495–498(1975).

[49] B. I. Kidyarov, I. V. Nikolaev, E. V. Pestryakov, V. M. Tarasov: Aluminium iodate octahydrate as a new nonlinear-optical crystal. Izv. Ross. Akad. Nauk, Ser. Fiz. 58(2), 131–134(1994)[In Russian, English trans.: Bull. Russian Acad. Sci.: Physics 58(2), 294–296(1994)]

[50] M. Webb, S. P. Velsko: Temperature sensitivity of phase-matched second-harmonic generation in $LiIO_3$. IEEE J. Quant. Electr. 26(8), 1394–1398(1990).

[51] H. Gerlach: Difference frequency generation in $LiIO_3$ using two tunable dye lasers. Opt. Commun. 12(4), 405–408(1974).

[52] R. B. Andreev, V. D. Volosov, V. N. Krylov: Parametric generation of high-power nanosecond light pulses at 0.74—1.85 μm. Pisma Zh. Tekh. Fiz. 4(5–6), 256–258(1978) [In Russian, English trans.: Sov. Tech. Phys. Lett. 4(3), 105–106(1978)].

[53] R. Danielius, G. Dikchyus, V. Kabelka, A. Piskarskas, A. Stabinis, Y. Yasevichyute: Parametric excitation of light in the picosecond range. Kvant. Elektron. 4(11), 2379–2395(1977)[In Russian, English trans.: Sov. J. Quantum Electron. 7(11),1360–1368(1977)].

[54] A. Arutyunyan, G. Arzumanyan, R. Danielius, V. Kabelka, R. Sharkhatunyan, Y. Yasevichyute: Investigation of parametric superluminescence of $LiIO_3$ crystals in the picosecond band. Lit. Fiz. Sbornik 18(2), 255–263(1978)[In Russian, English trans.: Sov. Phys. -Collection 18(2),62–67(1978)].

[55] L. Armstrong, S. E. Neister, R. Adhav: Measuring CFP dye laser damage thresholds on UV doubling crystals. Laser Focus 18(12), 49–53(1982).

[56] V. G. Dmitriev, V. N. Krasnyanskaya, M. F. Koldobskaya, I. S. Rez, E. A. Shalaev, E. M. Shvom: Frequency multiplication in nonlinear lithium iodate crystals. Kvant. Elektron. No. 2, 64–66(1973)[In Russian, English trans.: Sov. J. Quantum Electron. 3(2),126–127(1973)].

[57] M. Bass, H. H. Barrett: Avalanche breakdown and the probabilistic nature of laser-induced damage. IEEE J. Quant. Electr. QE–8(3), 338–343(1972).

8.14 Ag_3AsS_3，硫砷银（淡红银矿）

负单轴晶：$n_o > n_e$

分子量：494.724

密度：5.49 g/cm³[1]；5.629 g/cm³[2]；在 T = 293 K 时，5.635 g/cm³[3]；5.65 g/cm³[4]

点群：$3m$

晶格常数：

a = 10.74 Å[5]；10.756 Å[6]；10.82 Å[4]

c = 8.64 Å[5]；8.652 Å[6]；8.69 Å[4]

莫氏硬度：2～2.5[3]

在水中的溶解度：不溶解[3]

熔点：769 K[6]

线性热膨胀系数

T/K	$\alpha_t \times 10^6 / K^{-1}$，$\parallel c$	$\alpha_t \times 10^6 / K^{-1}$，$\perp c$
300	12	16

热导率 κ[2]

T/K	$\kappa/(W \cdot m^{-1} \cdot K^{-1})$，$\parallel c$	$\kappa/(W \cdot m^{-1} \cdot K^{-1})$，$\perp c$
300	0.113	0.092

室温下间接跃迁的带隙能量：

E_g = 2.067 eV，$\boldsymbol{E} \perp c$[2]

E_g = 2.100 eV，$\boldsymbol{E} \parallel c$[2]

室温下直接跃迁的带隙能：

E_g = 2.177 eV，$\boldsymbol{E} \perp c$[2]

E_g = 2.235 eV，$\boldsymbol{E} \parallel c$[2]

以 α = 1 cm^{-1} 计的透明范围：

0.63～12.5 μm（$\boldsymbol{E} \perp c$）[5]

0.61～13.3 μm（$\boldsymbol{E} \parallel c$）[5]

线性吸收系数 α

$\lambda/\mu m$	α/cm^{-1}	T/K	参 考 文 献	备 注
0.593	0.89	77	[7]	e 光
0.632 8	0.81	77	[7]	o 光
	0.64	77	[7]	e 光
0.678 9	0.64	77	[7]	o 光
9.31	0.25	77	[7]	e 光
0.576	36 ± 1	300	[5]	e 光，⊥c
0.593	16.1	300	[7]	e 光
0.632 8	1.83	300	[7]	o 光
	1.59	300	[7]	e 光
0.635 8	1.88	300	[8]	e 光
0.676 4	0.95	300	[8]	o 光
0.678 9	0.83	300	[7]	o 光
0.694 3	0.1	300	[9]	
	0.2	300	[10]	o 光，∥c
1.06	0.1	300	[10]	o 光，∥c
1.064 2	0.02	300	[9]	
5.3	0.3	300	[11]	e 光
	0.32	300	[12]	e 光
9.2	0.29	300	[13]	o 光
9.3	0.53	300	[7]	e 光
9.55	<0.1	300	[14]	SHG 方向
10.2	1.2	300	[5]	o 光，⊥c
	1.3	300	[15]	o 光
10.6	0.16	300	[16]	o 光
	0.38	300	[12]	o 光
	0.45	300	[8]	o 光
	0.6	300	[11]	o 光
	0.8	300	[17]	o 光
	1	300	[18]	o 光
11.6	0.5	300	[15]	o 光
14.5	≈70	300	[15]	o 光
15.2 ~ 20.8	<20	300	[19]	o 光

双光子吸收系数 $\beta(\parallel c)$

$\lambda/\mu m$	τ_p/ns	$\beta \times 10^{11}/(cm \cdot W^{-1})$	参 考 文 献
0.694 3	~20	10 000	[10]
	25	2 000	[9]
1.06	~20	3 000	[10]
1.064 2	20	<300	[9]

$T = 293$ K 时折射率的实验值[5]

$\lambda/\mu m$	n_o	n_e	$\lambda/\mu m$	n_o	n_e
0.587 6		2.789 6	1.530	2.772 8	2.548 5
0.632 8	3.019 0	2.739 1	1.709	2.765 4	2.542 3
0.667 8	2.980 4	2.709 4	2.50	2.747 8	2.528 2
1.014	2.826 4	2.590 1	3.56	2.737 9	2.521 3
1.129	2.806 7	2.575 6	4.62	2.731 8	2.517 8
1.367	2.783 3	2.557 0			

最佳色散关系方程组(λ 以 μm 为单位,$T = 293$ K)[20]:

$$n_o^2 = 9.220 + \frac{0.445\ 4}{\lambda^2 - 0.126\ 4} + \frac{1\ 733}{\lambda^2 - 1\ 000}$$

$$n_e^2 = 7.007 + \frac{0.323\ 0}{\lambda^2 - 0.119\ 2} + \frac{660}{\lambda^2 - 1\ 000}$$

在文献[5]、[21]中给出了其他色散关系方程组。

室温下高频(远高于 Ag_3AsS_3 晶体声学共振频率,即对"受夹"晶体)下测量的线性电光系数[22]

$\lambda/\mu m$	$r_{22}^S/(pm \cdot V^{-1})$
0.632 8	1.1

普通情况下有效二阶非线性系数的表达式(Kleinman 对称条件成立,$d_{15} = d_{24} = d_{31} = d_{32}$)[23]:

$$d_{ooe} = d_{31}\sin(\theta + \rho) - d_{22}\cos(\theta + \rho)\sin 3\phi$$
$$d_{eoe} = d_{oee} = d_{22}\cos^2(\theta + \rho)\cos 3\phi$$

有效二阶非线性系数的简化表达式(Kleinman 对称条件成立,$d_{15} = d_{24} = d_{31} = d_{32}$)[24]:

$$d_{ooe} = d_{31}\sin\theta - d_{22}\cos\theta\sin 3\phi$$
$$d_{eoe} = d_{oee} = d_{22}\cos^2\theta\cos 3\phi$$

二阶非线性系数值[17,25]:

$$|d_{22}(10.6\ \mu m)| = (0.2 \pm 0.03) \times |d_{36}(GaAs)| = (16.6 \pm 2.5)\ pm/V$$
$$|d_{31}(10.6\ \mu m)| = (1.6 \pm 0.1)^{-1} \times |d_{22}(Ag_3AsS_3)| = (10.4 \pm 2.2)\ pm/V$$

相位匹配角的实验值($T = 293$ K)

相互作用的波长/μm	$\theta_{exp}/(°)$	参 考 文 献
SHG, $o + o \Rightarrow e$		
10.6 $\Rightarrow$ 5.3	23.6	[26]

续表

相互作用的波长/μm	$\theta_{\exp}/(°)$	参 考 文 献
10.59⇒5.295	21.5	[17]
9.2⇒4.6	19.9	[13]
2.13⇒1.065	29.5	[27]
2.128 4⇒1.064 2	29.4	[28]
SFG, o + o⇒e		
12.2 + 1.064⇒0.978 6	17.2	[29]
8.9 + 1.064⇒0.950 4	20.0	[29]
6.3 + 1.064⇒0.910 3	23.5	[29]
10.57 + 0.694 3⇒0.651 5	25.3	[30]
10.6 + 0.676 4⇒0.635 8	25.7	[8]
10.693 5 + 0.672 6⇒0.632 8	25.81	[31]
10.588 1 + 0.673 0⇒0.632 8	25.93	[31]
10.300 6 + 0.674 2⇒0.632 8	26.12	[31]
10.191 8 + 0.674 7⇒0.632 8	26.36	[31]
9.533 3 + 0.677 8⇒0.632 8	27.09	[31]
9.268 8 + 0.679 2⇒0.632 8	27.43	[31]
6.355 2 + 0.702 8⇒0.632 8	32.90	[32]
6.257 1 + 0.704 0⇒0.632 8	33.16	[32]
6.162 9 + 0.705 2⇒0.632 8	33.37	[32]
5.907 9 + 0.708 7⇒0.632 8	34.06	[32]
5.737 5 + 0.711 2⇒0.632 8	34.53	[32]
5.539 3 + 0.714 4⇒0.632 8	35.12	[32]
5.257 8 + 0.719 4⇒0.632 8	36.00	[32]
SFG, e + o⇒e		
10.59 + 1.064⇒0.967	20.0	[33]
10.59 + 0.694 3⇒0.651 6	27.7	[34]
9.31 + 0.678 9⇒0.632 8	29.0	[35]
SFG, o + e⇒e		
7.8 + 2.47⇒1.875 9	33	[36]

内角带宽的实验值

相互作用的波长/μm	$\Delta\theta^{\text{int}}/(°)$	参 考 文 献
SHG, o + o⇒e		
10.6⇒5.3	0.098	[11]
9.2⇒4.6	0.082	[13]
SFG, e + o⇒e		
10.6 + 0.694 3⇒0.651 6	0.031	[34]

激光诱导的表面损伤阈值

$\lambda/\mu m$	τ_p/ns	$I_{thr}/(GW \cdot cm^{-2})$	参 考 文 献
0.694 3	1 000 000	0.000 006	[37]
	14	0.003	[38]
1.064 2	CW	0.000 000 1	[37]
	18	>0.012	[38]
2.098	200	>0.01	[38]
9.55	30	0.18	[14]
10.6	190	>0.046	[38]
	150	0.053	[11]

关于这一晶体

  淡红银矿在1967年被介绍过[5]，并且在其后十年，被IR区的非线性频率转换和IR上转换广泛应用。然而，这种材料具有避免不了的缺点，诸如表面损伤阈值低并且难于获得均匀的大尺寸单晶。在1980年，这种晶体为其他材料（例如AGS和GaSe）所取代。

参考文献

[1] B. H. T. Chai: Optical Crystals. In: *CRC Handbook of Laser Science and Technology, Supplement 2: Optical Materials*, ed. by M. J. Weber (CRC Press, Boca Raton, 1995), pp. 3–65.

[2] *Physical-Chemical Properties of Semiconductors. Handbook* (Nauka, Moscow, 1979) [In Russian].

[3] A. A. Blistanov, V. S. Bondarenko, N. V. Perelomova, F. N. Strizhevskaya, V. V. Tchkalova, M. P. Shaskolskaya: *Acoustic Crystals* (Nauka, Moscow, 1982) [In Russian].

[4] D. M. Bercha, Y. V. Voroshilov, V. Y. Slivka, I. D. Turyanitsa: *Complex Chalcogenides and Chalcohalogenides*, ed. by D. V. Chepur (Vishcha Shkola, Lvov, 1983) [In Russian].

[5] K. F. Hulme, O. Jones, P. H. Davies, M. V. Hobden: Synthetic proustite (Ag_3AsS_3): a new crystal for optical mixing. Appl. Phys. Lett. 10(4), 133–135(1967).

[6] M. I. Holovey, I. D. Olexeyuk, M. I. Gurzan, I. S. Rez, V. V. Panko, Y. V. Voroshilov, M. Y. Rigan, I. G. Ganeyev, A. V. Bagdanova: Preparation and some properties of synthetic proustite single crystals. Kristall und Technik 6(5), 631–637(1971).

[7] N. Ito: Sum-frequency mixing of CO_2 and He-Ne lasers in proustite. Opt. Lett. 7(2), 63–65(1982).

[8] E. N. Antonov, V. R. Mironenko, D. N. Nikogosyan, M. I. Golovey: Conversion of CO_2 laser radiation to the visible range in a proustite crystal. Kvant. Elektron. 1(8), 1742–1746(1974) [In Russian, English trans.: Sov. J. Quantum Electron. 4(8), 963–965

(1974)].

[9] D. S. Hanna, A. J. Turner: Nonlinear absorption measurements in proustite (Ag_3AsS_3) and CdSe. Opt. Quant. Electron. 8(3), 213 – 217(1976).

[10] V. V. Berezovskii, Y. A. Bykovskii, S. N. Potanin, I. S. Rez: Two-photon absorption in proustite. Kvant. Elektron. No. 2, 74 – 75 (1973) [In Russian, English trans. : Sov. J. Quantum Electron. 3(2),134 – 135(1973)].

[11] D. N. Nikogosyan, A. P. Sukhorukov, M. I. Golovey: Saturation of second harmonic generation of TEA CO_2 laser radiation in a proustite crystal. Kvant. Elektron. 2(3), 609 – 612(1975)[In Russian,English trans. :Sov. J. Quantum Electron. 5(3),344 – 346(1975)].

[12] V. V. Berezovskii, Y. A. Bykovskii, M. I. Goncharov, I. S. Rez: Nonlinear polarization coefficients of proustite and tellurium. Kvant. Elektron. No. 2, 105 – 107 (1972) [In Russian,English trans. :Sov. J. Quantum Electron. 2(2),180 – 182(1972)].

[13] G. J. Ernst, W. J. Witteman: Second-harmonic generation in proustite with a CW CO_2 laser. IEEE J. Quant. Electr. QE – 8(3), 382 – 383(1972).

[14] Y. M. Andreev, V. V. Badikov, V. G. Voevodin, L. G. Geiko, P. P. Geiko, M. V. Ivashchenko, A. I. Karapuzikov, I. V. Sherstov: Radiation resistance of nonlinear crystals at a wavelength of 9.55 μm. Kvant. Elektron. 31(12), 1075 – 1078(2001) [In Russian, English trans. : Quantum Electron. 31(12),1075 – 1078(2001)].

[15] L. O. Hocker, C. F. Dewey: Difference frequency generation in proustite from 11 to 23 μm. Appl. Phys. 11(2), 137 – 140(1976).

[16] N. P. Barnes, R. C. Eckardt, D. J. Gettemy, L. B. Edgett: Absorption coefficients and the temperature variation of the refractive index difference of nonlinear optical crystals. IEEE J. Quant. Electr. QE – 15(10), 1074 – 1076(1979).

[17] D. S. Chemla, P. J. Kupeček, C. A. Schwartz: Redetermination of the nonlinear optical coefficients of proustite by comparison with pyrargyrite and gallium selenide. Opt. Commun. 7(3), 225 – 228(1973).

[18] P. J. Kupeček, C. A. Schwartz, D. S. Chemla: Silver thiogallate ($AgGaS_2$)—Part I: nonlinear optical properties. IEEE J. Quant. Electr. QE – 10(7), 540 – 545(1974).

[19] D. Cotter, D. C. Hanna, B. Luther-Davies, R. C. Smith: Backward-wave medium infra-red down-conversion in proustite. Opt. Commun. 11(1), 54 – 56(1974).

[20] M. V. Hobden: The dispersion of the refractive indices of proustite (Ag_3AsS_3). Opto-electron. 1(3), 159(1969).

[21] R. A. Andrews: IR image parametric up-conversion. IEEE J. Quant. Electr. QE – 6(1), 68 – 80(1970).

[22] G. W. C. Kaye, T. H. Laby: *Tables of Physical and Chemical Constants*(Longman Group Ltd. ,London,1995).

[23] I. Shoji, H. Nakamura, K. Ohdaira, T. Kondo, R. Ito, T. Okamoto, K. Tatsuki, S. Kubota:

Absolute measurement of second-order nonlinear-optical coefficients of β-BaB_2O_4 for visible to ultraviolet second-harmonic wavelengths. J. Opt. Soc. Am. B 16(4), 620 – 624(1999).

[24] J. E. Midwinter, J. Warner: The effects of phase matching method and of uniaxial crystal symmetry on the polar distribution of second-order non-linear optical polarization. Brit. J. Appl. Phys. 16(11), 1135 – 1142(1965).

[25] D. A. Roberts: Simplified characterization of uniaxial and biaxial nonlinear optical crystals: a plea for standardization of nomenclature and conventions. IEEE J. Quant. Electr. 28(10), 2057 – 2074(1992).

[26] Y. A. Gorokhov, D. P. Krindach, D. N. Nikogosyan, A. P. Sukhorukov: Influence of thermal self-actions on second harmonic generation of continuous-wave radiation. Kvant. Elektron. 1(3), 679 – 683 (1974) [In Russian, English trans. : Sov. J. Quantum Electron. 4(3), 382 – 384(1974)].

[27] D. C. Hanna, B. Luther-Davies, H. N. Rutt, R. C. Smith: Reliable operation of a proustite parametric oscillator. Appl. Phys. Lett. 20(1), 34 – 36(1972).

[28] T. Elsaesser, A. Seilmeier, W. Kaiser: Parametric generation of tunable picosecond pulses in proustite between 1.2 and 8 μm. Opt. Commun. 44(4), 293 – 296(1983).

[29] J. Falk, J. M. Yarborough: Detection of room-temperature blackbody radiation by parametric upconversion. Appl. Phys. Lett. 19(3), 68 – 70(1971).

[30] D. N. Nikogosyan: The up-conversion efficiency of CO_2 laser radiation in a proustite crystal pumped by ultrashort light pulses. Kvant. Elektron. 2(11), 2524 – 2525(1975) [In Russian, English trans. : Sov. J. Quantum Electron. 5(11), 1378 – 1379(1975)].

[31] E. K. Pfitzer, H. D. Riccius, K. J. Siemsen: Fabry-Perot photographs of CO_2 laser lines up-converted in proustite. Opt. Commun. 3(4), 277 – 278(1971).

[32] H. D. Riccius, K. J. Siemsen: Up-conversion of CO laser lines by difference-frequency mixing in proustite(Ag_3AsS_3). Phys. Lett. A 45(5), 377 – 378(1973).

[33] A. J. Alcock, A. C. Walker: Fast linear detection system for TE CO_2 lasers. Appl. Phys. Lett. 23(8), 467 – 468(1973).

[34] J. Warner: Photomultiplier detection of 10.6 μm radiation using optical up-conversion in proustite. Appl. Phys. Lett. 12(6), 222 – 224(1968).

[35] N. Ito: Sum-frequency mixing of CO_2 and He-Ne lasers in proustite. Opt. Lett. 7(2), 63 – 65(1982).

[36] G. C. Bhar, D. C. Hanna, B. Luther-Davies, R. C. Smith: Tunable down-conversion from an optical parametric oscillator. Opt. Commun. 6(4), 323 – 326(1972).

[37] A. F. Milton: Upconversion—a systems view. Appl. Opt. 11(10), 2311 – 2330(1972).

[38] D. C. Hanna, B. Luther-Davies, H. N. Rutt, R. C. Smith, C. R. Stanley: Q-switched laser damage of infrared nonlinear materials. IEEE J. Quant. Electr. QE – 8(3), 317 – 324(1972).

8.15 HgGa$_2$S$_4$，硫镓汞

负单轴晶：$n_o > n_e$

分子量：468.180

密度：4.95 g/cm^3[1]

点群：$\bar{4}$

晶格常数[2]：$a = 5.506$ Å；$c = 10.299$ Å

莫氏硬度：3~3.5

$p = 0.101\ 325$ MPa 时的比热容 c_p[3]

T/K	$c_p/(\text{J}\cdot\text{kg}^{-1}\cdot\text{K}^{-1})$
293	350~490

热导率 κ[3]

T/K	$\kappa/(\text{W}\cdot\text{m}^{-1}\cdot\text{K}^{-1})$，∥$c$	$\kappa/(\text{W}\cdot\text{m}^{-1}\cdot\text{K}^{-1})$，⊥$c$
293	2.5~2.9	2.3~2.4

室温下的带隙能量：$E_g = 2.84$ eV[4]

透明范围：0.55~11 μm[5]；0.55~12.4 μm[6]；0.55~13 μm[7]

UV 透过截止边对"黄"晶体为 0.51 μm，而对"橙"晶体为 0.55 μm[8]

线性吸收系数 α

$\lambda/\mu\text{m}$	α/cm^{-1}	参考文献	备注
0.53	8	[9]	e 光，SHG 方向
	11	[7]	
0.9~8.5	0.1~0.2	[8]	
0.96	0.25	[10]	e 光，SFG 方向
1.06	0.1	[9]	o 光，SHG 方向
1.064	0.25	[10]	o 光，SFG 方向
4.34	0.15	[3]	橙色晶体
9.55	<0.2~0.3	[11]	SHG 方向
10.6	1.2	[10]	o 光，SFG 方向
	0.44	[3]	橙色晶体

$T = 293$ K 时折射率的实验值[2]

$\lambda/\mu m$	n_o	n_e	$\lambda/\mu m$	n_o	n_e
0.549 5	2.659 2	2.597 9	2.650 0	2.444	2.403
0.574 7	2.633 4	2.574 8	3.540 0	2.439	2.398
0.600 9	2.611 2	2.554 9	7.150 0	2.414	2.372
0.632 8	2.589 0	2.534 9	8.730 0	2.400	2.358
0.650 0	2.579 6	2.526 4	10.400	2.380	2.337
1.076 0	2.477	2.432	11.000	2.369	2.329
1.150 0	2.472	2.428			

最佳色散关系方程组(λ 以 μm 为单位,$T = 293$ K)[6]:

$$n_o^2 = 5.940\ 5 + \frac{0.236\ 1}{\lambda^2 - 0.092\ 9} - 0.002\ 57\lambda^2$$

$$n_e^2 = 5.741\ 2 + \frac{0.213\ 8}{\lambda^2 - 0.089\ 7} - 0.002\ 47\lambda^2$$

在文献[2]、[3]、[8],[12]中给出了其他色散关系。

有效二阶非线性系数的表达式(Kleinman 对称条件成立,$d_{15} = d_{31}$ 和 $d_{14} = d_{25} = d_{36}$)[13]:

$$d_{ooe} = d_{36}\sin\theta\sin 2\phi + d_{31}\sin\theta\cos 2\phi$$

$$d_{eoe} = d_{oee} = d_{36}\sin 2\theta\cos 2\phi - d_{31}\sin 2\theta\sin 2\phi$$

二阶非线性系数值:

$$|d_{36}(1.064\ \mu m)| = 80 \times d_{11}(SiO_2) \pm 30\% = (24.0 \pm 7.2)\ pm/V^{[7,14]}$$

$$|d_{36}(1.064\ \mu m)| = 1.8 \times d_{36}(AgGaS_2) \pm 15\% = (24.7 \pm 7.6)\ pm/V^{[9,15]}$$

$$|d_{36}(1.064\ \mu m)| = 1.8 \times d_{36}(AgGaS_2) \pm 15\% = (31.5 \pm 4.7)\ pm/V^{[9,14,16]}$$

$$|d_{31}(1.064\ \mu m)| = 0.33 \times |d_{36}(HgGa_2S_4)| = (8.1 \pm 2.5)\ pm/V^{[9,15]}$$

$$|d_{31}(1.064\ \mu m)| = 0.33 \times |d_{36}(HgGa_2S_4)| = (10.4 \pm 1.6)\ pm/V^{[9,14,16]}$$

相位匹配角的实验值($T = 293$ K)[8]

相互作用的波长/μm	$\theta_{exp}/(°)$
SHG, o + o ⇒ e	
9.55 ⇒ 4.775	67.5

激光诱导的表面损伤阈值

$\lambda/\mu m$	τ_p/ns	$I_{thr}/(GW \cdot cm^{-2})$	参考文献	备 注
0.82	0.000 22	>170	[4]	1 kHz
1.064	30	0.04	[3]	
		~0.06	[10]	12.5 Hz, 10 个脉冲

续表

$\lambda/\mu m$	τ_p/ns	$I_{thr}/(GW\cdot cm^{-2})$	参 考 文 献	备 注
1.25	0.000 16	>160	[4]	1 kHz
9.55	30	0.3	[11]	
10.6	CW	>0.000 000 016	[10]	

关于这一晶体

尽管硫镓汞于20世纪70年代被引入[2,7],由于其光学质量相当低,它的应用很少。最近,HgGa$_2$S$_4$的性能得到很大的改善,并且利用这一种非线性材料首次实现了OPO[17]。

参考文献

[1] *Physical-Chemical Properties of Semiconductors. Handbook* (Nauka, Moscow, 1979) [In Russian].

[2] V. V. Badikov, I. N. Matveev, V. L. Panyutin, S. M. Pshenichnikov, T. M. Repyakhova, O. V. Rychik, A. E. Rozenson, N. K. Trotsenko, N. D. Ustinov: Growth and optical properties of mercury thiogallate. Kvant. Elektron. 6(8), 1807 – 1810 (1979) [In Russian, English trans.: Sov. J. Quantum Electron. 9(8), 1068 – 1069 (1979)].

[3] V. Badikov, K. Mitin, A. Seryogin, E. Ryabov, V. Laptev, A. Malinovsky: HgGa$_2$S$_4$ crystals for mid-infrared optical parametric oscillators pumped by Nd: YAG lasers. Proc. SPIE 4972, 131 – 138(2003).

[4] V. Petrov, F. Rotermund, V. V. Badikov, G. S. Shevyrdyaeva: Mixed nonlinear crystal Cd$_x$Hg$_{1-x}$Ga$_2$S$_4$ used for optical parametric amplification. In: *Conference on Lasers and Electrooptics CLEO/QELS 2003, Technical Digest* (OSA, Washington DC, 2003), paper CMA5.

[5] P. G. Schunemann, P. G. Pollak: Synthesis and growth of HgGa$_2$S$_4$ crystals. J. Cryst. Growth 174(1 – 4), 278 – 282(1997).

[6] E. Takaoka, K. Kato: Second-harmonic generation in HgGa$_2$S$_4$. In: *CLEO/Europe 1998, Technical Digest* (OSA, Washington DC, 1998), p. 387, paper CFH7.

[7] B. F. Levine, C. G. Bethea, H. M. Kasper, F. A. Thiel: Nonlinear optical susceptibilities of HgGa$_2$S$_4$. IEEE J. Quant. Electr. QE – 12(6), 367 – 368(1976).

[8] Y. M. Andreev, P. P. Geiko, V. V. Badikov, G. C. Bhar, S. Das, A. K. Chaudhury: Nonlinear optical properties of defect tetrahedral crystals HgGa$_2$S$_4$ & AgGaGeS$_4$ and mixed chalcopyrite crystal Cd$_{0.4}$Hg$_{0.6}$Ga$_2$S$_4$. Nonl. Opt. 29(1), 19 – 27(2002).

[9] V. V. Badikov, I. N. Matveev, S. M. Pshenichnikov, O. V. Rychik, N. K. Trotsenko, N. D. Ustinov, S. I. Shcherbakov: Growth and nonlinear properties of HgGa$_2$S$_4$. Kvant. Elektron. 7(10),

2235 – 2237(1980)[In Russian, English trans.: Sov. J. Quantum Electron. 10(10), 1300 – 1301(1980)].

[10] S. A. Andreev, N. P. Andreeva, V. V. Badikov, I. N. Matveev, S. M. Pschenichnikov: Frequency up-conversion in a mercury thiogallate crystal. Kvant. Elektron. 7(9), 2003 – 2006(1980)[In Russian, English trans.: Sov. J. Quantum Electron. 10(9), 1157 – 1158(1980)].

[11] Y. M. Andreev, V. V. Badikov, V. G. Voevodin, L. G. Geiko, P. P. Geiko, M. V. Ivashchenko, A. I. Karapuzikov, I. V. Sherstov: Radiation resistance of nonlinear crystals at a wavelength of 9.55 μm. Kvant. Elektron. 31(12), 1075 – 1078(2001)[In Russian, English trans.: Quantum Electron. 31(12), 1075 – 1078(2001)].

[12] G. G. Matvienko, Y. M. Andreev, V. V. Badikov, P. P. Geiko, S. G. Grechin, A. I. Karapuzikov: Wide band frequency converters for lidar systems. Proc. SPIE 4546, 119 – 126 (2002).

[13] J. E. Midwinter, J. Warner: The effects of phase matching method and of uniaxial crystal symmetry on the polar distribution of second-order non-linear optical polarization. Brit. J. Appl. Phys. 16(11), 1135 – 1142(1965).

[14] D. A. Roberts: Simplified characterization of uniaxial and biaxial nonlinear optical crystals: a plea for standardization of nomenclature and conventions. IEEE J. Quant. Electr. 28 (10), 2057 – 2074(1992).

[15] J.-J. Zondy, D. Touahri, O. Acef: Absolute value of the d_{36} nonlinear coefficient of Ag-GaS_2: prospect for a low-threshold doubly resonant oscillator-based 3:1 frequency divider. J. Opt. Soc. Am. B 14(10), 2481 – 2497(1997).

[16] F. Rotermund, V. Petrov: Mercury thiogallate mid-infrared femtosecond optical parametric generator pumped at 1.25 μm by a Cr:forsterite regenerative amplifier. Opt. Lett. 25 (10), 746 – 748(2000).

[17] V. V. Badikov, A. K. Don, K. V. Mitin, A. M. Seregin, V. V. Sinaiskii, N. I. Shchebetova: A $HgGa_2S_4$ optical parametric oscillator. Kvant. Elektron. 33(9), 831 – 832(2003)[In Russian, English trans.: Quantum Electron. 33(9), 831 – 832(2003)].

8.16 $CdGeAs_2$，砷锗镉(CGA)

正单轴晶：$n_e > n_o$

分子量：334.753

密度：5.60 g/cm^3 [1]

点群：$\bar{4}2m$

莫氏硬度：3.5 ~ 4

诺氏(或维氏)硬度：485，压痕负荷为 50 g [2]

熔点：933 K[3]

线性热膨胀系数的平均值[2]

ΔT/K	$\alpha_t \times 10^6/K^{-1}$, $\parallel c$	$\alpha_t \times 10^6/K^{-1}$, $\perp c$
293～673	1.0	11.4

热导率[1]：$\kappa = 4.18$ W/(m·K)[或 6.69 W/(m·K)]

室温下带隙能量：$E_g = 0.52$ eV[4]；0.54 eV[5]

以"0"透过计的透明范围：2.3～18 μm[3]；2.4～18 μm[6]；2.45～18.1 μm[3]；多光子吸收峰存在于 12.5 μm 和 13.5 μm 处[3]

线性吸收系数 α

λ/μm	T/K	α/cm^{-1}	参考文献	λ/μm	T/K	α/cm^{-1}	参考文献
2.8	300	1.5	[7]	9～11	300	0.23	[9]
3.39	300	5.7	[8]	9.2	300	0.30	[11]
4～18	300	<0.9	[9]	9.27	77	0.002(?)	[12]
4.5	300	0.7	[10]		300	0.44	[12]
4.6	300	0.17	[11]	9.55	80	0.1	[13]
4.635	77	0.32	[12]		295	0.66	[13]
	300	1.03	[12]		300	0.42	[17]
4.755	80	0.3	[13]	10	300	0.2	[7]
	295	0.99	[13]	10.6	77	0.1	[14]
5.3	77	0.4	[14]		300	0.4	[6]
	300	1.3	[8]		300	0.5	[8]
5.5	300	0.46	[3]		300	2.4	[18]
5.85	77	0.42	[15]	10.6～11.7	77	0.14	[15]
	300	1.5	[15]		300	0.5	[15]
8.6～12	77	<0.2	[16]	11	300	<0.2	[10]
	300	<0.5	[16]	12.3	300	0.4	[4]

双光子吸收系数 β[7]

λ/μm	τ_p/ns	$\beta \times 10^{11}$/(cm·W^{-1})	备注
2.8	0.1	25 000	o 光

折射率的实验值[19]

$\lambda/\mu m$	n_o	n_e	$\lambda/\mu m$	n_o	n_e
2.3	3.607 6		4.8	3.535 4	3.627 3
2.4	3.597 3	3.754 5	5.0	3.533 6	3.624 9
2.5	3.589 5	3.731 6	5.5	3.528 5	3.617 8
2.6	3.582 3	3.715 6	6.0	3.525 1	3.613 4
2.7	3.577 3	3.703 0	6.5	3.522 3	3.610 4
2.8	3.572 1	3.692 6	7.0	3.520 0	3.607 3
2.9	3.568 4	3.684 6	7.5	3.517 5	3.605 0
3.0	3.564 5	3.677 5	8.0	3.515 7	3.603 0
3.1	3.561 5	3.671 4	8.5	3.514 0	3.600 9
3.2	3.558 1	3.666 1	9.0	3.512 0	3.598 8
3.4	3.553 6	3.657 4	9.5	3.509 8	3.596 6
3.6	3.550 3	3.650 8	10.0	3.507 8	3.594 2
3.8	3.546 8	3.645 4	10.5	3.505 4	3.592 2
4.0	3.544 0	3.640 2	11.0	3.503 1	3.589 6
4.2	3.541 5	3.636 8	11.5	3.500 4	3.587 1
4.4	3.539 1	3.632 9	12.0	3.497 7	
4.6	3.537 2	3.629 9	12.5	3.495 0	

最佳色散关系方程组(λ 以 μm 为单位,$T = 293$ K)[20]:

$$n_o^2 = 12.400\,8 + \frac{2.160\,3}{\lambda^2 - 2.061\,7} - 0.001\,33\lambda^2$$

$$n_e^2 = 13.007\,9 + \frac{3.261\,3}{\lambda^2 - 2.838\,2} - 0.001\,26\lambda^2$$

文献[10]、[21]的作者两次列出这些方程时对 n_e 的最后一个系数作了细微改变,即将 0.001 26 改为 0.001 25。

文献[6]、[9]、[22]给出了其他 Sellmeier 方程组。

光谱范围 2.65~10.6 μm 之间从室温到 T[K]时折射率的温度微商[20]:

$$dn_o/dT = (26.51/\lambda^2 - 6.45/\lambda + 19.17) \times 10^{-5} \times [1 + 9.0 \times 10^{-4}(T-293)]$$

$$dn_e/dT = (22.62/\lambda^2 - 5.35/\lambda + 15.20) \times 10^{-5} \times [1 + 6.1 \times 10^{-4}(T-293)]$$

普通情况下有效二阶非线性系数的表达式(Kleinman 对称条件成立,$d_{14} = d_{25} = d_{36}$)[23,24]:

$$d_{eeo} = 2d_{36}\sin(\theta+\rho)\cos(\theta+\rho)\cos 2\phi$$

$$d_{oeo} = d_{eoo} = -d_{36}\sin(\theta+\rho)\sin 2\phi$$

有效二阶非线性系数的简化表达式(Kleinman 对称条件成立,$d_{14} = d_{25} = d_{36}$)[24]:

$$d_{eeo} = d_{36}\sin 2\theta\cos 2\phi$$

$$d_{oeo} = d_{eoo} = -d_{36}\sin\theta\sin 2\phi$$

二阶非线性系数值:

$$d_{36}(10.6\ \mu m) = (4.7 \pm 0.4) \times d_{36}(AgGaSe_2) = (186 \pm 16)\ pm/V^{[20,25]}$$

相位匹配角的实验值($T = 293$ K)

相互作用的波长/μm	$\theta_{exp}/(°)$	参考文献	相互作用的波长/μm	$\theta_{exp}/(°)$	参考文献
SHG, e + e⇒o			SHG, o + e⇒o		
9⇒4.5	32.2	[21]	10.6⇒5.3	48.4	[9]
10⇒5.0	32	[21]		49	[8]
10.6⇒5.3	32	[8]		49.3	[20]
	32.6	[20]		50.7	[9]
	33.8	[15]		51.6	[19]
	34	[16]		52	[6]
	35	[19]	SFG, o + e⇒o		
11⇒5.5	32.5	[21]	16.4 + 9.54⇒6.03	47	[9]
11.7⇒5.85	35.7	[15]	12.9 + 9.59⇒5.5	46.1	[9]

内角带宽和温度带宽的实验值

相互作用的波长/μm	$\Delta\theta^{int}/(°)$	$\Delta T/K$	参 考 文 献
SHG, e + e⇒o			
10.6⇒5.3	0.84		[16]
SHG, o + e⇒o			
10.6⇒5.3	0.29		[6]
		41	[20]
DFG, o - e⇒e			
5.3 − 9.3⇒12.32	0.98		[4]

激光诱导的表面损伤阈值

$\lambda/\mu m$	τ_p/ns	$I_{thr}/(GW \cdot cm^{-2})$	参考文献	备 注
2.8	0.1	2.7	[7]	
5	0.0006	>6	[21]	10 Hz, 100 个脉冲成串列
9.55	30	0.16	[17]	
10.6	CW	>0.000 13	[14]	$T = 77$ K
	CW	>0.000 001	[6]	
	160	>0.004	[6]	
	160	0.038	[9]	
	150	0.033 ~ 0.04	[26]	

关于这一晶体

尽管砷锗镉具有一个最大的二阶非线性系数,然而这种材料很难生长出高光学质量的晶体。最近,Schunemann 成功地生长了在 4.6 μm 和 9.2 μm 处吸收系数小于 0.1 cm^{-1} 的 CGA 晶体[11],这晶体被 Vodopyanov 用于 DFG[10] 和 OPG[21]。

■ **参考文献**

[1] *Physical-Chemical Properties of Semiconductors. Handbook* (Nauka, Moscow, 1979) [In Russian].

[2] I. I. Kozhina, A. S. Borshchevskii: High-temperature x-ray investigations of A^{II}B^{IV}C$_2^V$ compounds. Vestnik LGU No. 22, 87 – 92 (1971) [In Russian].

[3] P. G. Schunemann, T. M. Pollak: Ultralow gradient HGF-grown ZnGeP$_2$ and CdGeAs$_2$ and their optical properties. MRS Bulletin 23(7), 23 – 27 (1998).

[4] R. G. Harrison, P. K. Gupta, M. R. Taghizadeh, A. K. Kar: Efficient multikilowatt mid infrared difference frequency generation in CdGeAs$_2$. IEEE J. Quant. Electr. QE – 18 (8), 1239 – 1242 (1982).

[5] G. C. Bhar, R. C. Smith: Optical properties of II-IV-V$_2$ and I-III-VI$_2$ crystals with particular reference to transmission limits. Phys. Stat. Solidi A 13(1), 157 – 168 (1972).

[6] R. L. Byer, H. Kildal, R. S. Feigelson: CdGeAs$_2$—a new nonlinear crystal phasematchable at 10.6 μm. Appl. Phys. Lett. 19(7), 237 – 240 (1971).

[7] K. L. Vodopyanov, S. B. Mirov, V. G. Voevodin, P. G. Schunemann: Two-photon absorption in GaSe and CdGeAs$_2$. Opt. Commun. 155(1 – 3), 47 – 50 (1998).

[8] D. S. Chemla, R. F. Begley, R. L. Byer: Experimental and theoretical studies of third harmonic generation in chalcopyrite CdGeAs$_2$. IEEE J. Quant. Electr. QE – 10(1), 71 – 81 (1974).

[9] H. Kildal, J. C. Mikkelsen: Efficient doubling and CW difference frequency mixing in the infrared using the chalcopyrite CdGeAs$_2$. Opt. Commun. 10(4), 306 – 309 (1974).

[10] K. L. Vodopyanov, P. G. Schunemann: Efficient difference-frequency generation of 7 – 20 μm radiation in CdGeAs$_2$. Opt. Lett. 23(14), 1096 – 1098 (1998).

[11] P. G. Schunemann, S. D. Setzler, T. M. Pollak: Crystal growth of low-loss CdGeAs$_2$. In: *Advanced Solid-State Lasers*, *OSA Trends in Optics and Photonics Series*, Vol. 50, ed. by C. Marshall (OSA, Washington DC, 2001), pp. 632 – 634.

[12] P. G. Schunemann, K. L. Schepler, P. A. Budni: Nonlinear frequency conversion performance of AgGaSe$_2$, ZnGeP$_2$, and CdGeAs$_2$. MRS Bulletin 23(7), 45 – 49 (1998).

[13] A. Zakel, J. L. Blackshire, P. G. Schunemann, S. D. Setzler, J. Goldstein, S. Guha: Temperature and pulse-duration dependence of second-harmonic generation in CdGeAs$_2$.

Appl. Opt. 41(12), 2299 – 2303(2002).

[14] N. Menyuk, G. W. Iseler, A. Mooradian: High-efficiency high-average-power second-harmonic generation with $CdGeAs_2$. Appl. Phys. Lett. 29(7), 422 – 424(1976).

[15] Y. M. Andreev, V. G. Voevodin, P. P. Geiko, A. I. Gribenyukov, A. P. Dyadkin, S. V. Pigulsky, A. I. Starodubtsev: Efficient generation of the second harmonic of NH_3 laser radiation in $CdGeAs_2$. Kvant. Elektron. 14(4), 784 – 786(1987) [In Russian, English trans.: Sov. J. Quantum Electron. 17(4), 491 – 493(1987)].

[16] V. E. Zuev, M. V. Kabanov, Y. M. Andreev, V. G. Voevodin, P. P. Geiko, A. I. Gribenyukov, V. V. Zuev: Applications of efficient parametric IR-laser frequency converters. Izv. Akad. Nauk SSSR, Ser. Fiz. 52(6), 1142 – 1148(1988) [In Russian, English trans.: Bull. Acad. Sci. USSR, Phys. Ser. 52(6), 87 – 92(1988)].

[17] Y. M. Andreev, V. V. Badikov, V. G. Voevodin, L. G. Geiko, P. P. Geiko, M. V. Ivashchenko, A. I. Karapuzikov, I. V. Sherstov: Radiation resistance of nonlinear crystals at a wavelength of 9.55 μm. Kvant. Elektron. 31(12), 1075 – 1078(2001) [In Russian, English trans.: Quantum Electron. 31(12), 1075 – 1078(2001)].

[18] N. P. Barnes, R. C. Eckardt, D. J. Gettemy, L. B. Edgett: Absorption coefficients and the temperature variation of the refractive index difference of nonlinear optical crystals. IEEE J. Quant. Electr. QE – 15(10), 1074 – 1076(1979).

[19] G. D. Boyd, E. Buehler, F. G. Storz, J. H. Wernick: Linear and nonlinear optical properties of ternary $A^{II} B^{IV} C_2^{V}$ chalcopyrite semiconductors. IEEE J. Quant. Electr. QE – 8 (4), 419 – 426(1972).

[20] E. Tanaka, K. Kato: Second-harmonic and sum-frequency generation in $CdGeAs_2$. In: *MRS Symposium Proceedings*, *Vol. 384*, *Infrared Applications of Semiconductors II*, ed by D. L. McDaniel, Jr., M. O. Manasreh, R. H. Miles, S. Sivananthan (Materials Research Society, Warrendale, PA, 1998), pp. 475 – 479.

[21] K. L. Vodopyanov, G. M. H. Knippels, A. F. G. van der Meer, J. P. Maffetone, I. Zweiback: Optical parametric generation in CGA crystal. Opt. Commun. 202(1 – 3), 205 – 208(2002).

[22] G. C. Bhar: Refractive index interpolation in phase-matching. Appl. Opt. 15(2), 305 – 307 (1976).

[23] R. C. Eckardt, H. Masuda, Y. X. Fan, R. L. Byer: Absolute and relative nonlinear optical coefficients of KDP, KD^*P, BaB_2O_4, $LiIO_3$, MgO: $LiNbO_3$, and KTP measured by phase-matched second-harmonic generation. IEEE J. Quant. Electr. 26(5), 922 – 933(1990).

[24] J. E. Midwinter, J. Warner: The effects of phase matching method and of uniaxial crystal symmetry on the polar distribution of second-order non-linear optical polarization. Brit. J. Appl. Phys. 16(11), 1135 – 1142(1965).

[25] A. Harasaki, K. Kato: New data on the nonlinear optical constant, phase-matching,

and optical damage of AgGaS$_2$. Jpn. J. Appl. Phys. 36(2), 700–703(1997).
[26] H. Kildal, G. W. Iseler: Laser-induced surface damage of infrared nonlinear materials. Appl. Opt. 15(12), 3062–3065(1976).

8.17 Tl$_3$AsSe$_3$，硒砷铊(TAS)

负单轴晶：$n_o > n_e$

分子量：784.002

密度：7.83 g/cm^3 [1]

点群：$3m$

晶格常数[2]：$a = 9.80$ Å；$c = 7.08$ Å

莫氏硬度：2~3 [2]

熔点：584 K[1]

线性热膨胀系数[3]

T/K	$\alpha_t \times 10^6$/K^{-1}, $\parallel c$	$\alpha_t \times 10^6$/K^{-1}, $\perp c$
300	18.2	28

对 6 mm 长晶体以 0.5 透过计的透明范围：1.28~17 μm[2]

线性吸收系数 α

λ/μm	α/cm^{-1}	参考文献	备 注
2~12	<0.02	[1]	
9.6	0.0005	[4]	
10.6	0.082	[5]	SHG 方向
	0.038	[2]	

300 K 时折射率的实验值[3]

λ/μm	n_o	n_e	λ/μm	n_o	n_e
2.056	3.419	3.227	7.854	3.345	3.162
3.059	3.380	3.190	9.016	3.340	3.158
4.060	3.364	3.177	9.917	3.336	3.155
5.035	3.357	3.171	10.961	3.331	3.152
5.856	3.354	3.168	12.028	3.327	3.147
6.945	3.349	3.164			

$\lambda = 2 \sim 10.6$ μm 时折射率的温度微商($T = 80 \sim 300$ K)[3]：

$$\frac{dn_o}{dT} = -4.52 \times 10^{-5} \text{ K}^{-1}$$

$$\frac{dn_e}{dT} = +3.55 \times 10^{-5} \text{ K}^{-1}$$

最佳色散关系方程组(λ 以 μm 为单位,$T = 300$ K)[3]:

$$n_o^2 = 1 + \frac{10.210\lambda^2}{\lambda^2 - 0.197\,136} + \frac{0.522\lambda^2}{\lambda^2 - 625}$$

$$n_e^2 = 1 + \frac{8.933\lambda^2}{\lambda^2 - 0.197\,136} + \frac{0.308\lambda^2}{\lambda^2 - 625}$$

在文献[2]中给出了其他色散关系方程组。

普通情况下有效二阶非线性系数的表达式(Kleinman 对称条件成立,$d_{15} = d_{24} = d_{31} = d_{32}$)[6]:

$$d_{ooe} = d_{31}\sin(\theta + \rho) - d_{22}\cos(\theta + \rho)\sin 3\phi$$
$$d_{eoe} = d_{oee} = d_{22}\cos^2(\theta + \rho)\cos 3\phi$$

有效二阶非线性系数的简化表达式(Kleinman 对称条件成立,$d_{15} = d_{24} = d_{31} = d_{32}$)[7]:

$$d_{ooe} = d_{31}\sin\theta - d_{22}\cos\theta\sin 3\phi$$
$$d_{eoe} = d_{oee} = d_{22}\cos^2\theta\cos 3\phi$$

I 类相互作用二阶非线性系数的极大值 $d_+ = |d_{31}\sin\theta| + |d_{22}\cos\theta|$:

$$d_+(10.6\ \mu m) = (3.47 \pm 1.04) \times d_+(Ag_3AsS_3) = (68 \pm 31)\text{ pm/V}^{[2,8,9]}$$
$$d_+(10.6\ \mu m) = (3.3 \pm 1.0) \times d_+(Ag_3AsS_3) = (37 \pm 13)\text{ pm/V}^{[2,9,10]}$$
$$d_+(10.6\ \mu m) = 20 \sim 30\text{ pm/V}^{[4]}$$

相位匹配角和内角带宽的实验值

相互作用的波长/μm	$\theta_{exp}/(°)$	$\Delta\theta^{int}/(°)$	参 考 文 献
SHG, o + o ⇒ e			
4.8⇒2.4	27		[11]
9.6⇒4.8	19	0.21	[11]
10.6⇒5.3		0.30	[12]
SFG, o + o ⇒ e			
9.6 + 4.8⇒3.2	21		[11]
9.6 + 2.4⇒1.92	28		[11]

激光诱导的表面损伤阈值

$\lambda/\mu m$	τ_p/ns	$I_{thr}/(GW \cdot cm^{-2})$	参 考 文 献	备 注
9.25	20	>0.00001	[4]	10 kHz
9.6	70	>0.01	[11]	
10.6	20	>0.27	[12]	
	150	0.01~0.017	[13]	
	200	0.016	[2]	

关于这一晶体

TAS 是一种毒性相当高的 IR 非线性晶体,它是在 20 世纪 70 年代发展起来的[1-3],自那时起,主要用于 CO_2 激光器辐射的 SHG。最近,用这种晶体通过 20 ns、30 kHz CO_2 激光脉冲实现了 4.625 μm 波长 6 W 的准连续输出[4]。

■ **参考文献**

[1] M. Gottlieb, T. J. Isaacs, J. D. Feichtner, G. W. Roland: Acousto-optic properties of some chalcogenide crystals. J. Appl. Phys. 45(12), 5145 - 5151(1974).

[2] J. D. Feichtner, G. W. Roland: Optical properties of a new nonlinear optical material: Tl_3AsSe_3. Appl. Opt. 11(5), 993 - 998(1972).

[3] M. D. Ewbank, P. R. Newman, N. L. Mota, S. M. Lee, W. L. Wolfe, A. G. DeBell, W. A. Harrison: The temperature dependence of optical and mechanical properties of Tl_3AsSe_3. J. Appl. Phys. 51(7), 3848 - 3852(1980).

[4] D. R. Suhre, L. H. Taylor: Six-watt mid-infrared laser using harmonic generation with Tl_3AsSe_3. Appl. Phys. B 63(3), 225 - 228(1996).

[5] N. P. Barnes, R. C. Eckardt, D. J. Gettemy, L. B. Edgett: Absorption coefficients and the temperature variation of the refractive index difference of nonlinear optical crystals. IEEE J. Quant. Electr. QE - 15(10), 1074 - 1076(1979).

[6] I. Shoji, H. Nakamura, K. Ohdaira, T. Kondo, R. Ito, T. Okamoto, K. Tatsuki, S. Kubota: Absolute measurement of second-order nonlinear-optical coefficients of β-BaB_2O_4 for visible to ultraviolet second-harmonic wavelengths. J. Opt. Soc. Am. B 16(4), 620 - 624(1999).

[7] J. E. Midwinter, J. Warner: The effects of phase matching method and of uniaxial crystal symmetry on the polar distribution of second-order non-linear optical polarization. Brit. J. Appl. Phys. 16(11), 1135 - 1142(1965).

[8] D. S. Chemla, P. J. Kupeček, C. A. Schwartz: Redetermination of the nonlinear optical coefficients of proustite by comparison with pyrargyrite and gallium selenide. Opt. Commun. 7(3), 225 - 228(1973).

[9] D. A. Roberts: Simplified characterization of uniaxial and biaxial nonlinear optical crystals: a plea for standardization of nomenclature and conventions. IEEE J. Quant. Electr. 28(10),

2057 – 2074(1992).

[10] J. H. McFee, G. D. Boyd, P. H. Schmidt: Redetermination of the nonlinear optical coefficients of Te and GaAs by comparison with Ag_3SbS_3. Appl. Phys. Lett. 17(2), 57 – 59 (1970).

[11] R. C. Y. Auyeung, D. M. Zielke, B. J. Feldman: Multiple harmonic conversion of pulsed CO_2 laser radiation in Tl_3AsSe_3. Appl. Phys. B 48(4), 293 – 297(1989).

[12] D. R. Suhre: Efficient second-harmonic generation in Tl_3AsSe_3 using pulsed CO_2 laser radiation. Appl. Phys. B 52(6), 367 – 370(1991).

[13] H. Kildal, G. W. Iseler: Laser-induced surface damage of infrared nonlinear materials. Appl. Opt. 15(12), 3062 – 3065(1976).

8.18　CdSe，硒化镉

正单轴晶：$n_e > n_o$

分子量：191.370

密度：5.81 g/cm^3，$T = 288$ K[1-3]

点群：$6mm$

晶格常数：

　　$a = 4.30$ Å[4]；4.298 5 Å[5]；4.299 9 Å[6]

　　$c = 7.01$ Å[4]；7.015 0 Å[5]；7.010 9 Å[6]

莫氏硬度：3.25[4]

努氏硬度：44 ~ 90[7]；71，压痕载荷为 20 g[4]；90，压痕载荷为 20 g[5]

熔点：1 512 K[1]；1 525 K[2]；1 531 K[2]

线性热膨胀系数的平均值[2]

T/K	$\alpha_t \times 10^6 / K^{-1}$, $\parallel c$	$\alpha_t \times 10^6 / K^{-1}$, $\perp c$
77 ~ 298	2.45	4.4

$p = 0.101\ 325$ MPa 时的比热容 c_p[1]

T/K	$c_p/(J \cdot kg^{-1} \cdot K^{-1})$
298	258

热导率 κ[8]

T/K	$\kappa/(\text{W}\cdot\text{m}^{-1}\cdot\text{K}^{-1})$, $\parallel c$	$\kappa/(\text{W}\cdot\text{m}^{-1}\cdot\text{K}^{-1})$, $\perp c$
293	6.9	6.2

室温下带隙能量(直接跃迁):

$E_\text{g} = 1.67 \text{ eV}^{[2]}$; $1.7 \text{ eV}^{[9,10]}$; $1.74 \text{ eV}^{[1,11-16]}$; $1.75 \text{ eV}^{[17]}$; $1.8 \text{ eV}^{[3]}$

以"0"透过计的透明范围: $0.7 \sim 24 \ \mu\text{m}^{[3]}$

线性吸收系数 α

$\lambda/\mu\text{m}$	α/cm^{-1}	参考文献	$\lambda/\mu\text{m}$	α/cm^{-1}	参考文献
0.75~20	<0.1	[18]	3.39	0.01	[24]
0.1~10	0.01~0.02	[19]	4	0.04	[25]
1.06	0.062±0.006	[20]	10.6	0.000 5	[26]
1.064 2	0.013±0.001	[21]		0.016	[25]
	0.02	[22],[23]		0.032	[27]
1.32	0.01	[22]			

双光子吸收系数 β

$\lambda/\mu\text{m}$	τ_p/ns	$\beta\times 10^{11}/(\text{cm}\cdot\text{W}^{-1})$	参考文献	备注
1.06	~20	90 000±9 000	[28]	$\boldsymbol{E}\perp c$
		39 000±4 000	[28]	$\boldsymbol{E}\parallel c$
	20	14 000	[29]	$\perp c, \boldsymbol{E}\perp c$
		14 000	[29]	$\parallel c, \boldsymbol{E}\perp c$
		6 000	[29]	$\perp c, \boldsymbol{E}\parallel c$
1.064 2	0.03	3 000±500	[23]	
	0.038	1 800	[30]	$\parallel c$
	0.040	3 500	[31]	
	11	2 500±800	[21]	
	15	8 000	[32]	$\perp c, \boldsymbol{E}\parallel c$
		16 000	[32]	$\perp c, \boldsymbol{E}\perp c$
		16 000	[32]	$\parallel c, \boldsymbol{E}\perp c$
	16	5 000±1 400	[20]	$\parallel c$
	~20	23 000	[33]	$\boldsymbol{E}\perp c$
		21 000	[33]	$\boldsymbol{E}\parallel c$
	26	3 800±1 100	[21]	
1.15	0.003	199	[34]	$\perp c, \boldsymbol{E}\parallel c$
1.22	0.003	215	[34]	$\perp c, \boldsymbol{E}\parallel c$
1.30	0.003	304	[34]	$\perp c, \boldsymbol{E}\parallel c$
1.318 8	80	6 700±2 000	[21]	

续表

$\lambda/\mu m$	τ_p/ns	$\beta \times 10^{11}/(cm \cdot W^{-1})$	参考文献	备注
1.37	0.003	355	[34]	$\perp c, \boldsymbol{E} \parallel c$
1.42	0.003	8	[34]	$\perp c, \boldsymbol{E} \parallel c$
1.44	0.000 083	20	[35]	$\parallel c$
1.46	0.003	5	[34]	$\perp c, \boldsymbol{E} \parallel c$
1.50	0.003	10	[34]	$\perp c, \boldsymbol{E} \parallel c$

折射率的实验值

$\lambda/\mu m$	n_o	n_e	参考文献	$\lambda/\mu m$	n_o	n_e	参考文献
0.8	2.644 8	2.660 7	[36]	3.2	2.453 2	2.472 6	[36]
0.9	2.582 6	2.602 7	[36]	3.4	2.451 8	2.471 4	[36]
1.0	2.550 2	2.569 6	[36]	3.6	2.450 9	2.470 2	[36]
1.013 9	2.548 1	2.567 7	[27]	3.8	2.449 8	2.469 4	[36]
1.128 7	2.524 6	2.544 4	[27]	4.0	2.449 1	2.468 5	[36]
1.2	2.513 2	2.533 1	[36]		2.449	2.470	[37]
1.367 3	2.497 1	2.517 0	[27]	5.0	2.446 4	2.465 7	[27]
1.4	2.492 9	2.513 3	[36]	6.0	2.445	2.466	[37]
1.529 5	2.486 1	2.505 9	[27]		2.443 4	2.462 5	[27]
1.6	2.481 8	2.500 8	[36]	7.0	2.439 8	2.458 6	[27]
1.710 9	2.477 6	2.497 4	[27]	8.0	2.436 7	2.455 2	[27]
1.8	2.473 2	2.493 0	[36]	9.0	2.433 3	2.451 4	[27]
2.0	2.468 2	2.487 3	[36]	10.0	2.431	2.452	[37]
	2.468	2.489	[37]		2.429 4	2.447 5	[27]
2.2	2.464 2	2.484 0	[36]	11.0	2.425 2	2.443 0	[27]
2.325 3	2.462 7	2.482 3	[27]	12.0	2.420 4	2.437 9	[27]
2.4	2.461 2	2.479 8	[36]	14.0	2.410	2.431	[37]
2.6	2.459 0	2.478 4	[36]	16.0	2.399	2.419	[37]
2.8	2.456 2	2.475 7	[36]	20.0	2.376	2.390	[37]
3.0	2.455 3	2.474 8	[27]	22.0	2.339	2.351	[37]
		2.474 1	[36]	24.0	2.291		[37]

最佳色散关系方程组(λ 以 μm 为单位,$T = 293$ K)[38]:

$$n_o^2 = 4.224\ 3 + \frac{1.768\ 0\lambda^2}{\lambda^2 - 0.227\ 0} + \frac{3.120\ 0\lambda^2}{\lambda^2 - 3\ 380}$$

$$n_e^2 = 4.200\ 9 + \frac{1.887\ 5\lambda^2}{\lambda^2 - 0.217\ 1} + \frac{3.646\ 1\lambda^2}{\lambda^2 - 3\ 629}$$

在文献[39]、[40]中给出了其他色散关系方程组。

文献[40]中给出了温度为 73 K、173 K、373 K、573 K 时的 Sellmeier 的方程。

非线性折射率 γ

$\lambda/\mu m$	$\gamma \times 10^{15}/(cm^2 \cdot W^{-1})$	参考文献	$\lambda/\mu m$	$\gamma \times 10^{15}/(cm^2 \cdot W^{-1})$	参考文献
1.064 2	−15	[14]	1.44~1.54	130	[35]

室温下高频(远高于 CdSe 晶体声学共振频率,即对"受夹"晶体)下测量的线性电光系数[41]

$\lambda/\mu m$	$r_{13}^S/(pm \cdot V^{-1})$	$r_{33}^S/(pm \cdot V^{-1})$
3.391 3	1.81	4.3

有效二阶非线性系数的表达式(Kleinman 对称条件成立,$d_{15} = d_{24} = d_{31} = d_{32}$)[42]:

$$d_{oeo} = d_{eoo} = d_{31}\sin\theta$$

二阶非线性系数值:

$$d_{33}(2.12\ \mu m) = 40\ pm/V^{[43,44]}$$
$$d_{31}(10.6\ \mu m) = -18\ pm/V^{[45]}$$
$$d_{33}(10.6\ \mu m) = 36\ pm/V^{[45]}$$

相位匹配角的实验值($T = 293$ K)

相互作用的波长/μm	$\theta_{exp}/(°)$	参考文献
SHG, o + e⇒o		
16.4 + 3.479⇒2.87	73.7	[26]
15.96 + 2.28⇒1.995	62.2	[46]
14.1 + 3.604⇒2.87	70.9	[26]
13.7 + 2.849 2⇒2.358 7	65	[47]
10.6 + 2.72⇒2.164 6	70.5	[24]
10.361 + 2.227⇒1.833	78	[48]
9.871 + 2.251⇒1.833	84	[48]
9.776 + 2.256⇒1.833	90	[48]
8.278 + 4.3⇒2.83	84	[25]
8.253 + 4.4⇒2.87	84	[25]
8.236 + 4.5⇒2.91	84	[25]
7.88 + 3.36⇒2.358 7	90	[47]
7.86 + 3.37⇒2.358 7	90	[18]

内角带宽的实验值[24]

相互作用的波长/μm	$\Delta\theta^{int}/(°)$
SFG, o + e⇒o	
10.6 + 2.72⇒2.164 6	1.24

光谱带宽的实验值[24]

相互作用的波长/μm	$\Delta\nu/cm^{-1}$
SFG, o + e⇒o	
10.6 + 2.72⇒2.164 6	15

激光诱导的表面损伤阈值

$\lambda/\mu m$	τ_p/ns	$I_{thr}/(GW \cdot cm^{-2})$	参 考 文 献
0.694 3	30	0.008	[49]
	500 000	<0.000 002	[50]
1.833	200	0.03	[27]
1.995	20	>0.05	[46]
2.29~2.52	0.005	>0.22	[51]
2.36	35	0.05	[18]
2.596	22	>0.009	[52]
2.79	50	>0.014	[53]
2.797	0.1	>4	[19]
9.55	30	0.13	[54]
10.6	200	0.06	[55]

关于这一晶体

这种晶体的主要特点是其 IR 透过可上达 24 μm。在 20 世纪 70 年代，CdSe 被广泛应用于 OPO、DFG 和上转换；到今天它仍然被人们所应用[19,51,53]。

■ **参考文献**

[1] Physical-Chemical Properties of Semiconductors. Handbook (Nauka, Moscow, 1979) [In Russian].

[2] Physical Quantities. Handbook, ed. by I. S. Grigoriev, E. Z. Meilikhov (Energoatomizdat, Moscow, 1991) [In Russian].

[3] B. H. T. Chai: Optical Crystals. In: CRC Handbook of Laser Science and Technology, Supplement 2: Optical Materials, ed. by M. J. Weber (CRC Press, Boca Raton, 1995), pp. 3–65.

[4] E. M. Voronkova, B. N. Grechushnikov, G. I. Distler, I. P. Petrov: *Optical Materials for Infrared Technique* (Nauka, Moscow, 1965) [In Russian].

[5] A. A. Blistanov, V. S. Bondarenko, N. V. Perelomova, F. N. Strizhevskaya, V. V. Tchkalova, M. P. Shaskolskaya: *Acoustic Crystals* (Nauka, Moscow, 1982) [In Russian].

[6] *Handbook of Optical Constants of Solids II*, ed. by E. D. Palik (Academic Press, Boston, 1991).

[7] M. J. Weber: Optical Crystals. In: *CRC Handbook of Laser Science and Technology*, Vol. IV, *Optical Materials: Part 2*, ed. by M. J. Weber (CRC Press, Boca Raton, 1987), pp. 5–14.

[8] J. D. Beasley: Thermal conductivities of some novel nonlinear optical materials. Appl. Opt. 33(6), 1000–1003 (1994).

[9] B. Jensen, A. Torabi: Refractive index of hexagonal II-VI compounds CdSe, CdS, and $CdSe_xS_{1-x}$. J. Opt. Soc. Am. B 3(6), 857–863 (1986).

[10] K. J. Bachmann: *The Materials Science of Microelectronics* (VCH Publishers, NewYork, 1995).

[11] E. W. van Stryland, M. A. Woodall, H. Vanherzeele, M. J. Soileau: Energy band-gap dependence of two-photon absorption. Opt. Lett. 10(10), 490–492 (1985).

[12] S. S. Mitra, L. M. Narducci, R. A. Shatas, Y. F. Tsay, A. Vaidyanathan: Nonlinear absorption in direct-gap semiconductors. Appl. Opt. 14(12), 3038–3042 (1975).

[13] A. Miller, G. S. Ash: Two-photon absorption and short pulse stimulated recombination in $AgGaSe_2$. Opt. Commun. 33(3), 297–300 (1980).

[14] M. Sheik-Bahae, D. C. Hutchings, D. J. Hagan, E. W. van Stryland: Dispersion of bound electron nonlinear refraction in solids. IEEE J. Quant. Electr. 27(6), 1296–1309 (1991).

[15] E. W. van Stryland, L. L. Chase: Two-Photon Absorption. Inorganic Materials. In: *CRC Handbook of Laser Science and Technology, Supplement 2: Optical Materials*, ed. by M. J. Weber (CRC Press, Boca Raton, 1995), pp. 299–328.

[16] C. Kittel: *Introduction to Solid State Physics*, Seventh Edition (John Wiley & Sons, NewYork, 1996).

[17] M. Schäffner, X. Bao, A. Penzkofer: Principal optical constants measurement of uniaxial crystal CdSe in the wavelength region between 380 and 950 nm. Appl. Opt. 31(22), 4546–4552 (1992).

[18] A. A. Davydov, L. A. Kulevskii, A. M. Prokhorov, A. D. Saveliev, V. V. Smirnov: Parametric generation with CdSe crystal pumped by $CaF_2:Dy^{2+}$. Pisma Zh. Eksp. Teor. Fiz. 15(12), 725–727 (1972) [In Russian, English trans.: JETP Lett. 15(12), 513–514 (1972)].

[19] K. L. Vodopyanov: Megawatt peak power 8–13 μm CdSe optical parametric generator pumped at 2.8 μm. Opt. Commun. 150(1–6), 210–212 (1998).

[20] M. Bass, E. W. van Stryland, A. F. Stewart: Laser calorimetric measurement of two-photon absorption. Appl. Phys. Lett. 34(2), 142 – 144(1979).

[21] A. F. Stewart, M. Bass: Intensity-dependent absorption in semiconductors. Appl. Phys. Lett. 37(11), 1040 – 1043(1980).

[22] D. S. Hanna, A. J. Turner: Nonlinear absorption measurements in proustite (Ag_3AsS_3) and CdSe. Opt. Quant. Electron. 8(3), 213 – 217(1976).

[23] J. H. Bechtel, W. L. Smith: Two-photon absorption in semiconductors with picosecond light pulses. Phys. Rev. B 13(8), 3515 – 3522(1976).

[24] A. Ferrario, M. Garbi: Efficient up-conversion in CdSe. Opt. Commun. 17(2), 158 – 159 (1976).

[25] J. A. Weiss, L. S. Goldberg: Singly resonant CdSe parametric oscillator pumped by an HF laser. Appl. Phys. Lett. 24(8), 389 – 391(1974).

[26] R. G. Wenzel, G. P. Arnold: Parametric oscillator: HF oscillator-amplifier pumped CdSe parametric oscillator tunable from 14.1 μm to 16.4 μm. Appl. Opt. 15(5), 1322 – 1326 (1976).

[27] R. L. Herbst, R. L. Byer: Efficient parametric mixing in CdSe. Appl. Phys. Lett. 19 (12), 527 – 530(1971).

[28] A. Z. Grasyuk, I. G. Zubarev, A. N. Menzer: Anisotropy of two-photon absorption upon optical excitation of CdSe semiconductor lasers. Fiz. Tverd. Tela 10(2), 543 – 549 (1968)[In Russian, English trans. : Sov. Phys. -Solid State 10(2),427 – 431(1968)].

[29] F. Brükner, V. S. Dneprovskii, V. U. Khattatov: Two-photon absorption in cadmium selenide. Kvant. Elektron. 1(6), 1360 – 1364(1974)[In Russian, English trans. : Sov. J. Quantum Electron. 4(6),749 – 751(1974)].

[30] E. W. van Stryland, H. Vanherzeele, M. A. Woodall, M. J. Soileau, A. L. Smirl, S. Guha, T. F. Boggess: Two photon absorption, nonlinear refraction, and optical limiting in semiconductors. Opt. Eng. 24(4), 613 – 623(1985).

[31] W. L. Smith: Two-Photon Absorption in Condensed Media. In: *CRC Handbook of Laser Science and Technology*, Vol. III, *Optical Materials*: Part 1, ed. by M. J. Weber(CRC Press, Boca Raton, 1986), pp. 229 – 258.

[32] V. S. Dneprovskii, S. M. Ok: Role of absorption by nonequillibrium carriers in determination of two-photon absorption of CdSe and GaAs crystals. Kvant. Elektron. 3(3), 559 – 562(1976)[In Russian, English trans. : Sov. J. Quantum Electron. 6(3),298 – 300 (1976)].

[33] J. M. Ralston, R. K. Chang: Nd: laser induced absorption in semiconductors and aqueous $PrCl_3$ and $NdCl_3$. Opto-electron. 1(4), 182 – 188(1969).

[34] I. B. Zotova, Y. J. Ding: Spectral measurements of two-photon absorption coefficients for CdSe and GaSe crystals. Appl. Opt. 40(36), 6654 – 6658(2001).

[35] G.-M. Schucan, R. G. Ispasoiu, A. M. Fox, J. F. Ryan: Ultrafast two-photon nonlinearities in CdSe near 1.5 μm studied by interferometric autocorrelation. IEEE J. Quant. Electr. 34(8), 1374 – 1379(1998).

[36] W. L. Bond: Measurement of the refractive indices of several crystals. J. Appl. Phys. 36(5), 1674 – 1677(1965).

[37] M. P. Lisitsa, L. F. Gudymenko, V. N. Malinko, S. F. Terekhova: Dispersion of the refractive indices and birefringence of CdS_xSe_{1-x} single crystals. Phys. Stat. Solidi 31(1), 389 – 399(1969).

[38] G. C. Bhar: Refractive index interpolation in phase-matching. Appl. Opt. 15(2), 305 – 307 (1976).

[39] G. C. Bhar, D. C. Hanna, B. Luther-Davies, R. C. Smith: Tunable down-conversion from an optical parametric oscillator. Opt. Commun. 6(4), 323 – 326(1972).

[40] G. C. Bhar, G. C. Ghosh: Temperature dependent phase-matched nonlinear optical devices using CdSe and $ZnGeP_2$. IEEE J. Quant. Electr. QE – 16(8), 838 – 843(1980).

[41] I. P. Kaminow: Tables of Linear Electrooptic Coefficients. In: *CRC Handbook of Laser Science and Technology*, Vol. III, *Optical Materials: Part 2*, ed. by M. J. Weber(CRC Press, Boca Raton, 1986), pp. 253 – 278.

[42] J. E. Midwinter, J. Warner: The effects of phase matching method and of uniaxial crystal symmetry on the polar distribution of second-order non-linear optical polarization. Brit. J. Appl. Phys. 16(11), 1135 – 1142(1965).

[43] M. M. Choy, R. L. Byer: Accurate second-order susceptibility measurements of visible and infrared nonlinear crystals. Phys. Rev. B 14(4), 1693 – 1706(1976).

[44] W. J. Alford, A. V. Smith: Wavelength variation of the second-order nonlinear coefficients of $KNbO_3$, $KTiOPO_4$, $KTiOAsO_4$, $LiNbO_3$, $LiIO_3$, $\beta-BaB_2O_4$, KH_2PO_4, and LiB_3O_5 crystals: a test of Miller wavelength scaling. J. Opt. Soc. Am. B 18(4), 524 – 533(2001).

[45] D. A. Roberts: Simplified characterization of uniaxial and biaxial nonlinear optical crystals: a plea for standardization of nomenclature and conventions. IEEE J. Quant. Electr. 28(10), 2057 – 2074(1992).

[46] D. Andreou: 16 μm tunable source using parametric processes in non-linear crystals. Opt. Commun. 23(1), 37 – 43(1977).

[47] A. A. Davydov, L. A. Kulevskii, A. M. Prokhorov, A. D. Saveliev, V. V. Smirnov, A. V. Shirkov: A tunable infrared parametric oscillator in a CdSe crystal. Opt. Commun. 9(3), 234 – 236(1973).

[48] R. L. Herbst, R. L. Byer: Singly resonant CdSe infrared parametric oscillator. Appl. Phys. Lett. 21(5), 189 – 191(1972).

[49] M. Birnbaum, T. L. Stocker: Reflectivity enhancement of semiconductors by Q-switched ruby

lasers. J. Appl. Phys. 39(13), 6032 – 6036(1968).

[50] M. Bertolotti, F. de Pasquale, P. Marietti, D. Sette, G. Vitali: Laser damage on semiconductor surfaces. J. Appl. Phys. 38(10), 4088 – 4090(1967).

[51] M. A. Watson, M. V. O'Connor, D. P. Shepherd, D. C. Hanna: Synchronously pumped CdSe optical parametric oscillator in the 9 – 10 μm region. Opt. Lett. 28(20), 1957 – 1959(2003).

[52] J. M. Fukumoto: Three-stage optical parametric oscillator conversion from 1 μm to the 8 – 12 μm region. In: *Advanced Solid-State Lasers, OSA Trends in Optics and Photonics Series, Vol. 68*, ed. by M. E. Fermann, L. R. Marshall (OSA, Washington DC, 2002), pp. 558 – 562.

[53] T. H. Allik, S. Chandra, D. M. Rines, P. G. Schunemann, J. A. Hutchinson, R. Utano: Tunable 7 – 12 μm optical parametric oscillator using a Cr, Er: YSGG laser to pump CdSe and $ZnGeP_2$ crystals. Opt. Lett. 22(9), 597 – 599(1997).

[54] Y. M. Andreev, V. V. Badikov, V. G. Voevodin, L. G. Geiko, P. P. Geiko, M. V. Ivashchenko, A. I. Karapuzikov, I. V. Sherstov: Radiation resistance of nonlinear crystals at a wavelength of 9.55 μm. Kvant. Elektron. 31(12), 1075 – 1078(2001) [In Russian, English trans. : Quantum Electron. 31(12), 1075 – 1078(2001)].

[55] D. C. Hanna, B. Luther-Davies, H. N. Rutt, R. C. Smith, C. R. Stanley: Q-switched laser damage of infrared nonlinear materials. IEEE J. Quant. Electr. QE – 8 (3), 317 – 324 (1972).

第 9 章
一些最新的应用

这一章包括了讨论常用的新非线性材料最新应用的 7 篇简短评述。

9.1 深紫外光的产生

在 1986 年,Kato 发现通过倍频(NCPM SHG)产生的最短波长为 204.8 nm[1]。这一纪录是在 BBO 中实现的,仅在 10 年后这一纪录就被打破了。在 1996 年,一个中国的研究组报道了新的非线性晶体氟硼铍酸钾(KBBF)[2],这种晶体可以直接实现短到 172.5 nm 的 SHG[3]。然而,KBBF 具有层状特性,生长超过 1 mm 的晶体极为困难。这使得实现角度调谐相位匹配困难。为了这种晶体的深 UV 应用,特别提出了一种通过两块 CaF_2 棱镜耦合的光学接触方法[3],这相当不方便,并且不能用于每一种有效非线性转换中。

产生非常短 UV 波长(短于 205 nm)的一个替代方法是利用和频产生。这一方法是于 20 世纪 70 年代中期发展起来的[4,5]。为了满足相位匹配条件,参与和频的两束光在频率上应该尽可能地相差大一些:即其中一个波应该处于透过范围的接近 UV 截止边处,而另一个波接近于 IR 截止边处。最近，一个德国研究组，利用由一台钛宝石飞秒激光泵浦 OPO 产生的近 IR 闲频光和同一台激光器的四次谐波紫外光的 SFG，在 CLBO 晶体中实现了 175 nm 波长[6]，在 LBO 晶体中为 172.7 nm[7]，在 LB4 晶体中为 170 nm[8]，以及在 KB5 晶体中为 166 nm[9]。这些结果的综述，可见文献[10]。

最近，将 CLBO 用于最后的和频阶段，制备了几种大功率的准 CW 深紫外光源。在文献[11]中，产生了 205 nm 波长、250 mW 平均功率，在文献[12]中，发展了 196.3 nm 波长、1 W 的光源，而在文献[13]中，这一波长的平均功率达到了 1.5 W。

在文献[14]中报道了纳秒宽调谐深紫外光源。利用一组 BBO 谐波发生器和带放大的宽调谐钛宝石激光器，作者获得了光谱范围为 193~233 nm、重复频率为 10 Hz、能量超过 1 mJ 的脉冲光。

参考文献

[1] K. Kato: Second-harmonic generation to 2048Å in β-BaB$_2$O$_4$. IEEE J. Quant. Electr. QE-22(7), 1013-1014(1986).

[2] B. Wu, D. Tang, N. Ye, C. Chen: Linear and nonlinear optical properties of the KBe$_2$BO$_3$F$_2$(KBBF) crystal. Opt. Mater. 5(1-2), 105-109(1996).

[3] T. Togashi, T. Kanai, T. Sekikawa, S. Watanabe, C. Chen, C. Zhang, Z. Xu, J. Wang: Generation of vacuum-ultraviolet light by an optically contacted, prism-coupled KBe$_2$BO$_3$F$_2$ crystal. Opt. Lett. 28(4), 254-256(2003).

[4] F. B. Dunning, R. E. Stickel, Jr.: Sum frequency mixing in potassium pentaborate as a source of tunable coherent radiation at wavelengths below 217 nm. Appl. Opt. 15(12), 3131-3134(1976).

[5] G. A. Massey, J. C. Johnson: Wavelength-tunable optical mixing experiments between 208 nm and 259 nm. IEEE J. Quant. Electr. QE-12(11), 721-727(1976).

[6] V. Petrov, F. Noack, F. Rotermund, M. Tanaka, Y. Okada: Sum-frequency generation of femtosecond pulses in CsLiB$_6$O$_{10}$ down to 175 nm. Appl. Opt. 39(27), 5076-5079(2000).

[7] F. Seifert, J. Ringling, F. Noack, V. Petrov, O. Kittelmann: Generation of tunable femtosecond pulses to as low as 172.7 nm by sum-frequency mixing in lithium triborate. Opt. Lett. 19(19), 1538-1540(1994).

[8] V. Petrov, F. Rotermund, F. Noack, R. Komatsu, T. Sugawara, S. Uda: Vacuum ultraviolet application of Li$_2$B$_4$O$_7$ crystals: generation of 100 fs pulses down to 170 nm. J. Appl. Phys. 84(11), 5887-5892(1998).

[9] V. Petrov, F. Rotermund, F. Noack: Generation of femtosecond pulses down to 166 nm by sum-frequency mixing in $KB_5O_8 \cdot 4H_2O$. Electron. Lett. 34(18), 1748 – 1750(1998).

[10] V. Petrov, F. Rotermund, F. Noack, J. Ringling, O. Kittelmann, R. Komatsu: Frequency conversion of Ti: sapphire-based femtosecond laser systems to the 200-nm spectral region using nonlinear optical crystals. IEEE J. Sel. Topics Quant. Electr. 5(6), 1532 – 1542 (1999).

[11] K. F. Wall, J. S. Smucz, B. Pati, Y. Isyanova, P. Moulton, J. G. Manni: A quasi-continuouswave deep ultraviolet laser source. IEEE J. Quant. Electr. 39(9), 1160 – 1169(2003).

[12] J. Sakuma, A. Finch, Y. Ohsako, K. Deki, M. Yoshino, M. Horiguchi, T. Yokota, Y. Mori, T. Sasaki: All-solid-state, 1-W, 5-kHz laser source below 200 nm. In: *Advanced Solid- State Lasers*, *OSA Trends in Optics and Photonics Series*, *Vol. 26*, ed. by M. M. Fejer, H. Injeyan, U. Keller(OSA, Washington DC,1999), pp. 89 – 92.

[13] J. Sakuma, K. Deki, A. Finch, Y. Ohsako, T. Yokota: All-solid-state, high-power, deep-UV laser system based on cascaded sum-frequency mixing in $CsLiB_6O_{10}$ crystals. Appl. Opt. 39(30), 5505 – 5511(2001).

[14] A. V. Kachynski, V. A. Orlovich, A. A. Bui, V. D. Kopachevsky, A. V. Kudryakov, W. Kiefer: All solid-state pulsed ultraviolet laser widely tunable down to 188.5 nm. Opt. Commun. 218 (4 – 6), 351 – 357(2003).

9.2 通过 DFG 产生太赫兹波

非线性光学晶体最普遍的应用之一是通过差频产生红外范围的激光,甚至在量子电子学发展的早期,就曾采用这一方法来尝试产生亚毫米辐射。在1971 年,Yajima 和 Takeuchi 在锁模钕玻璃激光器 1.06 μm 辐射带宽的两个光谱分量之间,通过在它们的 DFG,以实验证明了在铌酸锂(LN)晶体中产生远红外可调谐光的可能性[1]。利用铌酸锂晶体并将其相位匹配角从 18°改变到 16.2°,他们设法获得了在 521~645 μm(0.58~0.47 THz)范围可调远红外辐射。次年,美国 Bell 实验室的科研组在 $ZnGeP_2$ 中观察到两台 CO_2 激光器频率之间的相位匹配 DFG[2]。所获得的调谐范围是 91~143 μm(3.3~2.1 THz),在 120 μm 处的能量转换效率为 1.3×10^{-8}。

30 年后,当紧凑型高效 THz 光源在分子光谱学、射电天文学、医学影像学、电子学等的应用成为现实时,旧的 THz 波 DFG 技术又被人们所想起而复活了。一个日本研究组[3]利用由二组极化畴周期稍有差异的 PPLN 晶体构建的两台 OPO 激光器的两束信号光进行 DFG。具体过程是用调 Q Nd:YAG 激光器的激光($\lambda = 1.064\ \mu m$)泵浦 PPLN 晶体,产生了两束接近 $1.5\ \mu m$ 的波,继而采用有机非线性 DAST 晶体对这两束光进行 DFG,结果获得光谱范围为 $120 \sim 160\ \mu m(2.5 \sim 1.87\ THz)$ 的远红外辐射,其功率转换效率相当低,为 3.8×10^{-10}。美国科学家 Shi 和 Ding 决定采用发展更好的无机晶体 GaSe 和 $ZnGeP_2$ (ZGP) 用于 DFG。利用 1.5 cm 长的 GaSe 晶体(II 类相互作用 e - e⇒o),他们设法产生了在极宽范围 $56.8 \sim 1618\ \mu m (5.27 \sim 0.18\ THz)$ 内的相干 THz 辐射,在 196 μm 处功率转换效率为 1.8×10^{-4} [4]。在一块 1.2 cm 长的 ZGP 晶体中,I 类(o - e⇒e)及 II 类(o - e⇒o)的相互作用均已实现,调谐范围分别为 $66.5 \sim 300\ \mu m(4.51 \sim 1.0\ THz)$ 和 $72.7 \sim 237\ \mu m(4.13 \sim 1.26\ THz)$ [5]。所测得的对 o - e⇒e 及 o - e⇒o 及相互作用的功率转换效率分别为 6.7×10^{-5}(在 97 μm 处)以及 3.6×10^{-5}(在 123 μm 处)。

最近发表的综述[6]总结了适用于亚毫米波产生的非线性光学晶体。除了产生 THz 波的 DFG 方法以外,人们也考虑了前向和后向 OPO 方法。

参考文献

[1] T. Yajima, N. Takeuchi: Spectral properties and tunability of far-infrared differencefrequency radiation produced by picosecond light pulses. Jpn. J. Appl. Phys. 10 (7), 907 - 915(1971).

[2] G. D. Boyd, T. J. Bridges, C. K. N. Patel, E. Buehler: Phase-matched submillimeter wave generation by difference-frequency mixing in $ZnGeP_2$. Appl. Phys. Lett. 21 (11), 553 - 555 (1972).

[3] K. Kawase, T. Hatanaka, H. Takahashi, K. Nakamura, T. Taniuchi, H. Ito: Tunable terahertz-wave generation from DAST crystal by dual signal-wave parametric oscillation of periodically poled lithium niobate. Opt. Lett. 25(23), 1714 - 1716(2000).

[4] W. Shi, Y. J. Ding, N. Fernelius, K. Vodopyanov: Efficient, tunable, and coherent 0.18 - 5.27-THz source based on GaSe crystal. Opt. Lett. 27 (16), 1454 - 1456 (2002).

[5] W. Shi, Y. J. Ding: Continuously tunable and coherent terahertz radiation by means of phase-matched difference-frequency generation in zinc germanium phosphide. Appl. Phys. Lett. 83(5), 848 - 851(2003).

[6] Y. J. Ding, I. B. Zotova: Second-order nonlinear optical materials for efficient generation and amplification of temporally-coherent and narrow-linewidth terahertz waves. Opt. Quant. Electron. 32(4 - 5), 531 - 552(2000).

9.3 通过 SHG 的超短激光脉冲压缩

1990 年,澳大利亚的科学家提出了在 KDP 类型晶体中利用 II 类 SHG 来对波长、1 μm 脉宽为 1 ps 的脉冲进行脉宽压缩的新效应[1]。这一方法的基本思路是在寻常光和异常光相互作用的基波脉冲之间引进一个最佳的"预延时",具体做法是将一块具有相同切割方向的薄型 KDP 晶体(亦称之为"预延时"晶体)放置在相位匹配方向上,相对于 KDP 倍频晶体,其轴向旋转了 90°。由于在 SHG 晶体射入点 o 光和 e 光偏振基波光的群速之间的差异,应该实现一个长得多的非线性相互作用长度。这应该造成二次谐波脉冲的压缩,最大的压缩因子可达 5,并可提高功率转换效率。

这两种预言都被这个澳大利亚的科研组的实验工作证实了[2,3]。在文献[3]中,研究了 Nd:YLF 激光器 1.2 ps、1.053 μm 具有近高斯波型脉冲的压缩。引进一块 15 mm 厚的"预延迟"DKDP 晶体,在 25 mm 厚 II 类 SHG 的 KDP 类晶体波束入射处造成了 1.053 μm 脉冲异常光和寻常之间 1.4 ps 的延迟。在入射功率密度为 7 GW/cm^2 时,二次谐波绿光脉冲被压缩到 250 fs(大约为 5 倍),与没有预延迟的标准 40% 相比,功率转换效率提高 240%。在其后的工作[4]中获得了类似的结果,报道获得 2.5 倍以上的压缩。

在压缩 II 类 SHG 二次谐波脉冲的同时,如果选择基频波(o 光或 e 光)之间适当的强度比,这两个基频波都可以被压缩。在文献[5]中,报道了将 o 偏振基波从 1.3 ps 缩短到 280 fs。不过这一方法有其局限性,从整个非线性光学晶体范畴来看[2],只有 KDP 和其最相近的同构晶体 DKDP 能够提供基波和二次谐波脉冲之间群速的准确关系;而且只有在接近于 1 μm 波长(Nd 激光器)的 SHG 才可能做到。在文献[6]中,对于相对较长的 11 ps Nd:YAG 激光脉冲实现了 20 多倍的压缩。

如果对脉冲波前倾移实施群速调制(见前驱性工作[7,8]),就可能进一步发展这一方法。这就能对于 1.3 ps 钕玻璃激光器的 SHG 压缩,采用另一种非线性晶体,如报道用 BBO 晶体获得 9 倍的压缩[9]。此外,脉冲倾移压缩技术能用于 THG,甚至可用于 I 类相互作用。在文献[10]中无次脉冲的 351 nm 脉冲从 1.3 ps 被缩短到 140 fs*。

■ 参考文献

[1] Y. Wang, R. Dragila: Efficient conversion of picosecond laser pulses into second-harmonic frequency using group-velocity dispersion. Phys. Rev. A 41(10), 5645 – 5649 (1990).

[2] Y. Wang, B. Luther-Davies, Y. -H. Chuang, R. S. Craxton, D. D. Meyerhofer: Highly effi-

---

* 译者注:原文有误。根据文献[10],从实验中将 351 nm 的激光脉冲从 1.3 ps 压缩到 140 fs。

cient conversion of picosecond laser pulses with the use of group-velocity-mismatched frequency doubling in KDP. Opt. Lett. 16(23), 1862 – 1864(1991).

[3] Y. Wang, B. Luther-Davies: Frequency-doubling pulse compressor for picosecond high-power neodymium laser pulses. Opt. Lett. 17(20), 1459 – 1461(1992).

[4] T. Zhang, H. Daido, Y. Kato, L. B. Sharma, Y. Izawa, S. Nakai: Second-harmonic generation of a picosecond laser pulse at high intensities with time predelay. Jpn. J. Appl. Phys. 34(7A), 3546 – 3551(1995).

[5] A. Dubietis, G. Valiulis, R. Danielius, A. Piskarskas: Fundamental-frequency pulse compression through cascaded second-order processes in a type II phase-matched secondharmonic generator. Opt. Lett. 21(6), 1262 – 1264(1996).

[6] A. Umbrasas, J. -C. Diels, J. Jacob, G. Valiulis, A. Piskarskas: Generation of femtosecond pulses through second-harmonic compression of the output of a Nd: YAG laser. Opt. Lett. 20(21), 2228 – 2230(1995).

[7] M. R. Topp, G. C. Orner: Group dispersion effects in picosecond spectroscopy. Opt. Commun. 13(3), 276 – 281(1975).

[8] Z. Bor, B. Racz: Group velocity dispersion in prisms and its application to pulse compression and traveling-wave excitation. Opt. Commun. 54(3), 165 – 170(1985).

[9] A. Dubietis, G. Valiulis, G. Tamošauskas, R. Danielius, A. Piskarskas: Nonlinear second-harmonic pulse compression with tilted pulses. Opt. Lett. 22(14), 1071 – 1073(1997).

[10] A. Dubietis, G. Valiulis, G. Tamošauskas, R. Danielius, A. Piskarskas: Nonlinear pulse compression in the ultraviolet. Opt. Commun. 144(1 – 3), 55 – 59(1997).

9.4 自倍频晶体

自倍频(SFD)的想法是非常简单的。一种非线性光学晶体以三价稀土离子(通常是 Nd 或 Yb)掺杂,产生基频辐射并同时转换为它的二次谐波。这一思想首先在掺 Tm[1]和掺 Nd[2]的铌酸锂(LN)中实现。其后,利用其他基质晶体,诸如在 YAB[3]、MgO:LiNbO$_3$[4]、LaBGeO$_5$[5]和 GAB[6]中都尝试进行自倍频。从实用的观点来看,SFD 最重要的潜在应用是这些与激光二极管相关的应用,激光二极管自 20 世纪 90 年代以来就已实用化了。

首先,我们要列举一些自倍频的最佳结果,在 Nd:YAB(NYAB)中所取得的,以及在最近所发现的掺钕硼酸钙氧钆和硼酸钙氧钇(分别为 GdCOB[7-9]和 YCOB[9-11])晶体中实现的结果。在这些晶体中 Nd^{3+} $^4F_{3/2} \Rightarrow ^4I_{11/2}$ 跃迁的二次谐波波长相应于 530.5 nm。采用一块 0.5 cm 长的 NYAB 晶体(4 at.% Nd),以 I 类 SHG($\theta = 30.7°$)切割,用 807 nm 的 1.6 W 二极管泵浦,产生了 225 mW 的 CW 绿光辐射[12]。在文献[13]、[14]中,用一块 0.8 cm 长的 Nd:

GdCOB 晶体(7 at. % Nd)，以 I 类 SHG 切割($\theta = 90°, \phi = 46°$)，用 810 nm 波长的 1.25 W(吸收功率)二极管泵浦，获得 115 mW 的 530.5 nm 绿光。采用 Nd：YCOB 的类似实验(0.5 cm 长，5 at. % Nd，$\theta = 90°, \phi = 33.6°$)，在 3.8 W(吸收功率)、812 nm 二极管泵浦时，产生 245 mW 的绿光[15]。

澳大利亚 – 中国联合研究组研究了 I 类 Yb：YAB 晶体的 SFD(0.3 cm 长，8~10 at. % Yb，$\theta = 31°$)。使用中等的 1.4 W InGaAs 二极管 976 nm 泵浦，获得 160 mW 的 CW 绿光输出[16]。在高泵浦功率 11 W 的情况下，产生了 530.5 nm、1.1 W 绿光[17,18]，这是至今所报道所有二极管泵浦 SFD 的最高绿光功率。

在文献[19]中，在 Nd：YCOB 晶体中实现了 Nd^{3+} $^4F_{3/2} \Rightarrow {}^4I_{13/2}$ 离子跃迁($\lambda = 1332$ nm)。当 812 nm 吸收基频功率为 0.9 W 时产生约 16 mW CW 红光输出。

除了 SFD 外，在自倍频非线性晶体中也实现了和频及差频等混频过程[20,21]。在一种晶体中，SFD 和 SFM 通道的同时发生也是可能的[22,23]。在其后的工作中，利用两种不同泵浦波长(755 nm 和 807 nm)的组合，在 NYAB 晶体中实现了三基色红(669 nm)、绿(505 nm)及蓝(481 nm)波长的 CW 光输出。

参考文献

[1] L. F. Johnson, A. A. Ballman：Coherent emission from rare earth ions in electro-optic crystals. J. Appl. Phys. 40(1), 297 – 302(1969).

[2] V. G. Dmitriev, E. V. Raevskii, L. N. Rashkovich, N. M. Rubinina, O. O. Selichev, A. A. Fomichev：Simultaneous emission at the fundamental frequency and the second harmonic in an active nonlinear medium：neodymium-doped lithium metaniobate. Pisma Zh. Tech. Fiz. 5(21 – 22), 1400 – 1402(1979)[In Russian, English trans.：Sov. Tech. Phys. Lett. 5(11), 590 – 591(1979)].

[3] L. M. Dorozhkin, I. I. Kuratev, N. I. Leonyuk, T. I. Timchenko, A. V. Shestakov：Optical second-harmonic generation in a new nonlinear active medium：neodymium-yttriumaluminum borate crystals. Pisma Zh. Tekh. Fiz. 7(21), 1297 – 1300(1981)[In Russian, English trans.：Sov. Tech. Phys. Lett. 7(11), 555 – 556(1981)].

[4] T. Y. Fan, A. Cordova-Plaza, M. J. F. Digonnet, R. L. Byer, H. J. Shaw：Nd：MgO：$LiNbO_3$ spectroscopy and laser devices. J. Opt. Soc. Am. B 3(1), 140 – 147(1986).

[5] J. Capmany, D. Jaque, J. Garcia Sole, A. A. Kaminskii：Continuous wave laser radiation at 524 nm from a self-frequency-doubled laser of $LaBGeO_5$：Nd^{3+}. Appl. Phys. Lett. 72(5), 531 – 533(1998).

[6] C. Tu, M. Qiu, Y. Huang, X. Chen, A. Jiang, Z. Luo：The study of a self-frequency-doubling laser crystal Nd^{3+}：$GdAl_3(BO_3)_4$. J. Cryst. Growth 208(1 – 4), 487 – 492(2000).

[7] G. Aka, L. Bloch, J. M. Benitez, P. Crochet, A. Kahn-Harari, D. Vivien, F. salin,

P. Coquelin, D. Colin: A new non linear oxoborate crystal, characterized by using femtosecond broadband pulses. In: *Advanced Solid-State Lasers*, *OSA Trends in Optics and Photonics Series*, Vol. 1, ed. by S. A. Payne, C. Pollock (OSA, Washington DC, 1996), pp. 336 – 340.

[8] G. Aka, A. Kahn-Harari, D. Vivien, J.-M. Benitez, F. Salin, J. Godard: A new nonlinear and neodymium laser self-frequency doubling crystal with congruent melting: $Ca_4GdO(BO_3)_3$ (GdCOB). Eur. J. Solid State Inorg. Chem. 33(8), 727 – 736(1996).

[9] M. Iwai, T. Kobayashi, H. Furuya, Y. Mori, T. Sasaki: Crystal growth and optical characterization of rare-earth (Re) calcium oxyborate $ReCa_4O(BO_3)_3$ (Re = Y or Gd) as new nonlinear optical material. Jpn. J. Appl. Phys. 36(3A), L276-L279(1997).

[10] M. Yoshimura, T. Kobayashi, H. Furuya, K. Murase, Y. Mori, T. Sasaki: Crystal growth and optical properties of yttrium calcium oxyborate $YCa_4O(BO_3)_3$. In: *Advanced Solid-State Lasers*, *OSA Trends in Optics and Photonics Series*, Vol. 19, ed. by W. R. Bosenberg, M. M. Fejer (OSA, Washington DC, 1998), pp. 561 – 564.

[11] Q. Ye, B. H. T. Chai: Crystal growth of $YCa_4O(BO_3)_3$ and its orientation. J. Cryst. Growth 197(1 – 2), 228 – 235(1999).

[12] J. Bartschke, R. Knappe, K.-J. Boller, R. Wallenstein: Investigation of efficient self-frequency- doubling Nd: YAB lasers. IEEE J. Quant. Electr. 33(12), 2295 – 2300 (1997).

[13] F. Auge, S. Auzanneau, G. Lukas-Leclin, F. Balembois, P. Georges, A. Brun, F. Mougel, G. Aka, A. Kahn-Harari, D. Vivien: Efficient self-frequency-doubling Nd: GdCOB crystal pumped by a high brightness laser diode. In: *Advanced Solid-State Lasers*, *OSA Trends in Optics and Photonics Series*, Vol. 26, ed. by M. M. Fejer, H. Injeyan, U. Keller (OSA, Washington DC, 1999), pp. 77 – 81.

[14] D. Vivien, F. Mougel, F. Auge, G. Aka, A. Kahn-Harari, F. Balembois, G. Lucas-Leclin, P. Georges, A. Brun, P. Aschehoug, J.-M. Benitez, N. Le Nain, M. Jacquet: Nd: GdCOB: overview of its infrared, green and blue laser performances. Opt. Mater. 16(1 – 2), 213 – 220(2001).

[15] D. A. Hammons, M. Richardson, B. H. T. Chai, A. K. Chin, R. Jollay: Scaling of longitudinally diode-pumped self-frequency-doubling Nd: YCOB lasers. IEEE J. Quant. Electr. 36(8), 991 – 999(2000).

[16] P. Wang, P. Dekker, J. M. Dawes, J. A. Piper, Y. Liu, J. Wang: Efficient continuous-wave self-frequency-doubling green diode-pumped Yb: $YAl_3(BO_3)_4$ lasers. Opt. Lett. 25 (10), 731 – 733(2000).

[17] P. Dekker, J. M. Dawes, J. A. Piper, Y. Liu, J. Wang: 1.1 W CW self-frequency-doubled diode-pumped Yb: $YAl_3(BO_3)_4$ laser. Opt. Commun. 195(5 – 6), 431 – 436(2001).

[18] H. Jiang, J. Li, J. Wang, X.-B. Hu, H. Liu, B. Teng, C.-Q. Zhang, P. Dekker,

P. Wang: Growth of Yb: YAl₃(BO₃)₄ crystals and their optical and self-frequency-doubling properties. J. Cryst. Growth 233(1-2), 248-252(2001).

[19] Q. Ye, L. Shah, J. Eichenholz, D. Hammons, R. Peale, M. Richardson, A. Chin, B. H. T. Chai: Investigation of diode-pumped, self-frequency doubled RGB lasers from Nd: YCOB crystals. Opt. Commun. 164(1-3), 33-37(1999).

[20] F. Mougel, G. Aka, A. Kahn-Harari, D. Vivien: CW blue laser generation by self sum-frequency mixing in Nd: Ca₄GdO(BO₃)₃(Nd:GdCOB) single crystal. Opt. Mater. 13(3), 293-297(1999).

[21] A. Brenier, C. Tu, J. Li, Z. Zhu, B. Wu: Self-sum- and difference-frequency mixing in GdAl₃(BO₃)₄: Nd³⁺ for generation of tunable ultraviolet and infrared radiation. Opt. Lett. 27(4), 240-242(2002).

[22] Y. Chen, M. Huang, Y. Huang, Z. Luo: Simultaneous green and blue laser radiation based on a nonlinear laser crystal Nd: GdAl₃(BO₃)₄ and a nonlinear optical crystal KTP. Opt. Commun. 218(4-6), 379-384(2003).

[23] D. Jaque, J. Capmany, J. Garcia Sole: Red, green, and blue laser light from a single Nd: YAl₃(BO₃)₄ crystal based on laser oscillation at 1.3μm. Appl. Phys. Lett. 75(3), 325-327(1999).

9.5 周期性极化晶体

在过去10年中，非线性光学领域最重要的突破也许就是周期性极化晶体的引入。自从1991年以来，数以百计的研究工作致力于周期性极化非线性材料的应用。令人感到惊讶的是，直到现在，仍然没有人对这样的材料作一特别的专题，甚至在非线性光学的标准教科书中也没有提到它们，而这些晶体的历史可以回溯到40年前。

在1961年，在结晶石英中由Franken等人[1]首次观察到了SHG。由于在相互作用的波之间没有相位匹配，SH的功率是非常小的。两年后Giordmaine[2]和Maker等人[3]提出了双折射相位匹配(BPM)，应用了在基频和倍频不同偏振态的波之间相速度的差别。这一类的相位匹配在其后的30年被普遍地应用于非线性光学。然而，早在1962年，Bloembergen等人提出了称为准相位匹配(QPM)的另一种相位匹配方法[4]。QPM可以看作是在光波传播的方向上非线性材料的非线性极化率的周期性调制，这使得相位失配保持在零值的附近。30年后，与电场极化技术的结合使得铁电畴极化的翻转得以实现，这样就实际完成了QPM方法。现在周期性极化晶体PPLN、PPLT、PPKTP和PPRTA等是最经常使用的。关于QPM的综述，可见文献[5]、[6]、[7]。

准相位匹配与双折射匹配相比有几条优点。第一，对于材料或极性没有限制。第二，它可能应用最高的二阶非线性系数[例如，对于 LN, d_{33}(1.064 μm) = 25.2 pm/V，它比常常用于 BMP 的 d_{31}(1.064 μm) = 4.6 pm/V 要高得多。因此，沿 Z 轴所有相互作用波的极化可造成最高的有效非线性。应该强调的是，这样的相互作用在 BPM 中是不能实现的]。在 PPLN 的情况下，还有其他的优点，例如(比普通 LN)光折变效应的敏感度小；对于异常光波来说，IR 截止波长更长；这最后一个性质使得以 LN 为基础的 OPO 实现了可调谐，其闲频波的波长可以调至 6.6 μm[8]、6.8 μm[9]，甚至可以调到 7.3 μm[10]。

我们将很快列出在周期极化非线性材料应用中最重要的最新技术成就。在文献[11]中，研究了采用长 5.3 cm，6.5 μm 畴周期的 PPLN 晶体的一台 CW Nd：YAG 激光器(λ = 1.064 μm, P = 6.5 W)的 SHG，所测得的绿光输出功率为 2.7 W。另一台 Nd：YAG 激光器(λ = 0.946 μm, P = 2.6 W)的转换是用于二次谐波发生的[12]，采用一块长 0.9 cm，畴周期为 6.09 μm 晶体，CW 蓝光功率达到 0.74 W。在文献[13]中，研究了在长 1 cm、5 μm 畴周期的 PPKTP 晶体中，由一台 InGaAs MOPA(λ = 920 nm)发射的皮秒脉冲的 SHG，所得到的平均 SH 功率为 0.25 W。在文献[14]中研究了在一块 1.9 cm 长、9.5 μm 畴周期的 PPLN 晶体中，一台二极管泵浦 Nd：YVO$_4$ 激光器二条光束(λ_1 = 1.064 μm, P_1 = 1.2 W, λ_2 = 1.342 μm, P_2 = 1.0 W)之间的和频发生；所产生位于 593 nm 的 CW 黄光输出功率为 78 mW。在文献[15]中，一块 5 cm 长，畴周期为 30.3 μm 的 PPLN 晶体同时用于信号光(1.7 μm)和闲频光(2.8 μm)波长之间的 OPO(由 CW Nd：YAG 激光器泵浦，λ = 1.064 μm)和 DFG。CW IR 输出功率在 4.3 μm 处达到了 150 mW。

现在，我们将考虑用周期性极化晶体获得的 OPO 结果。在文献[16]中，一台基于 5 cm 长，具有从 28.5~29.9 μm 的 8 个畴周期 PPLN 的二极管泵浦 OPO 激光器，OPO 的调谐范围对于信号光是 1.461~1.601 μm 范围，对于闲频光为 3.173~3.917 μm，这可以通过改变倒格矢和/或通过升高晶体温度从 91 ℃ 到 173 ℃ 来实现。该系统产生的信号光脉宽为 34 ps，重复频率 235 MHz，输出平均信号光功率为 1 W。在文献[17]中，用了相似的泵浦激光器，但是重复频率较低(20 kHz)；采用了两块 1 cm 长的周期性极化晶体：PPKTP 晶体，畴周期为 37.8 μm 和 PPRTA 晶体，畴周期为 40.2 μm。在室温下，以 PPKTP 为基的 OPO 产生 1.72 μm 的信号光和 2.79 μm 的闲频光，总的输出功率为 2 W；以 PPRTA 为基的 OPO 产生 1.58 μm 的信号光和 3.26 μm 的闲频光，总的输出功率为 1.3 W。在文献[18]中，在一块长 5.5 cm 的 PPLN 晶体中，畴周期是 29.75 μm，产生 1.6 W 信号光功率(1.56~1.64 μm)以及 0.8 W 闲频光功率(3.34~3.03 μm)。基于 PPLN 的 OPG 由一台调 Q Nd：YVO$_4$ 激光器的 10 ns 脉冲泵浦，重复频率为 10 kHz。通过将 PPLN 温度从 140 ℃ 变到 250 ℃ 可实现调谐。在文献[19]中，最高 OPG 总功率达到 8.9 W，OPG 是由锁模 Nd：YVO$_4$ 振荡放大系统泵浦，这台系统产生重复频率为 82.3 MHz、7 ps 的脉冲，平均功率为 24 W。采用一块 5.5 cm 长，畴周期为 29.75 μm 的 PPLN 晶体，在文献[20]中实现了利用周期性极化非线性光学晶体 OPO 最大的调谐范围。OPO 是由 CW Nd：YVO$_4$ 激光器的二次谐波（λ = 532 nm, P = 0.8 ~ 3.3 W）泵浦。作为非线性元件，用了具有不同极化畴周期的 2.4 cm 长 PPKTP 晶体或 2.5 cm 长 PPLN 晶体：PPKTP 中 19 个周期从 8.96 μm 到 12.194 μm，PPLN 是 23 个周期从 6.51 μm 到 9.59 μm。通过改变倒格矢和/或改变非线性晶体的温度完成调谐，信号波调谐范围为 656 ~ 1 035 nm，闲频光调谐范围为 1 096 ~ 2 830 nm。对于 PPKTP，温度从 20 ℃ 变到 80 ℃，对于 PPLN，温度从 140 ℃ 变到 200 ℃。

改变周期性极化材料的温度是 OPO 调谐相当慢的办法，另一条 OPO 调谐的途径是通过改变畴周期（在多重倒格矢晶体中选取不同的超晶格），这需要机械平移设备，因此也是很慢的。在文献[21]中，提出了电光调制的方法。这想法是非常简单的：周期性极化的晶体分为三块相等的断片，第一块和第三块具有 50% 的占空比，而中间一段则保持未极化并且镀上一对电极。对中间一段施加高的电压以电光形式改变周期极化晶体的折射率，这会改进参量增益的光谱形状，因此导致增益极大值以及振荡波长的移动。改变施加的电压从 -180 V 到 +1 050 V，实现了在 1 562 ~ 1 664 nm 范围内 PPLN - OPO 快速调谐。文献[22]中报道了类似的结果。

有的作者采用双重倒格矢畴结构[23,24]或部分极化的非线性晶体[25]，文献[23]中在前一结构（PPLT，2 cm 长，11.9 μm 畴周期）中以皮秒 532 nm 泵浦（由 Nd：YAG 的 SH），产生光学参量振荡，而在第二个结构中（PPLT，1 cm 长，8.8 μm 畴周期），在 OPG 闲频光和泵浦辐射间产生 SFG，结果产生了 631 nm 的红光和 460 nm 的蓝光。在文献[24]中，第一个结构（PPLT，2 cm 长，14.9 μm 畴周期，工作温度 74.6 ℃）以 1 342 nm（二极管泵浦 Nd：YVO$_4$ 激光光器，35 ns，10 kHz）泵浦发生 SHG，而在第二个结构（PPLT，1 cm 长，4.9 μm 畴周期，工作温度 74.6 ℃）发生 THG。结果是产生了红光（671 nm）和蓝光（447 nm），平均功率分别为 0.75 W 和 0.15 W。最后，在文献[25]中报道了非线性晶体中仅有部分周期极化的工作（3.0 cm 长的 KTP 中 1.5 cm 被周期极化，畴周期为 16.46 μm），在这片晶体中产生了 1.327 μm 辐射的 QPM SHG，而在未极化的 1.5 cm 长的 KTP 晶体中，产生了通常的双折射 THG。

在文献[26]中，提出了一种特别的二组元准周期光学超晶格，能够对任何需要的波长实现三次谐波发生。相应的倒格矢包含了长度分别为 D_A 和 D_B 的两个基本组元 A 和 B 的序列。每一个组元由两个相逆的铁电畴组成，正畴

宽度等于 L。对于 LT 进行计算,基波波长为 1.44 μm,晶体温度为 30 ℃,对于 D_A = 13.12 μm,D_B = 18.65 μm,L = 9.31 μm,以及组元序列为 ABBBBA-BBBAB…显示了最高的转换效率。在实验中,在 1.5 cm 长的 PPLT 在 27.8 ℃ 时得到的 THG 转换效率为 27%;测量在 480 nm 处蓝光的输出功率等于 4 mW。

参考文献

[1] P. A. Franken, A. E. Hill, C. W. Peters, G. Weinreich: Generation of optical harmonics. Phys. Rev. Lett. 7(4), 118–119(1961).

[2] J. A. Giordmaine: Mixing of light beams in crystals. Phys. Rev. Lett. 8(1), 19–20 (1962).

[3] P. D. Maker, R. W. Terhune, M. Nisenoff, C. M. Savage: Effects of dispersion and focusing on the production of optical harmonics. Phys. Rev. Lett. 8(1), 21–22(1962).

[4] J. A. Armstrong, N. Bloembergen, J. Ducuing, P. S. Pershan: Interactions between light waves in a nonlinear dielectric. Phys. Rev. 127(6), 1918–1939(1962).

[5] M. M. Fejer, G. A. Magel, D. H. Jundt, R. L. Byer: Quasi-phase-matched second harmonic generation: tuning and tolerances. IEEE. J. Quant. Electr. 28(11), 2631–2654 (1992).

[6] L. E. Myers, R. C. Eckardt, M. M. Fejer, R. L. Byer, W. R. Bosenberg, J. W. Pierce: Quasiphase- matched optical parametric oscillators in bulk periodically poled LiNbO$_3$. J. Opt. Soc. Am. B 12(11), 2102–2116(1995).

[7] L. E. Myers, W. R. Bosenberg: Periodically poled lithium niobate and quasi-phase-matched optical parametric oscillators. IEEE. J. Quant. Electr. 33(10), 1663–1672 (1997).

[8] M. Sato, T. Hatanaka, S. Izumi, T. Taniuchi, H. Ito: Generation of 6.6-μm optical parametric oscillation with periodically poled LiNbO$_3$. Appl. Opt. 38(12), 2560–2563 (1999).

[9] P. Loza-Alvarez, C. T. A. Brown, D. T. Reid, W. Sibbett, M. Missey: High-repetition-rate ultrashort-pulse optical parametric oscillator continuously tunable from 2.8 to 6.8 μm. Opt. Lett. 24(21), 1523–1525(1999).

[10] M. A. Watson, M. V. O'Connor, P. S. Lloyd, D. P. Shepard, D. C. Hanna, C. B. E. Gavith, L. Ming, P. G. R. Smith, O. Balachninaite: Extended operation of synchronously pumped optical parametric oscillators to longer idler wavelengths. Opt. Lett. 27(23), 2106–2108(2002).

[11] G. D. Miller, R. G. Batchko, W. M. Tulloch, D. R. Weise, M. M. Fejer, R. L. Byer: 42%- efficient single-pass CW second-harmonic generation in periodically poled lithium niobate. Opt. Lett. 22(24), 1834–1836(1997).

[12] M. Pierrou, F. Laurell, H. Karlsson, T. Kellner, C. Czeranowsky, G. Huber: Generation

of 740 mW of blue light by intracavity frequency doubling with a first-order quasi-phasematched KTiOPO$_4$ crystal. Opt. Lett. 24(4), 205-207(1999).

[13] D. Woll, J. Schumacher, A. Robertson, M. A. Tremont, R. Wallenstein, M. Katz, D. Eger, A. Englander: 250mW of coherent blue 460-nm light generated by single-pass frequency doubling of the output of a mode-locked high-power diode laser in periodically poled KTP. Opt. Lett. 27(12), 1055-1057(2002).

[14] Y. F. Chen, S. W. Tsai, S. C. Wang, Y. C. Huang, T. C. Lin, B. C. Wong: Efficient generation of continuous-wave yellow light by single-pass sum-frequency mixing of a diodepumped Nd: YVO$_4$ dual-wavelength laser with periodically poled lithium niobate. Opt. Lett. 27(20), 1809-1811(2002).

[15] D.-W. Chen, K. Masters: Continuous-wave 4.3-μm intracavity difference frequency generation in an optical parametric oscillator. Opt. Lett. 26(1), 25-27(2001).

[16] T. Graf, G. McConnell, A. I. Ferguson, E. Bente, D. Burns, M. D. Dawson: Synchronously pumped optical parametric oscillation in periodically poled lithium niobate with 1-W average output power. Appl. Opt. 38(15), 3324-3328(1999).

[17] M. Peltz, U. Bäder, A. Borsutzky, R. Wallenstein, J. Hellström, H. Karlsson, V. Pasiskevicius, F. Laurell: Optical parametric oscillators for high pulse energy and high average power operation based on large aperture periodically poled KTP and RTA. Appl. Phys. B 73(7), 663-670 (2001).

[18] U. Bäder, T. Mattern, T. Bauer, J. Batschke, M. Rahm, A. Borsutzky, R. Wallenstein: Pulsed nanosecond optical parametric generator based on periodically poled lithium niobate. Opt. Commun. 217(1-6), 375-380(2003).

[19] B. Köhler, U. Bäder, A. Nebel, J.-P. Meyn, R. Wallenstein: A 9.5-W 82-MHz-repetitionrate picosecond optical parametric generator with CW diode laser injection seeding. Appl. Phys. B 75(1), 31-34(2002).

[20] U. Strößner, J.-P. Meyn, R. Wallenstein, P. Urenski, A. Arie, G. Rosenman, J. Mlynek, S, Schiller, A. Peters: Single-frequency continuous-wave optical parametric oscillator system with an ultrawide tuning range of 550 to 2830 nm. J. Opt. Soc. Am. B 19(6), 1419-1424 (2002).

[21] P. Gross, M. E. Klein, H. Ridderbusch, D.-H. Lee, J.-P. Meyn, R. Wallenstein, K.-J. Boller: Wide wavelength tuning of an optical parametric oscillator through electro-optic shaping of the gain spectrum. Opt. Lett. 27(16), 1433-1435(2002).

[22] S. Haidar, Y. Sasaki, E. Niwa, K. Masumoto, H. Ito: Electro-optic tuning of a periodically poled LiNbO$_3$ optical parametric oscillator and mixing its output waves to generate mid-IR tunable from 9.4 to 10.5μm. Opt. Commun. 229(1-6), 325-330 (2004).

[23] Z.-W. Liu, S.-N. Zhu, Y.-Y. Zhu, H.-T. Wang, G.-Z. Luo, H. Liu, N.-B. Min, X.-

Y. Liang, Z. -Y. Xu: Red and blue light generation in an LiTaO₃ crystal with a double grating domain structure. Chin. Phys. Lett. 18(4), 539 – 540(2001).

[24] J. -L. He, X. -P. Hu, S. -N. Zhu, Y. -Y. Zhu, N. -B. Min: Efficient generation of red and blue light in a dual-structure periodically poled LiTaO₃ crystal. Chin. Phys. Lett. 20(2), 2175 – 2177(2003).

[25] X. Mu, Y. J. Ding: Efficient third-harmonic generation in partly periodically poled KTiOPO₄ crystal. Opt. Lett. 26(9), 623 – 625(2001).

[26] C. Zhang, H. Wei, Y. -Y. Zhu, H. -T. Wang, S. -N. Zhu, N. -B. Ming: Thirdharmonic generation in a general two-component quasi-periodic optical superlattice. Opt. Lett. 26 (12), 899 – 901(2001).

9.6 光子带隙晶体

光子带隙晶体(或光子晶体,两种术语看来都不成功)简单说来就是非线性晶体,而其非线性是在二维变化。应该记住,周期性极化晶体是具有二阶非线性符号一维周期性改变的材料。最近,Berger 提出[1,2]将这一准相位匹配的思路扩展到多维空间中去。第一个二维周期极化非线性晶体是由英国的研究组在实验中实现的[3],他们在铌酸锂晶体中制备了具有六方对称性的周期结构(称为 HeXLN)。所获得的六方反转畴的六方格子的周期为 18.05 μm,整个反转区约占 30%,并设计作为在 ΓM 方向(X 轴)150 ℃时 1 531 nm 基频波的 QPM SHG。在这个方向的传播长度为 1.4 cm。HeXLN 放置在一个炉子内,升高温度以防止光折变损伤。在较低的入射强度(~ 0.2 GW/cm²)、4 ps 脉宽、1.531 μm 基波辐射,输出由不同颜色组成的多种光波,是从晶体中不同角度的输出所组合起来的。这些波对应于从基波方向(ΓM 方向)对称于 ±(1.1°±0.1°)角的 SH 辐射以及次级的 THG 和 FoHG 辐射。在更高的强度下,SH 点仍保持在同一位置,而 THG 光开始在一个大角度范围发射。最大的 SHG 外转换效率(强度为 ~ 0.2 GW/cm² 时)约为 60% 左右。

其后,在 HeXLN 中,仍由同一课题组进行了对 SHG 以及级联 THG 和 FoHG 更详细的研究,采用了功率较小的纳秒 IR 光源(1.520 ~ 1.560 μm,5 ns,2 kHz,5 ~ 16 MW/cm²)以及一块更短的 1 cm 长的 HeXLN 晶体[4]。在相对低的温度下,对于 1 536 nm 基波辐射,在 HeXLN 晶体中所获得的 SHG 温度带宽是 8.5 ℃,这比同一长度及同一周期的 PPLN(4.2 ℃)要大得多。在较高的辐射强度下(14 ~ 16 MW/cm²),除了在 768 nm 的 SHG 波束,文献[4]的作者们观察了从晶体中出射的绿光和蓝光,分别对应于级联 THG 和 FoHG,并发现了另一条绿光波束对应于双折射匹配Ⅱ类 THG。文献[4]的作者认为 HeXLN 是"极其适于同时满足相位匹配的多重非线性相互作用过程"。文献[5]的作者也给出了相似的说法,他们从理论上考虑了在非线性光子晶体中的谐波产生,并提出二维光子晶体"是实现观察几种谐波同时发生以及与多步级联过程相关的不同效应的理想对象"。遗憾的是,HeXLN 的这一显著特点将可能限制其在非线性光学中的实际应用。

在文献[6]中,研究了基于 HeXLN 的波导 1.536 μm 基波辐射的 SHG。对于 $TM_0(\omega) \Rightarrow TM_1(2\omega)$ SHG 过程,获得了内转换效率的最佳值(46%)。然而,也观察到了与三次谐波发生同时引起的损伤。

■ 参考文献

[1] V. Berger: Nonlinear photonic crystals. Phys. Rev. Lett. 81(19), 4136-4139(1998).

[2] V. Berger: From photonic band gaps to refractive index engineering. Opt. Mater. 11(2-3), 131-142(1999).

[3] N. G. R. Broderick, G. W. Ross, H. L. Offerhaus, D. J. Richardson, D. C. Hanna: Hexagonally poled lithium niobate: a two-dimensional nonlinear photonic crystal. Phys. Rev. Lett. 84(19), 4345-4348(2000).

[4] N. G. R. Broderick, R. T. Bratfalean, T. M. Monro, D. J. Richardson, C. M. de Sterke: Temperature and wavelength tuning of second-, third-, and fourth-harmonic generation in a two-dimensional hexagonally poled nonlinear crystal. J. Opt. Soc. Am. B 19(9), 2263-2272(2002).

[5] S. Saltiel, Y. S. Kivshar: Phase matching in nonlinear $\chi^{(2)}$ photonic crystals. Opt. Lett. 25(16), 1204-1206(2000).

[6] K. Gallo, R. T. Bratfalean, A. C. Peacock, N. G. R. Broderick, C. B. E. Gavith, L. Ming, P. G. R. Smith, D. J. Richardson: Second-harmonic generation in hexagonally-poled lithium niobate slab waveguides. Electron. Lett. 39(1), 75-76(2003).

9.7 通过 $\chi^{(3)}$ 非线性过程产生的 THG

到现在为止,在这本书中所描述的都是利用二阶非线性极化张量 $\chi^{(2)}$ 相关的所谓三波相互作用。利用 $\chi^{(3)}$ 非线性的四波相互作用可能也有实用意义,特别是 THG 的情况下(因为只用一块非线性晶体而不是用两块)。Midwinter 和 Warner[1] 在 1967 年推导了单轴和各向同性晶体的有效三阶非线性系数。其后,由一个中国科研组得到了双轴晶的相关表达式[2]。在文献[3]中,相对于 ADP 和 KDP 晶体的三阶非线性系数对碘酸锂的三阶非线性系数 c_{35} 和 c_{12} 进行了测量。

Qiu 和 Penzkofer[4] 研究了 BBO 晶体中 5 ps、1.054 μm 辐射的 THG,在输入强度为 50 GW/cm^2 时得到 0.8% 的转换效率。作者声称所观察到的三阶非线性辐射可能是由于直接三阶非线性过程或是由于级联二阶过程产生的，并认为两个过程具有相近的输出。10 年后，同时有两个研究组研究了 KTP 晶体中的 THG[5,6]。他们获得了相近的效率：在 0.49 cm 长晶体中，22 ps、28 GW/cm^2 入射强度的 1.618 μm 基波，其效率为 2.4%；在 20 GW/cm^2 的入射强度，30～40 ps、1.6～1.8 μm 基波辐射的效率为 1%。然而，两个研究组的结论是相互矛盾的：第一个研究组声称"二阶的贡献仅为 10%"。而第二个研究组证明"级联二阶过程是 KTP 晶体 THG 占优势的过程"。

最近，美国研究组在 0.3 cm 长 BBO 晶体中在 200 GW/cm^2 的入射强度（$\lambda = 1.055$ μm, $\tau_p = 350$ fs）时利用 I 类和 II 类相位匹配[7,8]达到 6% 的 THG 转换效率。他们的结论是"级联 SHG 以及 SFG 过程，甚至尽管非相位匹配，仍然能在具有二阶响应的非线性材料单晶 SHG 相位匹配中有重大贡献，甚至起决定性作用"。

参考文献

[1] J. E. Midwinter, J. Warner: The effects of phase matching method and of crystal symmetry on the polar dependence of third-order non-linear optical polarization. Brit. J. Appl. Phys. 16 (11), 1667-1674(1965).

[2] S.-W. Xie, X.-L. Yang, W.-Y. Jia, Y.-L. Chen: Phase-matched third-harmonic generation in biaxial crystals. Opt. Commun. 118(5-6), 648-656(1995).

[3] M. Okada: Third-order nonlinear optical coefficients of LiIO$_3$. Appl. Phys. Lett. 18 (10), 451-452(1971).

[4] P. Qiu, A. Penzkofer: Picosecond third-harmonic light generation in β-BaB$_2$O$_4$. Appl. Phys. B 45 (4), 225-236(1988).

[5] J. P. Feve, B. Boulanger, Y. Guillien: Efficient energy conversion for cubic third-harmonic generation that is phase-matched in KTiOPO$_4$. Opt. Lett. 25 (18), 1373-1375 (2000).

[6] Y. Takagi, S. Muraki: Third-harmonic generation in a noncentrosymmetrical crystal: direct third-order or cascaded second-order process? J. Lumunesc. 87-89, 865-867 (2000).

[7] P. S. Banks, M. D. Feit, M. D. Perry: High-intensity third-harmonic generation in beta barium borate through second-order and third-order susceptibilities. Opt. Lett. 24 (1), 4-6 (1999).

[8] P. S. Banks, M. D. Feit, M. D. Perry: High-intensity third-harmonic generation. J. Opt. Soc. Am. B 19(1), 102-118(2002).

第10章 闭卷的话

尽管在编写本书的工作中,我已尽所有的努力来减少错误及打印错误的数目,但把它们都排斥于书籍之外不是不可能的,起码说是困难的。因此,我为所有可能出现的错误向读者表示歉意,并请求读者一旦发现错误就通过邮件或 E-mail(niko@ phys. ucc. ie)通知我。我也非常感激对本书进行的评论,这将会在以后再版时予以考虑。

附录 A　所列举杂志的全称

Acta Crystallogr.
　　Acta Crystallographica
Appl. Opt.
　　Applied Optics
Appl. Phys.
　　Applied Physics
Appl. Phys. Lett.
　　Applied Physics Letters
Atmos. Oceanic Opt.
　　Atmospheric and Oceanic Optics(Russia)
Brit. J. Appl. Phys.
　　British Journal of Applied Physics
Bull. Mater. Sci.
　　Bulletin of Materials Science(India)
Bull. Acad. Sci. USSR, Phys. Ser.
　　Bulletin of USSR Academy of Sciences: Physical Series
Bull. Russian Acad. Sci. : Physics
　　Bulletin of the Russian Academy of Sciences: Physics
Chin. Phys. Lett.
　　Chinese Physics Letters
Cryst. Res. Technol.
　　Crystal Research and Technology
Doklady AN SSSR
　　Doklady Akademii Nauk SSSR(USSR)
Electron. Lett.
　　Electronics Letters
Exp. Techn. Phys.
　　Experimentelle Technik der Physik
Eur. J. Solid State Inorg. Chem.

European Journal of Solid State and Inorganic Chemistry

Fiz. Tekh. Poluprov.

Fizika i Tekhnuka Poluprovodnikov(USSR,Russia)

Fiz. Tverd. Tela

Fizika Tverdogo Tela(USSR,Russia)

IEEE J. Quant. Electr.

IEEE Journal of Quantum Electronics

IEEE J. Sel. Topics Quant. Electr.

IEEE Journal of Selected Topics in Quantum Electronics

IEEE Photon. Technol. Lett.

IEEE Photonics technology Letters

Int. J. Nonl. Opt. Phys.

International Journal of Nonlinear Optical Physics

Int. Mater. Rev.

International Materials Reviews

Izv. Akad. Nauk SSSR, Ser. Fiz.

Izvestiya Akademii Nauk SSSR, Seriya Fizicheskaya(USSR)

Izv. Ross. Akad. Nauk, Ser. Fiz.

Izvestiya Rossiiskoi Akademii Nauk, Seriya Fizicheskaya(Russia)

JETP Lett.

JETP Letters

J. Am. Ceram. Society

Journal of American Ceramic Society

J. Appl. Phys.

Journal of Applied Physics

J. Appl Spectrosc.

Journal of Applied Spectroscopy

J. Cryst. Growth

Journal of Crystal Growth

J. Korean Phys. Soc.

Journal of the Korean Physical Society

J. Luminesc.

Journal of Luminescence

J. Mat. Sci. Lett.

Journal of Materials Science Letters

J. Mater. Sci. Semicond. Process.
  Journal of Material Science in Semiconductor Proecssing
J. Mol. Struct.
  Journal of Molecular Structure
J. Opt. Soc. Am.
  Journal of Optical Society of America
J. Opt. Technol.
  Journal of Optical Technology(Russia)
J. Phys.
  Journal of Physics
J. Phys. Chem. Solids
  Journal of Physics and Chemistry of Solids
J. Phys. : Condens. Matter
  Journal of Physics: Condensed Matter
J. Phys. Soc. Japan
  Journal of the Physical Society of Japan
J. Synth. Cryst.
  Journal of Synthetic Crystals(China)
Jpn. J. Appl. Phys.
  Japanese Journal of Applied Physics
Kratkie Soobshch. Fiz.
  Kratkie Soobshcheniya po Fizike(USSR,Russia)
Kristallogr.
  Kristallografiya(USSR,Russia)
Kvant. Elektron.
  Kvantovaya Elektronika(USSR,Russia)
Laser Phys.
  Laser Physics(Russia)
Lit. Fiz. Sbornik
  Litovskii Fizicheskii Sbornik(Lithuania)
Mater. Lett.
  Materials Letters
MRS Bulletin
  Materials Research Society Bulletin
Mater. Res. Bull.

Materials Research Bulletin

Mater. Sci. Eng.
Materials Science and Engineering

Nonl. Opt.
Nonlinear Optics

Opt. Commun.
Optics Communications

Opto-electron.
Opto-electronics

Opto-Electron. Rev.
Opto-Electronics Review

Opt. Eng.
Optical Engineering

Opt. Laser Technol.
Optics & Laser Technology

Opt. Lett.
Optics Letters

Opt. Mater.
Optical Materials

Opt. Mekh. Promyshl.
Optiko-Mekhanicheskaya Promyshlennost(USSR, Russia)

Opt. Quant. Electron.
Optical and Quantum Electronics

Opt. Spectrosc. USSR
Optics and Spectroscopy USSR

Opt. Spektrosk.
Optika i Spektroskopiya(USSR, Russia)

Pisma Zh. Eksp. Teor. Fiz.
Pisma v Zhurnal Eksperimentalnoi i Teoreticheskoi Fiziki(USSR, Russia)

Pisma Zh. Tekh. Fiz.
Pisma v Zhurnal Tekhnicheskoi Fiziki(USSR, Russia)

Progr. Cryst. Growth Character. Mater.
Progress in Crystal Growth and Characterization of Materials

Phys. Lett.
Physics Letters

Phys. Rev.
  Physical Review
Phys. Rev. Lett.
  Physical Review Letters
Phys. Stat. Solidi
  Physica Status Solidi
Pure Appl. Opt.
  Pure and Applied Optics
Proc SPIE
  Proceedings SPIE
Quantum Electron.
  Quantum Electronics(Russia)
Rev. Laser Eng.
  Review of Laser Engineering
Russ. J. Inorgan, Chem.
  Russian Journal of Inorganic Chemistry
Solid State Commun.
  Solid State Communications
Sov. J. Opt. technol.
  Soviet Journal of Optical Technology
Sov. J. Quantum Electron.
  Soviet Journal of Quantum Electronics
Sov. Phys. -Crystallogr.
  Soviet Physics-Crystallography
Sov. Phys. -Doklady
  Soviet Physics-Doklady
Sov. Phys. -JETP
  Soviet Physics-JETP
Sov. Phys. -Semicond.
  Soviet Physics-Semiconductors
Sov. Phys. -Solid State
  Soviet Physics-Solid State
Sov. Phys. -Tech. Phys.
  Soviet Physics-Technical Physics
Sov. Tech. Phys. Lett.

Soviet Technical Physics Letters

Z. Kristallogr.

Zeitschrift für Kristallographie

Zh. Eksp. Teor. Fiz.

Zhurnal Eksperimentalnoi I Teoreticheskoi Fiziki(USSR. Russia)

Zh. Neorg. Khim.

Zhurnal Neorganicheskoi Khimii(USSR. Russia)

Zh. Prikl. Spektrosk.

Zhurnal Prikladnoi Spektroskopii(USSR. Russia)

Zh. Tekh. Fiz.

Zhurnal Tekhnicheskoi Fiziki(USSR. Russia)

附录B 在最后审稿时加入的最新参考文献

加入第2章 基本的非线性光学晶体

[1] H. Wang, A. M. Weiner: Efficiency of short-pulse type-I second-harmonic generation with simultaneous spatial walk-off, temporal walk-off, and pump depletion. IEEE J. Quant. Electr. 39(12), 1600 – 1618(2003).

[2] A. -Y. Yao, W. Hou, X. -C. Lin, Y. Bi, R. -N. Li, D. -F. Cui, Z. -Y. Xu: High power red laser at 671nm by intracavity -doubled Nd: YVO_4 laser using LiB_3O_5. Opt. Commun. 231(1 – 6), 413 – 416(2004).

[3] H. Q. Li, H. B. Zhang, Z. Bao, J. Zhang, Z. P. Sun, Y. P. Kong, Y. Bi, X. C. Lin, A. Y. Yao, G. L. Wang, W. Hou, R. N. Li, D. F. Cui, Z. Y. Xu: High-power nanosecond optical oscillator based on a long LiB_3O_5 crystal. Opt. Commun. 232(1 – 6), 411 – 415(2004).

[4] X. -C. Lin, Y. Zhang, Y. -P. Kong, J. Zhang, A. -Y. Yao, W. Hou, D. -F. Cui, R. -N. Li, Z. -Y. Xu, J. Li: Low-threshold mid-infrared optical parametric oscillator using periodically poled $LiNbO_3$. Chin. Phys. Lett. 21(1), 98 – 100(2004).

[5] M. V. Pack, D. J. Armstrong, A. V. Smith: Measurements of the $\chi^{(2)}$ tensors of $KTiOPO_4$, $KTiOAsO_4$, $RbTiOPO_4$ and $RbTiOAsO_4$ crystals. Appl. Opt. 43(16), 3319 – 3323(2004).

加入第3章 主要的红外材料

[6] W. Shi, Y. J. Ding, P. G. Schunemann: Coherent terahertz waves based on difference-frequency generation in an annealed zinc-germanium phosphide crystal: improvements on tuning ranges and peak powers. Opt. Commun. 233(1 – 3), 183 – 189(2004).

[7] P. Kumbhakar, T. Kobayashi, G. C. Bhar: Sellmeier dispersion for phasematched terahertz generation in $ZnGeP_2$. Appl. Opt. 43(16), 3324 – 3328(2004).

[8] R. S. Dubinkin, X. Mu, Y. J. Ding: Spectrum of two-photon absorption coefficients for $ZnGeP_2$. In: *International Quantum Electronics Conference CLEO/IQEC 2004, Technical Digest*(OSA, Washington DC 2004) paper IMD6.

加入第4章 常用晶体

[9] I. A. Begishev, M. Kalashnikov, V. Karpov, P. Nickles, H. Schönnagel, I. A. Kulagin,

T. Usmanov: Limitation of second-harmonic generation of femtosecond Ti: sapphire laser pulses. J. Opt. Soc. Am. B 21(2), 318 – 322(2004).

[10] J. Sakuma, Y. Asakawa, M. Obara: Generation of 5-W deep-UV continuouswave radiation at 266nm by an external cavity with a $CsLiB_6O_{10}$ crystal. Opt. Lett. 29(1), 92 – 94(2004).

[11] N. Pavel, I. Shoji, T. Taira, K. Mizuuchi, A. Morikawa, T. Sugita, K. Yamamoto: Room-temperature, continuous-wave 1-W green power by single-pass frequency doubling in a bulk periodically poled MgO: $LiNbO_3$ crystal. Opt. Lett. 29(8), 830 – 832(2004).

[12] K. Kato, N. Umemura: Sellmeier and thermo-optic dispersion formulas for $KTiOAsO_4$. In: *Conference on Lasers and Electrooptics CLEO/IQEC 2004, Technical Digest*(OSA, Washington DC 2004) paper CThT35.

[13] J. Hirohashi, K. Yamada, H. Kamio, S. Shichijyo: Embryonic nucleation method for fabrication of uniform periodically poled structures in potassium niobate for wavelength conversion devices. Jpn. J. Appl. Phys. 43(2), 559 – 566(2004).

[14] S. S. Saltiel, K. Koynov, B. Agate, W. Sibbett: Second-harmonic generation with focused beams under conditions of large group-velocity mismatch. J. Opt. Soc. Am. B 21(3), 591 – 598(2004).

加入第 5 章 周期性极化晶体及"衬底"材料

[15] I. Yutsis, B. Kirshner, A. Arie: Temperature-dependent dispersion relations for $RbTiOPO_4$ and $RbTiOAsO_4$. Appl. Phys. B 79(1), 77 – 81(2004).

[16] T. Skauli, P. S. Kuo, K. L. Vodopyanov, T. J. Pinguet, O. Levi, L. A. Eyres, J. S. Harris, M. M. Fejer, B. Gerard, L. Becouarn, E. Lallier: Improved dispersion relations for GaAs and applications to nonlinear optics. J. Appl. Phys. 94(10), 6447 – 6455(2003).

加入第 6 章 新发展及有前景的晶体

[17] P. Segonds, B. Boulanger, J.-P. Feve, B. Menaert, J. Zaccaro, G. Aka, D. Pelenc: Linear and nonlinear optical properties of the monoclinic $YCa_4O(BO_3)_3$* crystal. J. Opt. Soc. Am. B 21(4), 765 – 769(2004).

[18] P. Kumbhakar, T. Kobayashi: Nonlinear optical properties of $Li_2B_4O_7$(LB4) crystal for the generation of tunable ultra-fast laser radiation by optical parametric amplification. Appl. Phys. B 78(2), 165 – 170(2004).

[19] V. Petrov, A. Yelisseyev, L. Isaenko, S. Lobanov, A. Titov, J.-J. Zondy: Second harmonic generation and optical parmetric amplification in the mid-IR with orthorhombic

---

* 译者注:原书为"$Ca_4YO(BO_3)_3$",应为"$YCa_4O(BO_3)_3$"。

biaxial crystals LiGaS$_2$ and LiGaSe$_2$. Appl. Phys. B 78(5), 543-546(2004).

加入第 7 章　自倍频晶体

[20] A. Brenier, C. Tu, Z. Zhu, J. Li, Y. Wang, Z. You, B. Wu: Self-frequency tripling from two-cascaded second-order nonlinearities in GdAl$_3$(BO$_3$)$_4$: Nd^{3+}. Appl. Phys. Lett. 84(1), 16-18(2004).

[21] A. Brenier, C. Tu, Z. Zhu, B. Wu: Red-green-blue generation from a lone dual-wavelength GdAl$_3$(BO$_3$)$_4$: Nd^{3+} laser. Appl. Phys. Lett. 84(12), 2034-2036(2004).

加入第 8 章　很少用的文献晶体

[22] V. Petrov, V. Badikov, V. Panyutin, G. Shevyrdyaeva, S. Sheina, F. Rotermund: Mid-IR optical parametric amplification with femtosecond pumping near 800 nm using Cd$_x$Hg$_{1-x}$Ga$_2$S$_4$. Opt. Commun. 235(1-3), 219-226(2004).

[23] A. A. Mani, Z. D. Schultz, A. A. Gewirth, J. O. White, Y. Caudano, C. Humbert, L. Dreesen, P. A. Thiry, A. Peremans: Picosecond laser for performance of efficient nonlinear spectroscopy from 10 to 21 μm. Opt. Lett. 29(3), 274-276(2004).

加入第 9 章　一些最新的应用

[24] T. Kanai, T. Kanda, T. Sekikawa, S. Watanabe, T. Togashi, C. Chen, C. Zhang, Z. Xu, J. Wang: Generation of vacuum-ultraviolet light below 160 nm in a KBBF crystal by the fifth harmonic of a single-mode Ti: sapphire laser. J. Opt. Soc. Am. B 21(2), 370-375 (2004).

[25] W. Shi, Y. J. Ding: A monochromatic and high-power terahertz source tunable in the ranges of 2.7-38.4 and 58.2-3 540 μm for variety of potential applications. Appl. Phys. Lett. 84(10), 1635-1637(2004).

[26] P. Ni, B. Ma, S. Feng, B. Cheng, D. Zheng: Multiple-wavelength secondharmonic generations in a two-dimensional periodically poled lithium niobate. Opt. Commun. 233(1-3), 199-203(2004).

[27] E. H. G. Backus, S. Roke, A. W. Kleyn, M. Bonn: Cascading second-order versus direct third-order nonlinear optical processes in a uniaxial crystal. Opt. Commun. 234 (1-6), 404-417(2004).

名词索引

ADP
Ag$_3$AsS$_3$
AgGaS$_2$, see AGS
AgGa$_{1-x}$In$_x$Se$_2$, see AGISe
AgGaSe$_2$, see AGSe
AGISe
AGS
AGSe
Ammonium dihydrogen phosphate, see ADP

BaAlBO$_3$F$_2$, see BABF
BABF
β-BaB$_2$O$_4$, see BBO
Barium aluminum fluoroborate, see BABF
Ba$_2$NaNb$_5$O$_{15}$, see BNN
Barium sodium niobate, see BNN
Barium titanate, see BaTiO$_3$
BaTiO$_3$
BBO
Beta-barium borate, see BBO
BIBO
BiB$_3$O$_6$, see BIBO
Birefringent phase matching
Bismuth triborate, see BIBO
BNN

Cadmium germanium arsenide, see CGA

ADP 161—175, 479
Ag$_3$AsS$_3$ 437—443
AgGaS$_2$, 见 AGS
AgGa$_{1-x}$In$_x$Se$_2$, 见 AGISe
AgGaSe$_2$, 见 AGSe
AGISe 320—323
AGS 91—104
AGSe 104—115
磷酸二氢铵, 见 ADP

BaAlBO$_3$F$_2$, 见 BABF
BABF 267—269
β-BaB$_2$O$_4$, 见 BBO
氟硼酸铝钡, 见 BABF

Ba$_2$NaNb$_5$O$_{15}$, 见 BNN
铌酸钡钠, 见 BNN
钛酸钡, 见 BaTiO$_3$
BaTiO$_3$ 234—240
BBO 5—22, 466, 469, 480
偏硼酸钡, 见 BBO
BIBO 257—261
BiB$_3$O$_6$, 见 BIBO
双折射相位匹配 55, 205, 225, 473—474
三硼酸铋, 见 BIBO
BNN 414—419

砷锗镉, 见 CGA

Cadmium mercury thiocyanate, see CMTC 硫氰酸汞镉，见 CMTC
Cadmium selenide, see CdSe 硒化镉，见 CdSe
CBO CBO 380—383
CDA CDA 395—400
CdGeAs$_2$, see CGA CdGeAs$_2$，见 CGA
CdHg(SCN)$_4$, see CMTC CdHg(SCN)$_4$，见 CMTC
CdSe CdSe 456—464
Cesium dihydrogen arsenate, see CDA 砷酸二氢铯，见 CDA
Cesium lithium borate, see CLBO 硼酸锂铯，见 CLBO
Cesium titanyl arsenate, see CTA 砷酸钛氧铯，见 CTA
Cesium triborate, see CBO 三硼酸铯，见 CBO
CGA CGA 447—453
C$_4$H$_7$D$_{12}$N$_4$PO$_7$, see DLAP C$_4$H$_7$D$_{12}$N$_4$PO$_7$，见 DLAP
CLBO CLBO 186—193
CMTC CMTC 297—300
CO(NH$_2$)$_2$, see Urea CO(NH$_2$)$_2$，见尿素
CsB$_3$O$_5$, see CBO CsB$_3$O$_5$，见 CBO
CsD$_2$AsO$_4$, see DCDA CsD$_2$AsO$_4$，见 DCDA
CsH$_2$AsO$_4$, see CDA CsH$_2$AsO$_4$，见 CDA
CsLiB$_6$O$_{10}$, see CLBO CsLib$_6$O$_{10}$，见 CLBO
CsTiOAsO$_4$, see CTA CsTiOAsO$_4$，见 CTA
CTA CTA 410—414

DCDA DCDA 400—404
Deep UV light generation 深 UV 光发生 465—467
Deuterated L-arginine phosphate monohydrate, see DLAP 氘化左旋磷酸精氨酸一水合物，见 DLAP
Deuterated cesium dihydrogen arsenate, see DCDA 氘化砷酸二氢铯，见 DCDA
Deuterated potassium dihydrogen phosphate, see DKDP 氘化磷酸二氢钾，见 DKDP
DKDP DKDP 175—186, 469

DLAP	DLAP 383—387
Fluoroboratobeyllate, see KBBF	氟硼铍酸钾，见 KBBF
GaAs	GaAs 243—255, 491
Gadolinium calcium oxyborate, see GdCOB	硼酸氧钙钆，见 GdCOB
Gadolinium-yttrium calcium oxyborate, see GdYCOB	硼酸氧钙钇钆，见 GdYCOB
Gallium arsenide, see GaAs	砷化镓，见 GaAs
Gallium selenide, see GaSe	硒化镓，见 GaSe
GaSe	GaSe 130—137, 468
$GdCa_4O(BO_3)_3$, see GdCOB	$GdCa_4O(BO_3)_3$，见 GdCOB
GdCOB	GdCOB 271—277
$Gd_xY_{1-x}Ca_4O(BO_3)_3$, see GdYCOB	$Gd_xY_{1-x}Ca_4O(BO_3)_3$，见 GdYCOB
GdYCOB	GdYCOB 287—292
HeXLN	HeXLN 478—479
$HgGa_2S_4$	$HgGa_2S_4$ 444—447, 491
α-HIO_3	α-HIO_3 387—392
α-Iodic acid, see α-HIO_3	α-碘酸，见 α-HIO_3
KABO	KABO 261—265
$K_2Al_2B_2O_7$, see KABO	$K_2A1_2B_2O_7$，见 KABO
KB5	KB5 373—380, 466—467
KBBF	KBBF 265—267, 465—466, 491
$KBe_2BO_3F_2$, see KBBF	$BKe_2BO_3F_2$，见 KBBF
$KB_5O_8 \cdot 4H_2O$, see KB5	$KB_5O_8 \cdot 4H_2O$，见 KB5
KDP	KDP 139—161, 468—470, 480, 490
KD_2PO_4, see DKDP	KD_2PO_4，见 DKDP
KH_2PO_4, see KDP	KH_2PO_4，见 KDP
$K_3Li_2Nb_5O_{15}$, see KLN	$K_3Li_2Nb_5O_{15}$，见 KLN
KLN	KLN 419—422
KN	KN 208—219, 490

$KNbO_3$, see KN	$KNbO_3$,见 KN
KTA	KTA 201—208, 490
$KTiOAsO_4$, see KTA	$KTiOAsO_4$,见 KTA
$KTiOPO_4$, see KTP	$KTiOPO_4$,见 KTP
KTP	KTP 64—89, 480, 490
$La_2CaB_{10}O_{19}$, see LCB	$La_2CaB_{10}O_{19}$,见 LCB
Lanthanum calcium borate, see LCB	硼酸钙镧,见 LCB
LB4	LB4 292—295, 466, 491
LBO	LBO 22—42, 466, 489
LCB	LCB 269—271
LFM	LFM 392—395
LGS	LGS 317—319, 491
LGSe	LGSe 319—320, 491
$Li_2B_4O_7$, see LB4	$Li_2B_4O_7$,见 LB4
LiB_3O_5, see LBO	LiB_3O_5,见 LBO
$LiCOOH \cdot H_2O$, see LFM	$LiCOOH \cdot H_2O$,见 LFM
$LiGaS_2$, see LGS	$LiGaS_2$,见 LGS
$LiGaSe_2$, see LGSe	$LiGaSe_2$,见 LGSe
$LiInS_2$, see LIS	$LiInS_2$,见 LIS
$LiInSe_2$, see LISe	$LiInSe_2$,见 LISe
$LiIO_3$	$LiIO_3$ 426—436
$LiRbB_4O_7$, see LRB4	$LiRbB_4O_7$,见 LRB4
LIS	LIS 309—315
LISe	LISe 315—317
$LiTaO_3$, see LT	$LiTaO_3$,见 LT
Lithium formate monohydrate, see LFM	甲酸锂一水合物,见 LFM
Lithium gallium selenide, see LGSe	硒镓锂,见 LGSe
Lithium indium selenide, see LISe	硒铟锂,见 LISe
Lithium iodate, see $LiIO_3$	碘酸锂,见 $LiIO_3$
Lithium niobate, see LN	铌酸锂,见 LN
Lithium rubidium tetraborate, see LRB4	四硼酸铷锂,见 LRB4
Lithium tantalate, see LT	钽酸锂,见 LT
Lithium tetraborate, see LB4	四硼酸锂,见 LB4
Lithium thiogallate, see LGS	硫镓锂,见 LGS

Lithium thioindate, *see* LIS	硫铟锂，见 LIS
Lithium triborate, *see* LBO	三硼酸锂，见 LBO
LN	LN 42—64, 467—468
$LiNbO_3$, *see* LN	$LiNbO_3$，见 LN
LRB4	LRB4 295—297
LT	LT 221—228
Magnesium barium fluoride, *see* $MgBaF_4$	氟化镁钡，见 $MgBaF_4$
Magnesium-oxide-doped lithium niobate, *see* MgLN	氧化镁掺杂铌酸锂，见 MgLN
Mercury thiogallate, *see* $HgGa_2S_4$	硫镓汞，见 $HgGa_2S_4$
$MgBaF_4$	$MgBaF_4$ 240—243
MgLN	MgLN 55, 193—201, 470
$MgO:LiNbO_3$, *see* MgLN	$MgO:LiNbO_3$，见 MgLN
$Nb:KTiOPO_4$, *see* NbKTP	$Nb:KTiOPO_4$，见 NbKTP
$Nb_xK_{1-x}Ti_{1-x}OPO_4$, *see* NbKTP	$Nb_xK_{1-x}Ti_{1-x}OPO_4$，见 NbKTP
NbKTP	NbKTP 300—305
Nb:BNN	Nb:BNN 417
$Nd:GdAl_3(BO_3)_4$, *see* NGAB	$Nd:GdAl_3(BO_3)_4$，见 NGAB
$Nd_xGd_{1-x}Al_3(BO_3)_4$, *see* NGAB	$Nd_xGd_{1-x}Al_3(BO_3)_4$，见 NGAB
$Nd:GdCa_4O(BO_3)_3$, *see* Nd:GdCOB	$Nd:GdCa_4O(BO_3)_3$，见 Nd:GdCOB
Nd:GdCOB	Nd:GdCOB 341—347, 470—472
$Nd_xGd_{1-x}COB$, *see* Nd:GdCOB	$Nd_xGd_{1-x}COB$，见 Nd:GdCOB
$Nd:Gd_2(MoO_4)_3$, *see* NdGMO	$Nd:Gd_2(MoO_4)_3$，见 NdGMO
$Nd_{2x}Gd_{2-2x}(MoO_4)_3$, *see* NdGMO	$Nd_{2x}Gd_{2-2x}(MoO_4)_3$，见 NdGMO
NdGMO	NdGMO 355—359
$Nd:LaBGeO_5$, *see* NdLBGO	$Nd:LaBGeO_5$，见 NdLBGO
$Nd_xLa_{1-x}BGeO_5$, *see* NdLBGO	$Nd_xLa_{1-x}BGeO_5$，见 NdLBGO
NdLBGO	NdLBGO 351—355, 470—471
NdMgLN	NdMgLN 325—329, 470—471
$Nd:MgO:LiNbO_3$, *see* NdMgLN	$Nd:MgO:LiNbO_3$，见 NdMgLN
$Nd:YAl_3(BO_3)_4$, *see* NYAB	$Nd:YAl_3(BO_3)_4$，见 NYAB
$Nd_xY_{1-x}Al_3(BO_3)_4$, *see* NYAB	$Nd_xY_{1-x}Al_3(BO_3)_4$，见 NYAB
$Nd:YCa_4O(BO_3)_3$, *see* Nd:YCOB	$Nd:YCa_4O(BO_3)_3$，见 Nd:YCOB

Nd:YCOB

Nd$_x$Y$_{1-x}$COB, see Nd:YCOB

Neodymium-and magnesium-oxide-doped lithium niobate, see NdMgLN

Neodymium-doped gadolinium aluminum tetraborate, see NGAB

Neodymium-doped gadolinium calcium oxyborate, see Nd:GdCOB

Neodymium-doped gadolinium molybdate, see NdGMO

Neodymium-doped lanthanum borogermanate, see NdLBGO

Neodymium-doped yttrium aluminum tetraborate, see NYAB

Neodymium-doped yttrium calcium oxyborate, see Nd:YCOB

NGAB

NH$_4$H$_2$PO$_4$, see ADP

Niobium-doped KTP, see NbKTP

NYAB

Periodically poled crystals

Photorefractive effect

Potassium aluminum borate, see KABO

Potassium dihydrogen phosphate, see KDP

Potassium fluoroboratoberyllate, see KBBF

Potassium lithium niobate, see KLN

Potassium niobate, see KN

Potassium pentaborate tetrahydrate, see KB5

Potassium titanyl arsenate, see KTA

Potassium titanyl phosphate, see KTP

PPKTP

PPLN

PPLT

Nd:YCOB 347—351, 470—472

Nd$_x$Y$_{1-x}$COB, 见 Nd:YCOB

掺钕掺氧化镁铌酸锂, 见 NdMgLN

掺钕四硼酸铝钆, 见 NGAB

掺钕硼酸氧钙钆, 见 Nd:GdCOB

掺钕钼酸钆, 见 NdGMO

掺钕硼锗酸镧, 见 NdLBGO

掺钕四硼酸铝钇, 见 NYAB

掺钕硼酸氧钙钇, 见 Nd:YCOB

NGAB 337—341, 470—471, 491

NH$_4$H$_2$PO$_4$, 见 ADP

铌掺杂 KTP, 见 NbKTP

NYAB 329—337, 470—472

周期性极化晶体 473—478

光折变效应 55, 77

硼酸铝钾, 见 KABO

磷酸二氢钾, 见 KDP

氟硼铍酸钾, 见 KBBF

铌酸钾锂, 见 KLN

铌酸钾, 见 KN

五硼酸钾四水合物, 见 KB5

砷酸钛氧钾, 见 KTA

磷酸钛氧钾, 见 KTP

PPKTP 78, 216, 232, 473—475

PPLN 55, 216, 467, 473—475, 490, 492

PPLT 225, 473, 475—476

PPMgLN	PPMgLN 198, 490
PPRTA	PPRTA 232, 474—475
Predelay crystal	预延时晶体 469
Proustite, see Ag_3AsS_3	硫镓银,见 Ag_3AsS_3
Pulsewidth shortening	脉宽变短 469
Quasi-phase matching	准位相匹配 55, 78, 198, 225, 249, 308, 327, 473—474
RbH_2PO_4, see RDP	RbH_2PO_4,见 RDP
$RbTiOAsO_4$, see RTA	$RbTiOAsO_4$,见 RTA
$RbTiOPO_4$, see RTP	$RbTiOPO_4$,见 RTP
RDP	RDP 404—410
RTA	RTA 228—234, 490—491
RTP	RTP 305—309, 490—491
Rubidium dihydrogen phosphate, see RDP	磷酸二氢铷,见 RDP
Rubidium titanyl arsenate, see RTA	砷酸钛氧铷,见 RTA
Rubidium titanyl phosphate, see RTP	磷酸钛氧铷,见 RTP
Self-frequency doubling	自倍频发生 471
Self-sum-frequency generation	硒镓铟银,见 AGISe
Silver gallium-indium selenide, see AGISe	
Silver gallium selenide, see AGSe	硒镓银,见 AGSe
Silver thiogallate, see AGS	硫镓银,见 AGS
Submillimeter radiation	亚毫米波 467
Submillimeter wave generation	亚毫米波发生 468
自倍频 470—473	
TAS	TAS 453—456
Terahertz-wave generation	太赫兹波发生 467—468
Thallium arsenic selenide, see TAS	硒砷铊,见 TAS
Thallium mercury iodide, see Tl_4HgI_6	碘汞铊,见 Tl_4HgI_6
THI	THI 323—324
Tl_3AsSe_3, see TAS	Tl_3AsSe_3,见 TAS
Tl_4HgI_6, see THI	Tl_4HgI_6,见 THI

Ultrashort laser pulse compression 超短激光脉冲压缩 468—470

Urea 尿素 422—426

Yb:GdCa$_4$O(BO$_3$)$_3$, see Yb:GdCOB   Yb:GdCa$_4$O(BO$_3$)$_3$, 见 Yb:GdCOB

Yb:GdCOB   Yb:GdCOB 364—367

Yb$_x$Gd$_{1-x}$COB, see Yb:GdCOB   Yb$_x$Gd$_{1-x}$COB, 见 Yb:GdCOB

YbMgLN   YbMgLN 327

Yb:YAB   Yb:YAB 355—359

Yb:YAl$_3$(BO$_3$)$_4$, see Yb:YAB   Yb:YAl$_3$(BO$_3$)$_4$, 见 Yb:YAB

Yb$_x$Y$_{1-x}$Al$_3$(BO$_3$)$_4$, see Yb:YAB   Yb$_x$Y$_{1-x}$Al$_3$(BO$_3$)$_4$, 见 Yb:YAB

Yb:YCa$_4$O(BO$_3$)$_3$, see Yb:YCOB   Yb:YCa$_4$O(BO$_3$)$_3$, 见 Yb:YCOB

Yb:YCOB   Yb:YCOB 367—371

Yb$_x$Y$_{1-x}$COB, see Yb:YCOB   Yb$_x$Y$_{1-x}$COB, 见 Yb:YCOB

YCa$_4$O(BO$_3$)$_3$, see YCOB   YCa$_4$O(BO$_3$)$_3$, 见 YCOB

YCOB   YCOB 278—287, 491

Ytterbium-doped gadolinium calcium oxyborate, see Yb:GdCOB   掺镱硼酸氧钙钆, 见 Yb:GdCOB

Ytterbium-doped yttrium aluminum tetraborate, see Yb:YAB   掺镱四硼酸铝钇, 见 Yb:YAB

Ytterbium-doped yttrium calcium oxyborate, see Yb:YCOB   掺镱硼酸氧钙钇, 见 Yb:YCOB

Yttrium calcium oxyborate, see YCOB   硼酸氧钙钇, 见 YCOB

ZGP   ZGP 115—130, 468, 490

Zinc germanium phosphide, see ZGP   磷锗锑, 见 ZGP

ZnGeP$_2$, see ZGP   ZnGeP$_2$, 见 ZGP

ZnO-doped LN   ZnO 掺杂 LN 55

译者后记

功能晶体研究是功能材料研究中的一个热点和前沿领域，在我国晶体生长有着悠久的历史和传统。我国现代的晶体生长研究工作自20世纪50年代后期开始，从模仿和跟踪国际新晶体和有用晶体的生长起步。到80年代初，以偏硼酸钡（β-BBO）晶体的发现、生长和应用为标志，我国的功能晶体，特别是非线性光学晶体的探索和研究走上了独立自主、创新发展的道路。近30年来，各种新的、有实用意义的非线性光学晶体在我国涌现，并成为国际公认具有领先水平的材料领域。

非线性光学的基本参数是晶体研究工作的结果，同时又是晶体应用的基础。由于各种晶体的研究结果发表于全世界各种不同的学术刊物，同一晶体的性质，不同研究者获得的结果又各有差异（由于晶体质量和测试条件的差异），因此，获得系统、可靠的晶体基本性质资料是从事晶体研究和应用的科技工作者共同的愿望。

本人长期从事功能晶体的生长工作，非常了解朋友和同事们的需求。我于1998年在美国访问工作时，第一次看到刚出版的《非线性光学晶体手册》（第二版）时，就有马上购买的欲望。当回到中国，这本书就被同事和学生们辗转借阅，不知所终。好在1999年此书又有第三版，当我得到这一版后又有不少人借阅。于是我想，既然这本书在中国这么受欢迎，把它译成中文在国内出版，岂不是一方面扩大了这本书的流通范围，另一方面又为我国读者提供了更大方便？

自从有了这一想法，我也一直想抽时间动手将其译成中文。但无奈此心虽有，空闲却无。直到人过六十，一些杂事从身上卸下，终于有了动手的机会。而当翻译《非线性光学晶体手册》之时，又见到新版的《非线性光学晶体——一份完整的总结》一书。虽然两书出版时间有先后，但各有特色、相互补充，可谓珠联璧合。于是我将两本书一起译成了中文，又由老友吴以成院士进行校对。

虽然文字翻译工作完成，未曾想更为繁杂的工作接踵而来。看来手册类的著作不同于其他著作，手册中充满着各种数据，由各类不同符号和数字构成，这是手册的主体和精髓。打印、编辑、校对，处处小心，处处仔细，但极易出错！故从文字到数字，一遍遍地整理、校对。我和吴以成院士曾数遍重改全稿，激光的相关理论部分由何京良教授校读，我们的助手和学生姚淑华、张建秀、张素芳、郭永解、张娜娜、韩树娟、李真、秦海明、任娜等几次反复校对所有数据符号。我们希望将本书的错漏减到最少，尽管也许很难为零。

我们感谢蒋民华、陈创天、沈德忠和许祖彦院士，他们十分支持我们的翻译工作，陈创天院士还为译本写了序。

我们要感谢高等教育出版社的刘剑波编辑，她为这两本译著的出版做了大量工作。

我们还要感谢国家自然科学基金委员会工程与材料学部的高瑞平副主任和陈克新处长长期以来对我们研究工作的支持。这两本书的出版受到国家自然科学重大基金项目"功能晶体的低对称特性、生长和应用探索"和晶体材料国家重点实验室的支持。

我们感谢支持过我们工作和本书出版的所有朋友。也希望读者能不吝指出本书中仍然存在的错误。

郑 重 声 明

高等教育出版社依法对本书享有专有出版权。任何未经许可的复制、销售行为均违反《中华人民共和国著作权法》，其行为人将承担相应的民事责任和行政责任，构成犯罪的，将被依法追究刑事责任。为了维护市场秩序，保护读者的合法权益，避免读者误用盗版书造成不良后果，我社将配合行政执法部门和司法机关对违法犯罪的单位和个人给予严厉打击。社会各界人士如发现上述侵权行为，希望及时举报，本社将奖励举报有功人员。

反盗版举报电话：(010)58581897/58581896/58581879
反盗版举报传真：(010)82086060
E - mail: dd@hep.com.cn
通信地址：北京市西城区德外大街4号
　　　　　高等教育出版社打击盗版办公室
邮　　编：100120

购书请拨打电话：(010)58581118

图字：01-2009-0941号

Translation from the English language edition：
Nonlinear Optical Crystals：*A Complete Survey* by David N. Nikogosyan
Copyright © 2005 Springer Science + Business Media，Inc.
All Rights Reserved

图书在版编目（CIP）数据

非线性光学晶体：一份完整的总结/（俄罗斯）尼科戈相；
王继扬译. —北京：高等教育出版社，2009.11
书名原文：Nonlinear Optical Crystals：A Complete Survey
ISBN 978-7-04-027779-1

Ⅰ.非… Ⅱ.①尼…②王… Ⅲ.非线性光学晶体-
材料科学 Ⅳ.07

中国版本图书馆CIP数据核字（2009）第158926号

策划编辑	刘剑波	责任编辑	张海雁	封面设计	刘晓翔
责任绘图	尹文军	版式设计	张 岚	责任校对	王 超
责任印制	陈伟光				

出版发行	高等教育出版社	购书热线	010-58581118
社 址	北京市西城区德外大街4号	咨询电话	400-810-0598
邮政编码	100120	网 址	http://www.hep.edu.cn
总 机	010-58581000		http://www.hep.com.cn
		网上订购	http://www.landraco.com
经 销	蓝色畅想图书发行有限公司		http://www.landraco.com.cn
印 刷	涿州市星河印刷有限公司	畅想教育	http://www.widedu.com
开 本	787×1092 1/16	版 次	2009年11月第1版
印 张	32.5	印 次	2009年11月第1次印刷
字 数	600 000	定 价	80.00元

本书如有缺页、倒页、脱页等质量问题，请到所购图书销售部门联系调换。
版权所有　侵权必究
物料号 27779-00